ITALIAN PHYSICAL SOCIETY

PROCEEDINGS

OF THE

INTERNATIONAL SCHOOL OF PHYSICS « ENRICO FERMI »

COURSE XLVIII

edited by P. CALDIROLA
Director of the Course
and by H. KNOEPFEL

VARENNA ON LAKE COMO
VILLA MONASTERO
14th - 26th - JULY 1969

Physics of High Energy Density

1971

ACADEMIC PRESS • NEW YORK AND LONDON

SOCIETA' ITALIANA DI FISICA

RENDICONTI

DELLA

SCUOLA INTERNAZIONALE DI FISICA «ENRICO FERMI»

XLVIII Corso

a cura di P. Caldirola
Direttore del Corso
e di H. Knoepfel

VARENNA SUL LAGO DI COMO
VILLA MONASTERO
14-26 LUGLIO 1969

Fisica delle alte densità di energia

1971

ACADEMIC PRESS • NEW YORK AND LONDON

ACADEMIC PRESS INC.
111 Fifth Avenue
New York 3, N. Y.

United Kingdom Edition
Published by
ACADEMIC PRESS INC. (London) Ltd.
Berkeley Square House, London W. 1

Library of Congress Catalog Card Number: 72-119469

PRINTED IN ITALY

INDICE

ROBERT A. GROSS – The physics of strong shock waves in gases.

R. E. KIDDER – Interaction of intense photon and electron beams with plasmas.

Introduction.

P. CALDIROLA

Istituto di Scienze Fisiche dell'Università - Milano

Ladies and Gentlemen,

the 3rd course of this year of the International Summer School organized in Varenna by the «Società Italiana di Fisica» has as its subject the Physics of high energy density.

Before saying a few introductory words about the content of the Course that I have been charged to organize, I have the pleasure of giving to all the participants the welcome of the President of the «Società Italiana di Fisica», Prof. Giuliano Toraldo di Francia.

On account of his heavy duties, Prof. TORALDO DI FRANCIA, is not here to say the few, traditional opening words.

At the same time I would like to express my deep gratitude to him and to the Società Italiana di Fisica for having accepted a course on the Physics of High Energy Density in the program of the Varenna Summer School.

I really think that this enterprise will be extremely useful, especially for our young students, because I know that there are no high-level Courses on this topic among the European Universities.

During the last two decades the Physics of Plasmas, or, in other words, the physics of fully ionized gases, has grown in a spectacular way. This grow is mainly due to the fact that people has become aware of the possibility of obtaining a controlled release of energy from the so-called nuclear fusion reactions, as, for instance, the hydrogen-helium transformation.

To this aim it is necessary to heat the material to a temperature of the order of several hundreds of millions of degrees. Obviously, one must concentrate an extremely high energy quantity in a relatively small portion of the material.

This Course is precisely intended to face the two basic problems relevant to the afore-mentioned process. The first is how one can practically obtain such concentrations of energy, and the second is how the material will react to such enormous energy densities.

The interest in this field of the modern physical research is also due to the fact that, in this way, in the laboratory we can be faced with physical situations in which matter acquires unusual and fascinating properties.

The same properties are also verified in cosmic spaces and this is the reason why plasma physics has attracted the interest of astrophysicists.

The thorough analysis of these properties will certainly be extremely useful for an understanding of the fundamental behaviour of matter and for the technical applications, which will surely come out of our enlarged knowledge. The program of the Course has been prepared by Dr. Knoepfel who, through his well-known personal competence in this field of physical research, has made a skilfull choice of all the relevant topics. The arguments chosen, though apparently diverse, are in reality strictly closely linked with the main subject of the Course, as will be evident at the end of the various groups of lectures.

A first set of lectures will be dedicated, as we have already said, to the study of the concentration of high quantities of energy in small portions of the material.

This topic will be treated in two series of lectures: in the first by Prof. R. E. KIDDER of the Lawrence Laboratory California University, and in the second one by Prof. O. N. KROKHIN of the Lebedev Institute of Moscow.

Both these lecturers will talk about the use of gigantic light pulses produced, during extremely short time intervals, by high-power lasers. The energy is focused upon solid materials in order to achieve conditions suitable to the triggering of thermonuclear reactions.

Prof. KIDDER will describe the properties of intense laser beams emphasizing their capability to concentrate energy in small portions of the material.

As a consequence the temperature and pressure will reach extremely high levels.

Prof. Krokhin's lectures will be devoted to the detailed study of all those physical transformations taking place in the material under the action of the high-energy laser radiation, and will thus complete, in a useful way, the first set of lectures.

All the other lessons will be likewise dedicated to a thorough and systematic analysis of the behaviour of matter when it is subjected to a strong energy concentration.

The most spectacular phenomenon is given by the rapid expansion of matter which, in turns, gives rise to an intense shock wave propagating in the material. Such a strong shock wave will be accompanied by release of heat and emission of electromagnetic radiation.

To all these problems will be devoted the lectures given by Prof. G. E. DUVALL, Washington State University, Prof. R. A. GROSS, Columbia University of New York, Prof. F. D. BENNETT, Maryland University and by Prof. R. N. KEELER and Prof. E. B. ROYCE both from California University.

Prof. KEELER will give a set of preliminary lectures about the state-equation of condensed media (particularly of solids) under conditions of extremely high pressure and temperature, with emphasis on the physical properties associated to the electron distribution in the material medium.

Prof. BENNETT will add a set of lectures dealing with vaporization waves and with phase transitions.

Prof. DUVALL will teach about shock-waves theory in dense media. He will also develop an extensive and detailed treatment of the dynamics and structure of a shock-wave propagating in a solid.

Particular attention will be given to the mechanical effects of shock-waves, which are relevant to some interesting technological applications.

Then Prof. GROSS will give a set of lectures with the principal aim of describing the ionization effects which take place in the shock-wave.

Furthermore, he will present an analysis of the shock-wave structure, when one has to take into account the radiation of electromagnetic waves and when relativistic effects modify the mathematical treatment of the process.

Prof. ROYCE will deal with a particular problem, describing the properties of a magnetic material under the strong compressions induced by the passage of a shock-wave, and finally Prof. SCHALL will illustrate some relevant topics in the detonation theory.

Two particular problems will also be treated by one of the most famous physicist of our time, Prof. E. TELLER of the California University. Prof. TELLER will present the problem of the shock-waves formation in stars and develop a model, based on the Thomas-Fermi statistical theory of the atom, which describes the electronic properties of a strongly compressed solid.

As usual, the Course will be completed by a series of Seminars on various topics of the physics of the high-energy density.

I have on my list the names of Dr. LINHART, Dr. KNOEPFEL, Dr. SOMON and Dr. CARUSO of the «Laboratori Gas Ionizzati di Frascati» and of Prof. WINTERBERG of Las Vegas University. They will speak about some special topics of great interest in our programme.

We plan also to have sufficient time available for free discussions. I hope that the formal and informal lecture and discussion meetings at this enchanting Villa Monastero will be useful and pleasant for all of you.

I cannot conclude this brief presentation, without saying that the major merit for arranging the whole Course goes to Dr. KNOEPFEL, who, with his authority, experience and unwielding work, has succeeded in gathering here such a number of distinguished lecturers. I am sure that all the guests of the Società Italiana di Fisica will join in thanking him for his precious and valuable collaboration.

I close my talk with the traditional expression: I open, here, this 3rd Course 1969, the 48th since the beginning of the School, and I hope it will be pleasant and successful.

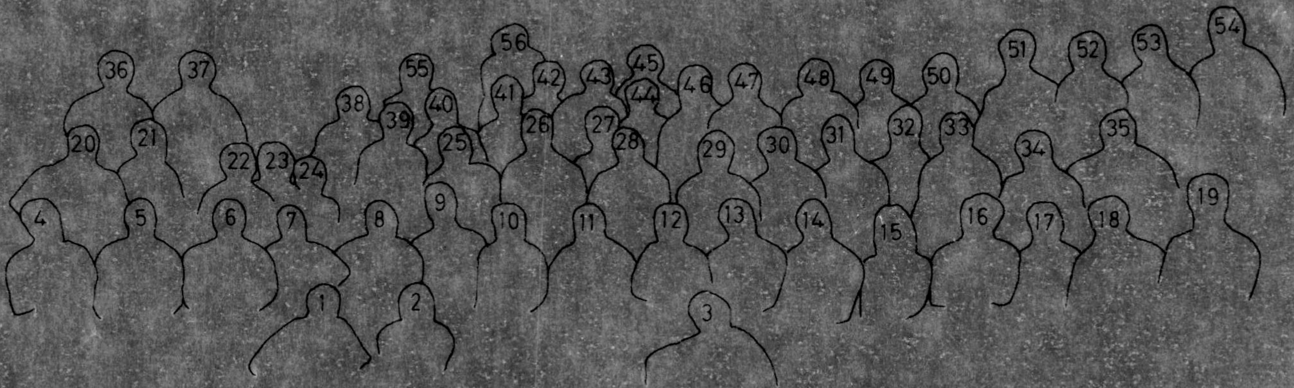

1. K. Hornung
2. M. Gambarelli
3. J. Mrikwiczka
4. N. Gylden
5. G. Iernetti
6. F. D. Bennett
7. E. B. Royce
8. P. Caldirola
9. R. N. Keeler
10. O. N. Krokhin
11. E. Teller
12. R. A. Gross
13. R. E. Kidder
14. H. Knoepfel
15. M. Pilo
16. A. Caruso
17. L. Saitta
18. G. Poletti
19. A. Gervat
20. J. Braun
21. J. G. Linhart
22. C. Rioux
23. C. Maisonnier
24. F. Grossetête
25. R. B. Oswald
26. R. H. Huddlestone
27. C. Summa
28. N. Cerullo
29. A. Di Giorgio
30. A. Fisher
31. J. Leorat
32. P. Lagus
33. G. Besançon
34. M. Samuelli
35. R. Luppi
36. A. Jaworowski
37. A. Lowrey
38. F. Gratton
39. R. Fortin
40. D. Schallhorn
41. W. Wölfli
42. E. Sindoni
43. R. Pozzoli
44. A. Brahme
45. C. Fauquignon
46. C. Chiosi
47. A. Saggion
48. G. Frigerio
49. Y. Avni
50. L. Egardt
51. L. Brun
52. P. Berling
53. B. Brunelli
54. R. Gratton
55. J. P. Somon
56. C. Di Gregorio

SOCIETÀ ITALIANA DI FISICA

SCUOLA INTERNAZIONALE DI FISICA « E. FERMI »

XLVIII CORSO - VARENNA SUL LAGO DI COMO - VILLA MONASTERO - 14-26 Luglio 1969

Some Thoughts about High Energy Densities (*).

E. TELLER

University of California, Lawrence Radiation Laboratory - Livermore, Cal.

It is always difficult to summarize a good conference on Physics. It is impossible to do so in the case of an invigorating mixture of old and neglected facts with ingenious, hopeful and uncertain ideas. Two weeks at the Lago di Como in a « school » established in the memory of Enrico Fermi were aimed at establishing a new discipline: The Physics of High Energy Densities. Our hosts, scientists from Frascati, have succeeded. We came with expectations and are leaving with reluctance carrying along the memories of a sunny lake and of exciting steps in establishing a new science.

We have two regrets. One is the absence of AL'TSHULER and ZEL'DOVICH, the two men who probably have done most in opening up this new field. The other is that no one from the excellent Laboratory of Los Alamos was with us. Much important work was not presented by the original authors.

The impetus came from Frascati. The hope of controlled fusion has generated the conviction that the one could succeed by concentrating on high energy densities. The enthusiasm generated in Frascati has given rise to impressive results. But when I try to look ahead I am reminded of examples of historic developments that take an unexpected turn.

When Columbus was engaged in the Medieval equivalent of fund raising his argument was that going westward he could establish better trade with China. Half a millenium later this aim is yet to be reached. But no one will say that Columbus was unsuccessful.

Controlled fusion may become really important only in the twenty-first century. But systematic fusion research and some ingenious European attempts

(*) Although this paper was presented as the summary of the Course, it is placed here at the beginning of the Proceedings as a stimulating scientific introduction (Editors' note).

to hit the « Jackpot » by finding a particularly effective approach already produced knowledge in plasma physics. This knowledge is applicable to astrophysics and to some branches of technology such as production of electricity in magnetohydrodynamic generators. The attempt to trigger fusion by concentrated pulses may give rise to the same mixture of practical applications and basic understanding. I will attempt to emphasize a few possibilities which are different from attaining the ambitious aim of creating and controlling fusion at the will of the experimenter.

The problem of high energy densities was actually approached in the conference from two sides: one old, the other new. The old approach is to use hydrodynamic shocks and obtain in this manner materials at high densities. With the help of high explosives, pressures of a few million atmospheres can be attained. These are similar to the pressures one finds near the center of the earth.

The new approach is to create high energy densities in electromagnetic radiation. The accelerating development of laser technology makes it easy to speculate and hard to predict what will happen next.

A considerable number of thorough and systematic lectures were given about shock phenomena by DUVALL of Washington State University. He laid great emphasis on the structure of the shock and the relation of this structure to various changes and re-adjustments in the shocked material. GROSS of Columbia University came closer to the frontiers when he talked about hydromagnetic shocks and radiative shocks. The only example of an apparent contradiction was proposed in the interesting discussion of detonation waves given by SCHALL. The discrepancy is small and might be explained by assuming that complete equilibrium of all molecular degrees of freedom is not attained in the detonation front. On the whole, shock phenomena give the impression of a closed science.

The same cannot be said about the effects produced by shocks. The work of compression on « incompressible » materials is comparable to chemical energies. Thus the compression alters the chemical properties and one is actually opening up a new field of physical chemistry.

This became quite clear in the lectures of ROYCE from the Lawrence Radiation Laboratory in Livermore. A systematic presentation of the results on the behavior of elements under high pressure shows that above a million atmospheres atomic volumes and compressibilities become much smoother functions of the atomic number than is observed under normal conditions. This is, of course, no great surprise. In the limit the behavior predicted by the simple statistical treatment of electrons (called the Thomas-Fermi model) should be valid. It is very amusing to observe the anomalously high compressibility of the rare-earth elements, due to the circumstance that as the metal atoms are compressed outer electrons are pushed into the partially filled f-shell. At strong compressions, the compressibility decreases as closed-shell core interactions become important.

One difficulty in obtaining truly high densities is due to the fact that the compressing shock heats the material and thus the compressibility is reduced. KEELER, from Livermore, discussed practical methods to circumvent this difficulty. This may be done if one approaches adiabatic conditions by using two consecutive shocks. An even more efficient method is to use magnetic fields as a cushion. The original shock compresses the region containing the the magnetic field thus increasing the magnetic field strength in a continuous manner. The magnetic field in turn compresses the sample which could either be a metal or could be surrounded by a skin of high conductivity. KNOEPFEL from Frascati gave an interesting review of the flux compression techniques.

The great importance of high magnetic fields was further emphasized by LINHART from Frascati. Several megaoersted can be reached. With the help of an inhomogenious field a test object can be accelerated to high velocities. The impact of this object can be used to produce strong shocks, great density and also high temperatures. All this is a part of the Frascati effort to obtain the extreme temperatures necessary to start the fusion reaction.

In private conversation LINHART mentioned to me some other, more unconventional uses of exceedingly strong magnetic fields. The question arose whether these fields might produce pairs of magnetic monopoles. The point is of a little interest since such pair production (even if the existence of monopoles is assumed) may be most difficult if the energy is made available in a conventional collision process.

Basically a simple argument is involved. One may assume that a particle and an antiparticle, with poles equal to μ and $-\mu$, are produced at a distance r from each other. Then the potential energy is $-\mu^2/r$. Due to the uncertainty relation the relative momentum will be of the order $\hbar/r$. In an appropriately simplified form the kinetic energy can be written as $\hbar c/r$. Thus the ratio of the absolute values of these energies is $\mu^2/\hbar c$ which according to DIRAC is 34.25. There is not sufficient energy for the two particles to come apart even if they should have been created. (One notes that for pair production in case of electrons or protons the ratio is $e^2/\hbar c = 1/137$ and the Coulomb attraction does not interfere with the pair production.) Actually the argument is not rigorous but at least there is a strong indication that the pair production could be ruled out by the strong interaction of the monopoles. Incidentally, the argument does not depend on the mass of the particle and antiparticle.

The situation changes, however, if an extremely strong magnetic field is introduced. In that case the magnetic field can pull the pole and the antipole apart after a potential barrier is surmounted. The Gamow factor, which governs this process, is given by e^{-G} where one can approximate the exponent by $G = (\mu^2/\hbar c)\int \mathrm{d}r/r$. The lower limit of the integration should be the Compton wavelength $\hbar/Mc$ whereas at the upper limit the magnetic field should overcome the attraction of the two particles. (This oversimplified discussion neglects the influence of the rest mass and of the temperature. It may suffice, however, to give a crude approximation.) In order to cut down the value G the upper limit of the integration cannot be a great multiple of the lower limit and one obtains for the magnetic field $H \approx M^2c^2/\hbar^2$. If we take for M the mass of the proton this gives almost 10^{20} Oe. Even in the hot core of a collapsing supernova one can reach 10^{18} Oe only by converting practically all gravitational energy into magnetic-field energy. Thus pair production of monopoles can occur at best in an incipient form. If, on the other hand, one wants to assume that the mass of the monopoles is equal to that of the electron's, magnetic fields of 10^{13} Oe will suffice. A stellar collapse probably could produce such fields.

This discussion is presented here because it shows the broad scope of the conference on the physics of high energy densities. So far, of course, all this occurred only around the dinner table (an integral part of the conference). The magnetic fields reached in the laboratory have not as yet exceeded 10^7 Oe. In fact, this field strength corresponds to an energy density not much greater than stored in chemical form under normal circumstances.

From the consideration of strong magnetic fields it is only one step to great energy concentration produced by hydromagnetic means. MAISSONIER from Frascati described experiments called « plasma focus ». In a small volume exceedingly intensive fields have been created. The violence of the processes in that small volume is evident. But whether temperature equilibria are attained and thus conditions required for thermonuclear processes are approached is less clear.

High energy densities in light are an even more fruitful field of research. This topic, laser physics, was discussed in the second major portion of the conference. That it took almost half a century to proceed from Einstein's induced emission process to functioning lasers is one of the peculiarities of contemporary physics. Once lasers have been discovered, there appeared to be no practical limit to their applicability.

Sometimes it is stated that Einstein's suggestion did not quite suffice since it is essential for the working of a laser that the induced emission should proceed in the same direction as the incident radiation. One may express this fact by saying that photons so emitted « join the crowd » of the incident photons, as is proper for particles obeying Bose statistics. Actually it is easy to see that this could not be otherwise. Induced emission is the thermodynamic twin of absorption. It must appear in the same location as the shadow. In fact one may call it a negative shadow. It follows immediately that as absorption leads to an exponential decrease in intensity, so induced emission leads to an exponential increase.

On an even more simple-minded level, that of classical electrodynamics, one can reach the conclusion that induced emission is, indeed, a negative shadow. To obtain this result one need only consider the interference of the amplitude of the spontaneous emission with the amplitude of the incident radiation. Depending on the relative phases one obtains absorption or induced emission. In both cases systematic interference can occur only in resonance and in the direction in which the wave-number vectors of the spontaneous radiation is the same as that of the incident radiation. This, of course, is the direction where under normal conditions the shadow appears. One should note that positive interference in the classical theory corresponds to a population inversion in quantum theory. That this has to be so follows from energy conservation: the detailed mechanism can be traced with the help of perturbation theory.

All this is simple. What a laser actually can do is not only impressive, it is also involved. The detailed explanation of these phenomena were given by KIDDER from the Lawrence Radiation Laboratory in Livermore. He included not only the explanation and operation of lasers but also the difficult subject of the interaction of laser light with matter. In this way a direct connection was established between quantum optics and the creation of a hot plasma.

Beautiful experiments utilizing lasers were presented by KROKHIN of the P.N. Lebedev Physical Institute, Moscow. Intense light beams were used to study the vaporization of metals and other substances such as LiD. He presented a careful discussion of these experiments. In this manner high-temperature phenomena can be explored which are otherwise not readily accessible to an experimental approach.

It is amusing to consider that the laser technique discussed by KROKHIN could be used to accelerate test objects to high velocities. Evaporation at the back of the object could give rise to a rocket effect. A homogeneous magnetic field could be used as a «rail» to guide the object and to keep it from tumbling. If the weight of the test object is chosen as one gram, velocities in the neighborhood of 10^7 cm/s could be reached without an excessive effort.

Interest had been directed toward even higher velocities, 10^8 cm/s. Such velocities would require massive equipment or, alternatively, would result in the accelearation of quite small objects. These high velocities were mentioned by the physicists of Frascati and also by WINTERBERG, Desert Research Institute, Reno. The reason for the interest is that impact at these high velocities could lead to thermonuclear reactions.

At the same time one should not forget that at the lower and more practical velocity of 10^7 cm/s one could learn a lot of physics and a lot of physical chemistry. A platelet moving at such a speed could produce stronger shocks than any on which studies have been reported. High temperatures would of course be generated. The point is that in these experiments one could explore the region in which relatively easy theoretical treatments similar to the Thomas-

Fermi model become applicable. Thus it would be possible to correlate present experience with less strong shocks and the basically simpler situation which prevails at the truly high pressures.

There is still another approach to high energy concentrations. One can study the behavior of matter in the vicinity of a nuclear explosion. Fortunately we have established the rule in the United States that effects of nuclear explosives connected with constructive uses or scientific research can be discussed in an open manner. This rule actually governs the operations which we like to call « Plowshare ».

It would be wonderful if we could make Plowshare explosions available to the international scientific community for the purpose of scientific co-operation and advancement. This will not be easy. But the first steps in this direction have been undertaken. Perhaps it is true that the first steps are the hardest.

In conclusion I want to mention an even more ambitious dream. International scientific co-operation has been hampered by secrecy. I like to hope that this situation could change. Niels BOHR has seen in the open nature of science its greatest effectiveness. Secrecy having been practiced for almost three decades will not disappear from science overnight. It can begin to disappear only if at least one government would take well-considered action for the purpose of its own long-range advantage and for the purpose of the long-range benefit of all.

During the conference I had occasion to mention this problem to several participants. In every case the answer was « It would be wonderful, but ... »

In spite of this, perhaps I should say because of this, I should be particularly happy if it would be the United States which would take the initiative to make science as open as it used to be. I see in openness a great hope for the future both for science and in a more general manner.

During the conference in Varenna the participants assembled to watch the landing of Apollo 11. That magnificent adventure, openly carried out before the eyes of the world, gave some hope that what seems impossible today need not remain impossible in the future.

Shock Waves in Condensed Media (*).

G. E. Duvall

Shock Dynamics Laboratory, Department of Physics
Washington State University, Pullman, Wash.

In this entire discussion we focus attention on a simple thought experiment. We consider a half-space with free surface normal to the x-axis and located at $x = 0$. The medium of interest lies in the region $x > 0$ or $x < 0$. A uniform pressure is applied to the surface or the surface is given an arbitrary velocity at an arbitrary time, and we inquire about the state of the medium at later times. This apparently restrictive model is in reasonable accord with the geometry and physics of most significant experiments and leads to a great variety of interesting problems.

1. – Basic shock relations.

The continuum differential equations of flow, independent of material properties, are, for one-dimensional plane flow:

$$\partial\varrho/\partial t + \partial\varrho u/\partial x = 0\,, \tag{1}$$

$$\varrho\, \mathrm{d}u/\mathrm{d}t = \varrho\, \partial u/\partial t + \varrho u\, \partial u/\partial x = -\,\partial p/\partial x\,, \tag{2}$$

$$\mathrm{d}E/\mathrm{d}t = -\,p\, \mathrm{d}V/\mathrm{d}t; \qquad V = 1/\varrho\,, \tag{3}$$

where t is time, x is Eulerian space co-ordinate, ϱ is density, u is particle or mass velocity, E is internal energy, and p is compressive stress in the x-direction, including all dynamic forces due to viscosity, stress relaxation, etc.

Application of a pressure to the surface of a half-space produces a region of change propagating out from the surface. If we suppose that a very long time has elapsed since the driving pressure was first applied at the free surface, and that pressure has been held at a constant value, p_1, then the region of change in the resulting flow may be supposed far removed from the driving surface. If the half-space fills the region $x > 0$, we may shift the origin of

---

(*) Research sponsored by the United States Office of Scientific Research under Contract No. AFOSR-69-1758.

co-ordinates to a point deep within the material and suppose that the region of change connects a uniform undisturbed state at $x = +\infty$ to a uniform compressed state at $x = -\infty$. To implement this model, we seek solutions of eqs. (1)-(3) of the form

$$\varrho = \varrho(\xi)\,, \qquad u = u(\xi)\,, \qquad p = p(\xi)\,, \qquad E = E(\xi)\,, \qquad V = V(\xi)\,,$$

where $\xi = x - Dt$ and D is a constant propagation velocity. Let the values of variables in the undisturbed state be designated by subscript « 0 » and those in the uniform state at $x = -\infty$ by subscript « 1 ». Then eqs. (1)-(3) can be integrated to yield the relations

(4) $$\varrho(D - u) = \varrho_0(D - u_0)\,,$$

(5) $$p - p_0 = \varrho_0(D - u_0)^2(V_0 - V) = \varrho_0(D - u_0)(u - u_0)\,,$$

(6) $$E - E_0 = \tfrac{1}{2}(p + p_0)(V_0 - V)\,.$$

Substitution of the final-state variables into eqs. (4)-(6) yields the *jump conditions*. They are particularly useful in the forms:

(7) $$\varrho_0/\varrho_1 = 1 - (u_1 - u_0)/(D - u_0)\,,$$

(8) $$(D - u_0)^2 = V_0^2[(p_1 - p_0)/(V_0 - V_1)]\,,$$

(9) $$E_1 - E_0 = \tfrac{1}{2}(p_1 + p_0)(V_0 - V_1)\,.$$

The undisturbed medium will usually be at rest. In deriving eqs. (4)-(9) it has been assumed to have a velocity u_0. The generality obtained by this assumption will at times be useful.

Any travelling wave which connects end states « 1 » and « 0 » and which satisfies eqs. (7)-(9) is called a *shock wave*. The locus of states (p_1, V_1) which satisfy eqs. (7)-(9) is called the « Rankine-Hugoniot (p, V) curve centered at (p_0, V_0) » or, more simply, the « Hugoniot » or « R-H curve ». It is also sometimes called the « dynamic adiabat » or « shock adiabat ». When the root of eq. (8) is taken, a duality of sign appears. If $D - u_0 > 0$ the compressed state lies to the left, the undisturbed state to the right, and the disturbance is a « forward-facing shock wave ».

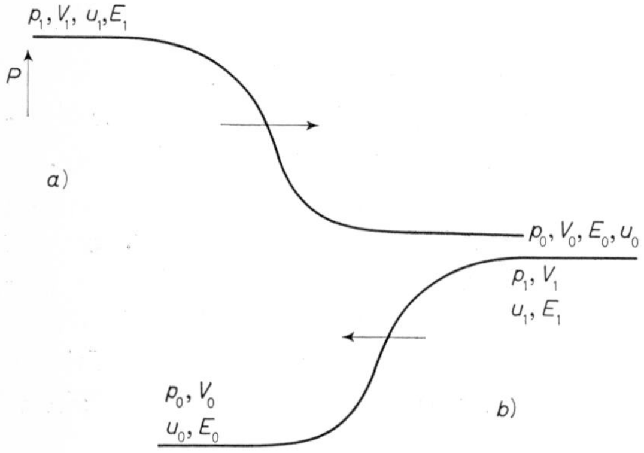

Fig. 1. – *a*) Forward-facing shock wave, $D - u_0 > 0$. *b*) Backward-facing shock wave, $D - u_0 < 0$.

If $D-u_0<0$, the undisturbed and compressed regions are interchanged and we have a « backward-facing shock wave ». These two cases are illustrated in Fig. 1.

Material is always accelerated in the direction of propagation of the shock wave. From eqs. (7)-(9) the change in mass flow or particle velocity can be calculated:

$$u_1-u_0=\pm[(p_1-p_0)(V_0-V_1)]^{\frac{1}{2}}. \tag{10}$$

If $D-u_0>0$, $u_1-u_0>0$; if $D-u_0<0$, $u_1-u_0<0$.

Equation (9) contains the thermodynamics of the shock transition and is called the « Rankine-Hugoniot equation ». For « normal » materials it can be satisfied only by compressive waves, $p_1>p_0$. We arbitrarily define « normal » materials as those for which the adiabatic (p, V) relation in one-dimensional compression, sometimes called « uniaxial strain », is concave upward. Most fluids are normal in this sense, and most solids are normal over a restricted range. For such materials the limitation of shock waves to compressive waves follows from both hydrodynamic and thermodynamic considerations. As to the latter, we find that by differentiating eq. (9) and combining it with the First and Second Laws of Thermodynamics an expression is obtained for entropy change along the Hugoniot:

$$\mathrm{d}S_1/\mathrm{d}p=[(V_0-V_1)/2T_1]\left[1-\frac{(p_1-p_0)/(V_0-V_1)}{|\mathrm{d}p/\mathrm{d}V|_{V_1}}\right], \tag{11}$$

where T_1 is temperature at (p_1, V_1). The bracket in eq. (11) is positive for all $p_1>p_0$ if the Hugoniot is concave upward. This is illustrated in Fig. 2, where it is obvious that the slope of the chord from 0 to A is less in magnitude than the slope of the tangent at A.

The increase in entropy in the shock front is produced by the presence of dynamic or irreversible forces associated with viscosity, stress-relaxation and the like. These forces are responsible for maintaining the linear relation between p and V in eq. (5): at any point of compression the total compressive stress p in the x-direction is the sum of an equilibrium and of a dynamic contribution. A physically unreal but mathematically interesting problem is to let the dynamic stress vanish and to represent the equilibrium stress by an equation of state,

$$p=\bar{p}(V, E).$$

In the absence of other irreversible processes, such as heat conduction, the shock front then becomes a mathematical discontinuity connecting states (p_0, V_0, u_0, E_0) and (p_1, V_1, u_1, E_1) [1].

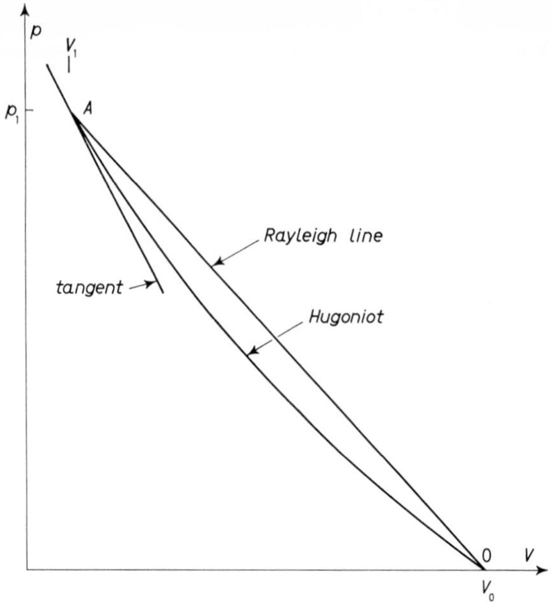

Fig. 2. – Entropy increases with pressure for a normal material.

A more realistic approximation is to add a viscous stress:

$$p = \bar{p}(V, E) - \alpha \, \mathrm{d}V/\mathrm{d}t \,. \tag{12}$$

If the dependence on E can be neglected, eqs. (5) and (12) together become a differential equation for the density profile in the shock front:

$$\alpha D \, \mathrm{d}V/\mathrm{d}\xi = p_0 + m^2(V_0 - V) - \bar{p}(V) \,. \tag{13}$$

In case the E-dependence of eq. (12) cannot be neglected, eqs. (5), (6) and (12) provide a description of the profile.

The sign of the entropy change in eq. (11) is directly related to the propagation process. For the shock wave of Fig. 1, moving into stationary material ($u_0 = 0$):

$$D^2 = V_0^2(p_1 - p_0)/(V_0 - V_1) > V_0^2 |\mathrm{d}p/\mathrm{d}V|_{V_0} \equiv c_0^2 \tag{14}$$

provided $(\partial^2 p/\partial V^2)_S > 0$ at (p_0, V_0). The velocity of sound is defined as $c = V(-\partial p/\partial V)_S^{\frac{1}{2}}$. The identity between $(\partial p/\partial V)_S$ and $\mathrm{d}p/\mathrm{d}V$ on the Hugoniot at (p_0, V_0) is possible because isentrope and Hugoniot have a second-order contact at the foot of the Hugoniot. This will be discussed later by ROYCE and KEELER. The inequality in (14) tells us that the shock wave overtakes any acoustic wave ahead of it. Under the same conditions a disturbance behind the shock front overtakes the shock, provided, of course, that both are travelling in the same direction. This can be seen in the following way: The propagation velocity of a forward-facing disturbance is $u_1 + c_1$; if $u_1 + c_1 > D$, the disturbance will overtake the shock. Equation (11) can be converted into a differential equation for the Hugoniot by assuming that $S = S(p, V)$ and proceeding to eliminate dS/dp. The result is:

$$\mathrm{d}p_1/\mathrm{d}V = \{(\partial p_1/\partial V_1)_S + \Gamma(p_1 - p_0)/2V_1\}/\{1 - \Gamma_1(V_0 - V_1)/2V_1\}\,, \tag{15}$$

where $\Gamma = V(\partial p/\partial T)_V/C_V$. It has already been shown that $-\mathrm{d}p/\mathrm{d}V > > (p - p_0)/(V_0 - V)$ in Fig. 2, and this with eq. (15) leads to the inequality:

$$c_1^2/(D - u_1)^2 - \Gamma_1(V_0 - V_1)/2V_1 > 1 - (\Gamma_1/2V_1)(V_0 - V_1)\,. \tag{16}$$

Then $c_1^2 > (D - u_1)^2$ or

$$u_1 + c_1 > D_1\,, \tag{17}$$

provided the Rayleigh line is less steep than the tangent to the Hugoniot at (p_1, V_1). We say that in this case the flow behind the shock is *subsonic*. A single shock connecting (p_0, V_0) and (p_1, V_1) is accordingly stable. If $D_1 > u_1 + c_1$, then (p_1, V_1) is a point of instability and the possibility of forming a second shock exists [2].

Experimentally produced shock waves seldom exactly satisfy the requirements of steady flow assumed in deriving the jump conditions. The states connected by the shock transition may not be precisely uniform or the shock wave has not propagated far enough to become steady. However, the experimental conditions may be very close to the theoretical assumptions, and it is quite likely that errors involved in applying the jump conditions to experiments are less than those originating from other sources. Such errors may be significant if gradients in adjacent regions are comparable to those in the shock or if time has not been sufficient for the flow to become steady and the curvature of the Hugoniot is large. The resolution of this question is a constant source of concern to experimentalists and no satisfactory resolution has been made. BLAND [3] has considered the development of a step change in pressure for a viscous material and concludes that the shock profile is essentially steady after travelling a distance of five shock thicknesses from the source. This is an interesting result. The difficulty in applying it is that, in general, the steady shock thickness is unknown.

2. – Rarefactions and characteristics.

Referring to our original model of a pressure on a half-space, we suppose that after being held at constant value p_1 while the shock was being formed, we then reduce the pressure to its ambient value p_0. A forward-facing rarefaction is produced, and we seek an appropriate method for describing the propagation of this rarefaction.

According to the discussion of Sect. **1**, waves of rarefaction cannot be steady in the sense of eqs. (4)-(6) for normal materials; *i.e.*, there are no solutions of the form $\varrho = \varrho(x - Dt)$, etc., for constant D. We do know, however, that waves of infinitesimal amplitude satisfy the simple wave equation, and that solutions of this are in the form

$$f(x - ct) + g(x + ct) .$$

That is, they consist of forward-facing and backward-facing waves. For such infinitesimal waves we know that

$$\mathrm{d}p = \varrho c \, \mathrm{d}u \tag{18a}$$

for forward-facing waves and

$$\mathrm{d}p = - \varrho c \, \mathrm{d}u \tag{18b}$$

for backward-facing waves. Here c is the velocity with which these infinitesimal disturbances travel into material at rest. Equations (18) are in accord with the jump conditions if we apply them to waves of infinitesimal amplitude.

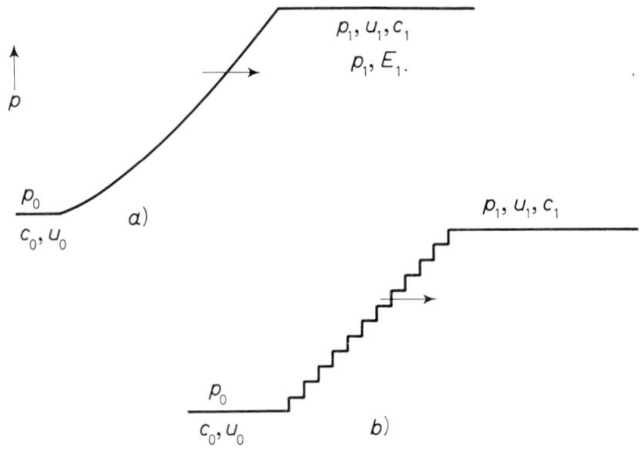

Fig. 3. – *a*) Continuous forward-facing rarefaction wave. The propagation velocity at each point is $u + c$. *b*) Representation of continuous rarefaction as sequence of small increments for which $\delta p = \varrho c \, \delta u$.

If we consider our forward-facing rarefaction to be represented by a sequence of jumps, $\mathrm{d}p$, to lower pressure, accompanied by a sequence of jumps, $\mathrm{d}u$, to smaller particle velocity, we can integrate eq. (18) to obtain the relation between p and u at any point in the rarefaction. This is illustrated in Fig. 3

and the result is

$$(19a)\qquad u - u_1 = \int_{p_1}^{p} \mathrm{d}p/\varrho c\,.$$

The equivalent relation for backward-facing rarefactions is

$$(19b)\qquad u - u_1 = -\int_{p_1}^{p} \mathrm{d}p/\varrho c\,.$$

Defining the new variable

$$(20)\qquad l = \int^{p} \mathrm{d}p/\varrho c\,,$$

Equations (19) can be written

$$(21)\qquad u - l = \text{const} = u_1 - l_1 \equiv -2s_1$$

for forward-facing rarefactions, and

$$(22)\qquad u + l = \text{const} = u_1 + l_1 = 2r_1$$

for backward-facing ones. The constants r_1 and s_1 are called « Riemann invariants ».

To understand the implications of eqs. (21) and (22) more fully, consider the special case of isentropic flow, for which eq. (3) reduces to $\mathrm{d}S/\mathrm{d}t = 0$. Then eqs. (1) and (2) can be combined to form an equivalent set of « characteristic equations »:

$$(23)\qquad [\partial/\partial t + (u + c)\,\partial/\partial x](u + l) = 0\,,$$

$$(24)\qquad [\partial/\partial t + (u - c)\,\partial/\partial x](u - l) = 0\,.$$

The curves on which $\mathrm{d}x/\mathrm{d}t = u \pm c$ are called « characteristic curves »; eqs. (23) and (24) are thus equivalent to the set:

$$(25)\qquad C+:\quad u + l = \text{const} = 2r\,;\qquad \mathrm{d}x/\mathrm{d}t = u + c\,,$$

$$(26)\qquad C-:\quad u - l = \text{const} = -2s\,;\qquad \mathrm{d}x/\mathrm{d}t = u - c\,.$$

In words, r is constant on the $C+$ characteristic curves for which $\mathrm{d}x/\mathrm{d}t = u + c$; s is constant on the $C-$ characteristics. Equations (25) and (26) are true when the flow is a mixture of forward- and backward-facing waves. For forward-facing waves, eq. (21) is also true; then

$$C+: \quad u + l = 2r, \tag{27}$$

$$C-: \quad u - l = 2s_1. \tag{28}$$

For backward-facing waves:

$$C+: \quad u + l = 2r_1, \tag{29}$$

$$C-: \quad u - l = -2s. \tag{30}$$

To summarize:

Forward-facing shock waves running into material with velocity u_0 and in the state (p_0, V_0) satisfy the relations

$$D - u_0 = V_0\sqrt{(p_1 - p_0)/(V_0 - V_1)}, \tag{31}$$

$$u_1 - u_0 = \sqrt{(p_1 - p_0)(V_0 - V_1)}. \tag{32}$$

Backward-facing shocks, the relations

$$D - u_0 = -V_0\sqrt{(p_1 - p_0)/(V_0 - V_1)}, \tag{33}$$

$$u_1 - u_0 = -\sqrt{(p_1 - p_0)(V_0 - V_1)}. \tag{34}$$

For forward-facing rarefactions running into material in the uniform state (p_0, V_0) with velocity u_0,

$$u - u_0 = l - l_0. \tag{35}$$

For backward-facing rarefactions moving into the same state,

$$u - u_0 = -(l - l_0). \tag{36}$$

For rarefactions there are no equations analogous to (31) and (33) because there is no single propagation velocity associated with a rarefaction. Waves for which eq. (35) or eq. (36) applies are called « simple waves ». In the general case of isentropic flow in which, for example, two rarefactions are interacting, eqs. (25) and (26) describe the flow.

In condensed materials compressed by shock waves to about 15 % of their initial volume or less, the shock wave is called « weak ». Then the entropy-change is small and the shock can be treated as a simple wave to which either eq. (35) or (36) applies. In this approximation the interactions of shock waves and rarefactions can be calculated from eqs. (25) and (26).

3. – Elementary wave interactions.

Equations (32), (34), (35) and (36) uniquely define and limit the values of particle velocity, u, which can be achieved by simple shock or rarefaction from a given state (p_0, V_0, u_0). This limitation on states which can be reached in a single wave transition supplies a powerful tool for thinking about and calculating the fields of high-amplitude waves. The problem is transformed into a «hodograph» plane in which the variables are (u, p), (u, l), (r, s) or some equivalent set. We shall use u, p here because of continuity conditions on u and p at an interface or boundary. The significance of this choice will appear later.

Various useful representations of a shock and of a rarefaction are shown in Fig. 4. In 4 *a*) is a cross-section of a half-space to which a pressure p_1 was applied at $t=0$ and released at $t=t_0$. The pressure profile at this particular $t>t_0$ is shown in 4 *b*). It consists of a forward-facing shock, designated $\mathcal{S}_+$, a region of uniform pressure p_1 and particle velocity u_1, and a rarefaction $\mathcal{R}_+$. The notations $\mathcal{S}$ and $\mathcal{R}$ are introduced here to denote shock and rarefaction waves, respectively. Forward-facing waves are denoted by the subscript «$+$», backward-facing by «$-$». In Fig. 4 *c*) the flow is shown in the (x, t) plane. Region I is the uniform initial state (p_0, V_0, u_0) with $u_0>0$. The shock front, $\mathcal{S}_+$, has constant slope until the following rarefaction overtakes it, reducing its amplitude and velocity. Region II is the uniform state (p_1, u_1, V_1) behind the shock. Region III is the rarefaction $\mathcal{R}_+$ in which pressure and particle velocity are diminishing. Region IV is again at the ambient pressure p_0 but volume and particle velocity now differ from V_0 and u_0. The path OAB is the trace of the half-space surface, sometimes called the «piston path», $\mathcal{P}$. The dashed curve is the path of a single particle or mass element traversed successively by $\mathcal{S}_+$ and $\mathcal{R}_+$. Figure 4 *d*) shows the wave process in the (p, V) plane. The initial shock compression is along the Rayleigh line to the state B on the Hugoniot. The rarefaction, assumed to be isentropic, expands the material along the dashed isentrope to the final state $C(V_0', p_0, u_0')$. In Fig. 4 *e*) the process is shown in the (p, u) plane. The straight line AB with slope $\mathrm{d}p/\mathrm{d}u=\varrho_0(D-u_0)$ is the image of the Rayleigh line. The compressed state B lies on the image of the Hugoniot curve and the dashed curve BC is the image of the isentrope of Fig. 4 *d*). Because the shock process is entropic and because most materials have positive thermal expansion coefficients, the final state (u_0', p_0) is normally to the left of (u_0, p_0) for forward-facing waves.

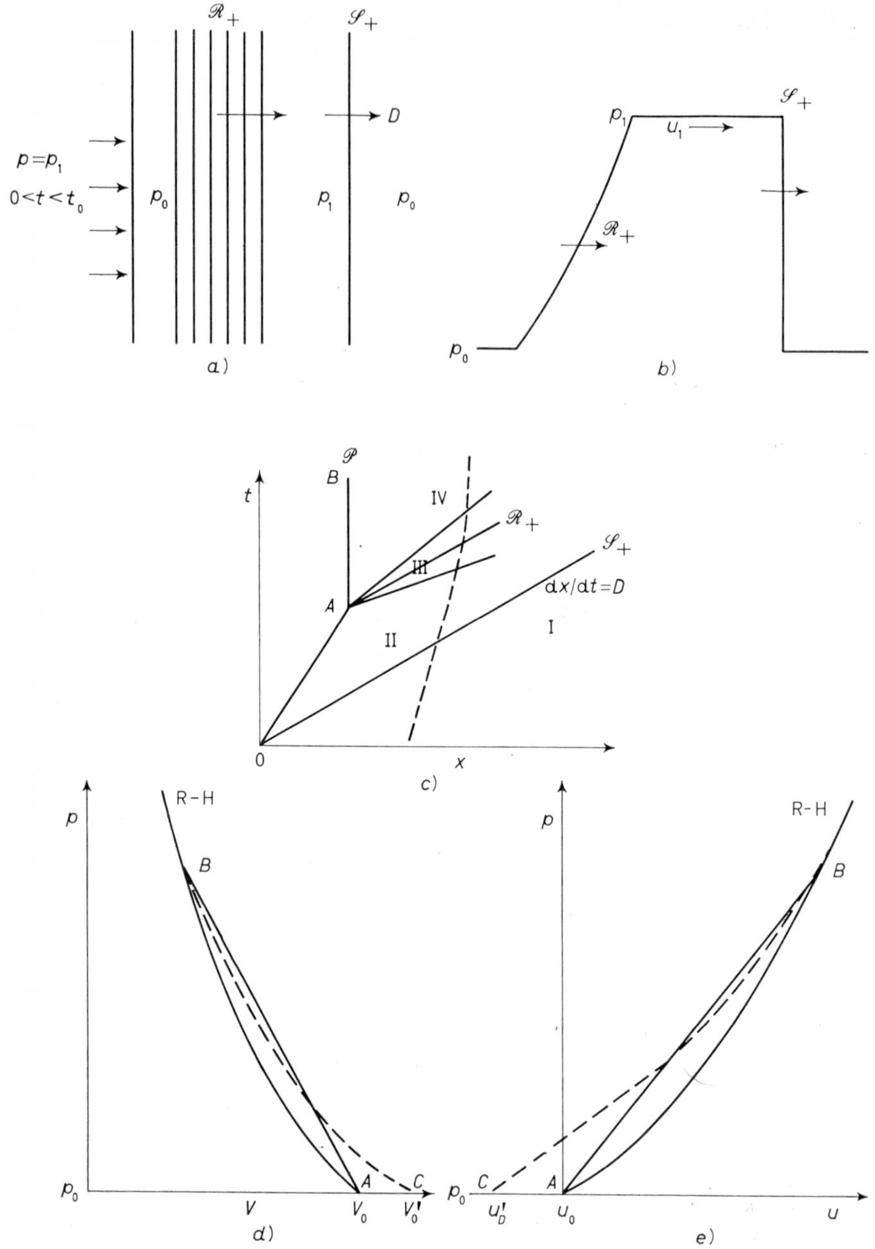

Fig. 4. – Forward-facing rarefaction overtaking a shock. *a*) planes of constant phase in half-space; *b*) pressure profile, $t > t_0$; *c*) $(x\text{-}t)$ diagram; *d*) $(p\text{-}V)$ diagram; *e*) $(p\ u)$-plane.

A set of similar figures can be drawn for backward-facing waves. Figure 4 *d*) is then unchanged. The others are produced by reflection in the vertical axis, except that appropriate adjustments need to be made depending upon the magnitude and sign of u_0.

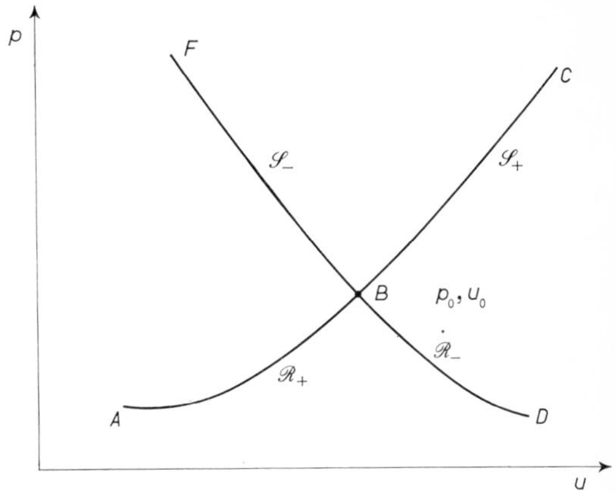

Fig. 5. – Wave transitions in the (p, u) plane.

The essential point of the following discussion is that states connected by shocks or by simple waves to an arbitrary state (p_0, u_0) must lie on fixed curves given by eqs. (32), (34), (35), (36), provided there are no time-dependent terms in the constitutive relations. This is illustrated in Fig. 5. AB is represented by eq. (35), BC by (32), BF by (34) and BD by (36). AB and BC have a second-order contact at B, arising from the second order contact between isentrope and Hugoniot at the initial state (p_0, V_0). So do BD and BF.

Some applications of the foregoing principles are illustrated in Fig. 6-13.

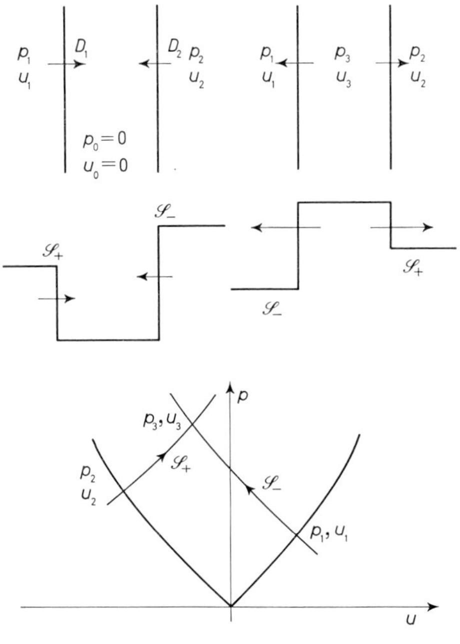

Fig. 6. – Collision of two shocks.

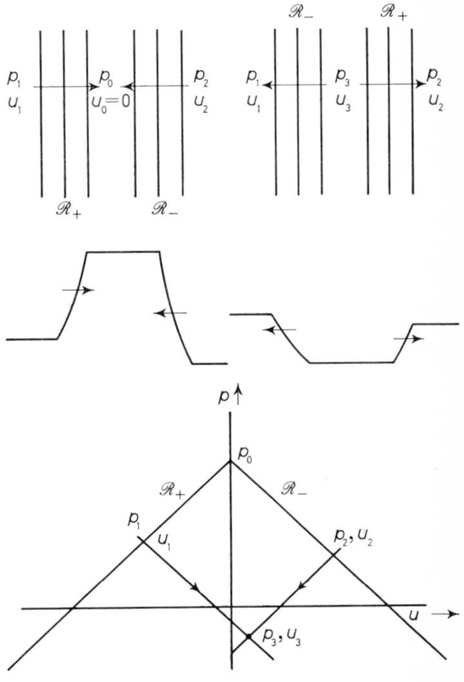

Fig. 7. – Collision of two rarefactions.

Figure 6: Two shocks approach one another, separated by a uniform state $p_0 = 0 = u_0$. They are represented by the points (p_1, u_1) and (p_2, u_2) in the (p, u) plane. After collision there are two waves drawing apart, separated by a new state (p_3, u_3) and running into the uniform states (p_1, u_1) and (p_2, u_2), respectively. The transition from (p_2, u_2) to (p_3, u_3) occurs across a forward-facing wave, that from (p_1, u_1) across a backward-facing wave. The state between the separating waves must be uniform in (p, u) because of continuity conditions on p and u. All these conditions are satisfied by the intersection of the $\mathscr{S}_-$ and $\mathscr{S}_+$ curves from (p_1, u_1) and (p_2, u_2), respectively.

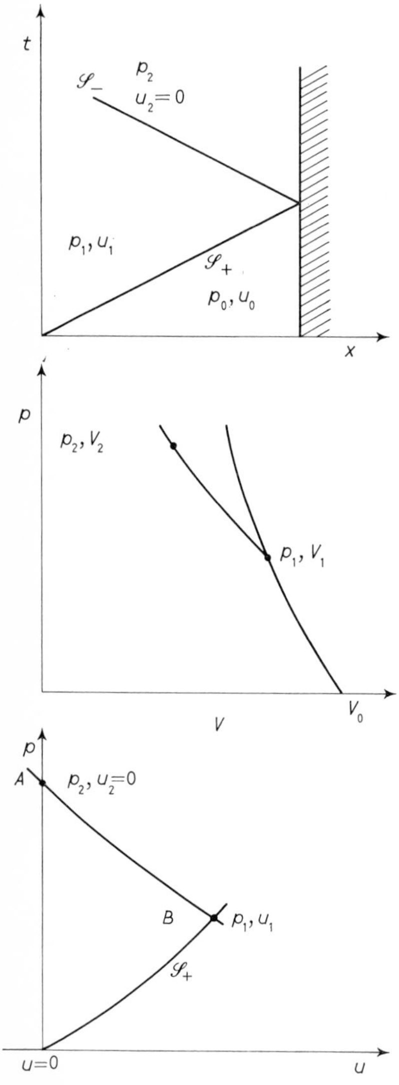

Fig. 8. – Reflection of a uniform shock at a rigid wall. $p_2/p_1 > 2$ if $\mathrm{d}^2p/\mathrm{d}V^2 > 0$.

Figure 7: Collision of two rarefactions: The argument is analogous to the previous one. The initial state p_0, $u_0 = 0$, is connected by rarefactions to (p_1, u_1) and (p_2, u_2). The final state reached from (p_1, u_1) and (p_2, u_2) must lie at the intersection of wave transition curves passing through these states and must be a state of uniform p and u. This is satisfied by (p_3, u_3).

Figure 8: Reflection of a uniform shock at a rigid wall: The initial shock, $\mathscr{S}_+$, carries material from $(u_0 = 0 = p_0)$ to (p_1, u_1). After reflection the state (p_1, u_1) is connected to the final state $(p_2, u_2 = 0)$ by a backward-facing wave. The final state must lie on the curve for backward-facing transitions in the (p, u) plane and also on the $u = 0$ axis. These conditions are satisfied at point A. Note particularly that if the $\mathscr{S}_+$ curve is concave upward in the (p, u) plane, $p_2 > 2p_1$. The $\mathscr{S}_-$ curve through B is the mirror image of the $\mathscr{S}_+$ curve in a vertical axis through B.

Figure 9: Reflection of a uniform shock at a free surface: This is similar to the preceding, but the final state must now lie on the $p = 0$ axis. For condensed

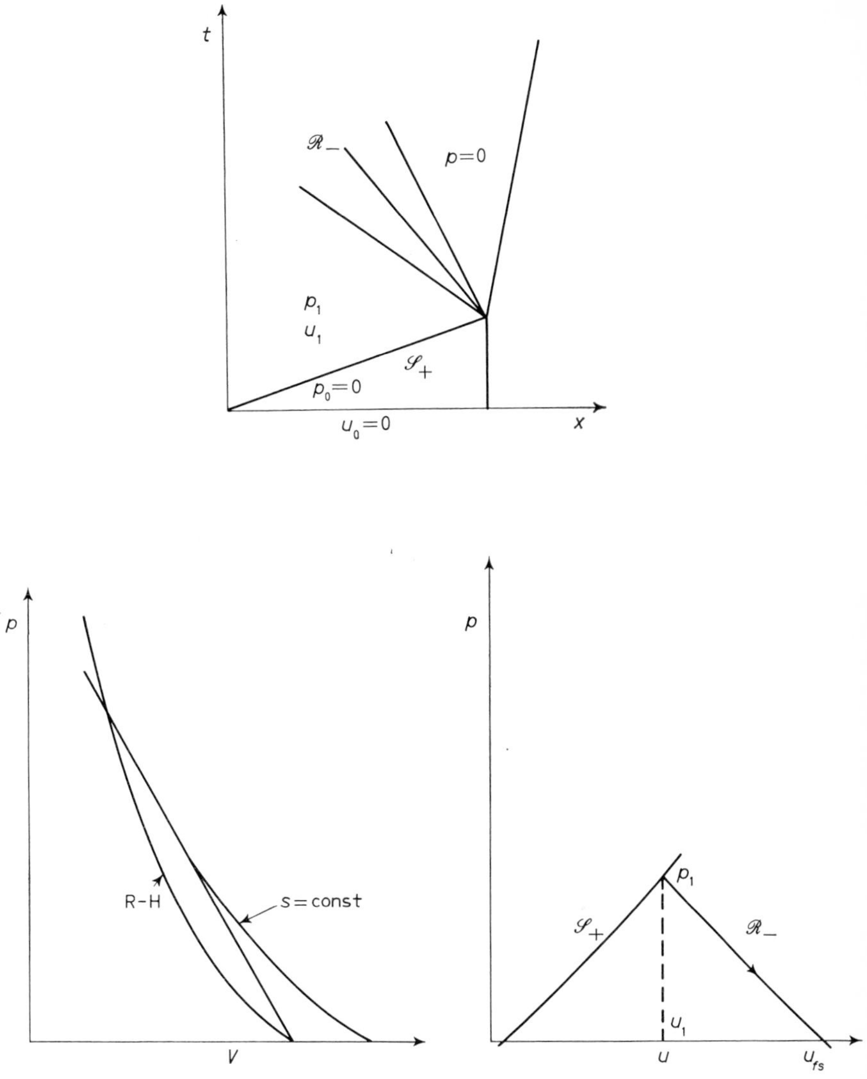

Fig. 9. – Uniform shock on a free surface.

material, u_{fs} is approximately equal to and usually slightly greater than $2u_p$.

Figure 10: Transmission of a uniform shock through an interface: The shock is incident from material I on material II. The (p, u) curves labelled I and II are images of the Hugoniot (p, V) curves of the respective materials. The final state (p_2, u_2) is common to both materials since p_2 and u_2 are to be continuous across the interface. This final state must be reached from (p_1, u_1)

by a backward-facing wave in material I and from ($p_0 = 0 = u_0$) by a forward-facing wave in II. The required state is as shown. Material II was arbitrarily chosen so that its Hugoniot image lies above I. In that case the wave reflected back into I is a shock wave. For II below I, the reflected wave is a rarefaction. It sometimes happens that the two curves cross; then the reflected wave is a shock or a rarefaction, depending on the amplitude of the incident shock. For an incident shock at the intersection there is no reflection. An analytic expression for the amplitude of a reflected shock can readily be obtained from the jump conditions written for the incident shock, the reflected shock, and the transmitted shock:

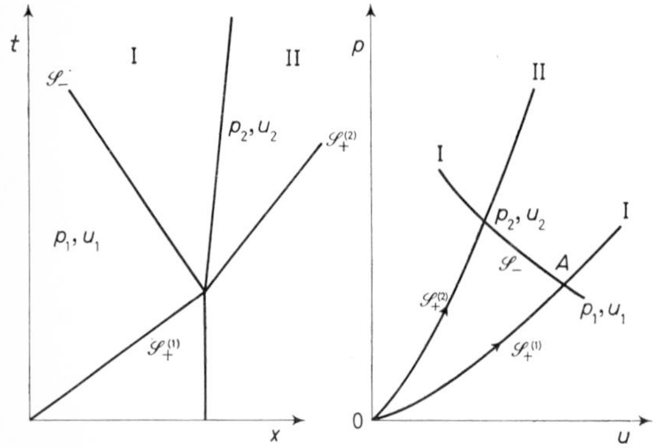

Fig. 10. – Transmission through an interface; $\varrho_{02} D_2 > \varrho_{01} D_1$.

$$\frac{p_2 - p_0}{p_1 - p_0} = \frac{1 + \varrho_0 D_1/\varrho_1 D'_{21}}{\varrho_0 D_1/\varrho_1 D'_{21} + \varrho_0 D_1/\varrho'_0 D_2}, \tag{37}$$

where

ϱ_0 = initial density of material I,

ϱ'_0 = initial density of material II,

D_1 = velocity of incident shock,

D'_{21} = velocity of reflected shock *relative to the material ahead* $= u_1 - D_{21}$,

D_{21} = velocity of reflected shock in laboratory co-ordinates,

D_2 = velocity of transmitted shock,

ϱ_1 = density behind initial shock.

The products ϱD appearing in eq. (37) are *shock impedances* corresponding to the various waves. Shock impedance for a given transition is equal to the magnitude of the slope of the corresponding chord in the (p, u) plane. For example, $\varrho_0 D_1$ is the slope of the chord OA in Fig. 10. In the limit of small amplitude waves, the shock impedance becomes equal to the acoustic impedance. Then eq. (37) reduces to the expression for acoustic reflection:

$$(p_2 - p_0)/(p_1 - p_0) = \delta p_2/\delta p_1 = 2\varrho'_0 c'_0/(\varrho_0 c_0 + \varrho'_0 c'_0). \tag{38}$$

For a rigid wall $\varrho_0' D_2 \to \infty$ and eq. (37) becomes

$$\frac{p_2 - p_0}{p_1 - p_0} = \frac{\varrho_1 D_{21}' + \varrho_0 D_1}{\varrho_0 D_1} \xrightarrow[p_1 \to p_0]{} 2 \,. \tag{39}$$

For a free surface, $\varrho_0' D_2 = 0$ and eq. (37) becomes

$$(p_2 - p_0)/(p_1 - p_0) = 0 \,. \tag{40}$$

Equation (37) applies exactly only when the reflected wave is a shock. Because of the second-order contact between the rarefaction branch and the shock branch of the cross curve passing through A, the formula provides a very good approximation for rarefactions in condensed materials.

Figure 11: Weak-shock approximation: Equations (32) and (35) yield very nearly the same curve when $(V_0 - V)/V_0 \lessdot 0.15$. Then either can be used and the (p, u) plane can be mapped with a set of identical curves and their mirror images. These curves can be regarded as transformations from the (x, t) plane. A curve along which transitions can be made by forward-facing waves is an image of a $C-$ characteristic and is called a Γ_- characteristic.

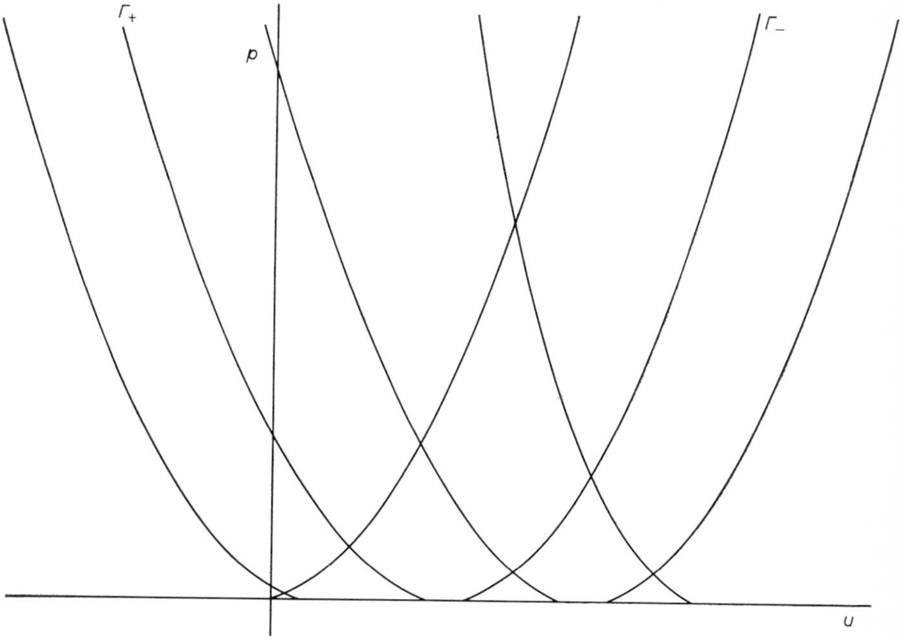

Fig. 11. – Weak shock approximation, $S = \text{const}$. No distinction between adiabat and Hugoniot; (p-u) plane can be mapped with images of the $C+$ and $C-$ characteristics. $u - l = \text{const}$ on Γ_-, $u + l = \text{const}$ on Γ_+.

Transitions through backward-facing waves are along Γ_+ characteristics. Transformation from the (x, t) to the (x, u) plane is not necessarily one-one. For example, a region of uniform state in the (x, t) plane maps into a single point in the (p, u) plane; a simple wave in the (x, t) plane maps into one of the Γ characteristics.

An example of application of the characteristic mapping of the (p, u) plane in the weak shock approximation is shown in Fig. 12. A forward-facing rarefaction travels into a uniform state $(p_0, u_0 = 0)$ bounded by a rigid wall, reducing it to the uniform state (p_1, u_1). The reflection process is represented in the (x, t) plane by drawing $C+$ and $C-$ characteristics at arbitrarily close intervals and labelling the fields between the characteristics as shown. In relating this flow to the (p, u) plane, each field in the (x, t) plane is considered a region of uniform state, represented by a single point in the (p, u) plane. Thus the regions 00, 01, 02, 03 map into the points 00, 01, 02, 03 on the Γ_- characteristic $(u - l = l_0)$ passing through $(p_0, u_0 = 0)$. Region 11 in (x, t) is reached from 01 across a backward-facing wave, therefore it lies on a Γ_+ characteristic passing through 01. The rigid boundary condition $u = 0$ must be satisfied in 11, therefore 11 in (p, u) lies at the intersection of the Γ_+ through 01 and the $u = 0$ axis. Similarly the transition from 02 to 12 to 22 takes place along a Γ_+ characteristic and terminates at $u = 0$. The intermediate state 12 is reached via a backward-facing wave from 02 and a forward-facing wave from 11, etc. This simple step by step procedure will succeed in unraveling the most complicated flow problems in (x, t) geometry provided only that the weak shock approximation is valid. With the aid of a large drawing board and considerable patience, the procedure is useful for graphical computation; and the solution of a few problems by this means is certain to establish the elements of finite amplitude wave propagation firmly in the mind of the student. It should be noted that the C characteristics are actually curved in the region of interaction, and the slopes of those characteristics

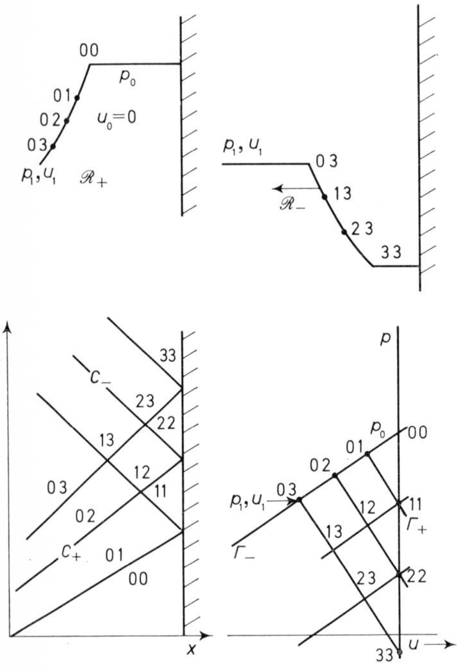

Fig. 12. – Rarefaction on a rigid wall.

bounding the fields 11, etc., are determined from the values of p and u at the intersections of the corresponding Γ characteristics.

A further application of the weak shock approximation is shown in Fig. 13, where a common experimental problem is described in the (x, t) and (p, u) planes. A pressure-free flyer plate with uniform velocity w collides at $t = 0$ with a stationary pressure-free target, producing shock waves which travel forward into the target and backward into the flyer. The shock in the flyer reflects from the back face as a rarefaction, and there is subsequently a succession of reflections between free surface and interface which ultimately bring the flyer to a stop. The sequence of states, preserving continuity of u and p, is shown in the figure. The time between reflections is twice the travel time through the flyer, so the time to effectively stop the flyer can be estimated.

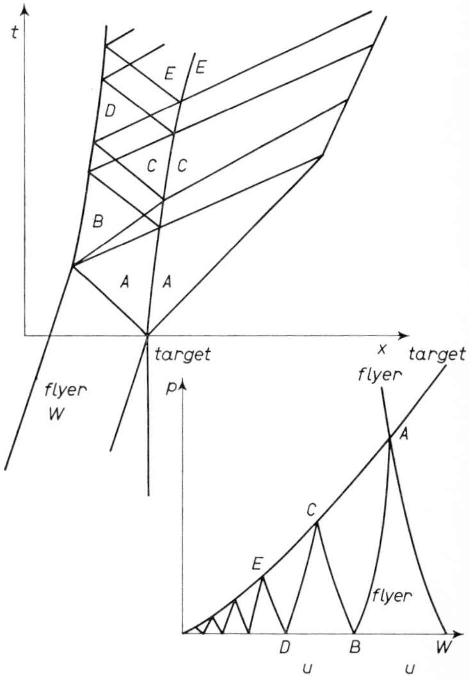

Fig. 13. – Flyer plate colliding with target.

4. – Elastic-plastic solids.

Much of the material in the preceding Sections has been couched in the language of fluids, though it applies equally well to solids, and has made but little reference to the explicit material properties involved. In this Section we become more specific about materials and examine more explicitly propagation effects in these models.

It must be recognized at the outset that there are no physically complete descriptions of the thermomechanical properties of solids. Hooke's law of elasticity is commonly used for small strains in metals and brittle solids, though there are materials to which it does not apply. Some materials are viscoelastic even at small strains, and the proper description of such materials is subject to current research. All solids fail through flow or fracture at some stress, and above this level, Hooke's law is totally improper. A satisfactory theory of fracture is far from realization; and the theory of plastic failure, while far advanced compared to fracture, is still logically incomplete and is often at odds with experimental results. However, the theory of plasticity is more completely formulated than other models of anelasticity, and its applications in shock propagation will be discussed here.

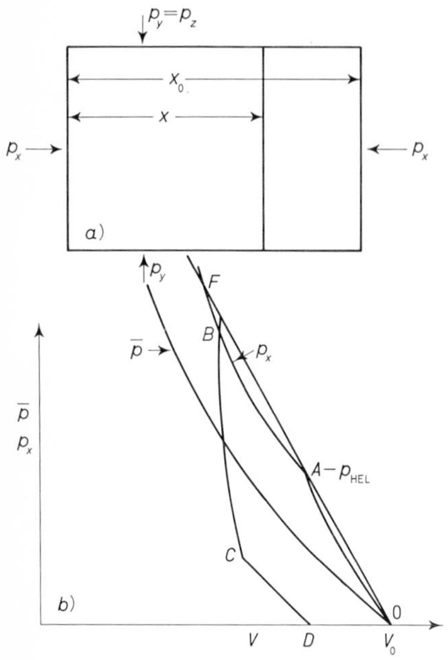

Fig. 14. – *a*) A parallelepiped of initial length x_0 has been compressed uniaxially to length x. *b*) Stress-strain relations for the sample of Fig. 14 *a*).

We restrict ourselves to the case of uniaxial strain existing in plane shock waves. We suppose a small element of volume to be compressed in the x-direction only and consider the relations between stress and strain. The notation used is shown in Fig. 14 *a*) and the expected stress-strain relation in a cycle of compression and rarefaction is shown in Fig. 14 *b*). Principal coordinates of the stress and strain matrices are (x, y, z) with x the direction of shock wave propagation. In order to maintain the condition of uniaxial strain while p_x is applied, p_y and p_z must be adjusted so as to maintain the lateral dimensions unchanged. Symmetry requires that $p_y = p_z$.

The most common assumptions of elasto-plasticity are:

i) Material response is elastic as long as deformation stresses do not exceed a characteristic value. The most commonly used criterion of failure is the von Mises condition

$$(41a)\qquad (p_x - p_y)^2 + (p_x - p_z)^2 + (p_y - p_z)^2 \leqslant 2Y^2 ,$$

where Y is the yield stress in simple tension. In uniaxial strain this becomes

$$(41b)\qquad |p_x - p_y| \leqslant Y .$$

If the inequality applies, the material is elastic and satisfies Hooke's law:

$$(42a)\qquad p_x = \lambda\theta + 2\mu\varepsilon_x ,$$

$$(42b)\qquad p_y = \lambda\theta + 2\mu\varepsilon_y ,$$

$$(42c)\qquad p_z = \lambda\theta + 2\mu\varepsilon_z ,$$

where $\theta = \varepsilon_x + \varepsilon_y + \varepsilon_z$. In uniaxial strain, $\varepsilon_x = (V_0 - V)/V_0$, $\varepsilon_y = \varepsilon_z = 0$. If the equality holds, the material is in the plastic state.

ii) In the plastic state, every increment in strain is the sum of an elastic and a plastic increment:

(43a) $$\mathrm{d}\varepsilon_x = \mathrm{d}\varepsilon_x^e + \mathrm{d}\varepsilon_x^p ,$$

(43b) $$\mathrm{d}\varepsilon_y = \mathrm{d}\varepsilon_y^e + \mathrm{d}\varepsilon_y^p ,$$

(43c) $$\mathrm{d}\varepsilon_z = \mathrm{d}\varepsilon_z^e + \mathrm{d}\varepsilon_z^p .$$

iii) There is no plastic dilatation:

(44) $$\mathrm{d}\varepsilon_x^p + \mathrm{d}\varepsilon_y^p + \mathrm{d}\varepsilon_z^p = 0 .$$

iv) The stress is supported solely by the elastic strain:

(45a) $$\mathrm{d}p_x = \lambda\,\mathrm{d}\theta + 2\mu\,\mathrm{d}\varepsilon_x^e ,$$

(45b) $$\mathrm{d}p_y = \lambda\,\mathrm{d}\theta + 2\mu\,\mathrm{d}\varepsilon_y^e ,$$

(45c) $$\mathrm{d}p_z = \lambda\,\mathrm{d}\theta + 2\mu\,\mathrm{d}\varepsilon_z^e ,$$

where λ and μ are, in general, functions of the density.

As p_x is increased from zero, the response is initially elastic and $\varepsilon_y = \varepsilon_z = 0$. Then

(46) $$p_x - p_y = (1-2\nu)p_x/(1-\nu) ,$$

where $\nu = \lambda/2(\lambda+\mu)$ is Poisson's ratio. The yield stress is reached at a value of p_x called the «Hugoniot elastic limit», denoted by p_{HEL}. From eqs. (41) and (46):

(47) $$p_{\mathrm{HEL}} = (1-\nu)\,Y/(1-2\nu) .$$

For further increases in p_x, the material is in the plastic state. Then

(48) $$p_x = \bar{p} + \tfrac{2}{3}(p_x - p_y) = \bar{p} + 2Y/3 ,$$

where $\bar{p} = (p_x + p_y + p_z)/3$, a function of density and internal energy alone. Referring to Fig. 14 *b*), eq. (48) applies to the segment *AB* of the p_x curve. The slope of the (p_x, V) curve in the elastic region is, from eqs. (42):

(49) $$\mathrm{d}p_x/\mathrm{d}V = -(\lambda + 2\mu)/V_0 = -(K + 4\mu/3)/V_0 ,$$

where K is bulk modulus. In the plastic region, AB, the slope is, for constant Y, from eq. (48)

$$\mathrm{d}p_x/\mathrm{d}V = \mathrm{d}\bar{p}/\mathrm{d}V = -K/V. \tag{50}$$

In accord with eq. (50), it is convenient to define the incremental dilatation as $\mathrm{d}V/V$. Bulk modulus normally increases with $\bar{p}$, so AB is normally concave upward. The yield stress, Y, is in general a function of plastic work and density. In such case eq. (50) is augmented by a $\mathrm{d}Y/\mathrm{d}V$ term. In any case the offset of p_x from the hydrostat, $\bar{p}$, is always $2Y/3$.

At point B in Fig. 14 *b*) we suppose that a change is made from monotonically increasing to monotonically decreasing p_x. Equation (41) must again be examined to determine whether the mass element is in the elastic or plastic state. During the initial compression process, p_x increased more rapidly than p_y until yield occurred. During unloading, p_x decreases more rapidly than p_y until yielding again occurs. Thus the portion BC of the unloading curve is elastic until $p_y - p_x = Y$ at C. From C to D, unloading is plastic and the unloading curve lies below the hydrostat by $\frac{2}{3}Y$.

Referring to the discussion following eq. (17), we see that point A of Fig. 14 *b*) may be a point of instability for single shock compressions. To see that this is indeed the case, suppose that a shock wave has been generated with amplitude p_{HEL}, traveling with speed

$$D_E = [V_0(\lambda + 2\mu)]^{\frac{1}{2}}.$$

The velocity of this shock front relative to the material behind it is

$$D_e - u_E = (V_A/V_0)D_E = V_A\sqrt{(\lambda + 2\mu)/V_0}\,. \tag{51}$$

If an additional compression of small amplitude is produced to follow the already established shock, it will travel with velocity c_A relative to the material ahead of it, where, according to eq. (50),

$$c_A = \sqrt{KV_A} = V_A\sqrt{(\lambda + 2\mu/3)/V_A}\,.$$

Comparing this with eq. (51) we find that

$$(D_E - u_E)^2/c_A^2 = (3V_A/V_0)(1-\nu)/(1+\nu) \simeq 3(1-\nu)/(1+\nu) = \tfrac{3}{2} \quad \text{for } \nu = \tfrac{1}{3}, \tag{52}$$

since $V_A/V_0 \simeq 1$ at the Hugoniot elastic limit. According to eq. (52), the second wave does not overtake the shock, so there is a region of the (p_x, V) curve above the point A which cannot be reached by a single shock from

(p_0, V_0). This region, AF in Fig. 14 b), is defined by extending the Rayleigh line OA until it intersects the Hugoniot AB at the point F. Pressures in this region will be reached by a double shock, illustrated in Fig. 15. The first

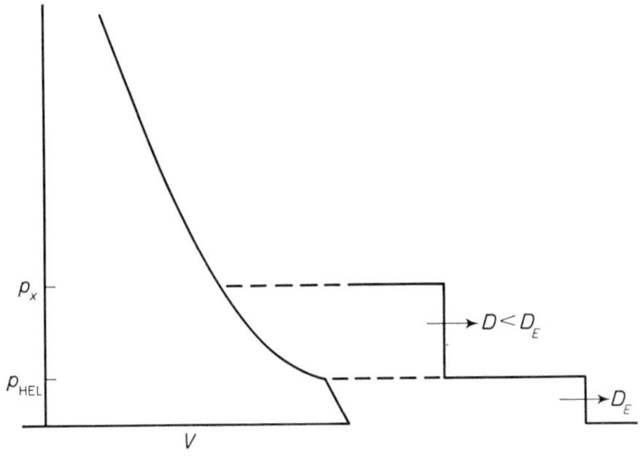

Fig. 15. – Double shock in elastic-plastic material.

shock is called the « elastic precursor »; it travels with elastic velocity, and its amplitude is p_{HEL}.

The unloading wave following an elastic-plastic shock also consists of two waves, but these are not always clearly distinguishable because of the spreading of the rarefactions. The decay process for an elastic-plastic shock is accelerated, however, by an elastic wave running back and forth between the shock front and the plastic rarefaction [4].

An example of an elastic-plastic shock in aluminum is shown in Fig. 16.

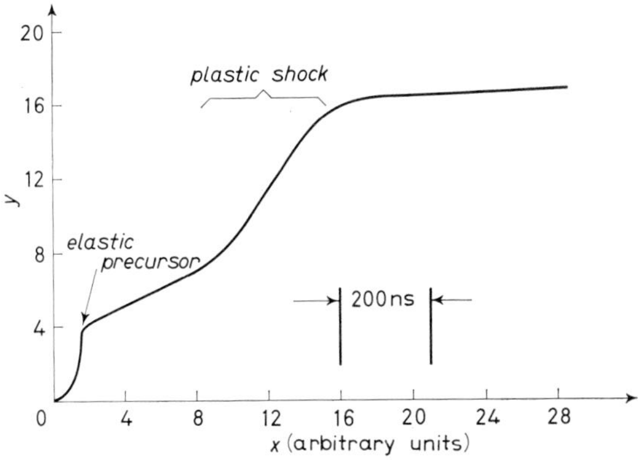

Fig. 16. – Shot no. 68-015. 6061-T6 Al target $\frac{1}{2}$ in. thick, recorded with 1 $\frac{1}{4}$ in. quartz gauge in Graham configuration. Peak pressure: 19 kb. Elastic precursor: 5.2 kb. (R. Mitchell, W.S.U.).

This record was obtained with a quartz gauge on the surface of a 6061-T6 Al target struck by a half-inch thick plate of the same material. The elastic precursor and plastic shock are readily distinguished.

A number of experiments on wave shape and shock decay have established that in aluminum, at least, and probably in other metals as well, the elastic-plastic model is substantially correct, though there are significant deviations due probably to Bauschinger effect and stress relaxation [5].

5. – Solid-solid phase transitions.

Another source of instability in shock waves is the solid-solid phase transition. A schematic diagram of a first-order transition in the (p, V) plane is shown in Fig. 17. An isotherm crosses the mixed phase region at constant pressure; an adiabat has, typically, a small negative slope. The change in slope of the adiabat on crossing the phase boundary can be readily calculated by straightforward application of thermodynamics [6]. In terms of sound velocity in phase 1, c_1, and equilibrium sound velocity in mixed phase, c_m, the change of slope is given by

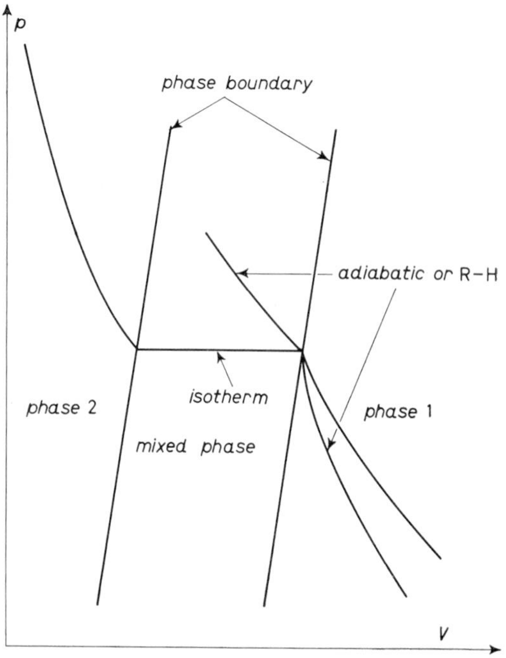

Fig. 17. – Reversible transition.

$$(53) \qquad (c_1^2 - c_m^2)/c_1^2 c_m^2 = (C_p/V^2 T)[(\partial T/\partial p)_S - \mathrm{d}T/\mathrm{d}p]^2 ,$$

where all quantities are evaluated at the phase boundary. Numerical computations for iron and bismuth show that the mixed phase adiabat is almost flat. Since the Hugoniot for the second shock and the mixed phase adiabat have a second order contact at the phase boundary, the Hugoniot is similarly flat.

In preparing a formal representation of a set of constitutive relations for use with the flow equations, eqs. (1)-(3), it is useful to keep in mind their mode of application. The most widely used procedure for integrating the flow equations is a staggered-difference scheme developed by VON NEUMANN and

RICHTMYER [7] in which time is advanced in increments of ΔT, and at each new time the positions, velocities, pressures, densities and energies of all the mass elements are computed. With this procedure in mind, we suppose that at a given point in the mixed phase region all the thermodynamic quantities are known and we then advance specific volume by a small amount $\mathrm{d}V$ and compute the other thermodynamic variables anew.

In an equilibrium phase change we assume that p and T are common to both phases and that extensive variables are given by a mass-weighted average over the two phases; *i.e.*, for example:

$$V = V_1(1-f) + V_2 f, \tag{54}$$

$$E = E_1(1-f) + E_2 f, \quad \text{etc.}, \tag{55}$$

where f is the mass fraction of material in phase 2. We now suppose that $V \to V + \Delta V$ and consequently $p \to p + \Delta p$, $f \to f + \Delta f$, $E \to E + \Delta E$, $T \to T + \Delta T$, etc., and proceed to calculate Δp, ΔT, ΔE and Δf. We must first of all assume that the equation of state is known for each phase individually, and it is most convenient to assume equations of state in the form

$$V_i = V_i(p, T), \tag{56a}$$

$$E_i = E_i(p, T). \qquad i = 1, 2, \tag{56b}$$

Equations (54)-(56) are now differentiated with the intent of expressing $\mathrm{d}V$ and $\mathrm{d}E$ in terms of $\mathrm{d}p$, $\mathrm{d}T$, and $\mathrm{d}f$ with coefficients that depend on p, T, f. The result is

$$\mathrm{d}V = l_1 \,\mathrm{d}p + m_1 \,\mathrm{d}T + n_1 \,\mathrm{d}f, \tag{57}$$

$$\mathrm{d}E = l_2 \,\mathrm{d}p + m_2 \,\mathrm{d}T + n_2 \,\mathrm{d}f, \tag{58}$$

where

$$l_1 = (1-f)V_{1,p} + fV_{2,p}, \tag{59a}$$

$$l_2 = -(1-f)(TV_{1,T} + pV_{1,p}) - f(TV_{2,T} + pV_{2,p}), \tag{59b}$$

$$m_1 = (1-f)V_{1,T} + fV_{2,T}, \tag{59c}$$

$$m_2 = (1-f)(C_{p1} - pV_{1,T}) + f(C_{p2} - pV_{2,T}), \tag{59d}$$

$$n_1 = V_2 - V_1,$$

$$n_2 = E_2 - E_1,$$

$$V_{1,T} \equiv (\partial V_1/\partial T)_p, \qquad V_{2,p} \equiv (\partial V_2/\partial p)_T, \quad \text{etc.},$$

C_{p1} = specific heat at constant pressure for phase 1, etc.

There is an additional relation between $\mathrm{d}E$ and $\mathrm{d}V$, *viz.* the first law. If there is no heat transfer,

$$\mathrm{d}E = -(p+q)\,\mathrm{d}V, \tag{60}$$

where q represents any irreversible forces in the compression process. Equations (57), (58) and (60) can now be solved for $\mathrm{d}p$ and $\mathrm{d}T$:

$$\mathrm{d}p = a_1\,\mathrm{d}V + a_2\,\mathrm{d}f, \tag{61}$$

$$\mathrm{d}T = b_1\,\mathrm{d}V + b_2\,\mathrm{d}f, \tag{62}$$

where

$$a_1 = [m_2 + m_1(p+q)]/D, \tag{63a}$$

$$a_2 = (m_1 n_2 - m_2 n_1)/D, \tag{63b}$$

$$b_1 = -[l_2 + l_1(p+q)]/D, \tag{63c}$$

$$b_2 = (l_2 n_1 - l_1 n_2)/D, \tag{63d}$$

$$D = l_1 m_2 - l_2 m_1. \tag{63e}$$

It's clear from eqs. (61) and (62) that, since only $\mathrm{d}V$ is given, one more relation is required before $\mathrm{d}p$ and $\mathrm{d}T$ can be computed. For reversible transitions this is the Clausius-Clapeyron relation:

$$\mathrm{d}p/\mathrm{d}T = \Delta S/\Delta V = \text{fcn. of } p \text{ or } T. \tag{64}$$

Combining eqs. (61), (62) and (64) yields

$$\mathrm{d}f = \chi(p, T, V, f)\,\mathrm{d}V. \tag{65}$$

This equation, together with eqs. (61) and (62), makes it possible to determine $\mathrm{d}p$ and $\mathrm{d}T$. When that is accomplished, another increment, $\mathrm{d}V$, is taken and the process is continued until the mass element enters the single phase region.

The above relations are also useful for computing the Hugoniot directly. To do this, replace eq. (60) by the differential form of the Rankine-Hugoniot equation and continue as before. To determine an isentrope, set $q=0$ in eq. (60).

We have seen in Sect. **4** how the cusp in the (p, V) compression curve gives rise to a double wave: the elastic precursor followed by a plastic shock. The cusp at the phase boundary, shown in Fig. 17, is also a point of instability which splits a shock to higher pressure into two. In fact, if the material undergoing a phase change is also elastic-plastic, a three wave structure may be produced, as in iron, illustrated in Fig. 18.

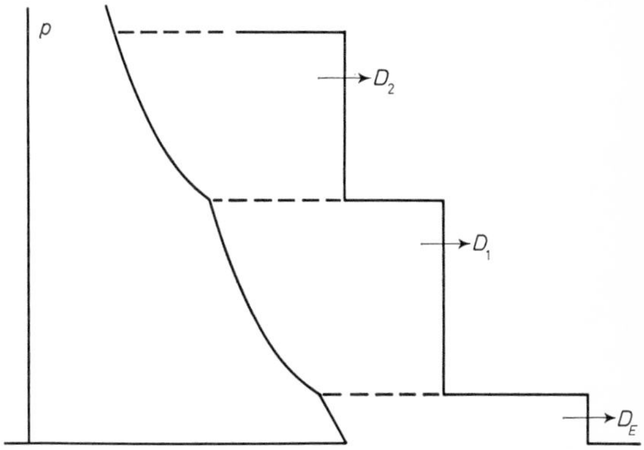

Fig. 18. – Three-wave structure in iron.

The phase transition is also responsible for a new phenomenon, the rarefaction shock. The unloading curve for an elastic-plastic material is always concave upward, the condition required for a rarefaction to spread as it progresses. But the unloading curve for an equilibrium phase transition of the kind shown in Fig. 17 has a convex upward region and this is responsible for producing a rarefaction shock. This is illustrated in Fig. 19.

The unloading curve is shown in Fig. 19 *a*) with two cusps, one at *B* and one at *D*. Suppose the material has been uniformly shocked to point $C(u_2, p_2)$,

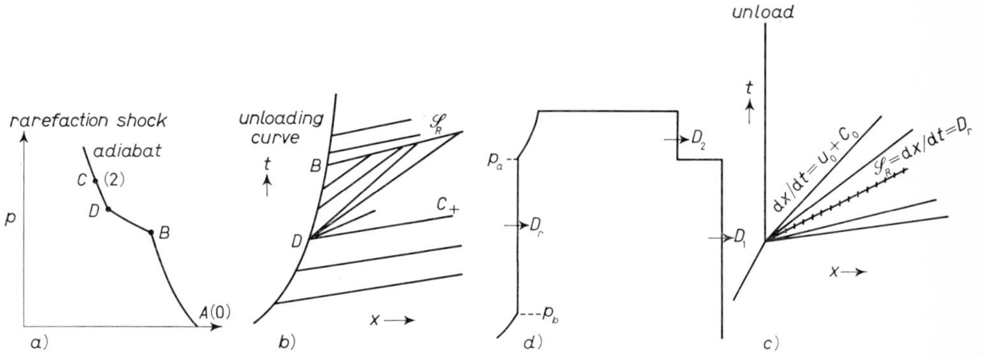

Fig. 19. – Rarefaction shock.

and that the pressure on the driving surface is slowly reduced so that the trace of the surface in the (x, t) plane is the curve *BD* shown in Fig. 19 *b*). At each decrement in driving pressure the free surface can be thought to send out a disturbance traveling along a $C+$ characteristic with velocity $u + c$. When the unloading process has progressed until the surface pressure has decreased from *C* to *D*, a singularity in the sound velocity *c* is encountered. At *D* we suppose *c* to take on all values from the second phase value on *CD* at *D* to the mixed phase value on *DB* at *D*. Then the point *D* on the unloading curve in Fig. 19 *b*) is the source of a fan of $C+$ characteristics as shown. The slower sound velocities correspond to the steeper lines. From D to B the sound velocities are all small and the $C+$ characteristics are steep. At B a second singularity in c is encountered, but now the characteristics, instead of forming a fan as at D, crowd into one another, intersecting one another, and ultimately intersecting some of the characteristics from D. An intersection of two characteristics means that the field variables carried along the characteristics are multiple-valued at that point. In Fig. 19 *b*) the build-up of shock amplitude occurs over some distance as more and more characteristics intersect. In Fig. 19 *c*) we suppose that the driving pressure is released suddenly and the fully developed shock, $\mathcal{S}_R$, radiates from the corner in the unloading path as shown. In Fig. 19 *d*) is shown schematically the structure of a wave including a double compression shock and a rarefaction shock. The pressures p_a and p_b are determined by the condition that

$$(u+c)_a = D_R = (u+c)_b \,, \tag{65}$$

where D_R is the velocity of the rarefaction shock.

Some mechanical effects of the rarefaction shock will be described in Sect. **8**.

6. – Stress-relaxation in elastic-plastic solids.

In order to provide a suitable framework for discussing the constitutive relations for a stress-relaxing solid, it is necessary to introduce some general concepts from continuum mechanics. We consider again an element of mass in a flow field, subject to forces of acceleration and compression transmitted through its immediate neighbors. We are interested in the response of this mass element to the stresses transmitted across its boundaries, and in order to discuss them we choose a co-ordinate system (x_1, x_2, x_3) which diagonalizes the stress and strain matrices. The principal stresses and strains are $\{\sigma_i\}$ and $\{\varepsilon_i\}$ respectively, with $i=1, 2, 3$. We define a set of «stress deviators», S_i, and «strain deviators», E_i:

$$S_i = \sigma_i + \overline{p} \,, \tag{66}$$

$$E_i = \varepsilon_i - \theta/3 \,, \tag{67}$$

where $\overline{p} = -(\sigma_1+\sigma_2+\sigma_3)/3$, $\theta = \varepsilon_1+\varepsilon_2+\varepsilon_3$.
The S_i incorporate all the stress of deformation; the E_i are the strains of deformation. For purely elastic strain we can write Hooke's law in incre-

mental form:

(68) $$\mathrm{d}\sigma_i = \lambda\,\mathrm{d}\theta + 2\mu\,\mathrm{d}\varepsilon_i\,.$$

The definitions of S_i, E_i, p and θ lead to a restatement of Hooke's law as two equations:

(69) $$\mathrm{d}\overline{p} = -K\,\mathrm{d}\theta\,,$$

(70) $$\mathrm{d}S_i = 2\mu\,\mathrm{d}E_i\,.$$

In this form we separate those stresses which produce deformation from those which merely alter density so that the two relations can be discussed independently. It is commonly assumed that when nonelastic behavior occurs, it will appear in the deviator relation, not in hydrostatic compression. A notable exception is the porous solid, but that is not considered here. Following common practice we can write the constitutive relations for an elastic-plastic solid as:

(71) $$\mathrm{d}E_i = \mathrm{d}E_i^e + \mathrm{d}E_i^p\,,$$

(72) $$\mathrm{d}S_i = 2\mu\,\mathrm{d}E_i^e\,,$$

(73) $$\mathrm{d}\overline{p} = -K\,\mathrm{d}\theta\,,$$

where E_i^e and E_i^p are the elastic and plastic components of the strain deviator, respectively.

For a viscoelastic-plastic material, eqs. (71) and (73) apply as before, but (72) is replaced by

(74) $$\mathrm{d}S_i = 2\mu\,\mathrm{d}E_i^e + 2\eta\,\mathrm{d}\dot{E}_i^p\,,$$

where the dot indicates convective derivative with respect to time.

For a stress-relaxing solid we make the assumption that the plastic strain increment, in response to a change in stress, does not take its final value immediately. Its change is inhibited by a relaxation mechanism, undefined at this point. This process is represented by a relation of the form

(75) $$\mathrm{d}E_i^p/\mathrm{d}t = F_i(S_i,\,\varrho)/2\,.$$

The right-hand side of eq. (75) depends upon the amount by which E_i^p differs from its equilibrium value. Combining eqs. (71), (72) and (75) we arrive at the relation

(76) $$\mathrm{d}E_i/\mathrm{d}t - (1/2\mu)\,\mathrm{d}S_i/\mathrm{d}t = F_i(S_i\,,\,\overline{p})/2\,.$$

Equations (76) and (73) comprise a set of constitutive relations for a stress-relaxing material. Note particularly that the stress deviator, S_i, is entirely supported by the elastic strain, eq. (72). This distinguishes it fundamentally from the viscoelastic solid, eq. (74). Equations (71)-(73) for the elastic-plastic solid and eqs. (71), (73), and (76) for the elastic-plastic relaxing solid must be supplemented by a yield condition, *e.g.* the von Mises condition, eq. (41*a*). In uniaxial strain the yield condition can be incorporated in $F(S_i, \overline{p})$ in eq. (76). For this geometry eq. (76) can be replaced by a single equation:

$$\mathrm{d}p_x/\mathrm{d}t = a^2(\mathrm{d}\varrho/\mathrm{d}t) - F(p_x, \varrho)\,, \tag{77}$$

where a is the elastic sound speed at density ϱ.

In Sect. 2 we combined the flow equations, eqs. (1)-(3), under the assumption $p = p(\varrho)$, to form a set of characteristic equations, eqs. (25) and (26). A similar procedure can be executed in the present case. Combining eqs. (1), (2) and (77) yields the characteristic set:

$$C+:\quad \mathrm{d}p_x + \varrho a\,\mathrm{d}u = -F\,\mathrm{d}t\,,\qquad \mathrm{d}x/\mathrm{d}t = u + a\,, \tag{78}$$

$$C-:\quad \mathrm{d}p_x - \varrho a\,\mathrm{d}u = -F\,\mathrm{d}t\,,\qquad \mathrm{d}x/\mathrm{d}t = u - a\,, \tag{79}$$

along with eq. (77), which applies along the particle path, sometimes called the « C_0 characteristic »:

$$\mathrm{d}p_x - a^2\,\mathrm{d}\varrho = -F\,\mathrm{d}t\,,\qquad \mathrm{d}x/\mathrm{d}t = u\,. \tag{80}$$

The characteristic equations are less useful for this and other time-dependent constitutive relations than for time-independent relations because there are now no quantities which remain constant on characteristics. This means that wave transitions are no longer limited to specific curves in the (p, u) plane, as described in Sect. 3, and that type of analysis loses most of its utility. The characteristic equations can still be used in numerical analysis, though it is almost always simpler to use a von Nuemann-Richtmyer procedure.

The principal observable effect of eq. (77) on the shock wave is decay of the elastic precursor. The nature of this decay can be seen by an approximate analysis. Suppose the precursor is never of such large amplitude that its speed of propagation differs significantly from the ambient elastic speed a_0. Then the jump condition for the precursor becomes (eq. (5)):

$$p_x = \varrho_0 a_0 u$$

and when the precursor amplitude decays by $\mathrm{d}p_x$, the particle velocity behind the precursor decays by an amount

$$du = dp_x/\varrho_0 a_0 \, . \tag{81}$$

Under the assumed conditions, eq. (78) also applies along the path of the precursor. Combining eqs. (78) and (81) yields the relation

$$dp_x/dt = -F/2 \, . \tag{82}$$

The function F is expected in general to be quite complicated. We can get a qualitative picture of its effect by assuming the form, for compression only,

$$F = (p_x^e - p_x^s)/T \, , \qquad p_x^e > p_x^s , \tag{83}$$

where $T =$ constant. Compression by the precursor is assumed to be elastic, so p_x of eq. (82) lies on a metastable extension of the elastic compression curve, $p_x^e(V)$. Above the yield point there is a stress $p_x^s(V)$ which will finally be reached for the given volume V after a very long time. This is curve AB of Fig. 14 *b*). According to eqs. (82) and (83), decay of the precursor amplitude, $p_x \equiv p_x^e(V)$ continues until $p_x^e(V) = p_x^s(V)$, which occurs at the static value of the Hugoniot elastic limit. To see the effect more explicitly, note that

$$(d/dt)(p_x^e - p_x^s) = (1 - c^2/a^2)(dp_x^e/dt) \, , \tag{84}$$

where $c^2 = K/\varrho$, $a^2 = (K + 2\mu/3)/\varrho$. If Poisson's ratio, ν, is independent of density, so is c^2/a^2. Then eqs. (82)-(84) can be integrated to yield

$$p_x^e(V) - p_x^s(V) = (p_x^e - p_x^s)_0 \exp[-x/x_0] \, , \tag{85}$$

where

$$x_0 = 2TD/(1 - c^2/a^2) \, . \tag{86}$$

Integrating eq. (84) under the assumption that $\nu =$ constant enables us to simplify eq. (85):

$$p_x^e - p_{\text{HEL}}^s = (p_x^e - p_{\text{HEL}}^s)_0 \exp[-x/x_0] \, , \tag{87}$$

where p_{HEL}^s is the static value of the Hugoniot elastic limit, related to the static yield strength by eq. (47).

Equation (82) was derived on the assumptions that the precursor follows a characteristic and that the energy equation, eq. (3), does not affect the propagation process. A more rigorous expression can be obtained by combining eq. (77) with eqs. (1)-(3) and specializing the result along the shock path [8]:

$$\frac{\mathrm{D}p_x}{\mathrm{D}x} = \left(1-\frac{u}{D}\right)\frac{(D-u)^2-a^2}{\frac{3}{2}(D-u)^2+a^2/2}\,\frac{\partial p_x}{\partial \mathrm{x}} - \frac{(D-u)^2}{D}\,\frac{F'}{\frac{3}{2}(D-u)^2+a^2/2}\,, \tag{88}$$

$$F' = (1-\alpha\Gamma y/2\mu)F\,. \tag{89}$$

Here the block derivative, $\mathrm{D}/\mathrm{D}x$, refers to differentiation along the shock path, $\partial p_x/\partial x$ is evaluated immediately behind the precursor front, and F' is a modification to F resulting from the assumption that a fraction α of plastic work goes into heat. In eq. (89), Γ is the Gruneisen parameter. F' and F differ by less than 10% for metals in which plastic flow occurs.

Under the assumptions that $D-u=a$ and $\alpha=0$, eq. (88) reduces to eq. (82).

Considerable effort in recent years has been devoted to attempts to relate the relaxation function F of eq. (75) to the motion and multiplication of dislocations. The basic relation is

$$\mathrm{d}E^p/\mathrm{d}t = hNbv = F/2\mu\,, \tag{90}$$

where N is the number of dislocations per unit area, b is the Burgers vector, h is a numerical constant the order of units, and v is the mean velocity of dislocations. Since $E_p = 2\varepsilon_1/3$ in uniaxial strain, eq. (90) becomes

$$\mathrm{d}\varepsilon_1/\mathrm{d}t = 3hNbv/2\,. \tag{91}$$

There are various models for multiplication and motion of dislocations. One which is frequently used is due to GILMAN:

$$N = N_{om}(1+A\varepsilon^p)\,, \tag{92}$$

$$v = v_{\max}\exp\left[-D/\tau\right]\,, \tag{93}$$

where

N_{om} = initial density of mobile dislocations,

$v_{\max}$ = maximum dislocation velocity $\sim v_{\text{shear}}$,

D = drag coefficient,

A = multiplication coefficient,

τ = resolved shear stress $=(p_x-p_y)/2$.

To the extent that eq. (82) applies, no multiplication occurs in the front of the elastic precursor, so the value of A is of no consequence in determining the precursor decay. A set of decay curves for single crystal tungsten are shown

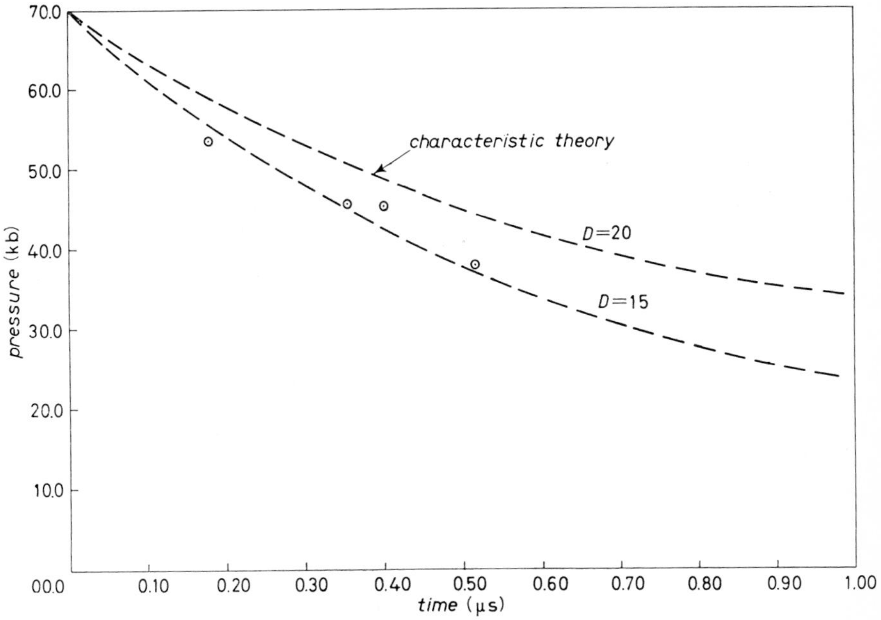

Fig. 20. – Precursor decay in single crystal tungsten, courtesy of T. E. MICHAELS, W.S.U. (unpublished).

in Fig. 20 for different values of the drag coefficient D. Here, as in work reported elsewhere, it is necessary to assume a value of N_{om} much higher than the measured values in order to get reasonable agreement with experiments.

In order to see the effect of dislocation multiplication, one must record the wave profile between the elastic precursor and the plastic shock. Such a profile obtained with a quartz gauge on LiF is shown in Fig. 21. By adjusting the parameter A in eq. (92), the sharp drop in amplitude immediately following the elastic peak can be explained.

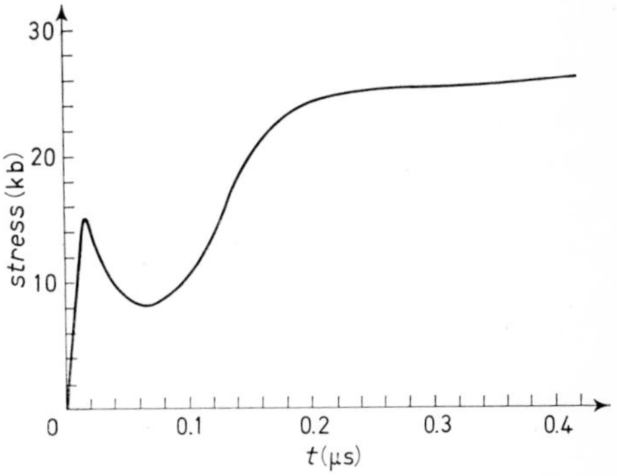

Fig. 21. – Precursor and shock in LiF. (J. ASAY, W.S.U.).

It is not yet clear whether critical tests of dislocation theory can be made from shock profile measurements. There is not yet any clear-cut evidence that dislocations bear any relation to elastic-plastic behavior in shock. It appears, however, that the subject is worth pursuing in depth for at least one case until a definitive answer appears.

7. – Irreversible phase transitions.

Equilibrium phase transitions were discussed in Sect. **5** and a formal procedure for integrating the flow equations through the mixed-phase region was described. The central relations are given in eqs. (61) and (62) together with the equilibrium assumption, eq. (64). When considering the irreversible case we need to re-examine the assumptions made in obtaining eqs. (61) and (62). In a general sense it is possible to have irreversible mass transfer, irreversible heat transfer and irreversible work transfer between the two phases. The first of these occurs if the reaction parameter f is out of equilibrium, the second if temperatures of the two phases are unequal and adiabaticity is violated, the third if pressures in the two phases are out of equilibrium. Of these three it is quite easy to imagine f out of equilibrium; this is in fact probably the usual case since the deviation of f from equilibrium corresponds to a value of the Gibbs energy above the minimum and this acts as a force to drive the reaction toward equilibrium.

If nucleation of the second phase occurs at very many points in every volume element so that dimensions of crystals in either phase are very small, it is unlikely that pressure will be significantly out of equilibrium. Temperature equilibration, however, takes place relatively slowly and in any particular case it may well call for a closer examination. However as an approximation at this stage it looks reasonably good and preferable to the other simple alternative that no heat exchange whatsoever occurs between phases. Equations (54) and (55) represent another possible source of error inasmuch as the interfacial energy between phases is ignored. Here again it seems unlikely that the effect will be large, and it seems appropriate to ignore it for the present. With assumptions unchanged from those previously made, we again arrive at eqs. (61) and (62) with coefficients the same as before. It's important to note that both here and in Sect. **5** it is assumed that both phases have the same particle velocity, u. This is appropriate for solid-solid transitions; it would not be appropriate for gas-liquid or gas-solid transitions.

The difference between the treatment of reversible and irreversible transitions then reduces to the computation of f. In Sect. **5** f was computed from the Clausius-Clapeyron equation (eqs. (64) and (65)). In the irreversible case we assume that

$$\mathrm{d}f = g(V, T, f)\,\mathrm{d}t\,. \tag{93}$$

Equations (61), (62) and (93) form the constitutive relations for the irreversible transition when $0<f<1$. If $f=0$ or 1, eqs. (56) apply. This formulation of the transition process is particularly appropriate for the von Neumann-Richtmyer integration since in that procedure $\mathrm{d}t$ is specified explicitly.

The above formulation represents the following physical model. Each phase has an equation of state surface in (p, V, T) space, and in equilibrium the surfaces are separated by the mixed phase region; V, for example, is never double-valued. We now suppose that each phase is defined for all p and T and that V may lie on either surface, one of which is metastable, or may take any value between the two surfaces. The value it has at any moment is determined by its previous value, by $\mathrm{d}t$, by the equation of continuity, and by previous values of p, T and f; and it seeks an equilibrium state through the operation of eq. (93) until it arrives at an equilibrium surface where $f=1$ or 0.

As an example of the application of these equations, consider the case of iron, which has a transition from *bcc* to *hcp* at 130 kb and approximately room temperature. For the transition we assume that:

$$V_2(p, T) - V_1(p, T) = -0.0059\ \mathrm{cm^3/g} \tag{94a}$$

$$\text{static transition pressure,}\ p_t = 130\ \mathrm{kb}\,, \tag{94b}$$

$$\mathrm{d}p_t/\mathrm{d}T = -0.065\ \mathrm{kb/^\circ K}\,, \tag{94c}$$

$$\mathrm{d}f/\mathrm{d}t = (f_{eq} - f)/\tau\,, \tag{94d}$$

$$f_{eq} = (V - V_1)/(V_2 - V_1)\ \text{at}\ p\ \text{and}\ T\,, \tag{94e}$$

$$\tau = \tfrac{1}{3}\ \mu\mathrm{s}\,. \tag{94f}$$

A constant pressure of 200 kb is applied to the surface of a half-space at $t=0$ and we seek the wave profile at subsequent times and the rate of decay of the precursor wave associated with the transition. The elastic precursor is ignored.

The integration was performed by a modification of the von Neumann-Richtmyer method [9]. The development of the pressure profile at early times is shown in Fig. 22 and the fully developed double wave is shown in Fig. 23. In each figure the effect of changing $V_2 - V_1$ is shown; it influences the shape of the profile but not the decay rate. The decay of the elastic precursor is shown in Fig. 24. The computation was also made with $\mathrm{d}p_t/\mathrm{d}T = 0$ with no substantial change in the results. The dashed line shown in Fig. 24 fits the decay curve obtained by numerical integration reasonably well. It was obtained analytically in the following way:

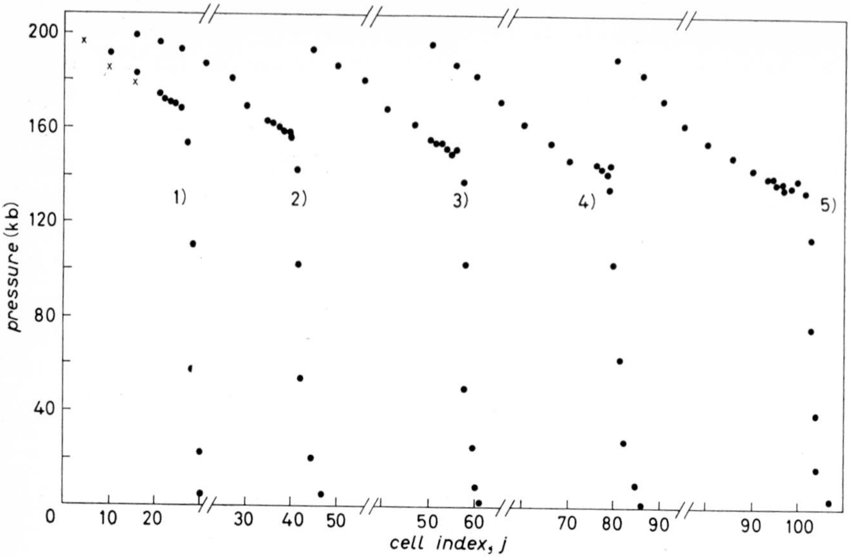

Fig. 22. – Pressure profiles at early times. (Ref. [9]). $p_1 = 0.200$ Mb; cell width = 0.01 cm; $\tau = \frac{1}{3}$ μs; 1) $t = 0.526$ μs; 2) $t = 0.812$ μs; 3) $t = 1.105$ μs; 4) $t = 1.554$ μs; • $\Delta V = -0.004$ cm³/g; × $\Delta V = -0.0059$ cm³/g.

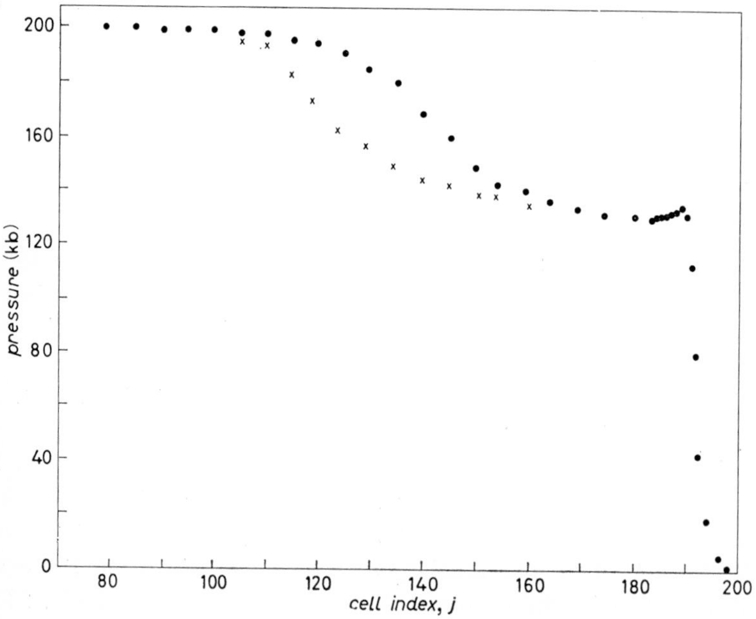

Fig. 23. – Double shock in iron. (Ref. [9]). $p_1 = 0.200$ Mb; cell width = 0.01 cm; • $\Delta V = -0.004$ cm³/g; × $\Delta V = -0.0059$ cm³/g; $\tau = \frac{1}{3}$ μs.

Assume the entire process to be temperature independent. Then eqs. (61) and (94*d*) can be combined in the form

$$\text{(95)}\qquad \mathrm{d}p/\mathrm{d}t = a^2\,\mathrm{d}\varrho/\mathrm{d}t + [(m_1 n_2 - m_2 n_1)/(l_1 m_2 - l_2 m_1)](f_{eq} - f)/\tau\,,$$

where m_1, n_1, etc., are as defined in Sect. **5** and a is the frozen sound speed, *i.e.*, $\sqrt{\mathrm{d}p/\mathrm{d}\varrho}$ with $f=$ constant. Equation (95) is then identical in form to the stress-relaxation equation, eq. (77). Following the procedure described there, we form the characteristic equations, assume the first shock velocity is equal to sound speed, and obtain

$$\text{(96)}\qquad \mathrm{d}p_1/\mathrm{d}t = -[(m_1 n_2 - m_2 n_1)/(l_1 m_2 - l_2 m_1)]\cdot(f_{eq} - f)/2\tau\,,$$

where p_1 is the amplitude of the transition shock. Assuming $\Delta V = V_2 - V_1 =$ constant as before and $C_{p1} = C_{p2}$, eq. (96) becomes

$$\text{(97)}\qquad \mathrm{d}p_1/\mathrm{d}t = -(\Delta V f_{eq}/2\tau)\,\mathrm{d}p_1/\mathrm{d}V.$$

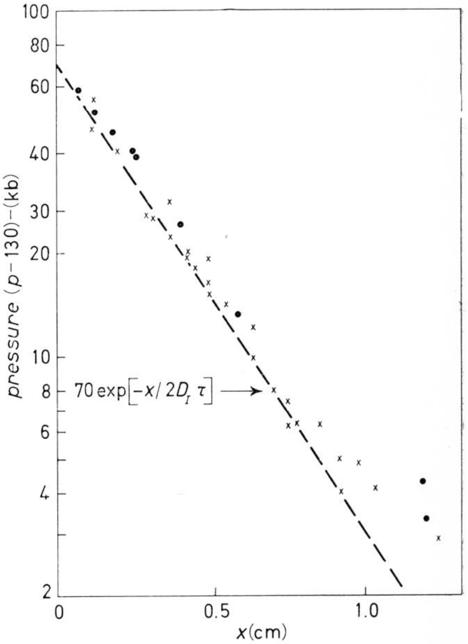

Fig. 24. – Decay of α-phase wave in iron. (Ref. [9]). Driving pressure = 200 kb; $\tau = \frac{1}{3}$ μs; $D_I\tau = 0.17$ cm; • temperature-independent; × temperature-dependent.

Here we have assumed that $f = 0$, *i.e.*, the material is in the metastable first phase at the peak of the first shock. If $\mathrm{d}p_1/\mathrm{d}V =$ constant, eq. (94*e*) gives

$$\text{(98)}\qquad f_{eq} = \begin{cases} 1\,, & V_1 \leqslant V_t) + \Delta V,\\ (V_1 - V_t)/\Delta V = [(p_1 - p_t)/\Delta V]\,\mathrm{d}V_1/\mathrm{d}p\,, & V_t + \Delta V \leqslant V_1 \leqslant V_t,\\ 0\,, & V_t \leqslant V_1, \end{cases}$$

where V_t, p_t represent the intersection of the phase 1 Hugoniot with the mixed-phase boundary. Combining eqs. (97) and (98) and integrating gives

$$\text{(99a)}\qquad p_1 = p_0 - (x\Delta V/2D_1\tau)\,\mathrm{d}p_1/\mathrm{d}v\,,\qquad V_1 \leqslant V_t + \Delta V,$$

$$\text{(99b)}\qquad = p_t + (p_0 - p_t)\exp[-x/2D_1\tau]\,,\qquad V_t + \Delta V \leqslant V_1 \leqslant V_t,$$

$$\text{(99c)}\qquad = p_t\,,\qquad V_1 = V_t,$$

where $x = D_1 t$, D_1 is the first shock velocity, and p_0 is the driving pressure. For $p_0 = 200$ kb, eq. (99*b*) applies. The result is shown in Fig. 24.

The graphs in Fig. 22-24 were calculated with $\tau = \frac{1}{3}\,\mu$s. In order to determine the relation between rise time in the second shock and τ, the constitutive relations can be combined with eqs. (4)-(6) to determine the steady profile in the second shock. The result of that integration is that the rise time $= 2.25\tau$. Unfortunately the theory of the rates at which solid-solid phase transitions occur is not developed to the state where predictions of experimental results can be made or where experimental results can be used to determine physical parameters. It is evident that careful experiments and creative theoretical work are both required in this area.

8. – Mechanical effects.

The most obvious terminal effects of shock experiments in solids are fractures produced by wave interactions. The geometry of fracture depends principally on the paths of intersecting tensile waves; limiting conditions for fracture are determined by dynamic strength of the material, which may depend on geometry and certainly depends upon stress rate or strain rate.

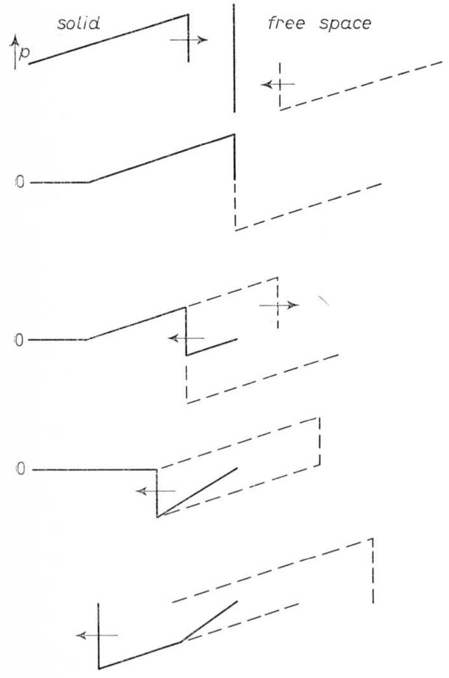

Fig. 25. – Reflection at a free surface, linear waves.

The geometry is indicated in Fig. 25 and 26 for the simplest case, plane spall. In Fig. 25 are shown a sequence of pressure profiles of a compressive pulse approaching and being reflected from a free surface. Propagation and interactions are assumed linear for purposes of illustration. The solid curve is the real pressure profile; the dashed curves represent the pulses from which it is composed. The interaction of the rarefaction in the incident pulse with the reflected rarefaction from the free surface produces tensile stresses in the specimen. If these produce the conditions required for fracture at any point and time, fracture is initiated at that point and time. This is illustrated more precisely in Fig. 26. In 26 *a*) is shown an (x, t) diagram for the process of Fig. 25. The incident shock, $\mathscr{S}_+$, is followed by a simple wave rarefaction represented by the fields 11, 12, 13, separated by C_+ characteristics. The free surface, ABC, is immediately accelerated at the incidence of $\mathscr{S}_+$ and a back-ward-facing rarefaction fan, $\mathscr{R}_-$, centered at B, is generated. The C_+ characteristics pass through the fan of the reflected rarefaction and each in turn is reflected from the free surface as shown. The map of the flow in the (p, u) plane is shown in Fig. 26 *b*).

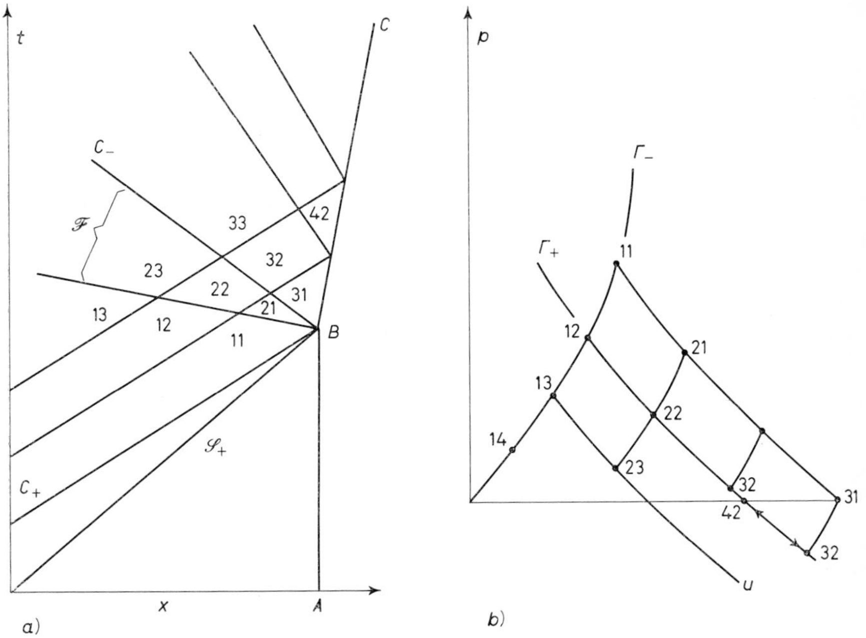

Fig. 26. – Reflection at a free surface. *a*) (x, t) plane; *b*) (p, u) plane.

The point to note here is that the transition from the field 12 to the free surface state takes place along a Γ_+ characteristic, and that the mapping goes into the negative pressure region and returns to the $p = 0$ state at 42. This is in qualitative agreement with the sketches of Fig. 25.

Whether or not a spall occurs depends upon the magnitude and duration of the stress. BREED, MADER and VENABLE have found [10] that spall occurs if the fracture stress, σ_f, satisfies the relation

(100) $$\sigma_f = A + F(\Delta\sigma/z)^{\frac{1}{2}},$$

where z is the distance from the free surface at which spall occurs, $\Delta\sigma = \sigma_f$, and $\Delta\sigma/z$ is the stress gradient. A and B are material constants. Equation (100) implies that fracture occurs when $\sigma\,\Delta t$ reaches a characteristic value.

BUTCHER, BARKER, MUNSON and LUNDERGAN have proposed a slightly more general relation which includes the above [11]; fracture occurs when

$$\Delta t \sigma^r = G \,, \tag{101}$$

where r and G depend on the material.

A still more general relation has been used by F. TULER [12] to describe experiments in 6061T6 Al; fracture occurs at time t_f defined by the relation

$$\int_0^{t_f} (\sigma_0 - \sigma)^r \, dt = G \,, \tag{102}$$

where $r = 2.02$, $G = 3.98 \cdot 10^{13}$, $\sigma_0 = -10^{10}$ dyn/cm^2. All of these rules apply to plane spall, described in Fig. 25 and 26. None have any particularly sound theoretical basis. Present efforts in this area are directed toward experiments which will provide information for nucleation and growth models similar to those discussed by McCLINTOCK [13].

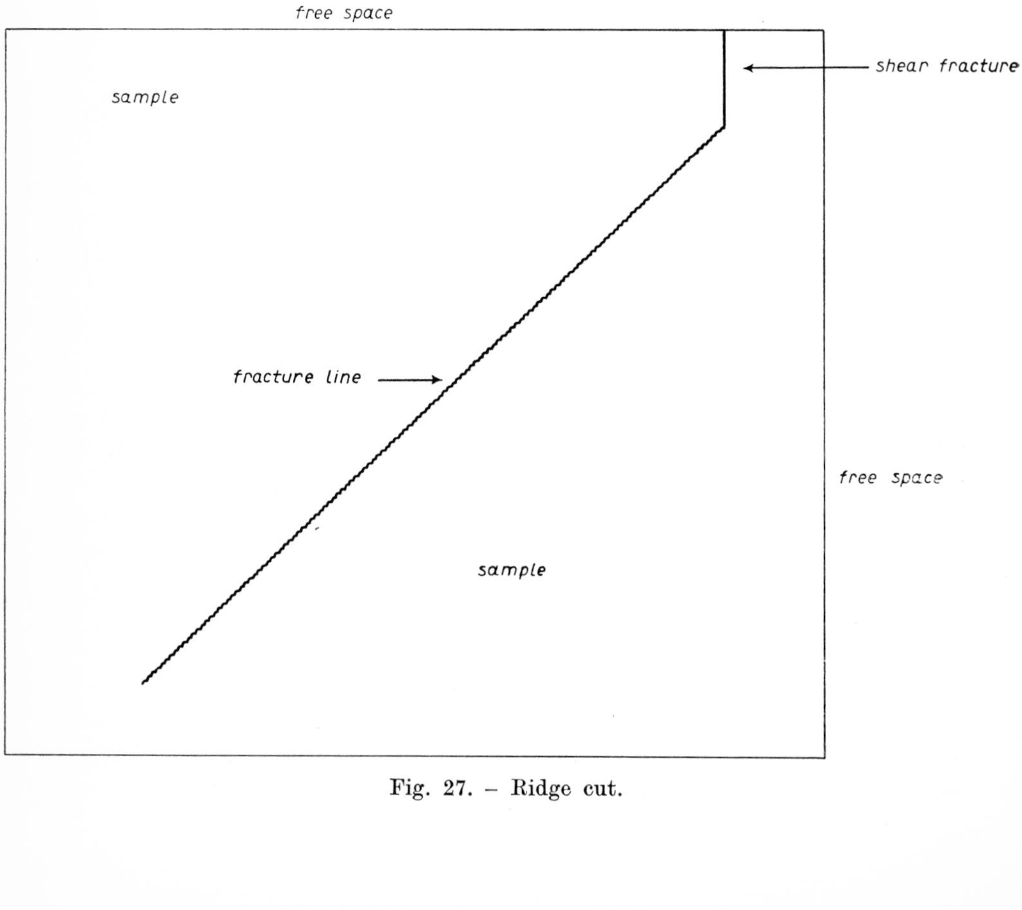

Fig. 27. – Ridge cut.

Another common phenomenon is the «ridge fracture»: any corner in a shocked sample may be neatly split, except near the outermost tip, where common shear fracture normally occurs, Fig. 27. These have been discussed in detail by RINEHART and PEARSON [14] and by FOWLES and ANDERSON [15]. The latter have performed a linear elastic analysis of the configuration shown in Fig. 28 and 29. In Fig. 28 the line labelled Φ represents a compressive wave

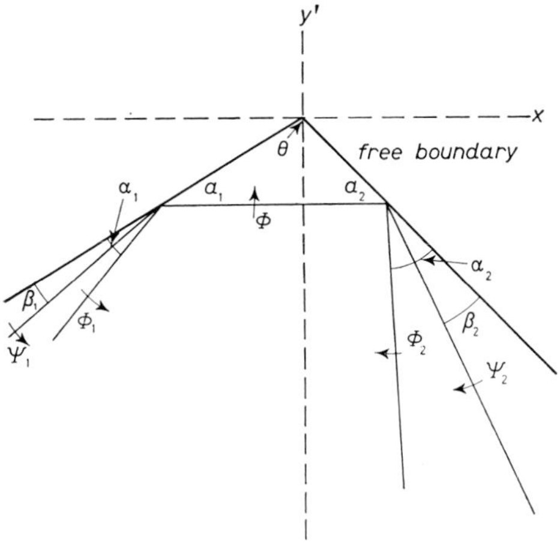

Fig. 28. – Wave incident upon corner. (From ref. [15]).

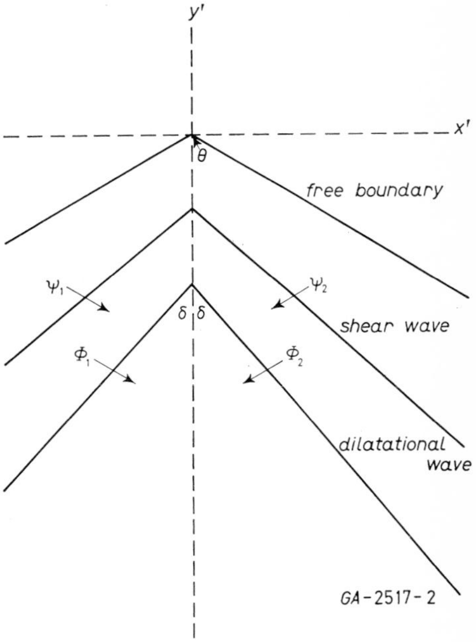

Fig. 29. – Interaction of reflected waves. (From ref. [15]).

running into a corner, generating reflected dilatation and shear waves Φ_1, Φ_2, and ψ_1, ψ_2, respectively. At a later time, shown in Fig. 29, only the reflected waves remain, and these meet along the dotted axis labelled y', the points of intersection running down as time progresses. The interacting tensile waves, Φ_1 and Φ_2 and/or shear waves ψ_1 and ψ_2 produce a fracture running along y' if the amplitude of the incident wave is sufficiently great. Their analysis, which agreed reasonably well with experiments, indicated that, for a given incident wave, the maximum fracture stresses were developed for $\theta \simeq 120°$. The shear fracture near the corner in Fig. 27 is very likely due to the finite time required for fracture, as in eqs. (100)-(102).

An interesting type of fracture occurs when a layer of explosive is detonated in contact with a metal plate. The situation is characterized in Fig. 30. In Fig. 30 *a*) the detonation front is travelling with speed D across the surface of the plate, inducing a trailing shock, $\mathscr{S}$. Pressure on the plate falls quickly behind the detonation front as indicated. This produces a corresponding rarefaction behind $\mathscr{S}$. The shock $\mathscr{S}$ is reflected from the bottom surface as a rarefaction, $\mathscr{R}$, and the interaction of $\mathscr{R}$ with the rarefaction behind $\mathscr{S}$ produces a tension stress field which may be strong enough to fracture the plate along

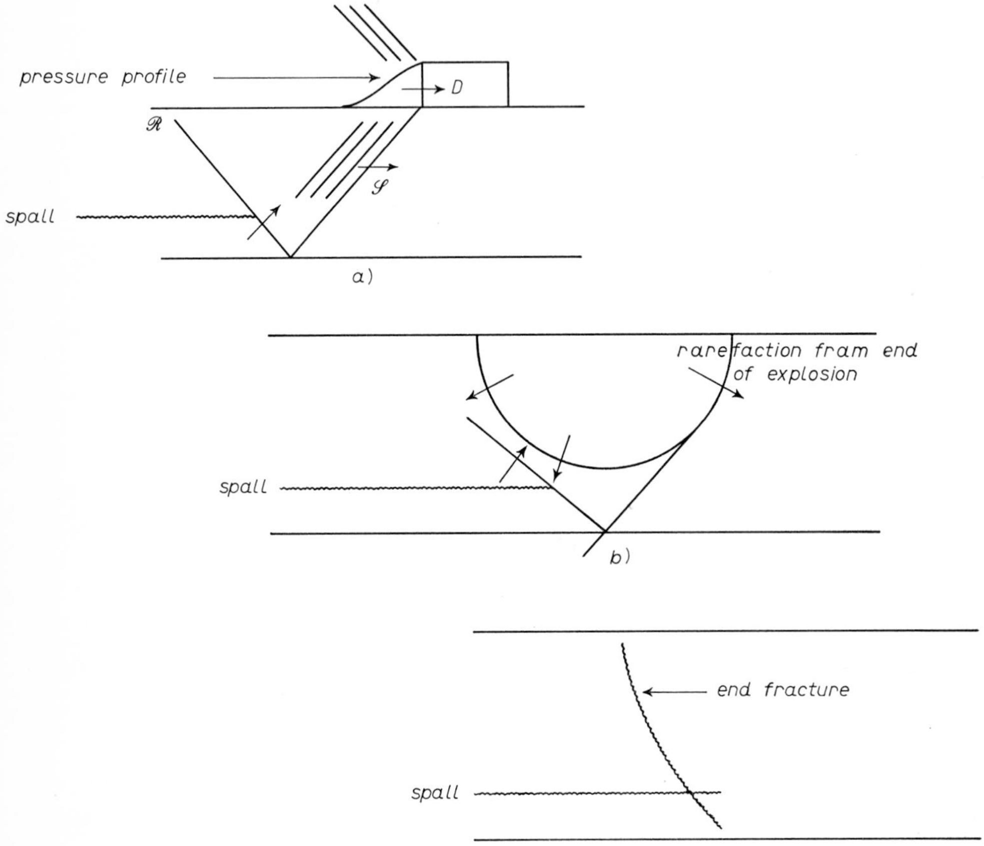

Fig. 30. – Generation of end fracture. *a*) Steady waves from detonating explosive; *b*) rarefaction generated at end of explosive intersects with bottom rarefaction; *c*) end fracture resulting from interaction.

the surface labelled « spall ». When the detonation reaches the end of the explosive charge, a rarefaction is generated with a more or less circular wave front as shown in Fig. 30 *b*). This rarefaction, interacting with $\mathscr{R}$, produces a stress field which results in an end fracture with orientation shown in Fig. 30 *c*). The geometry of this interaction has been worked out [16], but no one has yet calculated the stress fields and correlated them with experiments or with fracture formulae.

An interesting property of iron and carbon steels is their propensity to fracture under explosive or projectile impact so as to produce a very smooth surface. This fracture is called the « smooth spall » and has been described by RINEHART and PEARSON [14], by ERKMAN [17], by LETHABY and SKIDMORE [18], and by TUPPER [19]. In the last three works the smooth spall is attributed to the interaction of rarefaction shocks associated with the α-ε phase transition in iron.

9. – Shock waves on a one-dimensional lattice of mass points.

The mass points are connected by nonlinear springs, as in Fig. 31, and by dashpots. The forces exerted by the former on the N-th point mass are denoted

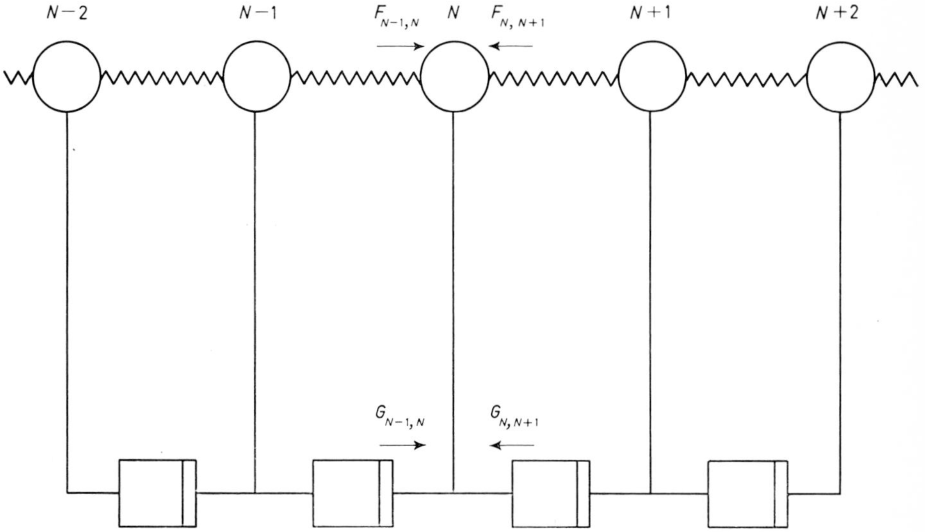

Fig. 31. – Dissipating lattice model. $\mathrm{d}^2X_N/\mathrm{d}T^2 = F_{N-1,N} - F_{N,N+1} + G_{N-1,N} - G_{N,N+1}$.

by $F_{N-1,N}$ and $F_{N,N+1}$; those associated with the dashpots are $G_{N-1,N}$ and $G_{N,N+1}$. The equation of motion of the N-th particle is

$$\mathrm{d}^2X_N/\mathrm{d}T^2 = F_{N-1,N} - F_{N,N+1} + G_{N-1,N} - G_{N,N+1}\,, \tag{103}$$

where X_N is its position at time T; X and T are reduced values corresponding to unit mass and unit equilibrium separation between mass points. The lattice is supposed to extend from $N=1$ to ∞ and we imagine the first mass point to be given a velocity u_1 at $T=0$. The entire lattice is assumed quiescent before that. For a parabolic force law,

$$(104)\quad F_{N,N+1}=-(S_{N+1}-S_N)+\alpha(S_{N+1}-S_N)^2\,,\qquad S_N=X_N-1,\qquad G_{N,N+1}=0\,,$$

the acceleration of the N-th particle is shown in Fig. 32 [20]. The broken line shown there is the result of a numerical integration, the dashed line is

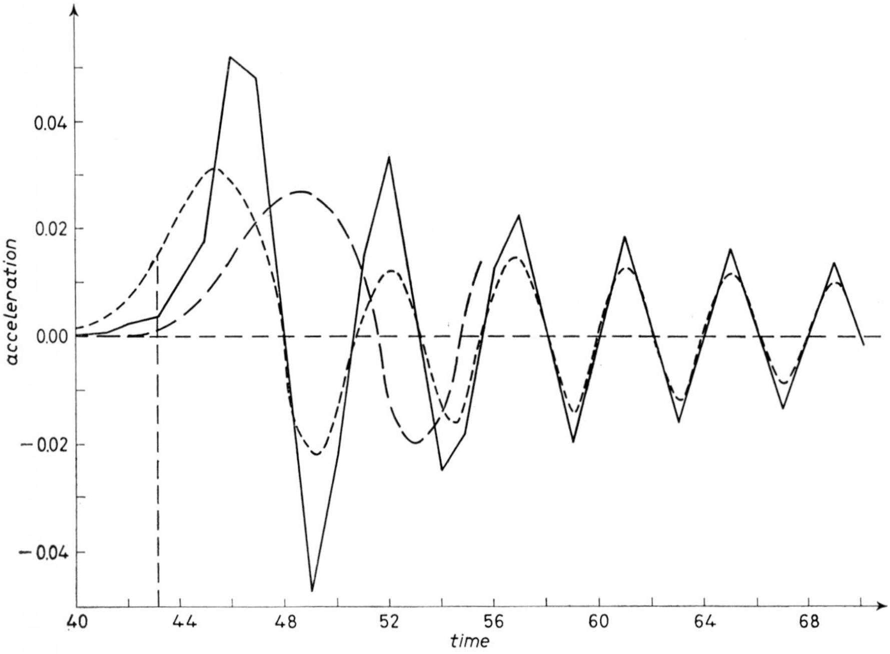

Fig. 32. – Acceleration of a mass point in a nonlinear lattice. (Ref. [20]).

an analytic approximation obtained by averaging the nonlinear force term over time, and the long-dashed line is the exact solution when the coefficient $\alpha=0$ in eq. (104). If solutions are compared for successively larger values of N, it is found that the initial acceleration increases monotonically with N and that the rate of decay of the oscillation decreases monotonically. This suggests that the solution may be approaching a permanent regime:

$$(105)\qquad S_N(T)=S(T-N\theta)\,,$$

where θ is a constant equal to the reciprocal wave speed. When eq. (105) is substituted into (103) with the force law of (104), we obtain

$$d^2S/dy^2 = [S(y-\theta) - 2S(y) + S(y+\theta)]\cdot\{1-\alpha[S(y-\theta) - S(y+\theta)]\}\,, \tag{106}$$

where

$$y = T - N\theta\,.$$

Equation (106) is an uncommon type of differential-difference equation and is not readily solved, even numerically. An approximate solution can be found by expanding $S(y\pm\theta)$ in powers of θ and retaining fourth order terms. If dissipation is nonzero, however small, nearly-steady oscillations of the type suggested by Fig. 32 are obtained. The frequencies of oscillation predicted by the approximate theory are, however, quite different from those obtained from numerical integration of the transient equations. It thus appears that this picture of the permanent regime in the lattice is qualitatively correct, but that eq. (106) must be solved if parameters of the permanent regime are to be calculated [21].

One interesting application of these results is to the von Neumann-Richtmyer integration of the flow equations. If artificial viscosity is too small, the results of such an integration oscillate wildly. This has sometimes been interpreted as instability of the numerical integration procedure; it is in fact the true physical behavior of the lumped-constant system used to model the continuum for purposes of numerical integration.

REFERENCES

[1] R. COURANT and K. O. FRIEDRICHS: *Supersonic Flow and Shock Waves* (New York, 1948). This contains much material basic to Sects. **1-3**.

[2] G. E. DUVALL: *Les ondes de détonation* (Paris, 1962), p. 337.

[3] D. R. BLAND: *Journ. Inst. Math. Appl.*, **1**, 56 (1965).

[4] J. O. ERKMAN and G. E. DUVALL: *Developments in Mechanics*, edited by T. C. HUANG and M. W. JOHNSON Jr., vol. **3**, Pt. 2 (New York, 1965), p. 179.

[5] W. HERRMANN: *Wave Propagation in Solids* (New York, 129), p. 129.

[6] G. E. DUVALL and Y. HORIE: *Proc. Fourth Symposium on Detonation, Oct. 12-15, 1965* (Washington, D. C., 1966), p. 248.

[7] R. D. RICHTMYER and K. W. MORTON: *Difference Methods for Initial-Value Problems*, 2nd Ed. (New York, 1967).

[8] T. J. AHRENS and G. E. DUVALL: *Journ. Geophys. Res.*, **71**, No. 18, 4349 (1966).

[9] Y. HORIE and G. E. DUVALL: *Proc. Army Symposium on Solid Mechanics, 1968*, p. 127; available from Applied Mechanics and Research Laboratory, Army Materials and Mechanics Research Center, Watertown, Mass.

[10] B. R. Breed, C. L. Mader and D. Venable: *Journ. Appl. Phys.*, **38**, 3271 (1967).
[11] B. M. Butcher, L. M. Barker, D. E. Munson and C. D. Lundergan: *Am. Inst. Aeronautics and Astronautics Journ.*, **2**, 977 (1964).
[12] F. W. Tuler: Sandia Laboratories, Albuquerque, N. Mex., private communication.
[13] F. A. McClintock: *International Journ. Fract. Mech.*, **4**, 101 (1968).
[14] J. S. Rinehart and J. Pearson: *Behavior of Metals Under Impulsive Loads* (New York, 1954).
[15] G. R. Fowles and G. D. Anderson: Poulter Laboratories Internal Report 032-59, Stanford Research Institute, Menlo Park, Calif. (1959).
[16] W. E. Drummond: *Comments on the cutting of metal plates with high explosive charges*, paper no. 57-A-89, American Society of Mechanical Engineers, 345 E. 47th St., New York, N. Y. 10017 (1957).
[17] J. O. Erkman: *Journ. Appl. Phys.*, **32**, 939 (1961).
[18] J. Lethaby and I. C. Skidmore: S.W.A. Branch Note No. 3/59, AWRE, Aldermaston, Berks. (March 1959).
[19] S. J. Tupper: *On the propagation of plane stress waves generated in a thick steel plate by a surface explosion*, A.R.D.E. Report (B) 12/61 (Sept. 1961).
[20] R. Manvi, G. E. Duvall and S. C. Lowell: *Int. Journ. Mech. Sci.*, **11**, 1 (1969).
[21] G. E. Duvall, R. Manvi and S. C. Lowell: *Journ. Appl. Phys.*, **40**, 3771 (1969).

Shock Waves in Condensed Media (*).

R. N. Keeler and E. B. Royce

University of California, Lawrence Radiation Laboratory - Livermore, Cal.

I. – Experimental Techniques.

R. N. Keeler

1·1. – Background.

The physical measurements required to accurately determine the state of material behind a shock front are characteristically similar to measurements made in other studies of the properties of matter at high energy density. They must be carried out in an extremely short time, since such high energy densities (up to 1 MJ/cm³) cannot be steadily maintained in the laboratory; experimental measurements must be read from a remote location, since the energy densities involved cannot be dissipated in a nondestructive manner; and the diagnostics must be as complete as possible, since the system cannot be returned to its original state for a verification of the measured results.

In his lectures Duvall has extensively reviewed the hydrodynamics of shock waves. He showed that the assumption of a one-dimensional, steady-state shock wave, that is, a shock wave whose properties do not vary with time, permits the derivation of the so-called Rankine-Hugoniot relationships [1.1]. Various schemes using spherically convergent or cylindrical implosion systems have been used in the past, but these are expensive, subject to instabilities, and produce data of degraded accuracy. In addition, convergent geometries are not steady, since pressure increases continuously. Because of the one-dimensionality of the Rankine-Hugoniot relationships, meaningful property measurements in shock waves can only be made in a one-dimensional planar geometry. Therefore, experiments using shock waves as a high-pressure–pro-

(*) Work performed under the auspices of the U.S. Atomic Energy Commission.

a flying plate, which subsequently impacts a sample. The second technique, ducing technique are limited, not only in terms of diagnostics, but also in their basic physical configuration.

The Rankine-Hugoniot relationships are normally written in terms of the variables p, $v = \varrho^{-1}$, U_s, U_p, and E.

$$v = v_0 \left(1 - \frac{U_p}{U_s}\right) \quad \text{conservation of mass.} \tag{1.1}$$

$$p = p_0 + \varrho_0 U_s U_p \quad \text{conservation of momentum.} \tag{1.2}$$

$$\Delta E = \tfrac{1}{2}(p + p_0)(v_0 - v) \quad \text{conservation of energy.} \tag{1.3}$$

The definition of these variables and their dimensions are as follows:

p = pressure, megabar (1 Mbar $= 10^{12}$ dyne/cm^2),

ϱ = density, g/cm^3,

v = specific volume, cm^3/g,

E = specific internal energy (Mbar-cm^3/g $= 10^{12}$ ergs/g),

U_p = particle velocity, or bulk material velocity behind the shock front, cm/μs,

U_s = shock wave velocity, cm/μs.

The subscript 0 refers to initial (unshocked) conditions.

Of the five variables, only two are independent. Therefore, in carrying out shock-wave experiments, it is customary to measure two of the five variables; from these measurements, the other variables can be calculated.

There are two basic techniques for achieving very high shock pressures in the laboratory. The first involves use of high explosives to drive a strong shock wave into a sample as shown in Fig. 1.1 or, as in Fig. 1.2, to accelerate

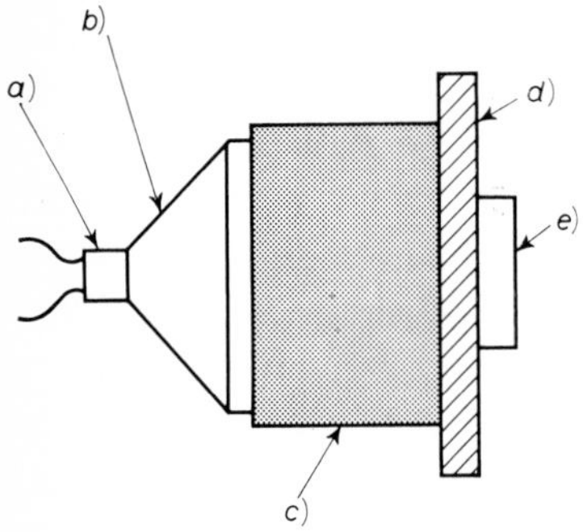

Fig. 1.1. – H.E. contact experiment. *a*) detonator; *b*) plane-wave lens; *c*) block of explosive; *d*) base plate; *e*) sample.

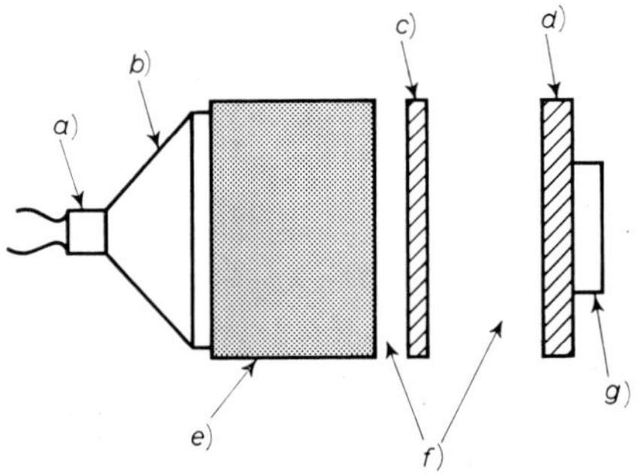

Fig. 1.2. – H.E. flying-plate experiment. *a*) detonator; *b*) plane-wave lens; *c*) flying plate; *d*) base plate; *e*) block of explosive; *f*) air gaps; *g*) sample.

shown in Fig. 1.3, involves the use of a one- or two-stage light gas or powder gun to accelerate a projectile, which subsequently impacts a target or sample. Following the arrival of the detonation wave or the impact of the accelerated projectile, the transit of the shock wave or desired physical property is measured in the sample.

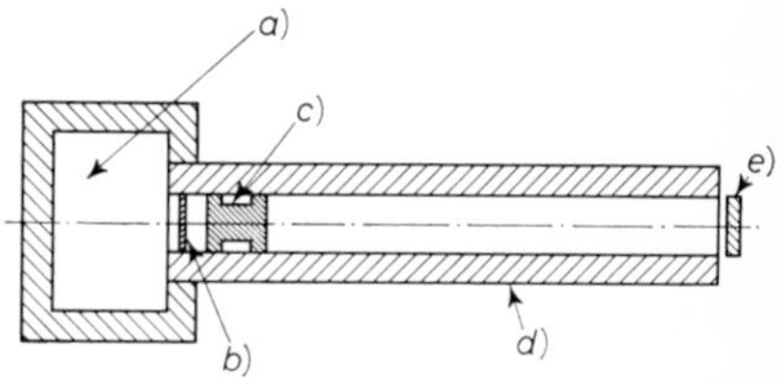

Fig. 1.3. – Light-gas gun. *a*) high-pressure chamber; *b*) diaphragm; *c*) projectile; *d*) barrel; *e*) sample.

It is instructive to consider the actual pressure limitations of these techniques. To do so, we consider the p-v plane. This plane with all its features is shown in Fig. 1.4 for a typical sample material: a rare gas, for example, or a metal. In this figure, polymorphic transitions in the solid state are ignored. It is important to note that the curves represented here are projections into the p-v plane from an equation-of-state surface in p-v-E or p-v-T space. For example, an isotherm in this plane is the projection onto the p-v plane, of the intersection of the $T=$ constant plane with the p-v-T equation-of-state surface.

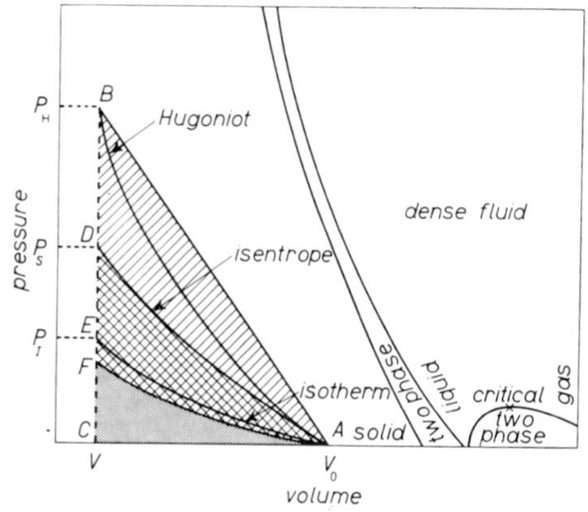

Fig. 1.4. – The p-v plane.

The first point of interest is the comparison of the isothermal curve, the isentrope, and the Hugoniot. The isentrope and isotherm represent a series of states which can be reals ized continuously; the Hugoniot represents the locus of all states which can be reached by shocking a material from a given initial state. It is a series of points for which the heat exchange between system and surroundings in going from the initial to the final state is zero, and for which the change in state is caused by the propagation of a one-dimensional steady-state discontinuous pressure disturbance. For these three processes,

(1.4) isotherm: $$\Delta E = \int_{S_0}^{S} T\,\mathrm{d}S_T - \int_{v_0}^{v} p\,\mathrm{d}v_T = \int_{v_0}^{v} T\left(\frac{\partial p}{\partial T}\right)_v \mathrm{d}v_T - \int_{v_0}^{v} p\,\mathrm{d}v_T\,,$$

(1.5) isentrope: $$\Delta E = -\int_{v_0}^{v} p\,\mathrm{d}v_s\,,$$

(1.3) Hugoniot: $$\Delta E = \tfrac{1}{2}(p + p_0)(v_0 - v)\,.$$

If we consider the energies required to compress a given substance from the same initial state to a given final volume, the energy required to carry out the shock compression process can be calculated from eq. (1.3) and is represented graphically by the area within the right triangle *ABC*. It is always in excess of the area under the Hugoniot curve. The energy for an isentropic process is given by eq. (1.5) and is represented by the area under the isentropic curve *AD*. Thus, the isentrope will always lie below the Hugoniot, since $(\partial p/\partial E)_v > 0$. The energy required to compress a sample isothermally is less than the area under the curve *AE* by the first term on the right-hand side of eq. (1.4), which is always negative. It is represented by the area under the curve *AF*. The increase in energy from isotherm through isentrope to Hugoniot leads to progressively higher temperatures for a given compression, and the accompanying upward displacement of curves as shown.

At this time, it is appropriate to introduce the concept of dynamic impedance. The physical meaning of dynamic impedance can be seen by observing what occurs when a shock wave crosses an interface between different materials.

In Fig. 1.5, a shock wave of pressure p_1 is progressing through material I. Upon impact with material II, shock waves of pressure p_2 are driven into material II and back into material I. When the second material has a

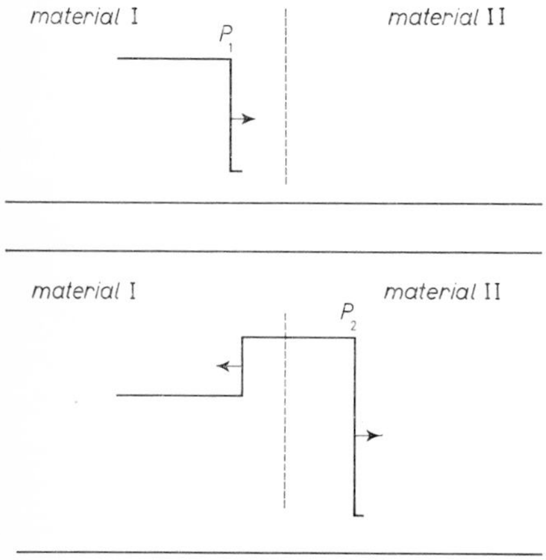

Fig. 1.5. – Shock wave into higher-impedance material.

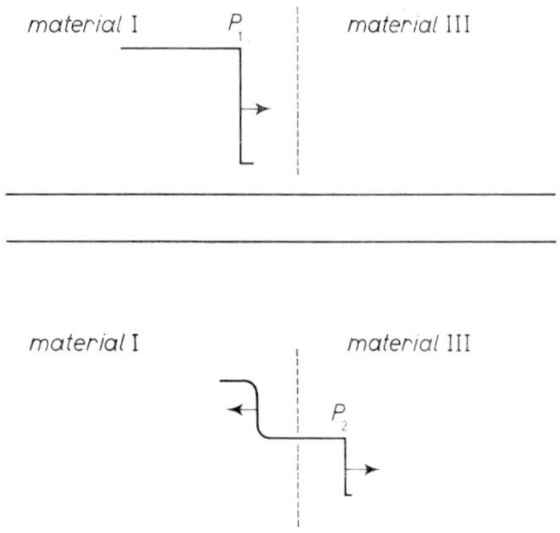

Fig. 1.6. – Shock wave into lower-impedance material.

lower dynamic impedance, as in Fig. 1.6, a rarefaction wave is propagated back into the first material, I, while a shock wave of reduced amplitude is driven into the second material, material III.

The relationships between materials of different dynamic impedance are best seen from a representation of Hugoniots in the p-U_p plane rather than the p-v plane. This choice of co-ordinates is convenient because in the passage of a shock wave through an interface between materials of different shock impedance, there must be continuity of pressure, p, and particle velocity, U_p, across the interface. The question of whether a shock wave of higher pressure than the initial shock wave and rarefaction wave of lower pressure than the initial wave is propagated back from the interface is a question of the relative dynamic impedance of the materials. The dynamic impedance is defined as $\varrho_0 U_s$, and is, in general, a function of pressure; its relation to reflected rarefaction and reflected transmitted shock waves will become clear later. A graphycal picture of shock interactions between materials of different dynamic impedance is shown in Fig. 1.7, where a number of Hugoniots and isentropes for various materials are plotted. In this figure, the momentum-conservation relationship $p = \varrho_0 U_s U_p$ is recalled, and the dynamic impedance at a given pressure is recognized as nothing more than the slope of the line connecting the origin and the particular pressure on the Hugoniot. Comparing Hugoniots for materials I, II, and III, it can be see that a shock proceeding from material I to material II will result in reflected shocks propagating in both directions. The locus of points which can be reached by driving a shock wave through a material of lower impedance into a material of higher impedance is represented by the curve AB. Similarly, the locus of points which can be reached by driving a shock wave into a material of lower impedance is the curve AC. The point C represents the shock wave propagating into a material of zero impedance, and therefore is equal to the free-surface velocity, U_{fs}, of a material from which a shock is emerging into a vacuum or into air. In order to obtain the curve BC, it is necessary to assume a functional form of the equation of state.

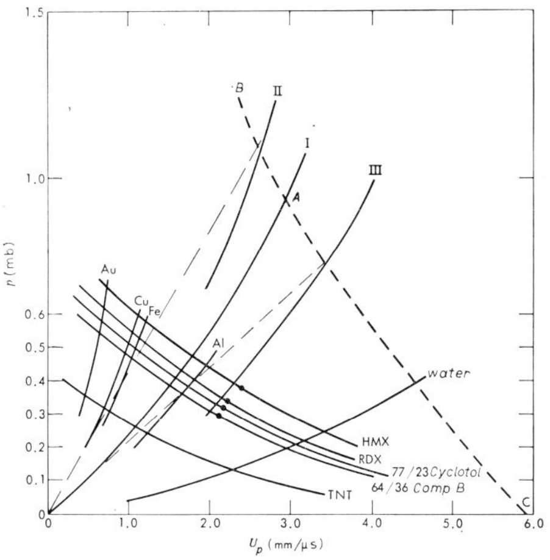

Fig. 1.7. – The p-U_p plane.

The expression for the free-surface velocity is

$$U_{fs} = U_p + U_r \,, \tag{1.6}$$

where U_r = the rarefaction velocity

$$U_r = \int_{p_0}^{p} \left(-\frac{\partial v}{\partial p}\right)_s^{\frac{1}{2}} dp \,. \tag{1.7}$$

At shock pressures up to 500 kbar, the assumption that $U_p = U_r$ (the « free-surface approximation ») is correct for most materials to about 1% [1.2].

The limits of pressure are also shown in Fig. 1.7. The Chapman-Jouguet pressures for several typical high explosives are shown [1.3]. The curve through the HMX point represents the pressure which can be reached in a material by impact of a detonation wave from a high explosive in contact with the material. Therefore, the pressure range attainable by use of contact high explosives extends downward from 0.8 Mbar.

Sound velocity in the hot detonation products is the limiting velocity for acceleration of a plate by high explosives. What actually takes place is a series of shock reflections and rarefactions, which serve to accelerate the plate to its final velocity. Acceleration stops, of course, when edgewise rarefactions and Taylor wave rarefactions from the detonation products reach the trailing edge of the plate. Acceleration of a 3.2 mm Monel plate to 5 mm/μs with a high explosive and subsequent impact upon a sample allows an extension of the maximum attainable pressure to 2.4 Mbar in tungsten.

The highest shock pressures attained to date, (5÷10) Mbar [1.4-1.6], have been generated by spherical implosion systems and two-stage gas guns. In these experiments, the Hugoniot pressures and volumes were obtained from measurements of flying plate velocity and shock velocity in the sample. Some higher-pressure experiments ($p = 33$ Mbar) [1.7] have been reported, but these experiments involve measurements of shock velocity only, and the calculated pressures and volumes are based on theoretical assumptions.

Of the four variables appearing in the first two Rankine-Hugoniot relationships, the shock velocity is easiest to measure directly, and it is normally measured in all shock-wave experiments. The pressure, volume, and particle velocity are much more difficult to obtain, since either the sensing elements used to make these measurements must be subject to the high pressures, temperatures, and shear stresses existing behind the shock front, or they must depend on the observation of radiation which must emanate from or pass through the shock front.

Experimental techniques used to make shock-wave equation-of-state measurements fall into four classes: 1) discrete, in which a signal is produced which refers to a definite event in time and space; 2) continuous, in which a continuous record is taken of the movement of a given surface; 3) internal, in which properties behind the shock front (*e.g.*, pressure or particle velocity) are measured; and 4) combination, in which the experimental technique involves two of the previously cited classes of experiments. Pioneering development of most of these techniques was carried out at the Los Alamos Scientific Laboratory, the Sandia Laboratories, the Stanford Research Institute, and various laboratories in the Soviet Union, Great Britain, and Australia [1.8].

The choice of which variables to measure is determined by the pressure range of the experiment, and by the phenomena which one expects to observe at the pressure of interest. For shock wave experiments carried out at pressures up to about half a megabar, it is customary to measure free-surface and shock velocity using either the «free-surface approximation» or the corrected particle velocity derived from the free-surface velocity and an evaluation of the Riemann integral [1.2]. The free surface measurement, of course, is dependent upon integrity of the free-surface, which begins to exhibit jetting, disintegration, and melting phenomena at pressures between 0.5 and 1 Mbar. Under these conditions, an alternative set of measurements can be made: the shock velocity in the sample, as before, and the measurement of the velocity of a flying plate or projectile just before impact on the sample. If the sample is made of the same material as the flying plate or projectile, we have a «symmetric impact». If a flying plate or projectile impacts a target of identical material, then from conservation of momentum, the particle velocity behind the shock front driven into the target material will be half the velocity of the flying plate. In this case, the measurement of projectile velocity and the measurement of shock velocity in the target are sufficient to characterize the pressure and volume behind the shock front. To develop a consistent set of equation-of-state data, various selected materials have been chosen, and experiments carried out in which shock velocity, flying plate velocity, and free-surface jump-off velocity were measured. These materials are then used as standards. When the equation of state of a given standard is known (or calculated using a particular equation-of-state model) over the region of interest, then the Hugoniot of an unknown material may be obtained by a measurement of shock velocities only. Again, we refer to the p-U_p plane in Fig. 1.7 to see how this is done. The shock wave in the standard material is represented by the intersection of the Rayleigh line, DA, with the previously determined Hugoniot. Then, the state in the target which is obtained by driving the shock wave into the target is determined by the intersection of the Rayleigh line for the target and the rarefaction (off Hugoniot) curve from point A. Of course a standard is essential when conducting experiments in liquids.

1·2. – Experimental equipment.

1·2.1. *Plane-wave lenses.* – Certain items of unusual experimental equipment find wide application in shock-wave physics. Because these devices are unfamiliar to many physicists, we will discuss them separately from the specific experimental techniques in which they are used. The first of these is the high-explosive lens [1.9]. The development of these lenses makes it possible to obtain plane-wave detonation fronts and essentially one-dimensional planar shock fronts in samples. The Snell's-Law plane-wave lens is fabricated with a rapidly burning high explosive on the outside of a cone, with the interior of the cone formed from a slower-burning explosive. A typical design of such a lens is shown in Fig. 1.8. One combination which has been used in explosive lenses is Baratol (a barium nitrate-loaded TNT) and Composition *B*. The detonation velocities of these explosives are 5 and 8 mm/μs, respectively. The shock planarity, or simultaneity of shock-arrival time in such a system, is within about 30 ns across a diameter up to 30 cm. Further planarity is achieved by allowing the detonation wave to propagate through a cylindrical block of high explosive before entering the experimental sample. This procedure also reduces the steepness of the rarefaction behind the shock front.

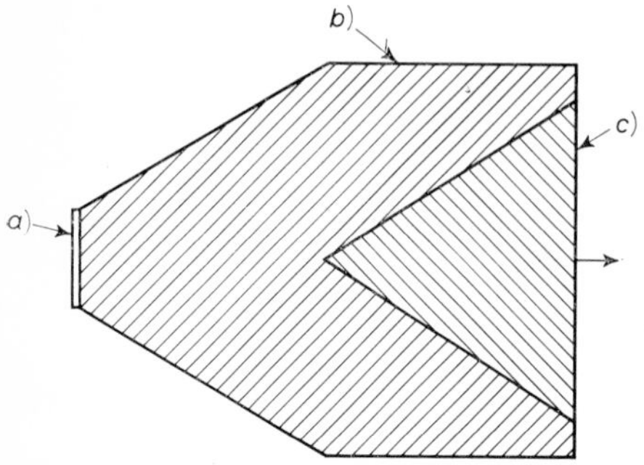

Fig. 1.8. – Plane-wave lens. *a*) detonator position; *b*) comp. *B*; *c*) baratol.

Where cost is an important factor, large amounts of high explosive are not desirable, and high precision is not mandatory, the so-called « mouse-trap » lens is frequently used. This lens consists of two flat blocks of high explosive placed at an angle with one another. The upper slab of high explosive is initiated at one end, and as the detonation wave travels down the length of a metal plate, the plate is turned until it is parallel with the second high-explosive pad. It then impacts the pad simultaneously over its entire area, creating a plane detonation wave. The time from firing signal to completion of a high-explosive experiment is of the order of 10 to 50 μs.

1˙2.2. *Gas and powder guns.* – In recent years, much of the shock-wave work which was formerly done with high explosives has been carried out with one- and two-stage gas and powder guns. The advantages of these guns are manifold. First, they enable high-pressure shock-wave experiments to be conducted under laboratory conditions, if proper precautions are taken. Of course, the same limitations on diagnostics prevail as were stated in Subsection 1˙1; however, it is no longer necessary to carry out the experiments in a remote location adapted for large-scale high-explosive detonations. These guns produce shock waves of a very controlled nature. The profiles behind the shock front are flat, since they are achieved by the impact of a projectile on a sample, rather than by detonation followed by a steadily decreasing pressure. It is much easier to interpret complex wave interactions in such a system. It is possible to tailor the shock duration and pressure with much greater precision, and finally, with a two-stage gun, to achieve pressures up to 10 Mbar with much greater accuracy and convenience than with high-explosive systems.

The design and use of one-stage [1.10, 1.11] and two-stage [1.6] guns have been discussed in some detail in the literature.

1·2.3. *Streaking camera* [1.12]. – In a streaking camera, an image is rapidly swept across a film, permitting observation of very fast events. The object of interest can be stationary or moving, and the sweep is accomplished with a rapidly rotating mirror which is mounted between the objective lens and the film. The simplest possible arrangement includes a target, objective lens, rotating mirror, and film. In practice, a more elaborate scheme is usually used to remove the camera to a remote location in order to avoid the near-range damaging effects of high explosives. With a mirror rotation speed of 2000 r.p.s., and a camera drum radius of 40 cm, the image traverses the film at about 1 cm/μs, with a resolution of approximately 1%.

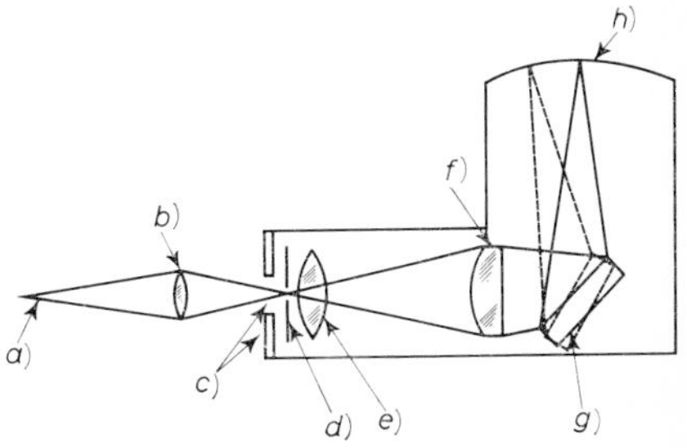

Fig. 1.9. – Streaking camera. *a*) target; *b*) objective lens; *c*) aperture; *d*) slit; *e*) collimating lens; *f*) telescope lens; *g*) rotating mirror; *h*) film.

A schematic drawing of a typical camera configuration is shown in Fig. 1.9. After the rotating mirror attains operating speed, high-explosive experiments are initiated by a signal triggered from the camera. When operating with gas guns, it is necessary to use a 360-degree streaking camera or an electronically triggered image-converter camera operated in the streaking mode because of the time interval between firing signal and experiment completion (~1 ms).

1·2.4. *Argon candle.* – The requirement for an intense light source of (10÷20) μs duration can be met with the argon «candle», which is merely a cardboard cylinder filled with argon gas and backed with a high-explosive charge [1.13]. The high explosive drives an intense shock wave into the argon gas, shock-heating the gas to high temperatures (2 eV) and producing a brilliant flash. A typical argon candle is shown in Fig. 1.10; a more elaborate light source currently used for spectroscopic studies is depicted in Fig. 1.11. The output of this source has been measured experi-

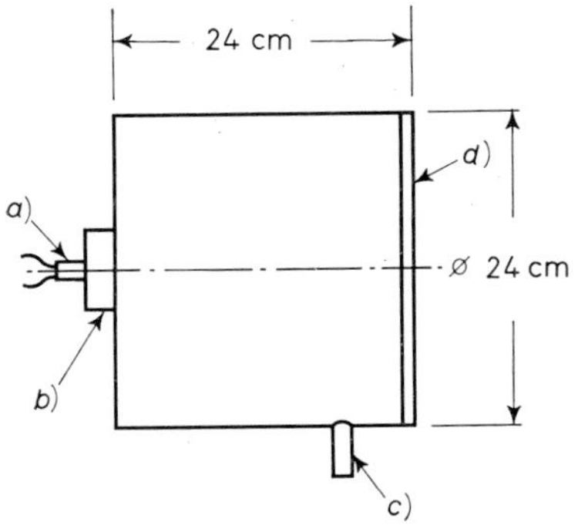

Fig. 1.10. – Argon candle. *a*) detonator; *b*) high-explosive pad; *c*) argon gas inlet; *d*) Lucite window.

mentally. Argon flows continuously through the explosive cylinder. Upon detonation of the explosive, a converging shock-wave system is driven into the argon gas, heating it to about 3.5 eV. This system provides an average

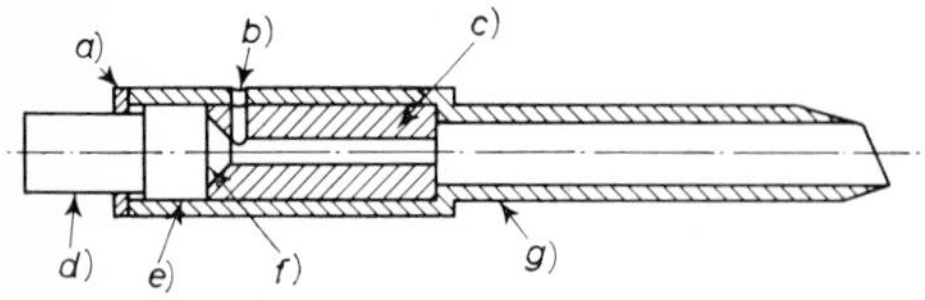

Fig. 1.11. – Argon candle for spectroscopy. *a*) cap; *b*) gas inlet; *c*) high-explosive cylinder; *d*) detonator, 0.95 cm diam.; *e*) tetryl pellet, 1.27 cm diam.; *f*) brass plug; *g*) aluminum tube.

power density of about 12 erg/cm^2 μs in the spectral range (4.200 ÷ 5400) Å over a circular area 10 cm in diameter at a distance of 600 cm. This device is currently being used in absorption spectroscopy experiments in shock-compressed materials [1.14].

1˙2.5. *The argon flash gap.* – The brilliant light flash generated by shock-heated argon can also be used to diagnose shock-wave arrival [1.2]. The argon flash gap technique involves the use of a very thin, $\sim$ 50-micron, argon-filled gap between a Lucite block and the sample. Gaps can be used to indicate either shock-arrival time or free-surface arrival time. Upon arrival of the shock wave or free surface, the argon is rapidly compressed by multiple shocks to a high temperature and becomes incandescent within $\sim 10^{-12}$ s after passage of the shock front. The flash duration is about 20 ns.

1˙3. – Discrete techniques.

The two major discrete techniques currently used in shock-wave research involve the use of electrical discharging pins and flash gaps. The timed optical or electrical signal is related to a known location, and from the distance and time information obtained, a velocity is computed. These techniques are useful in obtaining shock velocity, free-surface velocity, and the velocity of an impacting projectile or flying plate.

Only two other techniques are worthy of mention. The first is the flash X-ray technique in which X-ray tubes at known locations are discharged at a given time and a flash radiograph obtained of a projectile in flight. This technique has been used in conjunction with two-stage gas gun diagnostics [1.6]. The second technique is the interrupted optical beam technique, which is used for the same purpose. It is usually set up with lasers, photodiodes, and fast oscilloscopes. Both of these techniques suffer from decreased spatial resolution because of tilt, surface inhomogeneities, and free-run limitations caused by projectile tumbling. Some preliminary work has recently been reported in which an electromagnetic pickup technique was used to measure projectile velocity, but the results are too scanty to permit a full assessment.

1·3.1. *Pin techniques.* – The most common discrete technique in shock-wave measurements involves the use of shorting pins. A typical coaxial self-

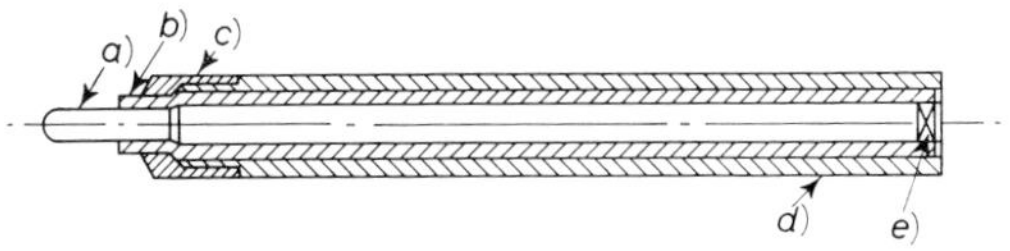

Fig. 1.12. – Coaxial self-shorting pin. *a*) brass rod; *b*) PVC insulation; *c*) end cap; *d*) pin body; *e*) mica washer.

shorting pin is shown in Fig. 1.12. The gap between the center conductor and the base of the pin is the open part of the pin circuit. Upon impact of the shock wave, the contact is closed, and current flows through the circuit shown in Fig. 1.13. Reproducibility of closure time depends upon the details of pin construction: surface uniformity, precision in locating the center coaxial lead, the accuracy in machining the pins, and the precision in placing the pins in the individual experiments. A small mica shim or washer is usually inserted between the center coaxial lead and the base of the outer pin body. Because

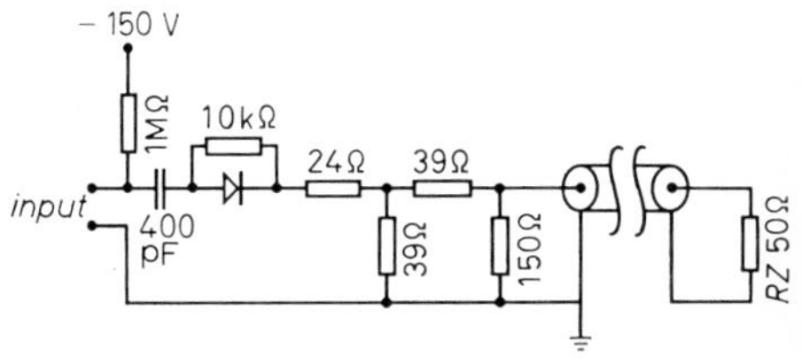

Fig. 1.13. – Pin circuit.

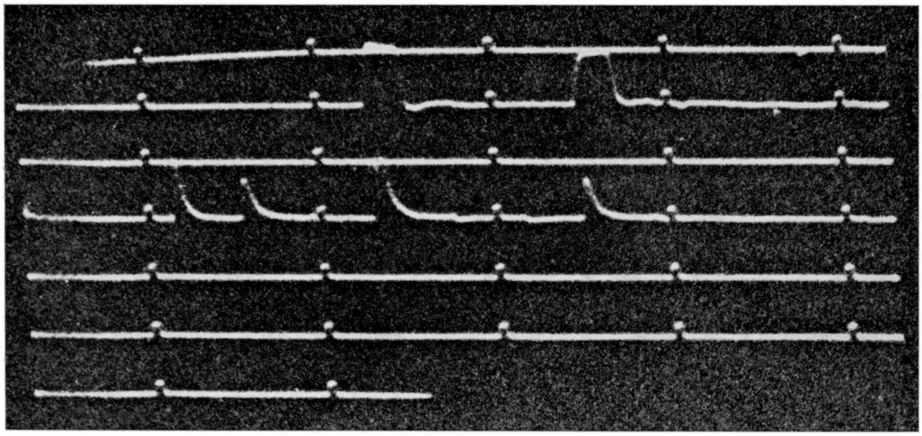

Fig. 1.14. – Self-shorting pin record.

of the details of breakdown and hydrodynamic flow across the gap after shock arrival, these pins are not normally used at pressures below 75 kbar. A typical pin record is shown in Fig. 1.14; these pins were spaced about 3 mm apart. The maximum jitter in discharge time was about 5 ns.

Some of the earlier papers by MINSHALL [1.15] are classic examples of the use of bare pins to determine free-surface position and observe multiple-wave profiles. An example of his data is shown in Fig. 1.15. From these data, MINSHALL was able to separate the elastic precursor wave, the 131-kbar wave, and the final wave which corresponds to the high-pressure (ε-phase) form of iron. It is important to avoid premature discharge of these pins by multiply shocked air. The pins and sample surface should be kept under vacuum.

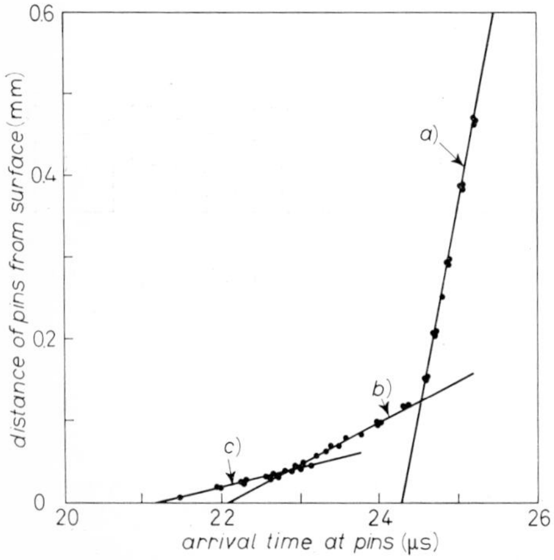

Fig. 1.15. – Pin data showing 3-wave structure in iron (MINSHALL). *a*) driving wave; *b*) first plastic wave; *c*) elastic wave.

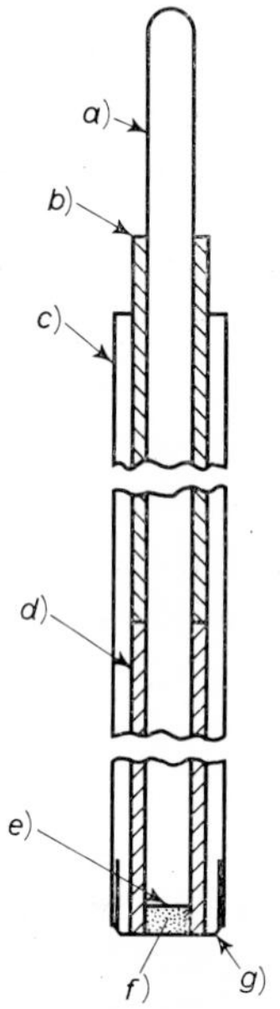

Fig. 1.16. – Coaxial piezoelectric pin. *a*) brass wire; *b*) teflon tubing; *c*) brass tube; *d*) epoxy resin; *e*) silver epoxy; *f*) piezoelectric crystal; *g*) copper plate.

Piezoelectric pins were developed at Livermore and elsewhere for application at lower pressures, where self-shorting pin closure is erratic. A coaxial piezoelectric pin developed for use in some low-pressure ((50 ÷ 100) kbar) experiments is shown in Fig. 1.16. These transducers require no external voltage source and can be either of a spring-loaded construction, or a solid assembly such as that shown in Fig. 1.16. Arrival time is accurate to within 5 ns. The

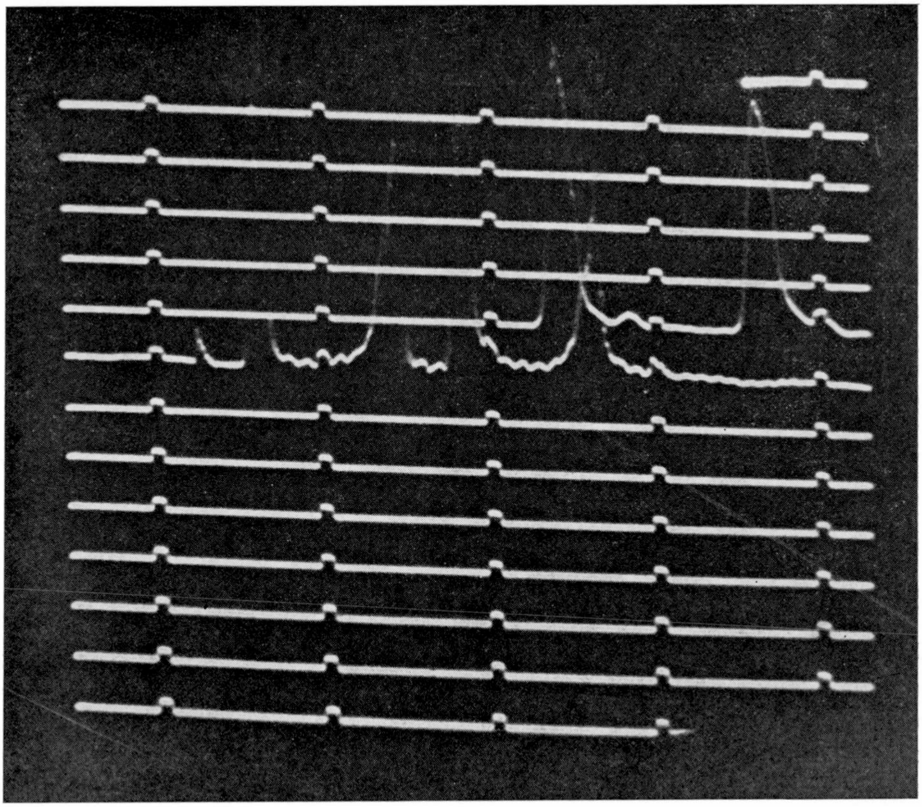

Fig. 1.17. – Piezoelectric pin record.

passage of a shock wave through liquid xenon, discharging a series of piezoelectric pins, is represented in Fig. 1.17.

1·3.2. *Streaking camera equation-of-state experiments using flash gap techniques.* – The standard equation-of-state measurement technique, which involves the use of flash gaps, is shown in Fig. 1.18 [1.16]. In this arrangement, the free surface and shock velocity of a number of samples are measured simultaneously. The slits are placed directly over the experiment, and the entire image is swept along the film. The horizontal straight lines in the photographic record, Fig. 1.19, are events occurring simultaneously within the error of shock tilt. They represent the arrival of the shock wave at the gaps located directly on the base plate.

The flashes to the right of the center flashes represent the arrival of the shock wave at the surface of the sample. To the left of the center flash, one sees first two short horizontal lines. These lines result from light leaks through the

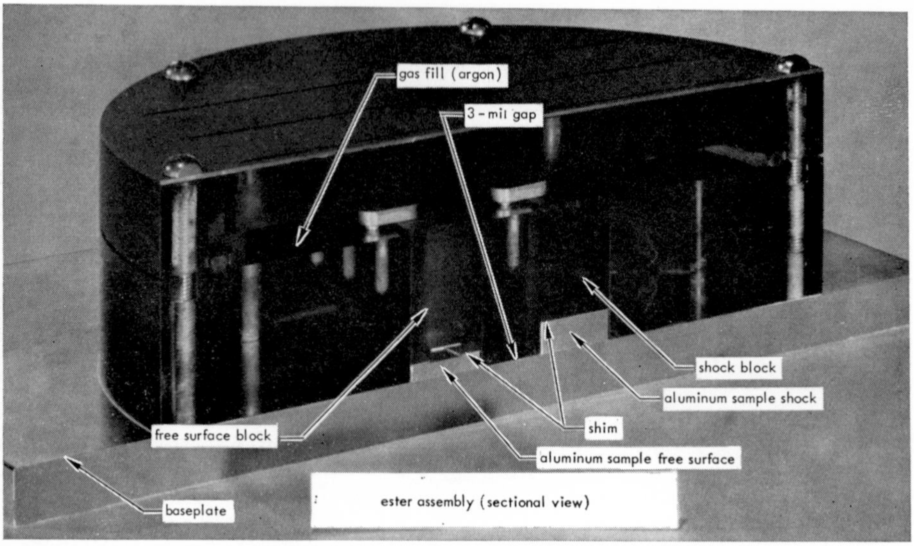

Fig. 1.18. – Ester assembly (sectional view).

Lucite which emanate from the shock emerging from the surface of the free-surface sample. The small horizontal plateau at the peak of the pulse represents the arrival of the free surface at the free-surface flash gap. The flashes of light at the top of the film represent the destruction of the sample and occur several microseconds later than the initial flashes. The particular shot shown was typical, using a 20-cm diameter explosive lens and a 10-cm thick high-explosive block.

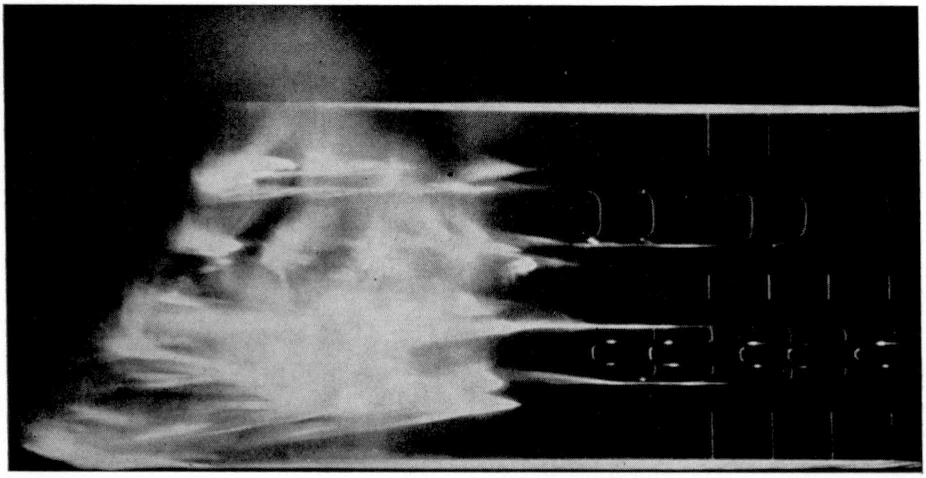

Fig. 1.19. – Flash-gap experimental record.

1·4. – Continuous methods.

In continuous methods, a continuous record is made of the movement of a free surface. Therefore, these methods depend upon preservation of integrity of the free surface as it accelerates under shock impact. Continuous methods are also applicable to measuring shock and particle velocities in transparent media. However, since under these conditions the record is at least partially internal to the shock front, these techniques will be discussed later in a section on combination techniques. Continuous techniques can be used to monitor the movement of a projectile or flying plate prior to impact with a sample. However, problems are created by the introduction of a hydrodynamic disturbance in the impacting surface, caused by the prism or mirror required to obtain a continuous record, and by the accompanying perturbation of the shock front as it propagates into the sample. Hence, continuous techniques have not yet been applied to such measurements. These techniques are appropriate for examining details of wave structure on breakout. It is possible to diagnose multiple-wave structure by recording the details of free-surface jump-off. There are numerous continuous methods for making free-surface measurements. Only two will be discussed here: the capacitor technique [1.17, 1.18] and the inclined-prism technique [1.19]. Other techniques commonly used are: the inclined wire [1.20], the inclined mirror [1.21, 1.22], the inclined flash gap [1.2] (which has generally been abandoned because of spurious preflashing caused by jetting into the flash gap itself), the knife-edge technique [1.23], and the optical-lever technique [1.22, 1.24]. The last two techniques are based on the observation of the reflection of stationary references in the moving free surface.

1·4.1. *Capacitor technique*. – The capacitor technique was developed simultaneously in the Soviet Union [1.17] and the at Los Alamos Scientific Laboratory [1.18]. This technique uses the capacitive charge between a plate suspended horizontally above a flat conducting surface as the surface is accelerated by the emerging shock wave and the subsequent rarefaction wave to provide a direct and continuous measurement of the free-surface velocity. The physical configuration and equivalent circuitry are shown in Fig. 1.20.

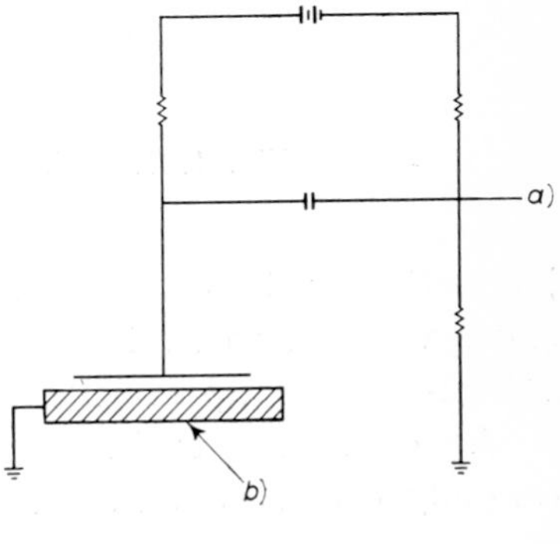

Fig. 1.20. – Capacitor technique. *a*) oscilloscope read-out; *b*) conducting sample.

Free-surface velocity is given as a function of time by the relationship

$$U_{fs}(t) = V_i(t)\,\frac{4\pi X_0^2}{\varepsilon S_0 E R}\,, \tag{1.8}$$

where

X_0 = spacing between capacitor plate and free surface at $t = t_0$,

S_0 = area of capacitor plate,

ε = dielectric constant of the medium between free surface and capacitor plate (m.k.s. units),

V_i = voltage drop measured across resistor,

E = charging voltage,

R = resistance, ohm.

Equation (1.8) is derived on the assumption that the time constant associated with the changing capacitance, $t_c = C(\mathrm{d}C/\mathrm{d}t)^{-1}$ is much greater than the RC time constant of the circuit. Thus, the circuit is in quasi-equilibrium at all times with respect to the changing capacitance, and the resistance of the circuit is so low that relatively little of the potential drop occurs across it. Equation (1.8) is also based on the assumption that during measurement, the displacement of the free surface does not exceed a small fraction of the initial separation of free surface and capacitor plate. Otherwise, for a constant free-surface velocity, the term X_0^2 is replaced by $(X_0 - U_{fs}t)^2$.

Experimentally, it is necessary for the capacitor plate and the free surface to be aligned as parallel as possible. Jetting and tilt which cause nonuniform

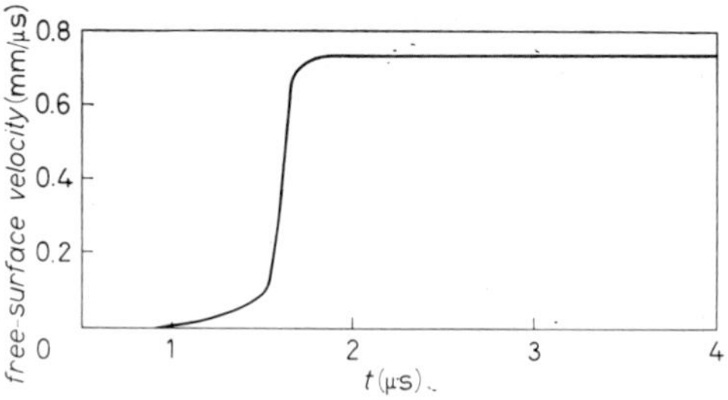

Fig. 1.21. – Free-surface record, capacitor technique.

jump-off must be eliminated, and the capacitor and free surface should be under vacuum, to avoid the effect of conducting gaseous precursors which, in high-pressure experiments, are driven in front of the accelerating free surface.

A record taken by TAYLOR of the Los Alamos Scientific Laboratory under carefully controlled conditions is shown in Fig. 1.21. The shock wave was generated by a gas-gun–accelerated projectile which impacted a sample. It clearly shows the elastic-plastic behavior in beryllium up to 120 kbar.

1·4.2. *The inclined-prism technique.* – This technique utilizes a prism with a lower surface slanted with respect to the free surface of the sample and reflecting upward [1.19]. The inclined-prism technique is based on the optical principle of total internal reflection of light passing through a medium and is dependent on the extinction of this reflection when there is an external surface in immediate proximity to the reflecting surface. When total internal reflection takes place from the interior of an optical element, a surface brought into contact with the reflecting surface will cause an optical disturbance such that there is complete transmission through the interface. This phenomenon occurs for the following reasons [1.25]: in total internal reflection, the amplitude of the light does not drop to zero just at the reflecting surface, but rather electromagnetic theory predicts a disturbance which decays exponentially beyond the surface but does not transfer net energy through it. When a dense medium is brought to within a small fraction of a wavelength from the surface, energy is drained off as light. The disappearance of total internal reflection in a prism inclined to the shock front indicates the arrival of the free surface at the totally reflecting interface. This technique is very valuable for making measurements in the pressure range of 0 to 150 kbar, since it does not depend on the destruction of a reflecting surface but only on the arrival of the sample free surface at the reflecting interface. An example of an inclined-prism experiment is shown in Fig. 1.22. This assembly must be evacuated to a pressure less than 50 microns due to air shocks. The light source is an argon candle, and the progress of the free surface is observed through the slit of a streaking camera. In designing the experiment, it is necessary to insure that the free surface of the sample travels along the inclined prism at a velocity faster than sound velocity in the mirror. The lower the angle, the higher is the velocity of arrival of the free surface along the prism. The largest angle that can be tolerated is given by the relationship

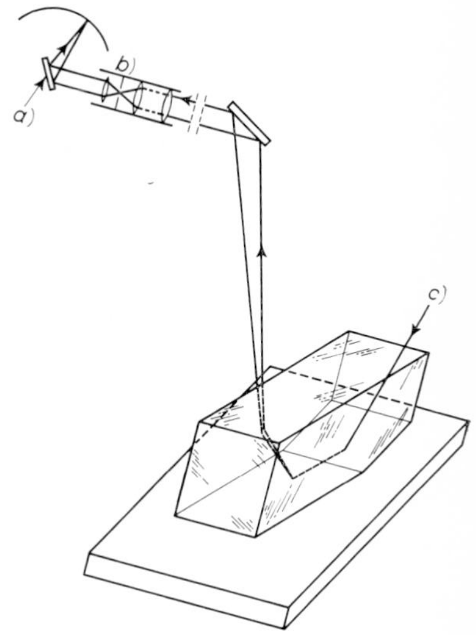

Fig. 1.22. – Inclined-prism experimental arrangement. *a*) rotating mirror; *b*) slit; *c*) light source.

$$\alpha_{\max} = \sin^{-1}\left(\frac{U_{fs}}{c}\right). \tag{1.9}$$

It is also necessary to insure that no hydrodynamic disturbance is propagated along the free surface of the sample. The free-surface velocities are given by the relationship

$$U_{fs} = \frac{W}{M}\frac{\mathrm{tg}\,\theta}{\mathrm{tg}\,\beta}, \tag{1.10}$$

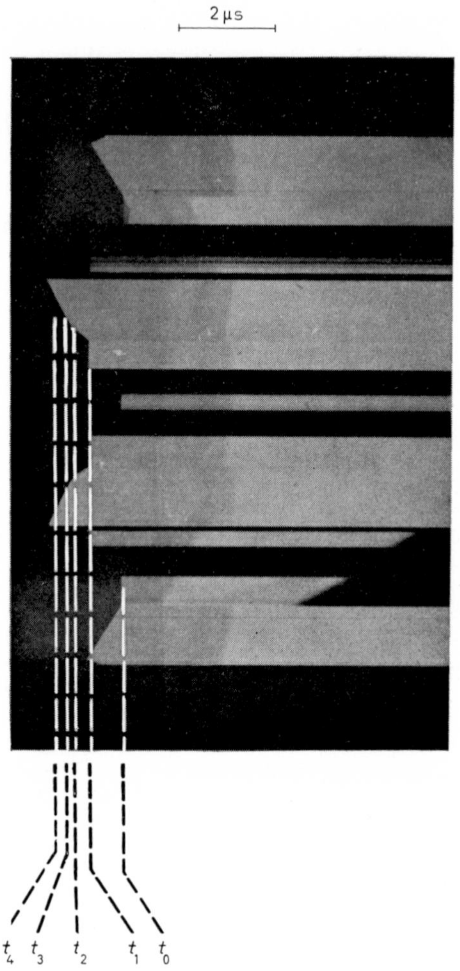

Fig. 1.23. – Inclined-prism experimental record.

where θ is the angle of the inclined prism with respect to the freesurface, β is the angle of the trace on the film, M is the magnification ratio between the film and the experiment, and W is the writing speed of the camera. Corrections can be made for shock tilt and alignment errors. An inclined-prism record is shown in Fig. 1.23; it is the response of a single crystal of 111- and 110-oriented silicon to a 200-kbar driving pressure. Four distinct waves can be seen: the elastic precursor and a three-wave structure associated with two successive phase transitions occurring at higher pressure. It is interesting to note the comparison of this record with the record (Fig. 1.25) made by Soviet investigators using the magnetic technique. The Soviets did not claim to have observed two polymorphic high-pressure transitions, even though both records clearly show the existence of two transitions.

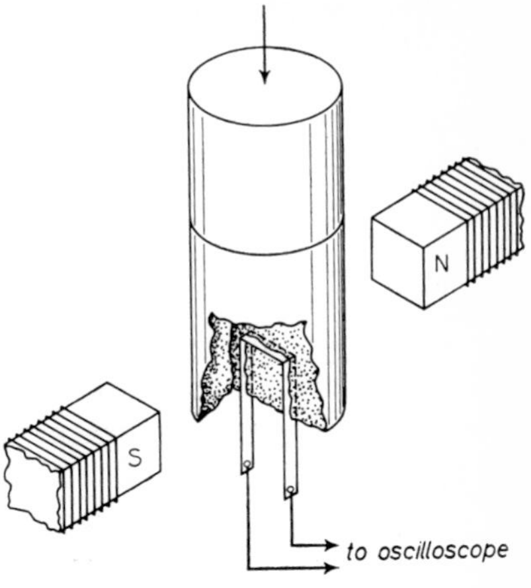

Fig. 1.24. – Electromagnetic technique, physical configuration.

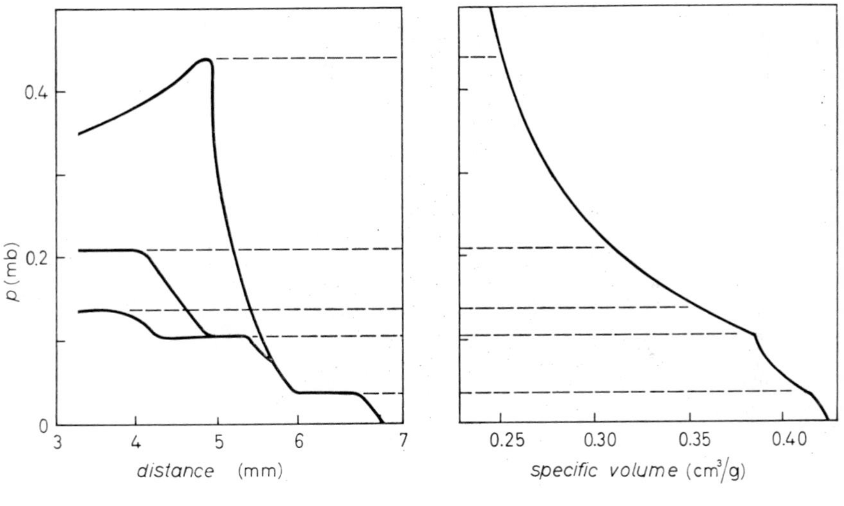

Fig. 1. 25. – Multiple-wave structure in silicon.

1·5. – Internal techniques.

Those diagnostic techniques which require the sensing element to be behind or internal to the shock front are classified as internal techniques. These techniques have the advantage that they provide direct read-outs of either pressure or particle velocity behind the shock front. These quantities are the most difficult to obtain experimentally. This is contrasted with the continuous external techniques which provide information on free-surface velocities. The free-surface information must be interpreted and unfolded in some detail to obtain the pressure profile of the original shock wave because of the multiple interactions and reflections generated by shocks moving back into the sample and by shocks reflected from the forward moving shock waves in the sample. The disadvantage of these techniques is that the sensing element is subject to the violent passage of the shock front and to the temperatures, pressures, and shear stresses behind the shock front. These techniques are valuable for providing qualitative information on multipleshock structure and rarefaction, but to date have provided only semi-quantitatively accurate data, since separate calibration measurements are required.

1.5.1. *The electromagnetic method.* – If an electrical conductor of length l is moved with a velocity, U_p, normal to the lines of a magnetic field of $\boldsymbol{B}$ gauss, an e.m.f. will be developed. This relationship is given by

$$V = U_p B l \cdot 10^{-8}, \tag{1.11}$$

with V in volt, B in gauss, l in cm, and U_p, particle velocity in cm/s. A typical experimental arrangement is shown in Fig. 1.24. The wire is embedded in a nonconducting sample, and after the arrival of the shock front, starts to move with the particle velocity characteristic of the shocked state. The method fails, of course, when the shocked material becomes highly conducting, and it is not applicable to metals. Apparently, circuit integrity can be maintained even after the passage of the shock front. This method was first used by DREMIN and SHREDOV [1.26] to study the behavior of detonation products. AL'TSHULER and PAVLOVSKII [1.27] were able to observe three waves in silicon at a driving pressure of 220 kbar. The first wave was an elastic precursor, and the next two waves were associated with a phase transition at 100 kbar, which was supposed to be from the diamond structure to the gray tin structure. The experimental record is shown in Fig. 1.25.

In determining the dynamic behavior of detonation products, one normally observes the response of metals of varied thickness in contact with the high explosive. However, the electromagnetic technique provides a direct measurement, limited only by the physical size of the wire. Similarly, multiple-wave structure is normally diagnosed by observation of the irregularities in the free surface jum-poff on emergence of a multiple-wave system. In order to reconstruct the original wave profile, account must be taken of the very complex system of wave interactions. With this technique, a continuous record of the passage of the multiple-wave system can be obtained.

1.5.2. *Manganin wire transducers.* – Manganin is an alloy which has an extremely low temperature coefficient of resistivity, and is commonly used in static high-pressure work. It was first used for dynamic high-pressure applications in England [1.28], and later extensively developed by the Stanford Research Institute [1.29], and has been used to diagnose multiple-wave structure in metals and ceramics in an epoxy encapsulation attached to the top of the material being studied. More recently, it has been fabricated as a ribbon and sandwiched between insulation directly inside the material of interest [1.30], as shown in Fig. 1.26. This technique is not applicable in metals above about 350 kbar, because the integrity of the encapsulating insulation breaks down at roughly this pressure. At pressures in metals much above 150 kbar, the thickness of insulation required seriously degrades the time resolution of the gauge. Hysteresis, work-hardening effects, and temperature effects on resistance at high pressure all combine to reduce the reproducibility of manganin-

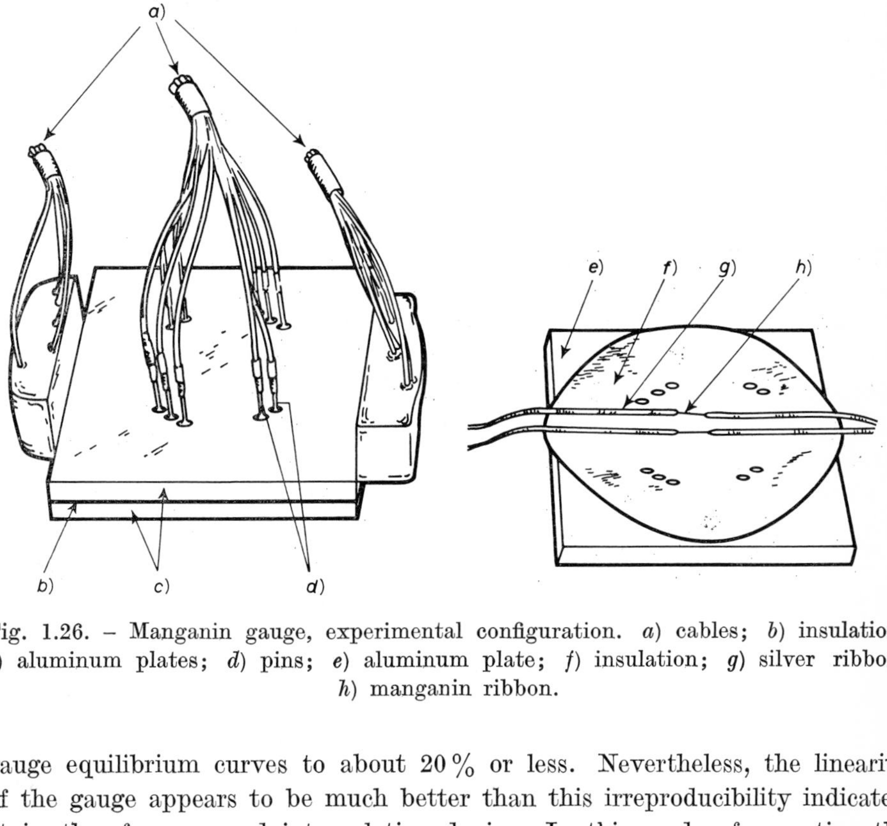

Fig. 1.26. – Manganin gauge, experimental configuration. *a*) cables; *b*) insulation; *c*) aluminum plates; *d*) pins; *e*) aluminum plate; *f*) insulation; *g*) silver ribbon; *h*) manganin ribbon.

gauge equilibrium curves to about 20% or less. Nevertheless, the linearity of the gauge appears to be much better than this irreproducibility indicates. It is, therefore, a good interpolation device. In this mode of operation the failure of the ideal elastic-plastic model was first shown conclusively with a manganin gauge [1.31].

A picture of a record of the passage of two shock waves is shown in Fig. 1.27, demonstrating the continued integrity of the gauge after shock and rare-

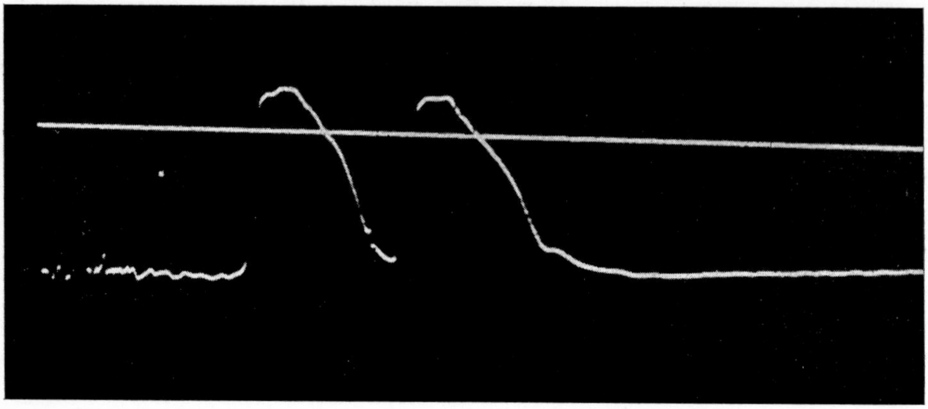

Fig. 1.27. – Manganin gauge, experimental record.

faction. In the future, the manganin gauge can be expected to provide useful qualitative and semiquantitative information on the behavior of matter under conditions of combined elastic and plastic loading.

1'6. – Combination techniques.

There are a number of techniques which provide information after passage of the shock front, but which have certain common characteristics with the preceding three classes of techniques. Three of these will be discussed in detail here: the immersed-foil technique, the quartz-crystal technique, and the laser interferometer technique. Other techniques have also been developed. One is utilization of the « detonation electric effect » [1.32], the electrical signal which is generated on impact of a shock or detonation wave with a metallic surface. This technique can be used to measure the velocity of the shock wave reflected back into hot dense detonation products. Another technique involves the use of a dielectric (*e.g.*, sapphire) to which electrostatic potential is applied [1.33]. The resulting current can be interpreted to obtain the interface velocity between the dielectric and the sample as a function of time.

1'6.1. *The immersed-foil technique.* – The immersed-foil technique [1.34] is similar in concept to the inclined-mirror and the inclined-prism techniques. The principal difference is that instead of a reflecting mirror or prism surface, a 0.01-mm aluminized Mylar foil is immersed at an angle in a transparent liquid covering the top of a sample, and measurements are made of the particle velocity and refractive index in the liquid material behind the shock front. Because of interface continuity conditions, this measured particle velocity is the same for both sample and fluid after shock-wave passage. A schematic is shown in Fig. 1.28. This technique has been quite useful in determining release adiabat points for minerals and has provided some information about the variation of refractive index of various fluids under pressure. With a

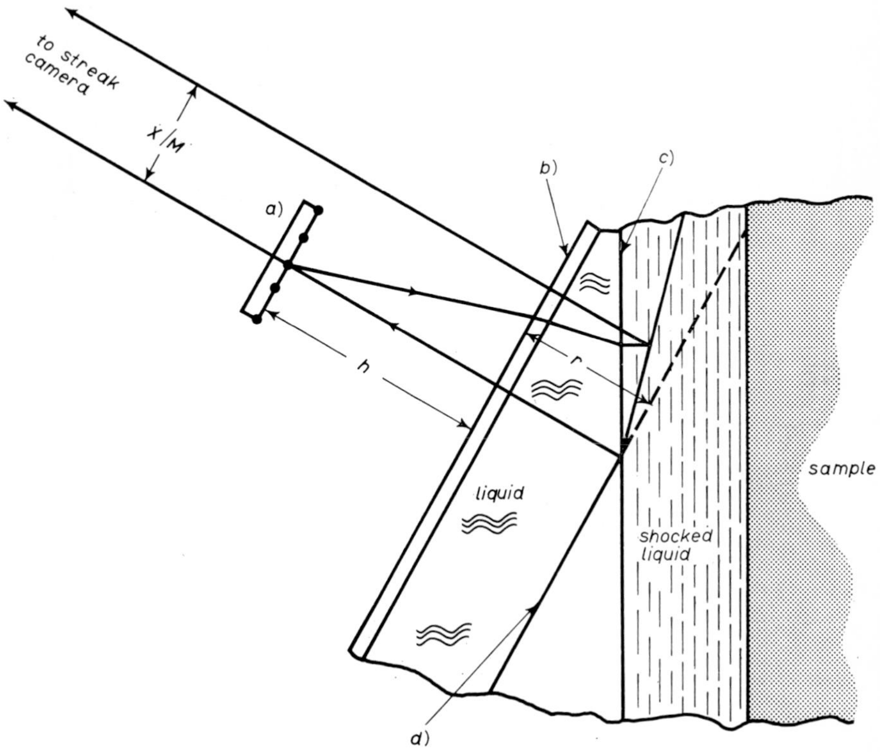

Fig. 1.28. – Immersed-foil technique. *a*) grid light source; *b*) plate glass; *c*) shock front; *d*) reflecting foil.

choice of different transparent fluids as anvil material, a set of rarefaction isentropes can be obtained, and the entire release adiabat can be mapped out. One unexpected windfall was the observation of a weak pulse propagating back from the thin immersed foil into the initially shocked material. This pulse apparently was a weak sonic disturbance, and was correlated with the predicted sound velocity, $c_s = \sqrt{(\partial p/\partial \varrho)_s}$. However, it is doubtful whether the accuracy of this measurement is better than about 5%.

1·6.2. *The quartz transducer.* – The piezoelectric effect in quartz has been utilized in the development of a stress gauge which is now used widely, particularly in the study of phase transitions [1.35]. The quartz crystal is cut cylindrically with faces perpendicular to the x-axis, and placed on top of the sample. A schematic of an experiment designed to look for multiple-wave structure in bismuth is shown in Fig. 1.29 [1.36]. One of the experimental records is shown in Fig. 1.30. In this experiment it was possible to determine the 25.4-kbar phase transition in bismuth to ± 0.8 kbar, and to correct for strength of material effects.

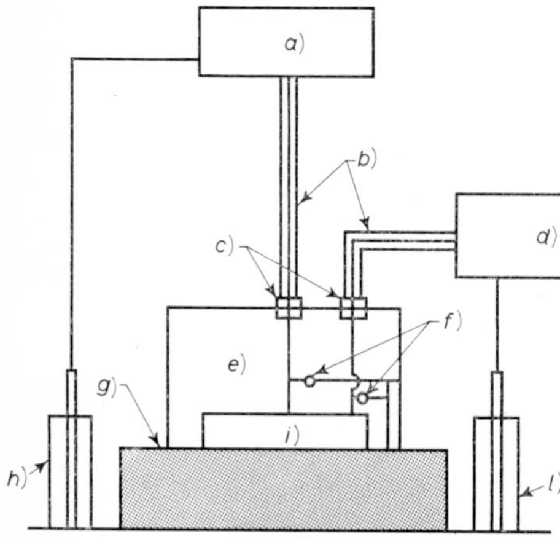

Fig. 1.29. – Quartz-gauge experimental configuration. *a*) Tektronix oscilloscopes; *b*) coaxial cable; *c*) connectors; *d*) Denver raster oscilloscopes; *e*) epoxy or oil; *f*) resistors; *g*) sample; *h*) barium titanate trigger pin; *i*) gauge; *l*) barium titanate shock-velocity pin.

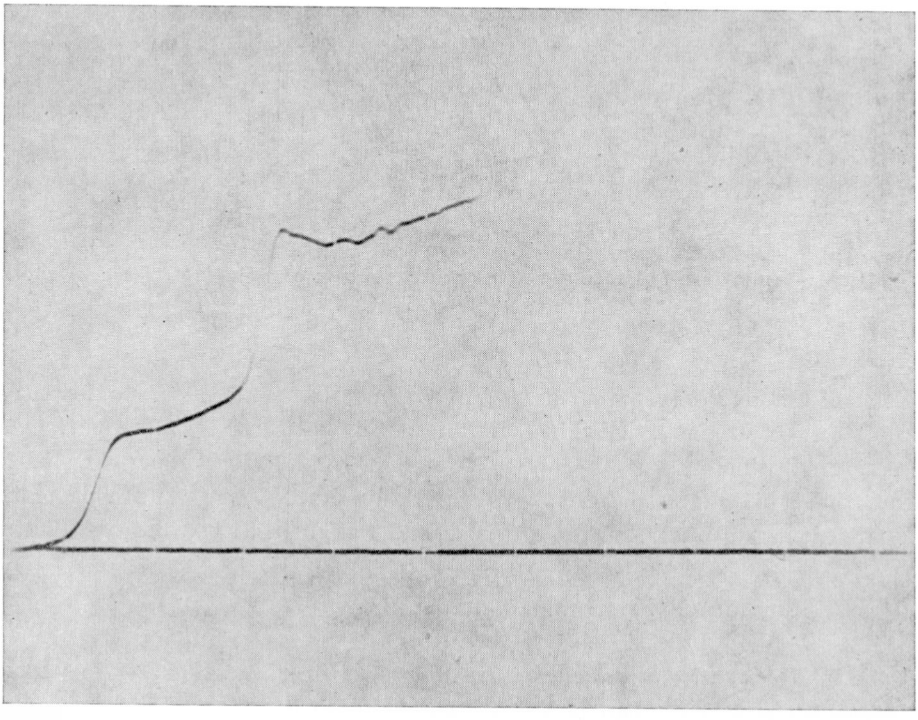

Fig. 1.30. – Three-wave structure in bismuth obtained from a quartz gauge.

The current produced by the piezoelectric effect is given by

$$i = A\left(\frac{\mathrm{d}D}{\mathrm{d}t}\right) = \left[kA\,\frac{U_s}{l}\,(P_0 - P_e)\right], \tag{1.12}$$

where A is the area of the electrode, $\mathrm{d}D/\mathrm{d}t$ the displacement current, U_s the wave velocity in quartz, l the gauge thickness, and P_0 and P_e, the pressures at front and rear of the gauge. The coefficient k is a function of pressure only, and for quartz, $k = 2.04 \cdot 10^{-8}$ C/cm² kbar. The gauge response is linear to 20 kbar and is usable up to 50 kbar. Because of impedance-matching considerations, this upper limit corresponds to a much higher pressure in dense materials. The quartz gauge is accurate to within 3 to 5% at pressures up to 50 kbar.

1'6.3. *The laser interferometer technique.* – The recent availability of single-frequency continuous-wave lasers has enabled high-precision measurement of complicated surface motion in shock-wave experiments. The first application of this technique was made by BARKER and his co-workers, [1.37] who developed the system shown in Fig. 1.31. The laser beam is directed at the target surface by a mirror and focused directly on the surface by a lens. This minimizes the area on which the light is incident, and helps to eliminate the effects of shock tilt. The light is directed back along the line of sight by the mirror, and recollimated by the second lens. Part of the light is reflected into the delay leg by a beam splitter, while the rest continues into the photomultiplier, which is read onto a fast-sweeping oscilloscope.

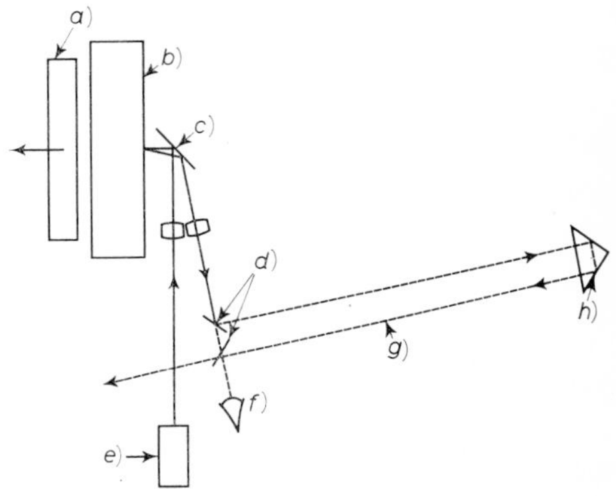

Fig. 1.31. – Experimental configuration. *a*) projectile plate; *b*) target specimen; *c*) mirror; *d*) beam splitters; *e*) laser; *f*) photomultiplier; *g*) delay leg; *h*) 90° reflecting prism.

The delay leg is the heart of the apparatus. This optical element makes it possible to beat light reflected from a surface at a given time with light reflected from the surface at a time L/c later. The delay leg is actually an acceleration-sensing element which produces fringes with a frequency proportional to the acceleration of the surface.

Just how this beating process provides the desired information can be explained in the following way. Light of a given wavelength is reflected from the surface of the target. The delay is equal to the path from the first beam splitter through the 90° prism to the second beam splitter, minus the distance between the beam splitters. If the length of this delay leg is an integer number of wavelengths, say $N\lambda$, then, the light arriving at the photomultiplier from both sources is in phase. Suppose the surface commences moving with a velocity such that λ is decreased slightly, with an increase in N of one half, then the two beams will be out of phase, and the photomultiplier will sense no signal. Thus, the photomultiplier actually registers as light fringes or oscillations the total change in the number of wavelengths in the delay leg caused by change in wavelength.

The actual Doppler shift for light is given by

$$\Delta\lambda = -\frac{2\lambda}{c}\,u(t)\,, \tag{1.13}$$

with $\Delta\lambda$ the change in wavelength, λ the wavelength of the light incident on the surface, c the velocity of light, and $u(t)$ the velocity of the surface. Now the length of the delay leg is given by

$$N\lambda = c\tau\,, \tag{1.14}$$

with τ the time required for the light to traverse the delay leg. Therefore, the number of waves of wavelength λ in the cavity is given by

$$N = \frac{c\tau}{\lambda}\,. \tag{1.15}$$

Differentiating, we obtain

$$\mathrm{d}N = -\frac{c\tau}{\lambda^2}\,\mathrm{d}\lambda\,, \tag{1.16}$$

or written in different form

$$\Delta N(t) = -\frac{c\tau}{\lambda^2}\,\Delta\lambda(t)\,. \tag{1.17}$$

λ, of course, is also a function of t but varies only slightly; the term $\Delta\lambda(t)$ dominates the right-hand side. Equation (1.17) can be expressed in terms of the Doppler formula eq. (1.13), which gives

$$u(t) = \left(\frac{\lambda}{2\tau}\right)\Delta N(t)\,. \tag{1.18}$$

This formula shows that it is possible to obtain the velocity as a function of time simply by counting the fringes observed on the oscilloscope trace, with $\lambda/2\tau$ velocity units per count. An idealized record is shown in Fig. 1.32. One difficulty here, of course, is the inability to discriminate between acceleration and deceleration simply by counting fringes. Some qualitative features of the free-surface behavior must be known.

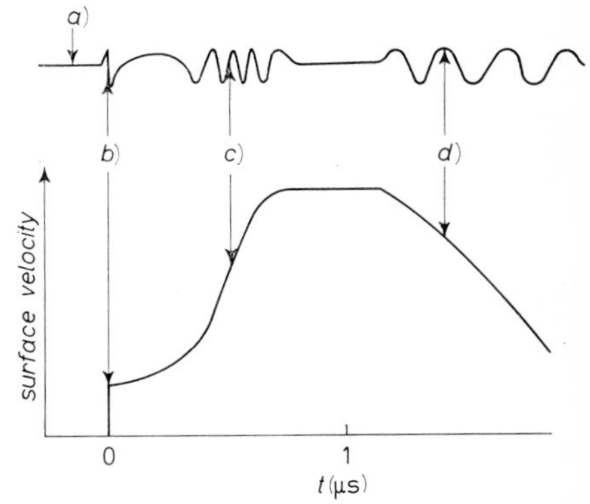

Fig. 1.32. – Idealized experimental record. *a*) oscilloscope trace; *b*) elastic shock front; *c*) plastic wave; *d*) release wave.

Free-surface measurements are limited, as discussed previously, by the complexity of analysis required to unfold the complex interactions between oncoming multiple waves and reflected rarefactions, and by the possibility of spallation obscuring details of release and rarefaction behind the shock front. This limitation can be partially overcome, using as an anvil a transparent material (*e.g.*, quartz or sapphire) which closely matches the impedance of the sample material. Light is then reflected from the interface between the two materials. This permits a more direct interpretation of the results in terms of the driving-wave system, since we are now looking at a surface where movement is more characteristic of particle velocity than free-surface velocity. In this application, however, it is necessary to know the dynamic behavior and refractive index of the transparent anvil material.

An alternate technique which is superior in some respects involves the use of a Fabry-Perot interferometer to vary transmission as a function of frequency of incident light [1.38]. This technique includes lens-free and tilt-insensitive optics; however, its accuracy is probably limited to approximately 3 %.

1·7. – Property of materials measurements.

In addition to the equation-of-state measurements discussed in this lecture, many properties-of-materials measurements have also been carried out in shock-wave experiments. Four of these will be reviewed in subsequent lectures: magnetic properties, electrical conductivity, acoustic velocity by Brillouin scattering, and flash X-ray diffraction. A great deal of optical work has been done in the Soviet Union, including measurements of temperature, surface reflectivity, and refractive index. A complete description of the techniques and experimental results is given in the excellent review article by KORMER [1.39]. There has also been work on the properties of shocked ferroelectrics [1.40], microwave measurements of dielectric coefficients [1.41, 1.42], and conductivity and depolarization of ionic salts under shock compression [1.43], optical spectroscopy [1.44], and Hall-effect measurements [1.45]. None of these investigations, however, has been carried out in depth, and therefore they will not be discussed in this lecture.

1'8. – **Conclusions.**

The pressure range from 0 to 150 kbar is fairly well characterized experimentally. Numerous techniques are available for high-precision measurements of single- and multiple-wave structures emerging from a free surface or into a transparent medium. However, one serious deficiency in this pressure range is our lack of understanding of release waves following shock fronts, and our ability to measure release phenomena only semiquantitatively. More work needs to be done on the development of highly accurate transducers such as the manganin gauge, which can withstand the effects of the passage of a shock wave.

Russian workers have carried out temperature measurements on a number of transparent shocked materials, and successfully observed melting transitions in several alkali halides. Unfortunately, the scatter in their data ((300 ÷ 500) °K) is too great to permit meaningful comparisons with equation-of-state models. Accurate temperature measurements are needed, both in transparent media and in shocked metals.

It appears that a major effort would be required to increase the accuracy of a shock- and particle-velocity measurements to 0.1 %, where a critical test of various equations of state might be made. Such precision is useless, however, until the effects of pressure and temperature on the Hugoniot elastic limit are understood. A more reliable technique for quantitatively identifying the melting transition along the Hugoniot must also be developed.

Finally, we must achieve a much deeper understanding of the mechanism of yielding under shock compression. Whatever the mechanism, dislocation or slip, it is important to know how the shocked material reforms itself behind the shock front. Should a shocked single crystal reform itself as a single crystal or even as a large number of crystallites of identical orientation, it will be possible to obtain much more useful information on materials under shock compression.

REFERENCES

[1.1] R. Courant and K. O. Friedrichs: *Supersonic Flow and Shock Waves* (New York, 1947), p. 121.

[1.2] J. M. Walsh and R. H. Christian: *Phys. Rev.*, **97**, 1544 (1955).

[1.3] G. E. Duvall: in *Response of Metals to High Velocity Deformation* (New York, 1960), p. 165.

[1.4] I. C. Skidmore and E. Morris: *Thermodynamics of Nuclear Materials* (Vienna, 1962), p. 173.

[1.5] L. V. Al'tshuler, A. A. Bakanova and R. F. Trunin: *Žurn. Ėksp. Teor. Fiz.*, **42**, 91 (1962); English translation in *Sov. Phys. JETP*, **15**, 65 (1962).

[1.6] A. H. Jones, W. M. Isbell and C. M. Maiden: *Journ. Appl. Phys.*, **37**, 3493 (1966).

[1.7] L. V. Al'tshuler, B. N. Moiseyev, L. V. Popov, G. V. Simakov and R. F. Trunin: *Žurn. Ėksp. Teor. Fiz.*, **54**, 785 (1968).

[1.8] See, for example, the following reviews: M. H. Rice, R. G. McQueen and J. M. Walsh: *Solid State Physics*, edited by Seitz and Turnbull, vol. **6** (New York, 1958), p. 1; G. E. Duvall: in *Response of Metals to High Velocity Deformation*, edited by Shewmon and Zackay (New York, 1960), p. 165; S. D. Hamann: in *Advances in High Pressure Research*, edited by R. S. Bradley, Vol. **1** (New York, 1966), p. 85; B. J. Alder: in *Solids Under Pressure*, edited by Paul and Warschauer (New York, 1963), p. 385; G. E. Duvall and G. R. Fowles: in *High Pressure Chemistry and Physics*, edited by R. S. Bradley, Vol. **2** (London, 1963), p. 209; L. V. Al'tshuler: *Usp. Fiz. Nauk*, **85**, 197 (1965), English translation in *Sov. Phys. Usp.*, **8**, 52 (1965); I. C. Skidmore: *Appl. Material Res.*, **4**, 131 (1965); D. G. Doran and R. K. Linde: *Solid State Physics*, edited by Seitz and Turnbull, Vol. **18** (New York, 1966), p. 229.

[1.9] J. R. Cook: *Research (London)*, **1**, 471 (1948).

[1.10] S. Thunborg, G. E. Ingram and R. A. Graham: *Rev. Sci. Instr.*, **35**, 11 (1964).

[1.11] R. K. Linde and D. N. Schmidt: *Rev. Sci. Instr.*, **37**, 1 (1966).

[1.12] B. Brixner: *Proceedings of the Third International Congress on High Speed Photography*, edited by R. B. Collins (New York, 1957), p. 289.

[1.13] I. Sh. Model: *Žurn. Ėksp. Teor. Fiz.*, **32**, 714 (1957); English translation in *Sov. Phys. JETP*, **5**, 589 (1957).

[1.14] M. van Thiel and A. C. Mitchell: unpublished work.

[1.15] F. S. Minshall: *Journ. Appl. Phys.*, **26**, 463 (1955).

[1.16] M. van Thiel and B. J. Alder: *Journ. Chem. Phys.*, **44**, 1056 (1966).

[1.17] A. G. Ivanov and S. A. Novikov: *Prib. Tekn. Eksp.*, **1**, 135 (1963).

[1.18] M. H. Rice: *Rev. Sci. Instr.*, **32**, 449 (1961).

[1.19] G. Eden and P. W. Wright: *Proceedings of the Fourth International Symposium on Detonation*, edited by S. J. Jacobs (Washington, D. C., 1966), p. 573.

[1.20] L. M. Barker, C. D. Lundergan and W. Hermann: *Journ. Appl. Phys.*, **35**, 1203 (1964).

[1.21] R. G. McQueen: in *Metallurgy at High Pressures and High Temperatures*, edited by K. A. Gschneidner jr., M. T. Hepworth and N. A. D. Parlee (New York, 1964), p. 76.

[1.22] D. G. Doran: *Journ. Appl. Phys.*, **34**, 844 (1968).

[1.23] W. C. DAVIS and B. G. CRAIG: *Rev. Sci. Instr.*, **32**, 579 (1961).
[1.24] G. R. FOWLES: *Journ. Appl. Phys.*, **32**, 1475 (1961).
[1.25] See, for example, F. A. JENKINS and H. E. WHITE: *Fundamentals of Optics*, III ed. (New York, 1957), p. 519.
[1.26] A. N. DREMIN and K. K. SHREDOV: *PMTF*, **2**, 154 (1964).
[1.27] L. V. AL'TSHULER and M. N. PAVLOVSKII: work cited in [1.8] (AL'TSHULER).
[1.28] P. J. A. FULLER and J. H. PRICE: *Nature*, **193**, 262 (1962); *British Journ. Appl. Phys.*, **15**, 751 (1964).
[1.29] D. BERNSTEIN and D. D. KEOUGH: *Journ. Appl. Phys.*, **35**, 1471 (1964).
[1.30] M. VAN THIEL and A. S. KUSUBOV: *Proceedings of the Symposium on Accurate Characterization of the High Pressure Environment* (Gaithersburg, Md., 1968), to be published.
[1.31] A. S. KUSUBOV and M. VAN THIEL: *Journ. Appl. Phys.*, in press.
[1.32] B. HAYES: *Journ. Appl. Phys.*, **38**, 507 (1967).
[1.33] R. A. GRAHAM and G. E. INGRAM: *Behaviour of Dense Media Under High Dynamic Pressure, Symposium HDP, IUTAM, Paris*, **1967** (New York, 1968), p. 470.
[1.34] T. J. AHRENS and M. H. RUDERMAN: *Journ. Appl. Phys.*, **37**, 4758 (1966).
[1.35] R. A. GRAHAM, F. W. NEILSON and W. B. BENEDICK: *Journ. Appl. Phys.*, **36**, 1775 (1965).
[1.36] D. B. LARSON: *Journ. Appl. Phys.*, **38**, 1541 (1967).
[1.37] L. BARKER: *Behaviour of Dense Media Under High Dynamic Pressure, Symposium HDP, IUTAM, Paris*, **1967** (New York, 1968), p. 484.
[1.38] P. M. JOHNSON and T. J. BURGESS: *Rev. Sci. Instr.*, **39**, 1100 (1968).
[1.39] S. B. KORMER: *Usp. Fiz. Nauk*, **94**, 641 (1968); (English translation in *Sov. Phys. Usp.*, **11**, 229 (1968).
[1.40] R. K. LINDE: *Journ. Appl. Phys.*, **38**, 4839 (1967).
[1.41] R. S. HAWKE, A. C. MITCHELL and R. N. KEELER: *Rev. Sci. Instr.*, **40**, 632 (1969).
[1.42] R. S. HAWKE, R. N. KEELER and A. C. MITCHELL: *Appl. Phys. Lett.*, **14**, 229 (1969).
[1.43] R. K. LINDE, W. J. MURRI and D. G. DORAN: *Journ. Appl. Phys.*, **37**, 2527 (1966).
[1.44] A. H. EWALD: work cited by S. D. HAMANN: *Advances in High Pressure Research*, edited by R. S. BRADLEY (New York, 1966), p. 86; M. VAN THIEL and A. C. MITCHELL: unpublished work.
[1.45] J. D. KENNEDY and W. B. BENEDICK: *Solid State Communications*, **5**, 53 (1967).

II. – High-Pressure Equations of State from Shock-Wave Data.

E. B. ROYCE

2·1. – The Grüneisen equation of state.

The interpretation of shock-wave data and the development of a more complete equation of state require the use of an equation-of-state model to furnish additional information, since the normal Hugoniot is a single line on an equation-of-state surface. If the Hugoniot lies only in the solid phase, it is common to utilize the Grüneisen form of the equation of state [2.1-2.4],

$$P(V,E) = P_{0\,^\circ\mathrm{K}}(V) + \frac{\gamma(V)}{V}\,(E - E_{0\,^\circ\mathrm{K}}(V)), \tag{2.1}$$

where Grüneisen's assumption is that gamma, $\gamma(V,\,T)$, is a function of volume only and is independent of the temperature or thermal energy, $(E - E_{0\,^\circ\mathrm{K}}(V))$. Since neither entropy nor temperature can be calculated explicitly, such an equation of state is still incomplete.

The meaning of such an assumption is most easily seen in terms of a quasi-harmonic model of the thermal excitations of a crystal lattice. The lattice vibrations are first treated as completely harmonic for a given specific volume, and a normal-mode analysis is carried out. The anharmonicity in the effective interatomic interaction is taken into account by allowing the various normal-mode frequencies to be volume-dependent. If ε_i is the energy in the i-th normal mode of natural frequency ν_i at some temperature T, the total energy of the system is

$$E(V,\,T) = \varphi(V) + \sum_i \varepsilon_i(V,\,T)\,, \tag{2.2}$$

$$\varepsilon_i(V,\,T) = \frac{1}{2}\,h\nu_i(V) + \frac{h\nu_i(V)}{[\exp[h\nu_i(V)/kT] - 1]}\,. \tag{2.3}$$

The potential $\varphi(V)$ is the energy of the crystal with the atoms at rest, the cohesive energy. The free energy is given by

$$A(V,\,T) = \varphi(V) + \sum_i \tfrac{1}{2}\,h\nu_i(V) + kT\sum_i \ln[1 - \exp[-h\nu_i(V)/kT]]\,, \tag{2.4}$$

and the pressure is given by

$$P(V,\,T) = -\frac{\mathrm{d}\varphi(V)}{\mathrm{d}V} + \frac{1}{V}\sum_i \gamma_i(V)\left\{\frac{1}{2}\,h\nu_i(V) + \frac{h\nu_i(V)}{[\exp[h\nu_i(V)/kT] - 1]}\right\} = -\frac{\mathrm{d}\varphi(V)}{\mathrm{d}V} + \frac{1}{V}\sum_i \gamma_i(V)\,\varepsilon_i(V,\,T)\,, \tag{2.5}$$

where the parameter $\gamma_i(V)$ is defined for each normal mode as

$$\gamma_i(V) = -\frac{\mathrm{d}\ln \nu_i(V)}{\mathrm{d}\ln V}\,. \tag{2.6}$$

The various terms in the Grüneisen equation (eq. (2.1)) may then be identified as

$$\gamma(V,T)=\frac{\sum_i \gamma_i(V)\,\varepsilon_i(V,T)}{\sum_i \varepsilon_i(V,T)}\,, \tag{2.7}$$

$$E_{0\,^\circ\mathrm{K}}(V)=\varphi(V)+\sum_i \tfrac{1}{2}h\nu_i(V)\,, \tag{2.8a}$$

$$P_{0\,^\circ\mathrm{K}}(V)=-\frac{\mathrm{d}\varphi(V)}{\mathrm{d}V}+\frac{\sum_i \gamma_i(V)h\nu_i(V)}{2V}\,. \tag{2.8b}$$

Except for materials with very light atoms, the zero-point terms in eqs. (2.8*a*), (2.8*b*) may be neglected, and one can identify $E_{0\,^\circ\mathrm{K}}(V)$ as the cohesive energy. The calculation of $P_{0\,^\circ\mathrm{K}}$ and $E_{0\,^\circ\mathrm{K}}$ is of fundamental interest in solid-state physics [2.4-2.6].

From the foregoing expressions it is clear that if all of the γ_i's are equal, gamma will be independent of temperature and dependent only on the volume. The converse is not true, however, since the independence of gamma on temperature will exist as long as the γ_i's yield averages independent of ν when averaged over a range of $h\nu$ of the order of kT. This is fortunate, since almost certainly, the γ_i's are not all equal.

Most metals have a Debye temperature below room temperature, and one can assume equipartition of energy between the normal modes. In that case, gamma is just the average value of the γ_i's taken over all of the normal modes and will, of course, be a function of the volume only. There is some evidence that one may consider gamma to be independent of temperature for all temperatures greater than 0.3 times the Debye temperature, based on lattice-dynamic calculations of BARRON [2.7].

It is worth noting that gamma is strictly defined thermodynamically as

$$\gamma(V,T)=V\left(\frac{\partial P}{\partial E}\right)_V=V\left(\frac{\partial P}{\partial T}\right)_V\Big/\left(\frac{\partial E}{\partial T}\right)_V\,, \tag{2.9}$$

$$\gamma(V,T)=\sum_i \gamma_i\left(\frac{\partial \varepsilon_i}{\partial T}\right)_V\Big/\sum_i\left(\frac{\partial \varepsilon_i}{\partial T}\right)_V\,. \tag{2.10}$$

This expression is equal to gamma defined previously in eq. (2.7), only under the same conditions that were required for gamma to be independent of temperature [2.4]. Thus, in general, gamma defined by the common expression

$$\gamma=V\left(\frac{\partial P}{\partial T}\right)_V\Big/C_V=-V\left(\frac{\partial P}{\partial V}\right)_T\left(\frac{\partial V}{\partial T}\right)_P\Big/C_V=-V\left(\frac{\partial P}{\partial V}\right)_S\left(\frac{\partial V}{\partial T}\right)_P\Big/C_P\,, \tag{2.11}$$

is equivalent to Grüneisen's gamma as defined in eq. (2.1) or (2.7) only at high temperatures. Of course $(\partial \ln V/\partial T)_P = 3(\partial \ln l/\partial T)_P = 3\alpha$, where α is the linear thermal expansion coefficient.

In summary, for practical manipulations, the Grüneisen form of the equation of state, with its assumption that gamma is a function of volume only, may be used for solids at temperatures higher than some fraction of the Debye temperature or at 0 °K. Such a gamma will be equal to the thermodynamic gamma. Depending on the distribution of the γ_i's with respect to ν_i, the Grüneisen form may also be a good approximation below the Debye temperature, but one cannot be sure of this.

To complete the solution of the equation-of-state problem, one must now estimate $\gamma(V)$ from experimental data. It is useful to consider the solid in the Debye, or acoustic, approximation, in which the material is assumed to be uniform and isotropic [2.1, 2.8, 2.9]. The thermal excitations can then be considered to be sound waves in a uniform continuum. The bulk, longitudinal, and transverse moduli are given in terms of the Lamé parameters (λ, μ) and Poisson's ratio $(\sigma = \lambda/2(\lambda + \mu))$ as follows:

(2.12) $$\beta_V = -V\left(\frac{\partial P}{\partial V}\right)_T = (3\lambda + 2\mu)/3\,,$$

(2.13) $$\beta_l = \lambda + 2\mu = 3\beta_V(1-\sigma)/(1+\sigma)\,,$$

(2.14) $$\beta_t = \mu = 3\beta_V(1-\sigma)/2(1+\sigma)\,.$$

The γ_i's are most conveniently written in terms of derivatives of the β_i's, under the assumption that the normal modes are all acoustic waves, traveling at velocities $C_i = (\beta_i V)^{\frac{1}{2}}$.

(2.15) $$\gamma_i = \frac{1}{3} - \frac{\mathrm{d}\ln C_i}{\mathrm{d}\ln V} = -\frac{1}{6} - \frac{1}{2}\frac{\mathrm{d}\ln \beta_i}{\mathrm{d}\ln V}\,.$$

Hence

(2.16) $$\gamma_V = -\frac{1}{6} - \frac{V\beta_V'}{2\beta_V}\,,$$

(2.17) $$\gamma_l = \gamma_V + V\sigma'/(1-\sigma^2)\,,$$

(2.18) $$\gamma_t = \gamma_V + 3V\sigma'/2(1+\sigma)(1-2\sigma)\,.$$

Primes denote differentiation with respect to volume.

But in the high-temperature limit, where there is equipartition of energy, eq. (2.7) yields

(2.19) $$\gamma = \tfrac{1}{3}\gamma_l + \tfrac{2}{3}\gamma_t\,.$$

Hence,

$$\gamma = \gamma_V + \frac{V\sigma'(4-5\sigma)}{3(1-\sigma^2)(1-2\sigma)}\,. \tag{2.20}$$

Now $\gamma_V(V)$ is just the gamma associated with a mode corresponding to a uniform volume expansion. Neglect of the second term in eq. (2.20) corresponds to the assumption that, on the average, all modes have the same gamma value as this mode, *i.e.*, that the volume expansion mode is « typical ». The parameter γ_V may be easily evaluated, since $\beta_V(V)$ is just the bulk modulus along the « cold compression curve, » the 0 °K isotherm or isentrope. Hence, $\gamma_V(V)$ may be evaluated simply from volume compression data. The problem is to make some estimate of the volume dependence of the Poisson ratio (σ).

Several models have been proposed to relate the total gamma to the curvature of the cold compression curve. These are described by the following formula [2.9-2.11],

$$\gamma_F = \left(\frac{\alpha}{2}-\frac{2}{3}\right) - \frac{V}{2}\,\frac{(\mathrm{d}^2/\mathrm{d}V^2)(P_{0\,^\circ\mathrm{K}}V^\alpha)}{(\mathrm{d}/\mathrm{d}V)(P_{0\,^\circ\mathrm{K}}V^\alpha)}\,. \tag{2.21}$$

For $\alpha = 0$, this formula yields a result (γ_S) first derived by SLATER [2.1] under the assumption that Poisson's ratio is independent of volume. Thus, our γ_V or « volume gamma » is just Slater's gamma. For $\alpha = \frac{2}{3}$, eq. (2.21) yields a result (γ_{DM}) first derived by DUGDALE and MACDONALD [2.12] and rederived for cubic materials by RICE, MCQUEEN, and WALSH [2.3], under the assumption that the logarithmic volume derivatives of all the interatomic force constants are equal. If $\alpha = \frac{4}{3}$, eq. (2.21) yields a result (γ_{FV}) derived by ZUBAREV and VASHCHENKO [2.9] from a free-volume theory.

The assumptions underlying eq. (2.21) may be more clearly seen if the equation is expanded in the variable $P_{0\,^\circ\mathrm{K}}/\beta_V$. Note that this parameter is small at low pressures and is always significantly less than one, since β_V increases with compression (ignoring phase transitions). The result of substituting eq. (2.12) and eq. (2.16) into eq. (2.21) is

$$\gamma_F = \frac{\gamma_V - (\alpha/6)(3 - P_{0\,^\circ\mathrm{K}}/\beta_V)}{(1-\alpha P_{0\,^\circ\mathrm{K}}/\beta_V)}\,. \tag{2.22}$$

Expansion yields

$$\gamma_F = \gamma_V - \frac{\alpha}{2} + \frac{\alpha P_{0\,^\circ\mathrm{K}}}{\beta_V}\left(\gamma_V - \frac{\alpha}{2} + \frac{1}{6}\right)\,. \tag{2.23}$$

Notice that at zero pressure, γ_{DM} is less than $\gamma_S = \gamma_V$ by $\frac{1}{3}$, and γ_{FV} is less than γ_S by $\frac{2}{3}$. Differentiation of eq. (2.1) at zero pressure shows that the first and second volume derivatives of the Hugoniot pressure (P_H) are equal to the same derivatives of $P_{0°K}$, the derivatives occurring in eq. (2.21) or eq. (2.16), provided that the thermal energy of the initial state can be ignored. If one assumes that the Hugoniot can be represented by a linear shock velocity-particle velocity plot, $U_s = C + SU_p$, [2.3, 2.13, 2.14], eq. (2.16) and eq. (2.23) yield the simple result

$$(\gamma_F)_0 = 2S - \frac{2}{3} - \frac{\alpha}{2}, \tag{2.24}$$

at zero pressure. The value of the zero-pressure parameter γ_0 so calculated from Hugoniot data may be directly compared with the value of the gamma calculated from eq. (2.11) and ordinary thermodynamic data [2.15]. Such a comparison shows that, on the average, the Dugdale-MacDonald gamma (γ_{DM}) is the gamma most appropriate for common metals [2.3], while the free volume gamma (γ_{FV}) may be more appropriate for alkali metals [2.11] and alkali halides [2.16].

However, eq. (2.23) shows that the various gammas become more nearly equal as the pressure is increased. Thus, we are free to select the gamma that best represents the low-pressure data, knowing that we will not significantly affect the high-pressure result.

It is interesting to note from eq. (2.20) that the assumption of the validity of the various models underlying eq. (2.21) is equivalent to the assumption that Poisson's ratio obeys the relation,

$$\sigma' = \frac{3\alpha(1-\sigma^2)(1-2\sigma)}{2V(4-5\sigma)} \left[\frac{P_{0°K}}{\beta_V} \left(2\gamma_V - \alpha + \frac{1}{3} \right) - 1 \right]. \tag{2.25}$$

Thus, in picking the appropriate value of α in eq. (2.21), one is merely adjusting σ' at zero pressure to give a reasonable value of γ_0. Equation (2.25) shows that all of the models implicitly assume that Poisson's ratio is independent of volume at high pressures.

Ultrasonic measurements of acoustic-wave velocities provide data on isentropic compressibilities. Such data can be converted to isothermal data by appropriate thermodynamic manipulation; for example,

$$\beta_T^{-1} = \beta_S^{-1} + \frac{9\alpha^2 TV}{C_P} = \beta_S^{-1} + \frac{C_P \gamma^2 T}{V \beta_S^2}, \tag{2.26}$$

where $\alpha = (\partial l/\partial T)_P/l$, the linear expansion coefficient. Such corrections are usually small. Ultrasonic data can be taken under conditions of static compression to a few kilobars and may be used to evaluate γ_l and γ_t from the volume derivatives of the longitudinal and transverse sonic velocities. Gamma then follows from eq. (2.19) [2.17]. This provides an alternative method of evaluating gamma at zero pressures. Such « acoustic » gammas are usually comparable to the « thermodynamic » gammas obtained from eq. (2.11) [2.17].

Measurements of the velocity of the leading edge of a rarefaction fan can be used to obtain a longitudinal sound-wave velocity C_l at high pressure. Since the « bulk » wave velocity C_B can be obtained from estimates of $(\beta_V)_S$ obtained from ordinary Hugoniot compression data, one may calculate the shear or transverse velocity C_t from the relation

$$C_B^2 = C_l^2 - \tfrac{4}{3} C_t^2 . \tag{2.27}$$

The volume dependence of these velocities, as evaluated from the shock-wave data, may then be used to evaluate gamma under compression. Such a procedure has been carried out for aluminum [2.18], where the various estimates of γ_0 are different. As one might expect, the resulting values of γ_0 and $(\mathrm{d}\ln\gamma/\mathrm{d}\ln V)$ are closest to the ultrasonic values.

2'2. – The treatment of shock-wave data.

With this rather lengthy background, we are now in a position to address the practical question of how to generate a Grüneisen-like equation of state, such as eq. (2.1), from an experimental Hugoniot. This may be accomplished by first substituting the experimental Hugoniot pressure and energy $(P_H(V), E_H(V))$ into eq. (2.1) and solving for gamma,

$$\gamma(V) = V\,\frac{P_H(V) - P_{0\,^\circ\mathrm{K}}(V)}{E_H(V) - E_{0\,^\circ\mathrm{K}}(V)} . \tag{2.28}$$

Equation (2.21) is a second expression for gamma in terms of $P_{0\,^\circ\mathrm{K}}(V)$. The elimination of $\gamma(V)$ between eqs. (2.21) and (2.28) and the substitution,

$$P_{0\,^\circ\mathrm{K}}(V) = -\left(\frac{\mathrm{d}E_{0\,^\circ\mathrm{K}}(V)}{\mathrm{d}V}\right) \tag{2.29}$$

then yield a third-order differential equation for $E_{0\,^\circ\mathrm{K}}(V)$, which may be integrated numerically. Appropriate differentiation yields $P_{0\,^\circ\mathrm{K}}$ and $\gamma(V)$ [2.2, 2.3, 2.13, 2.4-2.6, 2.19-2.21]. In some cases, a similar analysis has been used with the assumption of an analytic form for $P_{0\,^\circ\mathrm{K}}(V)$ [2.22-2.27, 2.16, 2.28].

The release path from a high-pressure state is an isentrope for normal materials, if one neglects effects of the finite yield strength. Such isentropes $(P_S(V), E_S(V))$ may be calculated from the Grüneisen equation of state by first rewriting eq. (2.1) in the form

$$E_S(V) - E_H(V) = \frac{V}{\gamma(V)}\left[P_S(V) - P_H(V)\right]. \tag{2.30}$$

This expression is then differentiated with respect to volume, and the substitution

$$P_S(V) = -\frac{\mathrm{d}E_S(V)}{\mathrm{d}V}, \tag{2.31}$$

yields a first-order differential equation for $P_S(V)$ that may be integrated numerically. The resulting $P_S(V)$ curve may then be integrated to yield $E_S(V)$. The object of calculating isentropes is to be able to calculate the Riemann integral for the particle velocity associated with the centered, simple rarefaction wave. Such a wave results when a shock wave is reflected from a free surface, and the resulting free-surface velocity is $U_{fs} = (U_p)_{shock} + (U_p)_{rarefaction}$.

$$(U_p)_{rarefaction} = \int_0^P \left(-\frac{\mathrm{d}V}{\mathrm{d}P_S}\right)^{\frac{1}{2}} \mathrm{d}P. \tag{2.32}$$

These calculations are necessary when measured free-surface velocities are to be converted to particle velocities. One starts out with the assumption that $(U_p)_{shock} = U_{fs}/2$; calculates an equation of state, a value of $(U_p)_{rarefaction}$, and a new value of $(U_p)_{shock}$; and iterates until a stable solution is obtained [2.2, 2.3, 2.13, 2.21]. Of course, if the particle velocity has been obtained directly through an impedance match to a standard or by a symmetric impact experiment, such a calculation is unnecessary.

As has been noted, the Grüneisen equation of state, eq. (2.1), is incomplete because the temperature, entropy, or any of the free energies cannot be calculated directly. In order to form a complete equation of state, further information, such as the specific heat, must be added. For solids, it is appropriate to take the specific heat in the Debye model,

$$C_V(V, T) = f(T_D(V)/T), \tag{2.33}$$

where f is the Debye function, and the volume dependence of the Debye temperature $T_D(V)$ is obtained by integrating the relation

$$\gamma(V) = -\frac{\mathrm{d}\ln T_D}{\mathrm{d}\ln V}. \tag{2.34}$$

The temperature on the Hugoniot may be calculated [2.29] with the use of the thermodynamic identity,

$$T\,\mathrm{d}S = C_V\,\mathrm{d}T + \left(\frac{\partial P}{\partial T}\right)_V T\,\mathrm{d}V\,, \tag{2.35}$$

combined with the first law of thermodynamics,

$$T\,\mathrm{d}S = \mathrm{d}E + P\,\mathrm{d}V\,. \tag{2.36}$$

If the differentiation in eqs. (2.35) and (2.36) is taken along the Hugoniot, one obtains

$$\frac{\mathrm{d}E_{\mathrm{H}}(V)}{\mathrm{d}V} + P_{\mathrm{H}}(V) = C_V(V,\,T)\,\frac{\mathrm{d}T_{\mathrm{H}}(V)}{\mathrm{d}V} + \left(\frac{\partial P}{\partial T}\right)_V T_{\mathrm{H}}(V)\,. \tag{2.37}$$

If one makes the substitution

$$\left(\frac{\partial P}{\partial T}\right)_V = \frac{\gamma(V)\,C_V(V,\,T)}{V}\,, \tag{2.38}$$

eq. (2.38) may be integrated along the Hugoniot to yield the temperature on the Hugoniot [2.21].

In a similar manner, eq. (2.35) may be integrated along an isentrope to yield temperatures [2.2, 2.3, 2.13, 2.29, 2.20, 2.21], once the temperature is known at some point on the isentrope. Thus, if the temperature is known on some curve that crosses the isentropes, a complete $P(V,\,T) - E(V,\,T)$ equation of state can be developed from the $P(V,\,E)$ Grüneisen equation of state [2.19].

The zero-degree Kelvin isotherm, or cold-compression curve, and room-temperature isotherm are probably the most scientifically interesting curves that can be calculated from shock-wave data. The energy and pressure on various isotherms may be calculated from the cold-compression curve with a knowledge of gamma and the specific heat [2.14].

$$E(V,\,T) = E_{0\,^\circ\mathrm{K}}(V) + \int_0^T C_V(V,\,T)\,\mathrm{d}T\,, \tag{2.39}$$

$$P(V,\,T) = P_{0\,^\circ\mathrm{K}}(V) + \frac{\gamma(V)}{V}\,[E(V,\,T) - E_{0\,^\circ\mathrm{K}}(V)]\,. \tag{2.40}$$

It is useful to note that such isotherms are probably more reliable than one might guess from the uncertainties in estimating gamma discussed earlier. At small compressions the off-set between the Hugoniot and isotherm is small enough that one does not need good values of gamma in order to make a good calculation of the isotherm from the Hugoniot. At large compressions, the off-set becomes considerably greater, and good estimates of gamma are needed to make good calculations of the isotherm. It is thus fortunate that all three models contained in eq. (2.21) give similar values of gamma at high pressures. The differences between the models are significant at small compressions, just where a large uncertainty in gamma is tolerable.

The convergence of the results from the three models for gamma lends considerable confidence to isotherms calculated from shock-wave data. Indeed, such isotherms are now replacing statically determined compression curves in the handbooks [2.30]. It is worth noting, however, that the gamma values to which the three models converge are not necessarily the correct values. All three models could be wrong!

It is instructive to compare gammas resulting from eq. (2.21) with results from other models at high compression [2.11]. Under strong compression we may represent $P_{0\,^\circ\mathrm{K}}(V)$ as V^{-n}, for a limited range of compressions. In this case, eq. (2.21) yields $\gamma_{\mathrm{F}} = (n - \frac{1}{3})/2$, independent of the value of α.

We should expect that at high compressions, the electrons in a metal would be essentially free. We shall show later that gamma for a free Fermi-Dirac gas is $\frac{2}{3}$. Thomas-Fermi calculations [2.31] indicate that the electronic pressure varies as $V^{-\frac{5}{3}}$ at high pressures, so that γ_{F} is also $\frac{2}{3}$.

The contribution of the nuclei to gamma, calculated in a Thomas-Fermi picture [2.11, 2.32] is described by a gamma decreasing slowly to one half as compression is increased. A classical model of a lattice of point nuclei in a uniform sea of electrons at high density also implies a value of one half for gamma [2.11].

Gammas calculated by the application of eq. (2.21) to the reduction of Hugoniot data lie in the range one half to one third at high compressions [2.11]. Thus, since these values are so similar to the predictions of the various theoretical models, considerable confidence may be placed in equations of state calculated from Hugoniot data in this manner.

Since gamma decreases with volume, it has become common to take γ/V constant [2.13]. Ross [2.33] has examined gammas calculated from Hugoniot data in the Dugdale-MacDonald approximation for a large number of metals and has found the results to be better represented by $\gamma/V^{\frac{1}{2}}$ constant over the experimental range. In any case, such relations should not be extrapolated to very high compressions, since their asymptotic behavior ($\gamma \to 0$ as $V \to 0$) is clearly wrong.

2·3. – Electronic and other corrections.

Several authors have introduced an equation-of-state correction for the nondegeneracy of the conduction electrons in metals at nonzero temperatures [2.10, 2.22-2.26]. Effects of electronic excitation have also been studied in insulators [2.16, 2.34]. It is useful to study such electronic effects in metals in terms of a free-electron gas [2.35]. The essential features of a free-electron Fermi-Dirac gas are that the density of states at the Fermi surface is given by

$$g(\varepsilon_F) = \frac{8\pi m V}{(2\pi\hbar)^3}\sqrt{2m\varepsilon_F}\,, \tag{2.41}$$

and that the Fermi energy is given by

$$\varepsilon_F = \frac{(2\pi\hbar)^2}{8m}\left(\frac{3N}{\pi V}\right)^{\frac{2}{3}}, \tag{2.42}$$

provided the gas is almost degenerate. Expansion of the Fermi-Dirac distribution function in powers of (kT/ε_F) then yields the thermal energy

$$E_e = \frac{\pi^2}{6}(kT)^2 g(\varepsilon_F)\,, \tag{2.43}$$

and an electronic heat capacity

$$(C_V)_e = \frac{\pi^2}{3} k^2 T g(\varepsilon_F)\,. \tag{2.44}$$

This electronic specific heat is normally observed under cryogenic conditions, since elsewhere it is dominated by the lattice specific heat. At low temperatures, the latter is proportional to T^3. The measurement of the electronic specific heat gives a direct evaluation of the density of states at the Fermi surface under the conditions where the measurements are made, and the result may be compared with the free-electron values given in eqs. (2.41) and (2.42). Thermodynamic consistency requires

$$\left(\frac{\partial E}{\partial V}\right)_T = T\left(\frac{\partial P}{\partial T}\right)_V - P\,. \tag{2.45}$$

Hence

$$P_e = \frac{\gamma_e}{V} E_e = \frac{1}{V}\left[\frac{\mathrm{d}\ln g(\varepsilon_F)}{\mathrm{d}\ln V}\right] E_e\,, \tag{2.46}$$

and

$$\gamma_e = \frac{\mathrm{d}\ln g(\varepsilon_F)}{\mathrm{d}\ln V} = \frac{2}{3}\,. \tag{2.47}$$

This result applies for any free electron gas, independent of its density, as long as the distribution is almost degenerate.

It is worth noting that this value of the electronic gamma is not very different from the lattice gamma under strong compression. Thus, the total effective gamma for the system will not be much changed by introducing the electronic contribution. In other words, for a given energy, calculated temperatures may be rather different, but calculated pressures will be the same, if the electronic terms are omitted. For this reason, low-temperature isotherms calculated with and without electronic corrections by different authors are similar [2.10], even though the electronic contribution to the pressure may be quite substantial.

Where the electronic contribution has been taken into account in the reduction of shock-wave data [2.10, 2.22-2.26], the procedure has been rather different from the procedure appropriate to the free-electron approximation. In the free-electron case, the density of states at the Fermi surface is just given by the substitution of eq. (2.42) into eq. (2.41), giving $g(\varepsilon_F)$ in terms of the specific volume V and number of valence electrons per mole N. On the other hand, the value of $g(\varepsilon_F)$ obtained from low-temperature specific-heat data and eq. (2.44) is often much larger than this free-electron value, particularly for transition metals, and it is the experimental value of $g(\varepsilon_F)$ that has been used in constructing the equation of state. Furthermore, the electronic gamma (γ_e) has been taken as one half, based on an examination of Thomas-Fermi results [2.31]. The large values of $g(\varepsilon_F)$ obtained at normal density for transition metals result from the fact that the Fermi surface splits a rather narrow d-electron band, where the d type conduction electrons are far from free. However, under compression, one should expect movement of the Fermi surface coupled with broadening of the d bands, so that under compression $g(\varepsilon_F)$ should be rather different. Furthermore, eq. (2.47) shows that in this case γ_e will be different from $\frac{2}{3}$, possibly even negative. For these reasons, the procedures that have been used seem rather questionable. We believe that it would be safer to assume the conduction electrons to be free at high compressions. Such an approximation should become more and more valid as the compression is increased.

2·4. – Porous materials.

One method of experimentally evaluating gamma and looking for electronic contributions is through the use of Hugoniot data on initially porous samples [2.10, 2.20, 2.24, 2.36-2.38, 2.39, 2.40]. Such samples achieve much higher internal energies and temperatures at a given specific volume, than do samples shocked from normal density. This can be easily seen from the Rankine-Hugoniot energy condition.

$$E = E_{00} = \tfrac{1}{2}(P + P_0)(V_{00} - V)\,, \tag{2.48}$$

where V_{00} and E_{00} are the initial specific volume and energy of the sample. If gamma is independent of temperature, subtraction of the values of the pressure and energy on the normal Hugoniot from those on the porous Hugoniot will yield

$$\gamma = V\left(\frac{\Delta P}{\Delta E}\right). \tag{2.49}$$

Unfortunately, there are several problems associated with this kind of experiment. If the shocked sample is to be at a uniform temperature, thermal diffusion may be necessary between the inside and outside of individual grains. This requires characteristic grain or particle sizes of the order of a few microns in metals, if such diffusion and thermalization is to take place on the time scale of a shock compression. The samples that have been used are normally not this fine grained, and there is some question as to whether the states reached are states of thermal equilibrium. It appears, however, that this problem may not be as severe as theoretical considerations would indicate. KORMER *et al.* [2.24], varied grain size from 0.5 micron to much larger values and saw no effect on the shock velocity. In effect, the average compression in the possibly nonequilibrium state reached by large-grain samples was the same as would have existed if there had been thermal equilibrium. There are still serious questions open on this point, however.

If we assume that the problem of thermal equilibrium can be ignored, there is still a problem of error buildup in the data analysis. The problem is that at pressures below a megabar in typical samples, the two Hugoniots are relatively close to each other. When pressures and energies on the Hugoniots are subtracted, the percentage error in the differences is so large that values of gamma are only poorly determined. The determinations are not sufficiently accurate to distinguish between various theories, although they indicate that the calculations are not seriously wrong.

If one proceeds to multi-megabar shock work, the differences between the normal and porous Hugoniots are great enough that an effective gamma can be fairly well determined. The problem now is that the final states are far into the liquid range, and gamma is not independent of the temperature or thermal energy.

In our laboratory, GROVER [2.41] has developed an equation of state for metals that treats the thermodynamics of the melting transition explicitly under the assumptions that: 1) a corrected entropy of melting is constant; 2) the specific heat in the liquid decreases at a universal rate as a function of the ratio of the temperature to the melting temperature under pressure; and 3) the melting temperature is a function of volume given by the Lindemann relation,

$$T_{\mathrm{m}} = A\,T_{\mathrm{D}}^{2}\,V^{\frac{2}{3}}\,M\,. \tag{2.50}$$

Here M is the atomic weight, T_D, the Debye temperature, and A, a normalizing constant. The volume dependence of the Debye temperature is given by eq. (2.34). This melting law fairly well reproduces the known data on the volume dependence of the melting temperature [2.42], including recent work by KRAUT and KENNEDY [2.43]. For states well into the liquid range, this equation of state predicts effective gammas (eq. (2.49)) significantly lower than the gammas for the solid at the same volume, although at just above the melting temperature the effective gammas are slightly larger than the solid gammas. A similar decrease in gamma with temperature has been obtained by KORMER *et al.* [2.24], based on a phenomenological theory in which the specific heat decreases with increasing temperature. It is interesting to note that Grover's assumption of a constant entropy of melting is in good agreement with shock data on the melting of NaCl under pressure [2.44].

Grover's theory is quite successful in reproducing the published porous Hugoniot data on Al [2.24], Ni [2.24], Cu [2.24], W [2.38], and U [2.39] to several megabars without normalizing to any such data. Thus, it appears that the porous Hugoniot data provide a good test of such a theory of the liquid state and the effective gammas predicted by the theory. Unfortunately, such data cannot be used to obtain the volume-dependence of gamma in the solid state, since the effects of the material being in the liquid state are so large. Neither can it be used to check the electronic terms, since their effect on measured variables (P, V, E) is much smaller than the effect of the melting transition.

* * *

This Section contains a number of ideas developed by R. GROVER of this laboratory. I am grateful for his contribution of these ideas and for criticism of the manuscript.

REFERENCES

[2.1] J. C. SLATER: *Introduction to Chemical Physics*, Chap. XIII and XIV (New York, 1939).
[2.2] J. M. WALSH, M. H. RICE, R. G. MCQUEEN and F. L. YARGER: *Phys. Rev.*, **108**, 196 (1957).
[2.3] M. H. RICE, R. G. MCQUEEN and J. M. WALSH: *Solid State Phys.*, **6**, 1 (1958).
[2.4] G. B. BENEDEK: *Phys. Rev.*, **114**, 467 (1959).
[2.5] F. E. PRIETO and C. RENERO: *Journ. Chem. Phys.*, **43**, 1050 (1965).
[2.6] E. B. ROYCE: *Phys. Rev.*, **164**, 929 (1967).
[2.7] T. H. K. BARRON: *Ann. of Phys.*, **1**, 77 (1957).
[2.8] D. J. PASTINE: *Phys. Rev.*, **138**, A 767 (1965).

[2.9] V. N. Zubarev and Y. Ya. Vashchenko: *Fiz. Tverd. Tela*, **5**, 886 (1963); English translation in *Sov. Phys. Solid State*, **5**, 653 (1963).
[2.10] L. V. Al'tshuler: *Usp. Fiz. Nauk*, **85**, 197 (1965); English translation in *Sov. Phys. Usp.*, **8**, 52 (1965).
[2.11] R. Grover, R. N. Keeler, F. J. Rogers and G. C. Kennedy: *Journ. Phys. Chem. Solids*, **30**, 2091 (1969).
[2.12] J. S. Dugdale and D. MacDonald: *Phys. Rev.*, **89**, 832 (1953).
[2.13] R. G. McQueen and S. P. Marsh: *Journ. Appl. Phys.*, **31**, 1253 (1960).
[2.14] L. V. Al'tshuler, K. Krupnikov and M. I. Brazhnik: *Žurn. Ėksp. Teor. Fiz.*, **34**, 886 (1958); English translation in *Sov. Phys. JETP*, **34** (7), 614 (1958).
[2.15] K. A. Gschneidner jr.: *Solid State Phys.*, **16**, 275 (1964).
[2.16] L. V. Al'tshuler, M. M. Pavlovskii, L. V. Kuleshova and G. V. Simakov: *Fiz. Tverd. Tela*, **5**, 279 (1963); English translation in *Sov. Phys. Solid State*, **5**, 203 (1963).
[2.17] R. Grover: Lawrence Radiation Laboratory, private communication (1968).
[2.18] A. C. Holt and R. Grover: paper presented at the *Symposium on Accurate Characterization of the High-Pressure Environment, Gaithersburg, Md., 1968*, to be published.
[2.19] R. E. Duff: *Fundamental Data Obtained from Shock-Tube Experiments*, edited by A. Ferri, Chap. 8 (New York, 1961).
[2.20] I. C. Skidmore: *Appl. Materials Res.*, **4**, 13 (1965).
[2.21] In our laboratory, this entire procedure is carried out automatically in a computer code written by F. J. Rogers; see LRL Report UCRL-50500 (1968).
[2.22] L. V. Al'tshuler, S. B. Kormer, A. A. Bakanova and R. F. Trunin: *Žurn. Ėksp. Teor. Fiz.*, **38**, 790 (1960); English translation in *Sov. Phys. JETP*, **11**, 573 (1960).
[2.23] L. V. Al'tshuler, A. A. Bakanova and R. F. Trunin: *Žurn. Ėksp. Teor. Fiz.*, **42**, 91 (1962); English translation, *Sov. Phys. JETP*, **15**, 65 (1962).
[2.24] S. B. Kormer, A. I. Funtikov, V. D. Urlin and A. N. Kolesnikova: *Žurn. Ėksp. Teor. Fiz.*, **42**, 686 (1962); English translation in *Sov. Phys. JETP*, **15**, 477 (1962).
[2.25] S. B. Kormer, V. D. Urlin and L. T. Popova: *Fiz. Tverd. Tela*, **3**, 2131 (1961); English translation in *Sov. Phys. Solid State*, **3**, 1547 (1962).
[2.26] K. K. Krupnikov, A. A. Bakanova, M. I. Brazhnik and R. F. Trunin: *Dokl. Akad. Nauk SSSR*, **148**, 1302 (1963); English translation in *Sov. Phys. Dokl.*, **8**, 205 (1963).
[2.27] L. V. Al'tshuler, L. V. Kuleshova and M. N. Pavlovskii: *Žurn. Ėksp. Teor. Fiz.*, **39**, 16 (1960); English translation in *Sov. Phys. JETP*, **12**, 10 (1961).
[2.28] S. B. Kormer, M. V. Sinitsyn, A. I. Funtikov, V. D. Urlin and A. V. Blinov: *Žurn. Ėksp. Teor. Fiz.*, **47**, 1202 (1964); English translation in *Sov. Phys. JETP*, **20**, 811 (1965).
[2.29] J. M. Walsh and R. H. Christian: *Phys. Rev.*, **97**, 1544 (1955).
[2.30] R. N. Keeler: *American Institute of Physics Handbook*, Sect. **4** d, in press.
[2.31] R. Latter: *Journ. Chem. Phys.*, **24**, 280 (1956).
[2.32] V. P. Kopyshev: *Dokl. Akad. Nauk SSSR*, **161**, 1067 (1965); English translation in *Sov. Phys. Dokl.*, **10**, 338 (1965).
[2.33] M. Ross: Lawrence Radiation Laboratory, private communication (1967).
[2.34] M. Ross: *Phys. Rev.*, **171**, 777 (1968).

[2.35] C. Kittel: *Introduction to Solid State Physics*, Chap. 10 (New York, 1956).
[2.36] Ya. B. Zel'dovich: *Žurn. Ėksp. Teor. Fiz.*, **32**, 1577 (1958); English translation in *Sov. Phys. JETP*, **32** (5), 1103 (1958).
[2.37] L. V. Al'tshuler, K. K. Krupnikov, B. N. Ledenev, V. I. Zuchikhin and M. I. Brazhnik: *Žurn. Ėksp. Teor. Fiz.*, **34**, 874 (1958); English translation in *Sov. Phys. JETP*, **34** (7), 606 (1958).
[2.38] K. K. Krupnikov, M. I. Brazhnik and-V. P. Krupnikova: *Žurn. Ėksp. Teor. Fiz.*, **42**, 675 (1962); English translation in *Sov. Phys. JETP*, **15**, 470 (1962).
[2.39] I. C. Skidmore and E. Morris: *Thermodynamics of Nuclear Materials* (Vienna, 1962), p. 173.
[2.40] B. L. Hord: *Bull. Am. Phys. Soc.*, **5**, 42 (1960); and private communication.
[2.41] R. Grover: *Bull. Am. Phys. Soc.*, **13**, 1647 (1968); and private commuuication.
[2.42] S. E. Babb jr.: *Rev. Mod. Phys.*, **35**, 400 (1963).
[2.43] E. A. Kraut and G. C. Kennedy: *Phys. Rev.*, **151**, 668 (1966).
[2.44] S. B. Kormer, M. V. Sinitsyn, G. A. Kirillov and V. D. Urlin: *Žurn. Ėksp. Teor. Fiz.*, **48**, 1033 (1965); English translation in *Sov. Phys. JETP*, **21**, 689 (1965).

III. - Stability of the Electronic Configuration in Metals at High Pressures: The Rare Earths.

E. B. Royce

3·1. - Introduction.

In previous lectures we have discussed the determination of a high-pressure equation of state from the use of shock-wave techniques and theoretical models. In particular, we have outlined methods for experimentally determining a Hugoniot and for calculating a cold-compression curve, $P_{0\,^\circ\mathrm{K}}(V)$, and cohesive-energy curve, $E_{0\,^\circ\mathrm{K}}(V) = -\int P_{0\,^\circ\mathrm{K}}(V)\,\mathrm{d}V$, from this Hugoniot. These curves are the results that are most easily related to ordinary solid-state concepts, since they are independent of thermal effects, depending only on the static interatomic interaction.

In this lecture, these curves will be analysed in terms of a rather primitive theory of binding in metals in order to infer some information regarding electronic configurations in compressed metals [3.1]. The densities attained on shock-wave compression are sufficiently large that the physical properties of metals may no longer be considered to result exclusively from interactions between the outer electrons of the constituent atoms. We shall be concerned here with the transition between two conditions, 1) the usual metallic state, where both the attractive and repulsive parts of the cohesive energy curve are associated with the delocalization of the outermost electrons of the free atom into energy bands [3.2]. and 2) the very high-density condition described by Thomas-Fermi theory, in which all of the electrons play a part in determining the equation of state [3.3].

3.2. – General considerations.

If we bring metal atoms together to form a crystal lattice, the energy of the electron system has a behavior as shown in Fig. 3.1, where we show the energy of *s* and *d* electrons as a function of mean interatomic spacing *R*. The Figure shows the effect of the delocalization of the electrons and the resulting formation of bands. It is, of course, just the energy shifts of these conduction-band electrons that produce metallic binding and the equation of state of the metal. The most binding states in each band have a high overlap charge density between the atoms, and it is the attraction of this overlap charge to the nuclei or cores that gives the binding. The most antibonding states have nodes in the electron distribution between atoms, and their repulsion comes from electron-electron and core-core electrostatic repulsion, plus the high kinetic energy associated with wave functions having many nodes.

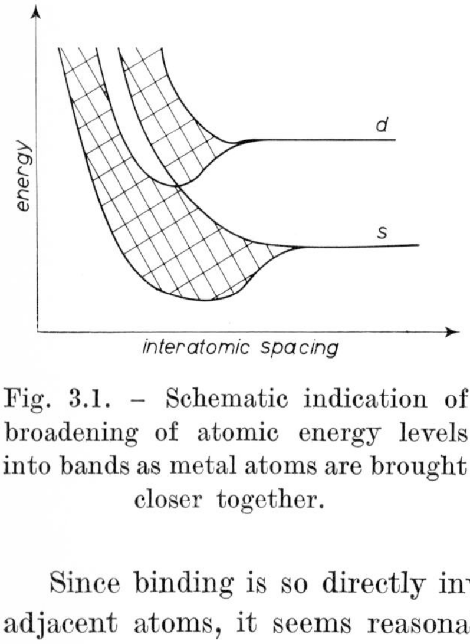

Fig. 3.1. – Schematic indication of broadening of atomic energy levels into bands as metal atoms are brought closer together.

Since binding is so directly involved with the overlap of the electrons on adjacent atoms, it seems reasonable to expect the minimum in the curve of energy *vs.* radius for the binding states to be associated with a particular degree of overlap. Thus, if only the binding states at the bottom of the conduction band were occupied, we might expect to find that the equilibrium mean interatomic spacing would be directly proportional to the mean radius of the electron distribution in the free atom.

The energy of states near the middle of the conduction band is rather insensitive to interatomic spacing, so that the population of these states should have relatively little effect on the equilibrium interatomic spacing. Thus, unless the band is almost completely filled, the equilibrium spacing should be relatively independent of the number of electrons in the conduction band. We should, then, expect to see the proportionality between interatomic spacing and free-atom radius in series of metals which appear horizontally in the periodic table.

It is interesting to note that the electrical properties of metals depend on electrons at the Fermi surface, while the equation of state or effective interatomic interaction depends, to a varying degree, on all of the conduction-band electrons. The superposition of contributions of many electrons makes the equation of state much less sensitive to details of the band structure than are the electrical properties. This is why we may use the primitive theory of binding outlined here.

As a measure of interatomic spacing, it is convenient to use the Wigner-Seitz radius, the radius of a sphere of volume equal to the volume of an atom. In terms of this radius $R_{WS}(P)$, the molar volume $V_m(P)$ is given by

$$V_m(P) = \left(\frac{4\pi}{3}\right) N_0 [R_{WS}(P)]^3 . \tag{3.1}$$

This radius has the advantage of being independent of crystal structure, a feature which is necessary since we often do not know crystal structures for shock-compressed states. The use of such a radius implies that the volume available to the conduction electrons is more important than the actual distance to neighboring atoms. The dependence $V_m(P)$ is obtained by inverting the $P_{0\,°K}(V)$ curve obtained from the shock-wave data [3.4].

A measure of the radius of the electronic distribution may be obtained from Hartree-Fock solutions to the problem of the electronic wave functions for a free atom [3.5, 3.6]. Where such wave functions are tabulated numerically [3.6], the radius for a particular orbital was taken as the radius at which the wave function falls to one-half its value at the outermost extremum. Where solutions were expanded in the form [3.5]

$$\psi = \sum_i P_i(r) \exp[-\alpha_i r] , \tag{3.2}$$

the radius was taken as α_m^{-1}, where α_m is the smallest of the α_i's; $P_i(r)$ is a polynomial. These two definitions gave the same Z-dependence for the radii $R_{nl}(Z)$.

Our basic assumption, now, will be a proportionality between these two kinds of radii as a function of Z. This is illustrated in Fig. 3.2, where the comparison is made at zero pressure for the $(3d, 4s, 4p)$ period of the periodic table, the first long period. The dots (•) represent the experimental data, and the solid lines the Hartree-Fock results. Since our assumption is one of proportionality rather than equality, the Hartree-Fock radii have been normalized to the experimental data by moving the curves up or down on the log plot. The shapes and slopes of the curves come from the free-atom Hartree-Fock solutions and cannot be changed in this fitting.

In K and Ca, the conduction band is primarily $4s$ in character, and the interatomic spacings are proportional to the radius of the $4s$ orbitals. Sc, Ti, V, and Cr show interatomic spacings proportional to the radius of the $3d$ orbital, even though their conduction bands are $3d$-$4s$ hybrids. This is the first of several pieces of evidence indicating that the d electrons are dominant in determining the equation of state of the transition metals. As more

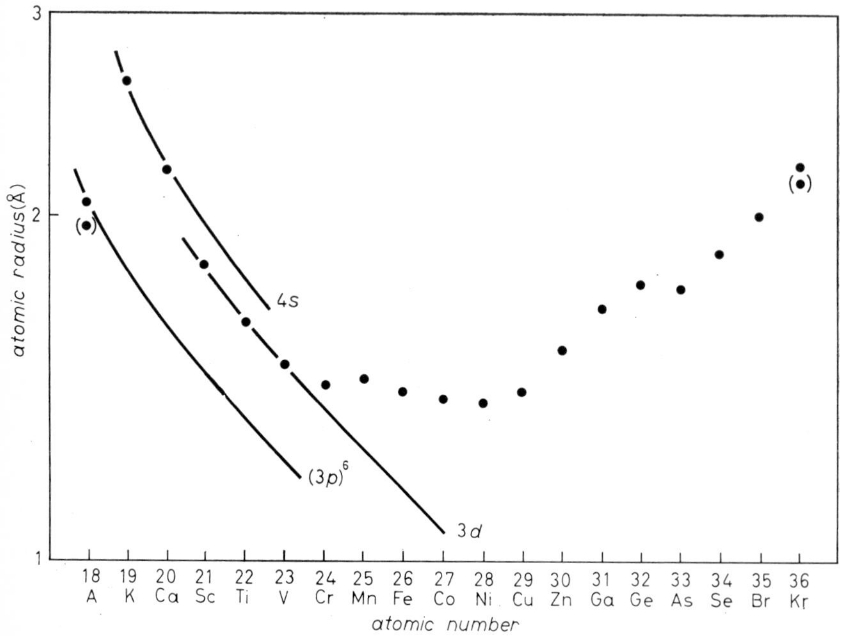

Fig. 3.2. – Atomic radius *vs.* atomic number for the first ($3d$, $4s$, $4p$) transition period at zero pressure. The dots are experimental data, and the lines show the Z-dependence obtained from the Hartree-Fock free-atom solutions as normalized to the experimental data.

$3d$ electrons are added to the conduction band, anti-bonding states become populated. For Mn and the metals with higher Z in the period, this causes the experimental interatomic spacings or atomic radii to be larger than the simple theory would predict.

Figure 3.3 shows the same comparison at 250 kbar. The broken lines reproduce the zero-pressure radii from Fig. 3.2 for comparison. At this high pressure, it continues to be possible to fit the experimental radii for K and Ca with the $4s$ curve from the Hartree-Fock calculations, and the radii for Sc, Ti, and V with the $3d$ curve. Cr is somewhat less compressible than Sc, Ti, and V, possibly due to its larger number of $3d$ electrons. The important feature to notice in Fig. 3.3 is the fact that, as a group, the metals with $3d$ electrons are much less compressible than the $4s$ metals. The large compressibility of metals of low density is well known; what is important here is the grouping. Again, we have evidence of the dominance of the *d* electrons in determining the equation of state, in this case making the materials very incompressible.

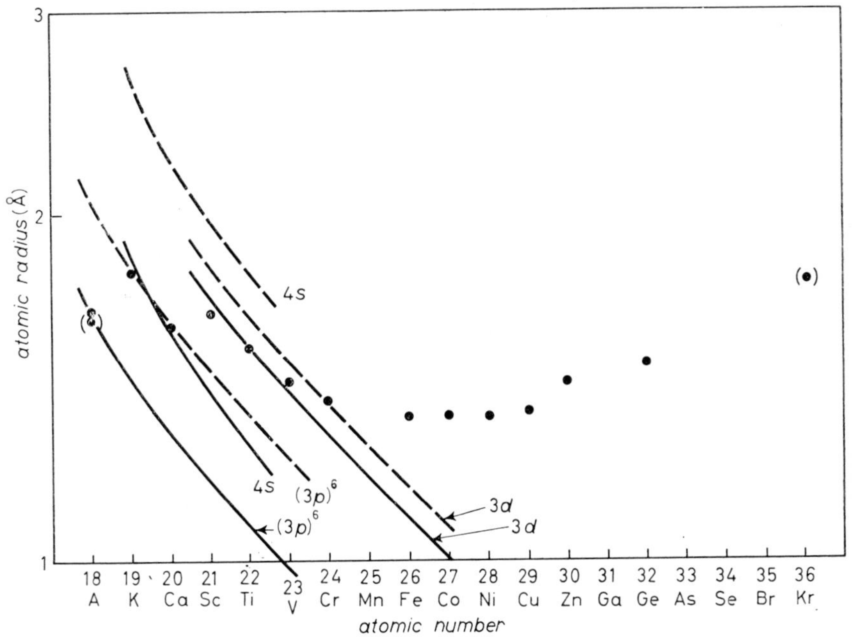

Fig. 3.3. – Atomic radius *vs.* atomic number for the first (3*d*, 4*s*, 4*p*) transition period at 250 kbar. The dots are experimental data at 0 °K. The solid lines show the *Z*-dependences obtained from the Hartree-Fock free-atom solutions as normalized to the experimental data. The broken lines are the zero-pressure Hartree-Fock *Z*-dependences taken from Fig. 3.2.

Figure 3.4 shows the same trend at 1 Mbar. These Figures show no evidence of any transitions in the electronic structure at experimental pressures.

On Fig. 3.2-3.4, we also show the radius of the argon $(3p)^6$ closed-shell core as a function of *Z*. These radii were normalized to equation-of-state results on A or isoelectronic KCl. Notice that the *s*-bonded metals are so compressed that their cores begin to overlap for volume compressions of the order of threefold.

In reality, of course, the conduction-band wave functions are all hybrids of the various atomic parent wave functions. When we discuss the *s* or *d* parts of the conduction band, we really mean those parts whose wave functions are primarily *s* or *d* in character. The present analysis shows, thus, that for the equation of state, the important feature of the conduction-band states populated is their atomic-state parentage.

Figures 3.5 and 3.6 are designed to emphasize the differences between s and d bonding and show the contributions to the pressure and cohesive energy from these different parts of the conduction band. Figure 3.5 was constructed as follows. The closed-shell contribution was taken from the rare-gas or alkali-halide $P_{0^\circ\mathrm{K}}(V)$ data. The s-electron contribution was obtained by subtracting

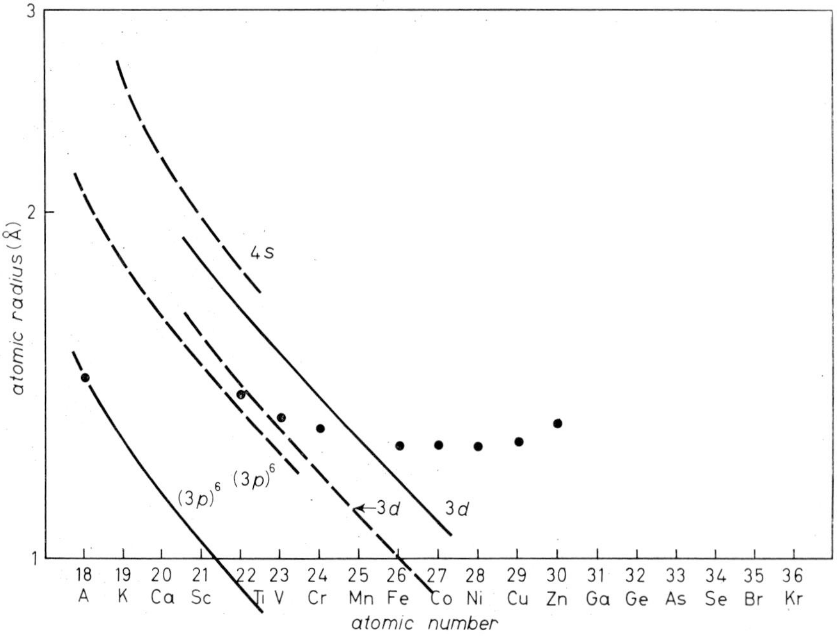

Fig. 3.4. – Atomic radius *vs.* atomic number for the first ($3d$, $4s$, $4p$) transition period at 1 Mbar. See Fig. 3.3.

the closed-shell contribution from the alkali-metal $P_{0^\circ\mathrm{K}}(V)$ data. The d-electron contribution was obtained from transition-metal data in an analogous way. Before these subtractions were performed, the $P_{0^\circ\mathrm{K}}(V)$ curves were converted to $P_{0^\circ\mathrm{K}}(R)$, and the R dependence was corrected for the variation of R with Z given by the dependence of the relevant Hartree-Fock radii. The energy curves in Fig. 3.6 were obtained by integrating the pressure curves. The reference energy is arbitrary for each curve. These Figures demonstrate that the d-electron contribution to the equation of state will dominate the s-electron contribution by an order of magnitude in both pressure and energy under compression.

The energy in Fig. 3.6 is the total binding energy per atom. Since the equation of state is mainly the result of conduction-band electron interactions, this energy is roughly the energy of the most binding electron states in the conduction band. The rapid rise of the d-electron energy on strong compres-

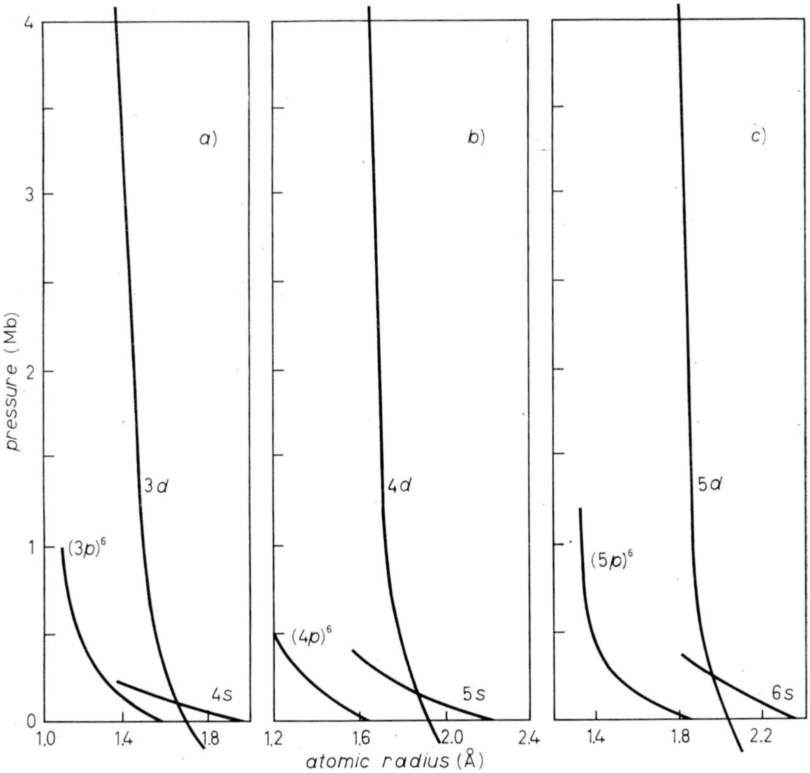

Fig. 3.5. – Contribution to the pressure from various orbitals as functions of atomic radius; *a*), *b*) and *c*) are for $Z = 21$, 39 and 57, respectively.

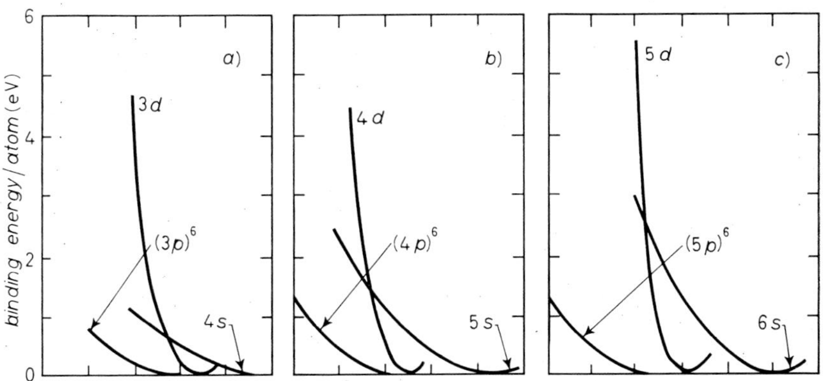

Fig. 3.6. – Contributions to the binding energy per atom from various orbitals as a function of atomic radius. The zero of energy is arbitrary for each orbital and has been taken at the minimum of each curve. *a*), *b*) and *c*) are for $Z = 21$, 39 and 57, respectively.

sion leads us to expect to find electronic transitions in which the *d* electron transfers to some other, less pressure-sensitive level when the material is strongly compressed. The transfer of electrons *into* the *d* states under strong compression appears very improbable.

3·3. – Rare-earth metals.

The analysis we have developed thus far provides a useful framework for the discussion of recent shock-compression data on rare-earth metals [3.7-3.9].

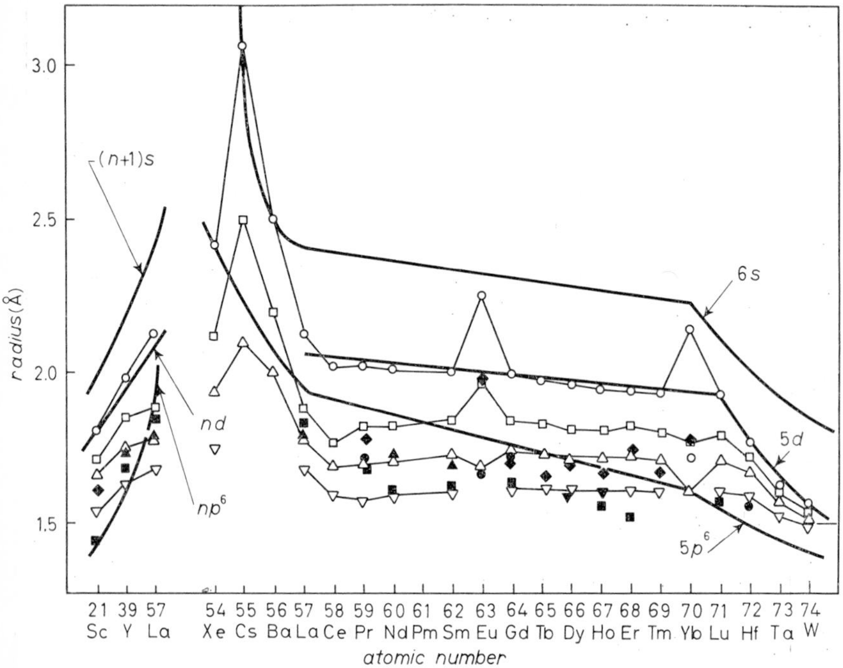

Fig. 3.7. – Atomic radius *vs.* atomic number for the rare-earth period and related metals. The heavy solid lines show the Z dependences obtained from the Hartree-Fock free-atom solutions as normalized to the zero-pressure experimental data. ○ 0 Mbar, □ 0.125 Mbar, △ 0.250 Mbar, ▽ 0.5 Mbar. ▼, ▲, ● Hugoniot stiffens, —— Gust fits; ▼, ■ Hugoniot stiffens, —— Al'tshuler fits; ▲, ◆ Hugoniot stiffens, —— McQueen fits.

Figure 3.7 shows atomic radii at 0 °K for several pressures as calculated from these experiments. Most of the rare-earth metals are normally divalent, with a $5d$-$6s$ hybrid conduction band. Their atomic radii at zero pressure fall on the $5d$ Hartree-Fock curve fitted to Hf, Ta, and W, would be expected. In

Eu and Yb, the stability of the half-filled and filled $4f$ shell, internal to the atom, results in a divalent metal, and the radii at zero pressure fall on the $6s$ curve fitted to Cs and Ba.

From the previous analysis, we should expect the trivalent metals to be incompressible, whereas in fact, all of the rare-earth metals exhibit the high compressibilities characteristic of s-bonded metals. Because of this high compressibility, we must conclude, that there is only a small amount of d bonding under pressure, *i.e.*, that all of the rare-earth metals are essentially divalent. This must be accomplished by a gradual transfer of the $5d$ electrons to conveniently vacant $4f$ levels as the pressure is increased. One of the polymorphic transitions observed statically at low pressures may take place simultaneously with the electronic transitions; these transitions are accompanied by large changes in resistivity.

The first-order $\alpha \to \gamma$ phase transition in cerium is thought to be the result of the reverse electronic transition ($4f \to 5d$). Apparently at higher pressures the $5d \to 4f$ transition takes place gradually.

Sc and Y are often considered similar to the rare-earth metals, since they have external electron configurations similar to La. However, they do not have the vacant internal f levels available to accept a d electron. Thus, we should expect and, in fact do find, that these metals are significantly less compressible than La. This lends further credence to the assumption of a $5d \to 4f$ transition in the rare-earth metals.

The $(5p)^6$ curve in Fig. 3.7 represents the radius at which the weak Van der Waals' attraction in solid Xe is just balanced by repulsion between overlapping $(5p)^6$ closed shells. The curve was obtained by fitting the Hartree-Fock $5p$ radii to the experimental radius for Xe. For compressions to radii significantly smaller than the $(5p)^6$ radii, we should expect to find a pressure contribution from core overlap. The various solid symbols on Fig. 3.7 indicate radii at which the Hugoniots for the rare-earth metals show an abrupt stiffening, as indicated by a break in the slope of the U_s-U_p plot. These breaks appear to be located at just those radii where the core contribution should become significant.

It appears that the rare-earth Hugoniots represent the first instance in which a core interaction has been identified in the compression of a metal. It is instructive to examine why this effect should be observed only in the rare earths. The alkali and alkaline-earth metals are very compressible but start out at such low zero-pressure densities that at least a fourfold compression would be needed before the core contribution would be a significant part of the pressure (see Fig. 3.5). The transition metals have much higher initial densities, but are very incompressible. From Fig. 3.5, we see that at least a fivefold compression would be needed for the core contribution to be a significant part of the total pressure. In the rare-earth metals, on the other hand the presence of d bonding at zero pressure results in a high density, and the absence of d bonding under pressure makes the material very compressible.

Two alternative explanations have been offered to account for the stiffening of the rare-earth metal Hugoniots. The first explanation [3.9] is based on the observation that d bonding makes a metal less compressible and attributes the stiffening of the Hugoniot to a $4f \rightarrow 5d$ transition. Since the $4f$ orbital is interior even to the $(5p)^6$ core, it should not be expected to show a significant energy variation with pressure. The zero-pressure equilibrium spacing is slightly greater than that where the d-electron contribution to the pressure would be just zero, due to the s-electron contribution. Thus, for very modest pressures, the energy of the d electrons will be lowered on compression. This is probably the reason for the $4f \rightarrow 5d$ transition in Ce. As the compression is increased, however, the energy of the d states rises very rapidly, and a $4f \rightarrow 5d$ transition appears impossible at the compressions where the Hugoniot stiffening is observed.

The stiffening of the Hugoniot has also been attributed to melting [3.8]. Some thermodynamic calculations appear to support this claim. However, studies of melting on the Hugoniot [3.10, 3.11], as well as studies with a model solid-liquid equation of state, generally indicate that the effect on the Hugoniot should be much less pronounced. Furthermore, a stiffening of the lattice might reasonably be expected to raise the melting temperature, according to the Lindemann law, so it would be surprising to find melting associated with a stiffening of the Hugoniot.

Several alkaline-earth metals and transition metals have been observed to exhibit a stiffening of the Hugoniot similar to the rare-earth metals. It has been claimed [3.9] that this stiffening, too, is associated with a transition in which more d states are occupied. In the light of the earlier discussion and the results shown in Fig. 3.6, it seems clear that this cannot be the case for these metals any more than for the rare-earth metals.

Since we have associated the stiffening of the rare-earth Hugoniots with a core effect, it is tempting to do the same for the stiffening of the alkaline-earth and transition-metal Hugoniots. However, in the latter cases, the stiffening is observed when the cores *just begin* to overlap, and experience with the rare-earth metals and examination of Fig. 3.5 indicate that this is not sufficient compression for there to be a significant core contribution to the pressure. The effect in the alkaline-earth and transition metals may be associated with core overlap, but it is not a *direct* core contribution to the pressures as in the rare-earth metals. That the stiffening of the transition-metal Hugoniots is not a general effect can be seen by noting that the effect is absent for Ti ($Z = 22$) but observed for Sc ($Z = 21$) and V ($Z = 23$). It seems more likely that the stiffening of the transition-metal Hugoniots is associated with some polymorphic phase transition, possibly associated with the onset of core overlap.

3·4. – Conclusions.

For the Thomas-Fermi model to be valid, the electronic shell structure must be unimportant in determining the equation of state. It seems unlikely that this can be true at compressions where the importance of the core interaction is changing relative to the conduction-electron interaction. Thus, agreement between Thomas-Fermi equation-of-state calculations and experimental results below a five- to tenfold volume compression at low temperatures must be considered fortuitous. The transition between the equation-of-state region explored by shock-wave techniques, which is governed by the conduction electrons, and the Thomas-Fermi region must involve a series of electronic transitions, as succeeding electronic shells begin to overlap and contribute to the equation of state. The core effects we have discussed here are the first such transitions.

* * *

I thank W. H. Gust of this laboratory for the use of his unpublished rare-earth data.

REFERENCES

[3.1] The general subject of this lecture is reviewed in: E. B. Royce: *Phys. Rev.*, **164**, 929 (1967).

[3.2] A. H. Wilson: *Theory of Metals* (New York, 1954); W. Hume-Rothery and B. R. Coles: *Adv. Phys.*, **3**, 149 (1954); N. F. Mott: *Adv. Phys.*, **13**, 325 (1964).

[3.3] R. Latter: *Journ. Chem. Phys.*, **24**, 280 (1956); R. D. Cowan and J. Ashkin: *Phys. Rev.*, **105**, 144 (1957); N. N. Kalitkin: *Žurn. Ėksp. Teor. Fiz.*, **38**, 1534 (1960); English translation in *Sov. Phys. JETP*, **11**, 1106 (1960); V. P. Kopyshev: *Dokl. Akad. Nauk SSSR, Ser. Mat. Phys.*, **161**, 1067 (1965); English translation in *Sov. Phys. Dokl.*, **10**, 338 (1965); H. M. Schey and J. L. Schwartz: *Phys. Rev.*, **137**, A 709 (1965).

[3.4] Sources of the data used are given in ref. [3.1], or in M. van Thiel, A. S. Kusubov and A. C. Mitchell: *Compendium of shock wave data*, Lawrence Radiation Laboratory, Report UCRL-50108, Rev. 1, 1967.

[3.5] R. E. Watson: *Technical Report No. 12, Solid State and Molecular Theory Group, MIT* (Cambridge, Mass., 1959); *Phys. Rev.*, **118**, 1036 (1960); **119**, 1934 (1960); **123**, 2027 (1961).

[3.6] F. Herman and S. Skillman: *Atomic Structure Calculations* (Englewood Cliffs, N. J., 1963).

[3.7] W. H. GUST: Lawrence Radiation Laboratory, private communication (1967). This work has been partially reported in: R. E. DUFF, W. H. GUST, E. B. ROYCE, M. ROSS, A. C. MITCHELL, R. N. KEELER and W. G. HOOVER: *Shock-wave studies in condensed media*, in *Behaviour of Dense Media Under High Dynamic Pressure* (New York, 1968), p. 397; *Bull. Am. Phys. Soc.*, **12**, 1128 (1967).

[3.8] R. G. MCQUEEN: Los Alamos Scientific Laboratory, Los Alamos, N. Mex., private communication (1968).

[3.9] L. V. AL'TSHULER, A. A. BAKANOVA and I. P. DUDOLADOV: *Žurn. Ėksp. Teor. Fiz. Pis. Red.*, **3**, 483 (1966), English translation in *JETP Lett.*, **3**, 315 (1966); L. V. AL'TSHULER, A. A. BAKANOVA and I. P. DUDOLADOV: *Žurn. Ėksp. Teor. Fiz.*, **53**, 1967 (1967); English translation in *Sov. Phys. JETP*, **26**, 1115 (1968); L. V. AL'TSHULER and A. A. BAKANOVA: *Usp. Phys. Nauk SSSR*, **96**, 193 (1968).

[3.10] V. D. URLIN and A. A. IVANOV: *Dokl. Akad. Nauk SSSR*, **149**, 1303 (1963); English translation in *Sov. Phys. Dokl.*, **8**, 380 (1963).

[3.11] S. B. KORMER, M. V. SINITSYN, G. A. KIRILLOV and V. D. URLIN: *Žurn. Ėksp. Teor. Fiz.*, **48**, 1033 (1965); English translation in *Sov. Phys. JETP*, **21**, 689 (1965).

IV. – Electrical Conductivity of Condensed Media at High Pressures.

R. N. KEELER

4·1. – Background.

Electrical conductivity measurements in shock-compressed solids and liquids have been carried out for the past twelve years [4.1-4.20]. It was originally anticipated that these measurements might provide considerable insight into the physical behavior of condensed media at high energy densities. However, many of the earlier expectations have not been realized. To begin with, these experiments, with one notable series of exceptions [4.6-4.8], did not thoroughly investigate all the experimental problems associated with carrying out the measurements. An additional complication is the fact that the passage of a shock wave with its associated sharp pressure gradients and high shear stress produces a number of artifacts in solids whose properties, in many cases, mask the intrinsic properties of the material under study [4.21]. In view of this, it is obvious that a great deal of careful experimentation will have to be done in the future before the techniques which are commonplace and routine in present-day solid-state physics can be meaningfully applied in shock-wave work. Nevertheless, a few significant investigations have either been recently carried out or are in progress, and these will be the subject of this lecture.

4·2. – Experimental techniques.

Because of the different experimental techniques which are required to carry out conductivity measurements, these measurements can be placed in three general categories, depending on the initial and final values of the conductivity as measured during the passage of the shock wave.

	Type of shock-induced conductivity change	Conductivity range (mho/cm)
1	insulator/semiconductor → semiconductor	$10 > \varkappa > 0 \rightarrow 10 > \varkappa > 10^{-8}$
2	metal → metal	$10 < \varkappa < 10^{6} \rightarrow 10 < \varkappa < 10^{6}$
3	insulator → metal	$\varkappa < 10^{-2} \rightarrow \varkappa > 10^{3}$

The three categories each demand different diagnostic techniques, but in general, the measurements may be said to be inductance-limited insofar as the basic electronic circuitry is concerned. That is, any abrupt change in current flow through a sample under shock compression is opposed by the inductive reactance of the sample and by the power supply. A certain minimum current flow is required, of course, to provide a measurable voltage drop. In the first type of measurement, where the initial conductivity is usually zero and the final conductivity is relatively low, a simple load resistor circuit can be utilized. An example of this kind of circuitry is shown in Fig. 4.1. This circuitry was originally developed by REIMERS [4.13] and has been used in one form or another in conductivity measurements since that time. The most comprehensive discussion of the use of this kind of circuitry has been given by MITCHELL and KEELER [4.22].

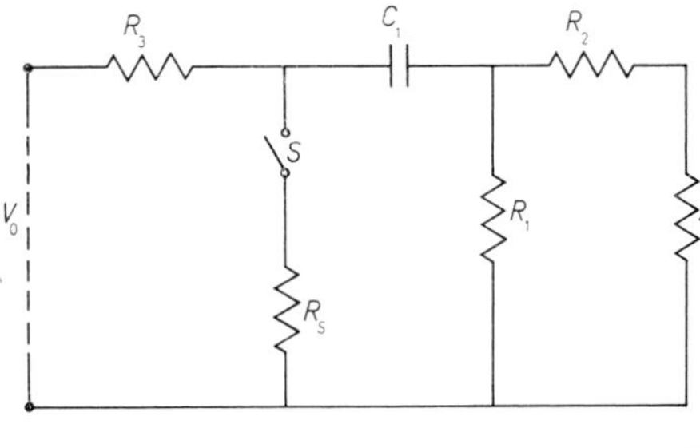

Fig. 4.1. – Circuit for measurement of electrical conductivity under shock conditions: the insulator-semiconductor range.

In the circuit diagram V_0 is the initial voltage; V, the measured voltage across the cable impedance; R_s, the resistance of the shocked sample; R_1, a parallel resistance; R_2, an attenuating resistance; I_s, the current through the sample; and Z_L, the load (cable) impedance. R_3 and C_1 are the charging resistance and capacitance. In actual use, R_3 is much greater than the other resistances. When the shock wave impacts the sample, the switch s is closed, and the circuit is complete.

The differential equation governing the response of this circuit is

$$I_s(t) = \frac{dQ}{dt} = \left(\frac{V_0}{R_s}\right) \exp\left[-\frac{t}{R_s C_1}\right]. \tag{4.1}$$

The circuit was designed in such a way that the time constant $R_s C_1$ was long with respect to the length of the experiment. At short times, the current through Z_L is given by

$$I_L = \frac{R_1 I}{R_1 + R_2 + Z_L}. \tag{4.2}$$

The voltage across the load (cable) impedance is

$$V = I_L Z_L = \frac{I R_1 Z_L}{R_1 + R_2 + Z_L}, \tag{4.3}$$

and the initial voltage

$$V_0 = I\left[R_s + \frac{R_1(R_2 + Z_L)}{R_1 + R_2 + Z_L}\right]. \tag{4.4}$$

Combining eqs. (4.3) and (4.4),

$$\frac{V}{V_0} = \frac{R_1 Z_L}{R_s(R_1 + R_2 + Z_L) + R_1(R_2 + Z_L)}. \tag{4.5}$$

The quantity V/V_0 is read from an oscilloscope trace such as that shown in Fig. 4.2.

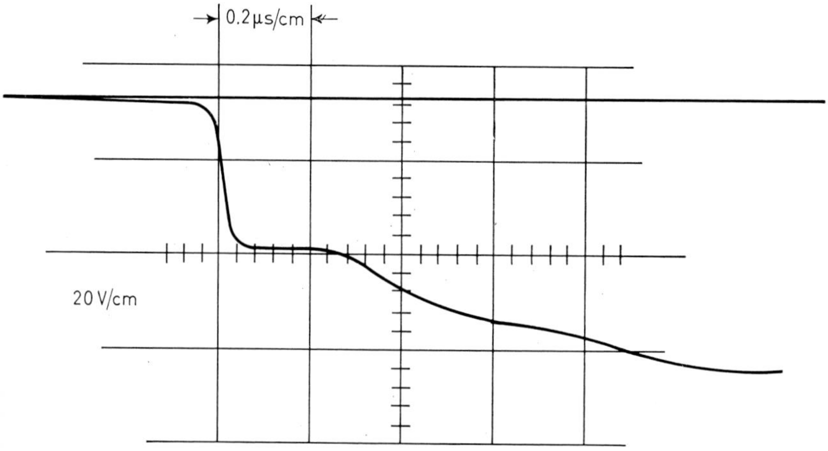

Fig. 4.2. – Oscilloscope trace of carbon tetrachloride conductivity measurement.

A cross-section of the experimental apparatus is shown in Fig. 4.3. The measurement is longitudinal, *i.e.* the current flows in a direction parallel to the direction of shock propagation with the anvil serving as one electrode and the base-plate as the other. The anvil and insulation are chosen to be a close dynamic impedance match to the material under investigation, to minimize the effects of reflected shocks and rarefactions. The apparatus shown was constructed for measuring the electrical conductivity of shock-compressed liquid carbon tetrachloride, but would be adaptable to solids as well. A picture of a typical experiment is shown in Fig. 4.4.

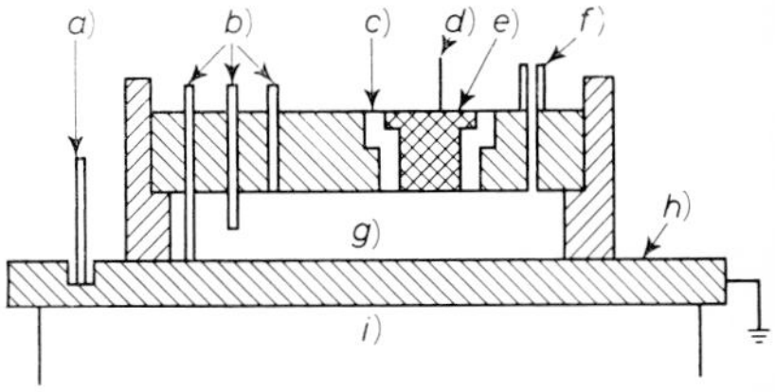

Fig. 4.3. – Experimental assembly for measuring electrical conductivity of shock-compressed fluids: semiconductor range. *a*) aluminum shock probes; *b*) CCl_4 shock probes; *c*) polyethylene insulation; *d*) to conductivity circuitry; *e*) conductivity electrode (calcium anvil); *f*) filling connection; *g*) CCl_4 sample; *h*) aluminum base plate; i) high explosive.

For the purpose of making measurements in the insulator→semiconductor range, the longitudinal geometry shown in Fig. 4.3 provides the most accurate data, since the effects of hydrodynamic interactions between anvil and sample are eliminated. The analysis is straightforward, but circuit behavior is not the main problem in conductivity measurements of the first type. Effects associated with shock propagation through a condensed material, hydrodynamic interaction with the measuring probes, and the extremely rapid change of conductivity with pressure together with pressure gradients behind the shock front are the major experimental difficulties that have been overcome. However, when the conductivity of a material does not change rapidly with pressure, transverse probes can be used with some effectiveness [4.23]. For liquids and nonpolar substances, this problem has been solved to the extent of permitting measurements to within 10% accuracy or better.

Since the geometry of the experiment is longitudinal and one-dimensional, the resistivity is calculated from the relationship

$$R_s = \varrho_s \left[\frac{4l}{\pi d^2}\right] f\,, \tag{4.6}$$

where R_s is the measured resistance, ϱ_s is the calculated specific resistivity, l is the distance between the anvil and baseplate, d is the anvil diameter, and f, the fringing field correction, to account for edge effects.

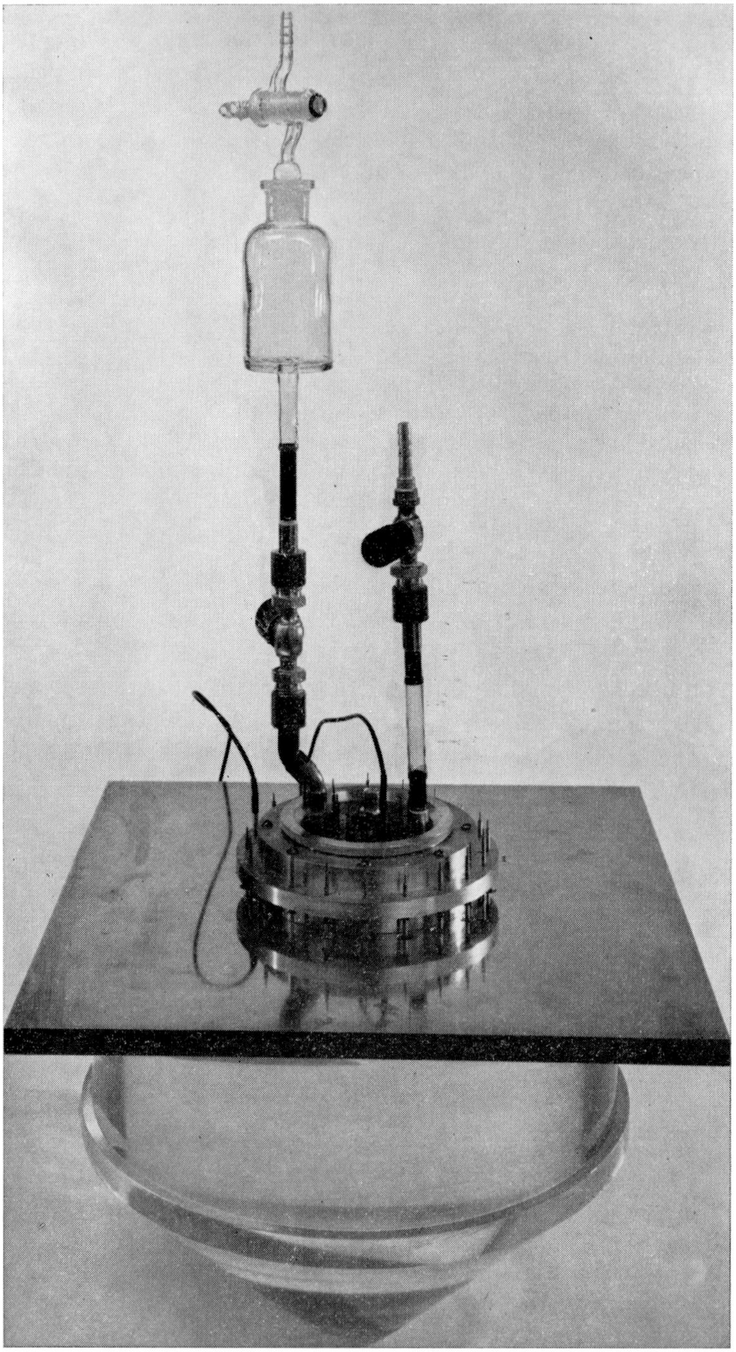

Fig. 4.4. – Assembled carbon tetrachloride conductivity experiment.

In the second category of measurements, a metal is shock-compressed to some pressure and temperature, and its conductivity, initially in the metallic range, may either increase or decrease; however, it is observed experimentally to remain very high at all times.

The high conductivity of metals requires that a sample be used whose resistance is sufficient to develop a measurable voltage drop across it. This dictates the use of a long, thin sample, and current-pulsing techniques. The initial experiments, carried out by FULLER and PRICE [4.18], utilized thin wires of iron and manganin embedded in epoxy resin. This technique has the disadvantage that the complex interactions between epoxy and wire arising from the two-dimensional geometry and large dynamic impedance mismatch leave the wire in a poorly characterized thermodynamic state. Furthermore, epoxy cannot be used as an insulator at pressures much above about 300 kbar, where it becomes a strong conductor. To avoid these problems, various insulators were tested as to their resistance under shock conditions and their dynamic impedance. The results are shown in Fig. 4.5. Wesgo 995 alumina

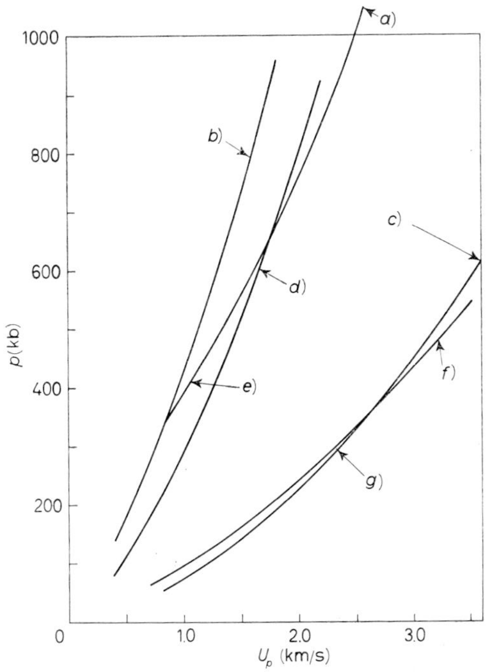

Fig. 4.5. – Equation of state and electrical conductivity of two insulators. *a*) alumina ceramic, $\varkappa = 1.57 \cdot 10^{-4}$ mho/cm, $p = 1100$ kbar; *b*) iron; *c*) Teflon, $\varkappa = 6.9 \cdot 10^{-3}$ mho/cm, $p = 630$ kbar; *d*) indium; *e*) alumina ceramic, $\varkappa = 7.9 \cdot 10^{-6}$ mho/cm, $p = 420$ kb; *f*) magnesium; *g*) Teflon, $\varkappa < 1 \cdot 10^{-7}$ mho/cm, $p = 300$ kbar.

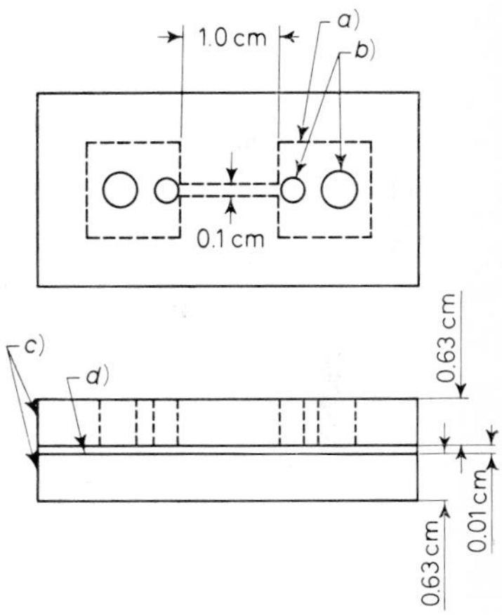

Fig. 4.6. – Schematic, metallic conductivity experiment. *a*) sample; *b*) indium electrodes; *c*) alumina blocks; *d*) epoxy or polyethylene filler.

ceramic (*) was the best insulator tested, both in terms of dynamic impedance relative to iron and as a high-resistance insulator. Teflon also performed well and is appropriate for use with lighter metals.

As was the case for the first type of measurement, it is necessary to maintain a one-dimensional geometrical configuration so that the hydrodynamic interactions can be understood, the effects of unrelieved stresses can be minimized, and the shocked state of the samples can be accurately characterized. A sample configuration for metallic conductivity experiments is shown in Fig. 4.6. The design in Fig. 4.6 is for use at the highest pressures ($\sim$ 1.4 Mbar). The leads to the iron foil are indium, a material whose dynamic impedance almost exactly matches that of alumina. This was done to minimize possible shearing effects at the iron foil lead-in contact caused by a dynamic impedance mismatch. The cross-sectional aspect ratios of the foils were 10:1 and 20:1, to achieve the desired one-dimensional configuration.

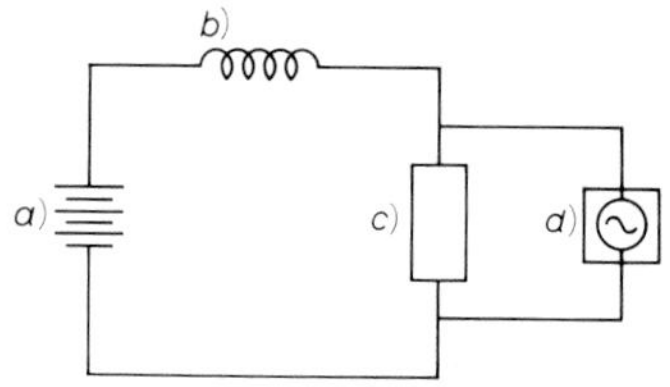

Fig. 4.7. – Electronic circuitry, metallic conductivity experiments. *a*) current supply; *b*) large inductance; *c*) foil sample; *d*) oscilloscope.

The circuitry used in the metallic conductivity experiments is straightforward and is shown in Fig. 4.7. Current is supplied by a current pulser, and the voltage is read across the sample on a fast-sweep oscilloscope. Two experimental records are shown in Fig. 4.8.

No satisfactory experimental solution presently exists for the third type of conductivity measurement: the insulator-metal transition. There are a number of suggested but untried possibilities which involve the interaction of the conducting medium with rapidly changing magnetic fields. However, there have been no successful measurements of this kind made to date, and it appears that most of the proposed solutions will provide data of considerably degraded accuracy. Since the problem of making this type of measurement is one of establishing current flow (or overcoming inductive effects), a flat pressure profile behind a shock front at least a microsecond in duration would be required to allow these effects to be overcome in circuitry such as that described previously. It is not clear that the integrity of the sample or leads

(*) Western Gold and Platinum Co., Belmont, Cal.

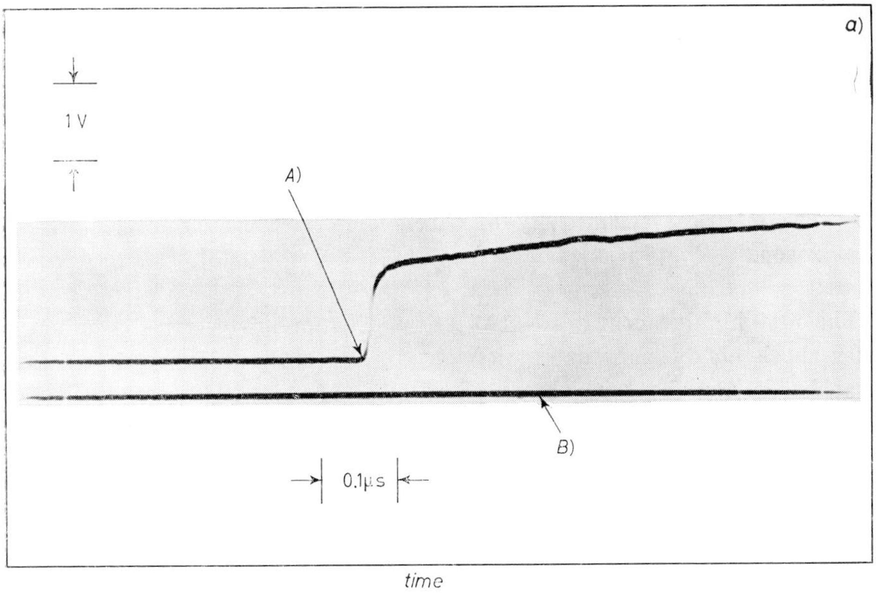

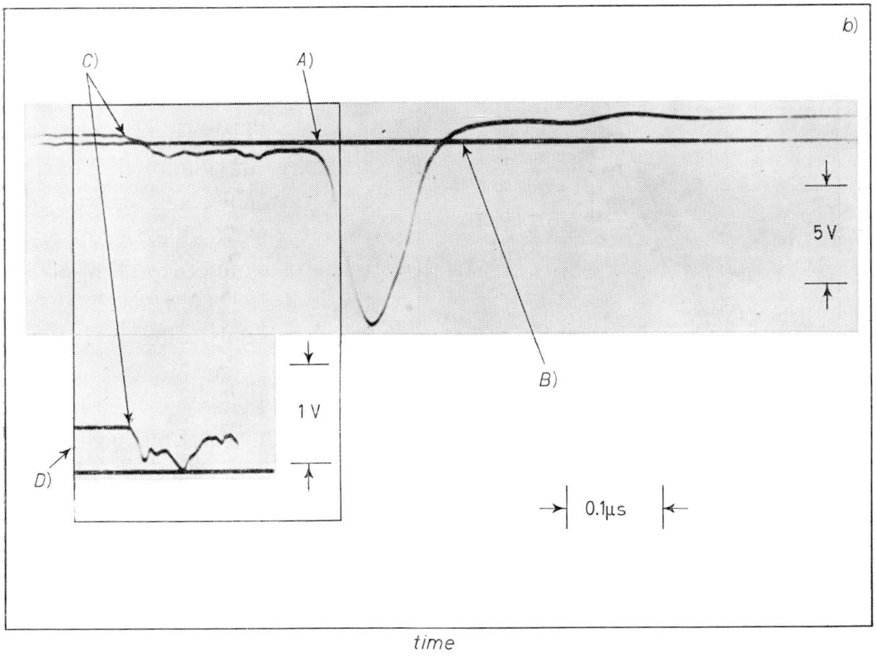

Fig. 4.8. – Oscilloscope traces, metallic conductivity experiments: *a*) copper, 1.1 Mbar; *b*) iron, 0.175 and 0.245 Mbar. *A*) shock-wave arrival; *B*) base-line; *C*) elastic wave precursor arrival; *D*) details of leading edge, 0.245 Mbar pulse, elastic-wave precursor.

could be preserved for as long as a microsecond, and unlike the first two classes of measurement, the data would not be taken with the arrival of the shock front as a reference point, but at some time later, when the state of the sample is not as well defined.

4·3. – The electrical conductivity of shock-compressed carbon tetrachloride and xenon.

The first highly accurate (∼10 %) investigation of electrical conductivity under shock conditions was carried out by MITCHELL. His work on carbon tetrachloride, together with the excellent optical work carried out in the Soviet Union, yields a rather interesting picture of this normally insulating liquid at high temperature and pressure.

Early work by WALSH and RICE [4.24] reported an attempt to observe the freezing of liquids under shock compression. They carried out a number of experiments in which they visually observed the transit of a shock wave through various fluids using an argon candle and a framing camera. They noted that at a pressure of about 60 kbar, carbon tetrachloride showed slight opacity; at 75 kbar the opacity became pronounced; and at 130 kbar the surface of the shock front became highly reflective. This reflectivity persisted to the highest pressure measured, 170 kbar. Although the authors suggest shock-

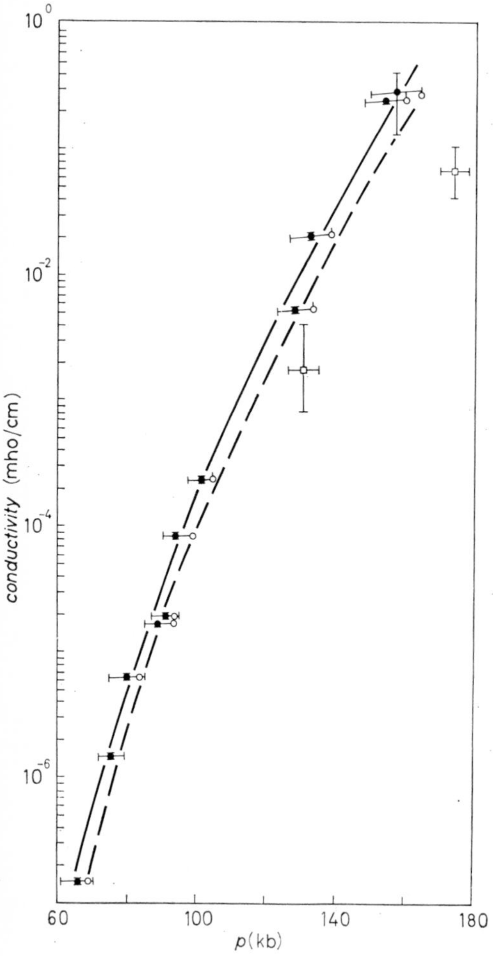

Fig. 4.9. – Electrical conductivity of CCl_4 *vs.* pressure—d.c. and micro-wave results. Mitchell and Keeler data: — — —, ○ without shock-wave attenuation correction behind shock front; ———, ● with shock-wave attenuation correction behind shock front; □ microwave data.

induced freezing as a possibility, the observed behavior is not completely consistent with freezing. One observation the authors made was that when opacity occurred, it occurred immediately; no kinetic effects were observed. If freezing were actually taking place, one might expect a finite nucleation time. The opacity increased monotonically with pressure; if freezing occurred along the Hugoniot at lower pressures, one would expect the Hugoniot to eventually emerge from the two-phase region. In this case, the opacity would increase, then decrease, and eventually disappear. Finally, there is no reason to associate a highly reflective surface with freezing. In view of these observations, one might seek an alternate explanation.

Figure 4.9 shows the variation of conductivity with pressure for carbon tetrachloride. The same data are shown in Fig. 4.10 in the familiar activation-energy plot. This plot shows a remarkable linearity of six decades in conductivity. Since the pressure, volume, and temperature are all changing, it might seem unlikely that the energy gap would remain constant along the Hugoniot. However, it should be recalled that over this entire range of pressure, (69÷170) kbar, the volume only changes about 7 or 8%, and the average separation distance of the atoms, only about 2 or 3%. Therefore, the average separation distance, to which the gap energy is most sensitive, is changing only slightly. The activation energy for this transport process is 4.3 eV. The question naturally arises, what is the mechanism of transport? To shed light on this question, a series of microwave experiments was carried out by R. S. Hawke [4.25]. In these experiments, he measured the phase shift and attenuation of 35 GHz microwaves reflected from the shock front and from the base plate behind the carbon tetrachloride. As can be seen in Fig. 4.9, the microwave results agree fairly well with the d.c. results up to a pressure of about 140 kbar. Since the relaxation time for ionic conductivity is around 10^{-7} s,

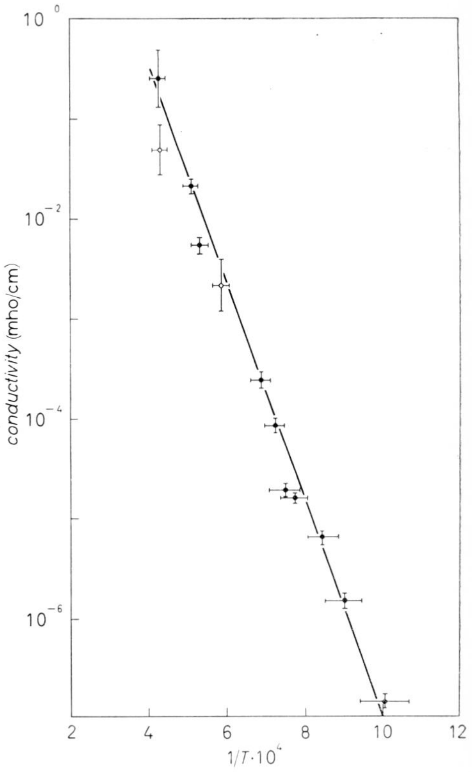

Fig. 4.10. – Determination of energy gap in CCl_4: $\log \varkappa$ *vs.* $1/T$. $\varepsilon_g = 4.30$ eV.

and the relaxation time characteristic of electronic conductivity is around 10^{-12} s, it is apparent that the response of the systems to the microwaves is characteristic of the electronic mechanism at pressures up to about 160 kbar. Above this pressure, there appears to be divergence which indicates a possibility of a ionic mechanism. If any appreciable portion of the conductivity at higher pressures is ionic, decomposition of the carbon tetrachloride is indicated. To verify whether or not this was the case, a series of recovery experiments was carried out by KUSUBOV [4.26]. He found that when carbon tetrachloride was shocked to pressures up to about 180 kbar, no evidence of decomposition was discovered on examination of the recovered products. However, at pressures slightly higher than 180 kbar, chlorine, carbon, and hexachloroethane were all found in the recovered samples. A high-pressure conductivity point was derived from the optical data of YUSHKO [4.27] ($\varkappa = 20$ mho/cm at 200 kbar). This point, which is consistent with an extrapolation of the d.c. measurements of MITCHELL is anomalously high compared with the microwave results and may be due to electronic effects directly within the shock front plus a very short skin depth.

Based on the indication that the mechanism for conductivity for carbon tetrachloride below 180 kbar is largely electronic, one may calculate the optical skin depth as a function of pressure and compare it with the previously quoted results of WALSH and RICE. At 65 kbar, the conductivity is $1.51 \cdot 10^{-7}$ mho/cm, from which a skin depth of 5.3 mm is calculated. This is consistent with the onset of opacity observed at this pressure. At 150 kbar the calculated skin depth is 5.3 μm, and the measured microwave dielectric coefficient is approximately 6.0. This is consistent with the approach to complete opacity and high reflectivity at the shock front. These calculations, therefore, are consistent with the observations of WALSH and RICE.

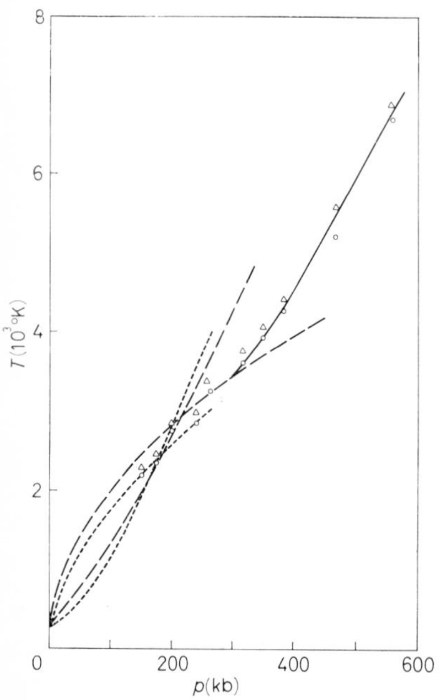

Fig. 4.11. – The Hugoniot and melting line of CCl_4. ——— Calculation by Ross.

To support their contention that the opacity of carbon tetrachloride was due to freezing, the Russians [4.28] carried out a series of temperature measurements in shocked carbon tetrachloride. Their results are shown in Fig. 4.11. However, they used the Simon equation to extrapolate the melting curve for carbon tetrachloride from 9 kbar up to 200 kbar. Such an extrapolation is highly questionable [4.29]. There is the additional complication that carbon tetrachloride exists in two polymorphic solid phases, and the liquid-solid equilibrium which exists at very high pressures may not be the same liquid-solid equilibrium which exists at the lower pressures. To clarify this point, a more accurate melting line is plotted on Fig. 4.11 using the melting law derived by Ross [4.30]. Some cell model temperature calculations [4.31] are also plotted and compared with the Russian experimental work. From this Figure, it can be seen that the temperature anomaly observed by the Russians starting at 200 kbar may be due to the same pyrolysis and decomposition observed by Kusubov in recovery experiments. The data available indicate, therefore, that carbon tetrachloride behaves in the following way: As a result of thermal excitation of electrons into the conduction band by the high temperatures achieved along the Hugoniot, carbon tetrachloride behaves as an intrinsic semiconductor up to about 170 kbar. Past this point, pyrolysis and decomposition take place. Melting and decomposition may be occurring simultaneously. These results, while quite interesting, are not immediately useful since the carbon tetrachloride molecule is difficult to treat theoretically, particularly in the liquid state.

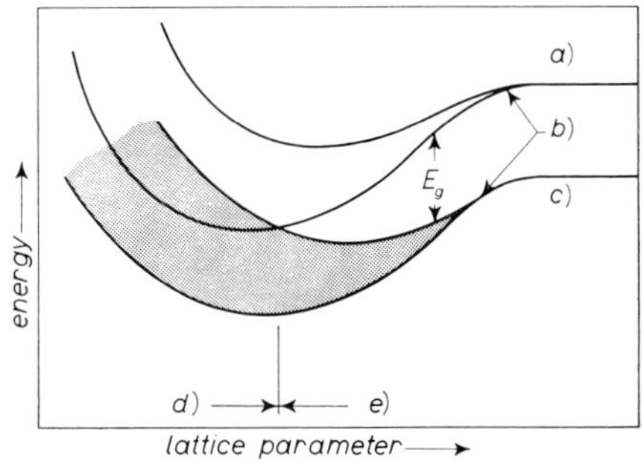

Fig. 4.12. – Conversion to the metallic state. *a*) first excited state; *b*) level splitting; *c*) ground state; *d*) metal; *e*) semiconductor ... insulator.

One of the goals of earlier experimentalists in shock-wave work was the direct observation of the conversion of an insulator to a metal. A very simplified view of the manner in which pressure creates such a transition can be understood by reference to Fig. 4.12. Consider an infinite number of atoms at infinite separation. Then, one may represent the energy states as a ground state and a first excited state. As the atoms are brought together to form a condensed system, their discrete energy spectra broaden into bands of allowed energies. In the case of an insulator, a gap exists between the highest full band and the lowest empty band. The separation between them, *e.g.*, is related to the activation energy characteristics of intrinsic semiconductors. If the insulator is compressed, the relative positions of the bands shift and their widths continue to increase until the top of the highest filled band overlaps the bottom of the lowest unfilled band and a metallic phase is formed.

The usual procedure in previous investigations has been the measurement of electrical conductivity at various shock pressures, calculation of the temperature of the shocked state, and preparation of a semi-logarithmic plot of conductivity against reciprocal temperature. This is based on the following well-known relationship for semiconductors:

$$\varkappa = A \exp\left[-\frac{\varepsilon_g}{2kT}\right], \tag{4.7}$$

where $\varkappa$ is the conductivity, A, a constant which varies slowly with temperature, and ε_g, the thermal gap between the valence and conduction band. The same type of functional relationship holds for a ionic mechanism, with $\varepsilon_g/2$ replaced by $E^{\pm}$, the activation energy for ionic conductivity.

AL'TSHULER *et al.* [4.13] made the first extensive measurements of electrical conductivity in shock-compressed solids. He chose NaCl as the subject of his investigation, and obtained an activation energy for conduction of 1.2 eV. Since the activation energy for ionic conduction of NaCl under normal conditions is 1.87 eV, AL'TSHULER concluded that the mechanism of conduction was ionic. However, it was pointed out later [4.28] that this was an unreasonable conclusion, since all available experimental data and theoretical predictions indicated that this activation energy should increase to about $(6 \div 7)$ eV at the pressure in question. ALDER [4.32] found a similar decrease in other alkali halides and suggested a band gap closure such as that shown in Fig. 4.11, although in view of the lack of theoretical support for such a suggestion, he proposed that a Hall-effect measurement should be made. Finally, KORMER argued rather conclusively [4.33] that the plastic deformation at the shock front creates a number of defects with associated electrons which act as donor states. Therefore, alkali halides under shock compression behave like semiconductors with donor levels whose number density is of the order of $10^{19}/cm^3$.

In view of these difficulties, the most fruitful area for fundamental investigation of physical properties has been liquids. Several earlier investigations by HAMANN and his co-workers [4.6-4.8] studied the properties of several weakly ionized fluids under pressure. He showed how the conductivity increase in water under shock compression was due to the large increase in the ionization constant of water caused by the increase in density and temperature behind the shock front [4.8]. His investigations have been, for the most part, concerned with the physicochemical properties of liquids and solutions.

Several years ago, a program was initiated at our laboratory to study the properties of the rare gases, argon and xenon, in the condensed state and at high pressure. These two substances are inert closed-shell atoms, and therefore most amenable to theoretical treatment. Because of the high temperature generated in the shock process, the atoms can make a close approach to one another, and thereby supply information about the repulsive part of the intermolecular potential. Theoretical interpretation of the high-pressure shock wave results on argon were best expressed in terms of an exponential-six potential and were in excellent agreement with the repulsive potential derived from Thomas-Fermi-Dirac theory and molecular beam-scattering experiments. To test the theorem of corresponding states, similar experiments were carried out on liquid xenon. When reduced by the critical constants, the two Hugoniot curves were in excellent agreement up to about 250 kbar; however, above this pressure, xenon showed anomalously high compressibility [4.34].

To test this idea, energy-band calculations were carried out by Ross [4.35]. These calculations were based on the Wigner-Seitz model, and employed the Thomas-Fermi-Dirac crystal potential. The results are shown in Fig. 4.13. The $6s$ band is initially the lowest conduction band, but it is replaced by the $5d$ band at about 27 cm³/mole. The $5d$ band eventually crosses the valence band at about 12 cm³/mole, at which volume xenon should become metallic. Ross showed that the combination of high temperatures (~ 1.5 eV), a large increase in the degeneracy of the conduction band, and a band gap that was decreasing with decreasing volume led to a state in which one in five xenon atoms contributed an electron to the conduction band. A calculation of the xenon Hugoniot fell within the experimental error of the measurements. These results and their implications have been discussed fully elsewhere [4.36]. Since

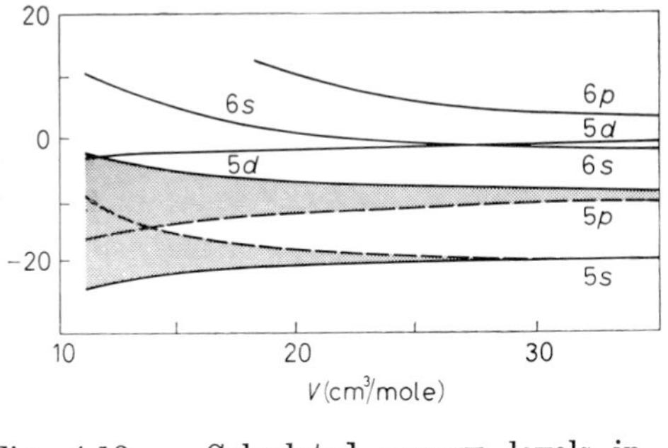

Fig. 4.13. – Calculated energy levels in xenon under compression.

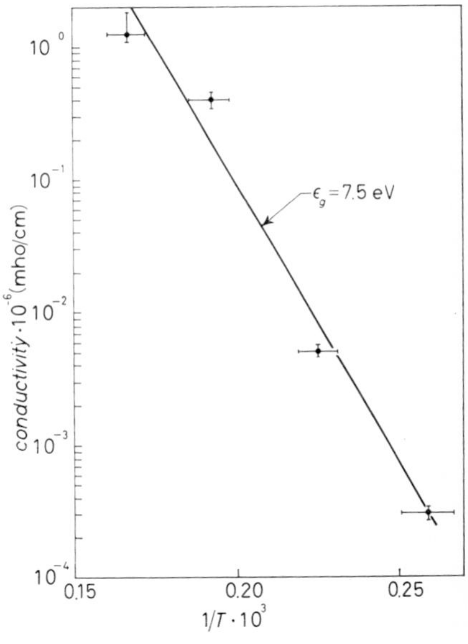

Fig. 4.14. – The electrical conductivity of shock-compressed xenon.

Ross' theory predicts a decreasing energy gap and a large increase in the electron population of the conduction band with compression, then electrical conductivity measurements should be expected to provide a verification of at least the qualitative correctness of the theory. Such a verification has been carried out, and is shown in Fig. 4.14. Utilizing the techniques developed by MITCHELL and KEELER [4.22] and VAN THIEL [4.37], the conductivity of shock-compressed liquid xenon has been measured over the pressure range 80 to 150 kbar. From the $\log \varkappa$ *vs.* $1/T$ plot shown, a band gap of (7.5 ± 0.7) eV was calculated. This is in excellent agreement with the prediction of ROSS. These developments mark only a modest start; much remains to be done. Whereas the theory of metals and liquid or solid semiconductors is, in principle, the same at high pressures as at 1 atm, the added complication of temperature makes the interpretation of shock data difficult. Temperature effects must be included in the band-gap calculations, and conductivity must be calculated for systems with large electron-phonon interactions. Finally, experimental techniques should be developed to circumvent the difficulties associated with measuring conductivity in solids, particularly for the insulator metal transition.

4'4. – Electrical conductivity of shock-compressed metals.

Shock-compressed semiconductors exhibit an exponential rise in conductivity with temperature. Measurements carried out under shock conditions can be simply correlated to provide a physically meaningful property—the energy gap. The energy gap calculated from shock-wave conductivity measurements can then be compared with theoretical predictions of the how the gap should vary with volume and temperature. The situation in metals, however, is not so clear. First-principles calculations of the electrical conductivity of metals are quite difficult even under normal conditions. These calculations generally only consider the electrons at the Fermi surface, which is adequate for small values of kT. However, at the temperatures generated in shock waves, it is necessary to also consider states considerably below the Fermi surface. An extreme example is potassium, which has a free-electron Fermi energy of 2.1 eV but can be shocked to about 3 eV. This lack of theoretical understanding of metallic conductivity has limited the use of conductivity measurements to the study of phase transitions and the generation of data of geophysical interest, rather than to the confirmation of theoretical models of conductivity.

The first conductivity measurements in metals at very high pressures ($p > 350$ kbar) were made at our laboratory, using demagnetization techniques described by ROYCE in the following Section, and the method described in the preceding Section on experimental techniques. The materials studied were copper, iron, iron-silicon alloy, and iron-nickel alloy. The results for copper are shown in Fig. 4.15 *a*), and the results for iron in Fig. 4.15 *b*). The iron results are compared with the static conductivity data of BALCHAN and DRICKAMER [4.38], and BRIDGMAN [4.39], the results of FULLER and PRICE [4.18] at lower shock pressures, and the earlier demagnetization experiments of ROYCE [4.40]. Some features of the iron results are worth discussing. The sharp decrease in the conductivity of iron at about 130 kbar is associated with the transition from the α-phase, which is body-centered cubic and ferromagnetic, to the ε-phase, which is hexagonal close-packed and nonmagnetic. The effect of demagnetization can be clearly seen in the experimental record shown in Fig. 4.8 *a*). The initial displacement above the baseline represents the re-

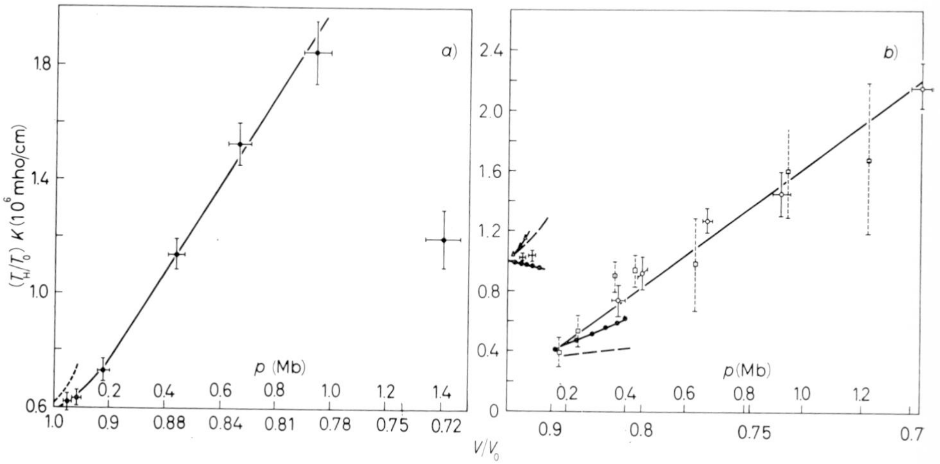

Fig. 4.15 *a*). – The electrical conductivity of copper at high pressure. •, this work; ——— BRIDGMAN [4.39]. *b*) The electrical conductivity of iron at high pressure. o, this work; □, ROYCE, LRL demagnetization [4.40]; •, FULLER and PRICE [4.18] (corrected for compression); ——— BALCHAN and DRICKAMER [4.38]; △ BRIDGMAN [4.39].

sistance of iron in the unshocked state. At a pressure of 175 kbar, a two-wave structure exists in alumina. The initial wave is an 83 kbar elastic wave, which is followed by the 175 kbar driving wave. At the arrival of the 83 kbar elastic wave, the resistivity appears to become negative and remains slightly negative until the arrival of the 175 kbar shock wave. The record then goes sharply negative.

This negative resistance can be explained in the following way: The arrival of the 175 kbar shock wave transforms the sample to the ε-phase, and magnetic susceptibility decreases by a factor about 1000. The original magnetic field can no longer be supported by the current in the sample; flux begins to diffuse from the sample, generating eddy currents which produce the sharp negative spike. The spike then decays with a time constant characteristic of the electrical conductivity. The idealized shape of the curve is given, to a first approximation, by [4.19].

$$\frac{V_2}{V_1}=\frac{R_2}{R_1}\left[1-2\frac{\mu_1}{\mu_2}\exp\left[-\frac{4\pi^2 t}{\mu_2\sigma_2 l^2}\right]\right], \tag{4.8}$$

where l is the thickness of the sample, σ_2, final sample conductivity, R_1 and R_2, initial and final sample resistances, μ_1 and μ_2, initial and final magnetic susceptibilities, t, time after shock arrival, and V_2/V_1, ratio of oscilloscope voltage to initial voltage. The finite rise time of the conductivity record at the arrival of the 175 kbar shock wave is a direct indication of the sluggishness of the iron transition. The closeness of the agreement of the conductivity measurement by current-pulsing techniques with the demagnetization results reported in the following Section is an excellent verification of the validity of the data for iron.

The continuous demagnetization occurring after the arrival of the 83 kbar elastic precursor is apparently a partial transition to the ε-phase. This conclusion is substantiated by other experimental data. A full discussion of this point can be found elsewhere [4.20] and in the Section on properties of magnetic materials under shock compression.

4.5. – Geophysical investigations.

The interior of the earth has three principal geophysical features: a crust, extending down for about 20 km; a solid mantle, presumably some mineral or combination of minerals; and a metallic core. The metallic core is liquid down to a radius of about 1400 km. The core-mantle interface is located at a radius of 3500 km. The estimated pressure at the core-mantle interface is 1.4 Mbar; at the inner core-outer core interface, 2.5 Mbar; and at the center of the earth, 3.5 Mbar. The maximum temperature encountered in the earth, at the center, is estimated to be between 5000 and 7000 °K. A schematic is shown in Fig. 4.16.

These ranges of pressures and temperatures are close to those which are attainable by Hugoniot measurements. Such measurements have, in fact, provided the principal equation-of-state guidance for geophysical theories in the last ten years. As an example, consider Fig. 4.17. BALCHAN and COWAN [4.41] measured the Hugoniot of a 19.8 wt% silicon-iron alloy and compared it with the limiting density-pressure curves calculated by BIRCH [4.42] from seismic data. According to these results, a core of 20% Si-80% Fe is much more probable than a core of pure iron. The equation of state of a proposed core material and its consistency with the seismic data are not the only boundary conditions which must be met. The electrical and magnetic properties of the proposed core material must be consistent with geomagnetic data. Therefore, it is of interest to see how measurements of electrical conductivity can be used in determining the composition of the earth's core.

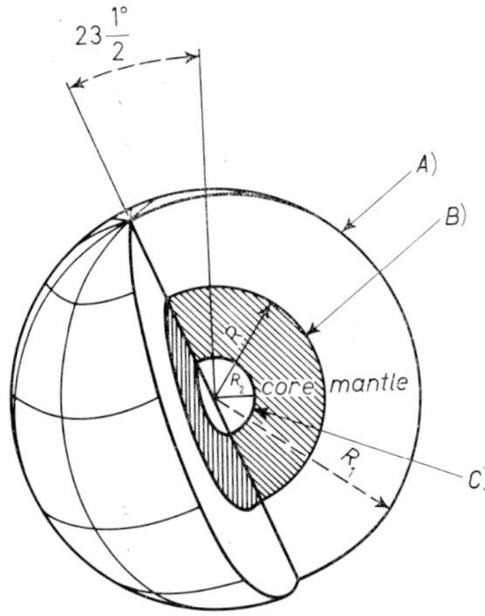

Fig. 4.16. – Major geophysical features of the earth. A) Radius $= 6730$ km; B) core mantle interface: radius $= 3800$ km, $p = 1.4$ Mbar, $\varrho_M = 5.7$, $\varrho_c = 10.0$, $T = 3000$ °K; C) outer core-inner core: radius $= 1400$ km, $p = 2.5$ Mbar, $\varrho_c = 12.5$, $T = 4500$ °K. Center: $p = 3.5$ Mbar; $\varrho_c = 13.0$, $T = 5000$ °K.

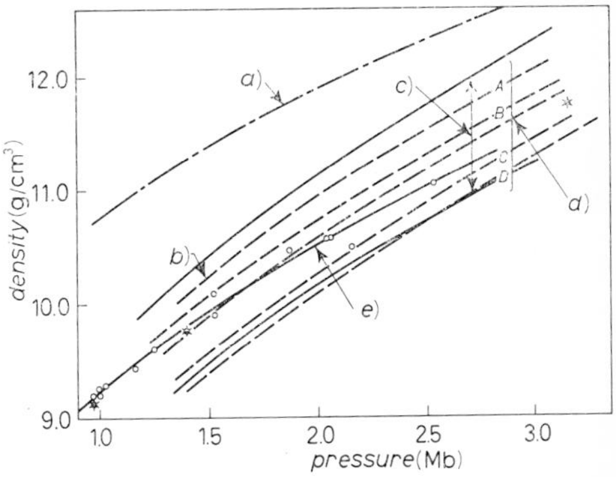

Fig. 4.17. – Equation of state of core materials (data due to BALCHAN) [4.40]. *a*) Hugoniot of pure iron, AL'TSHULER *et al.*; *b*) BIRCH, solution I; *c*) limits of earth's core, BIRCH; *d*) computed isentropes; *e*) Hugoniot for 80 Fe-20 Si alloy, $U_s = 5.444 + 1.235\, U_p$. • KORMER *et al.*, ○ Balchan results.

A qualitative requirement for the dynamo action is that the magnetic Reynolds number,

$$R_M = L v\, 4\pi\mu\sigma$$

should exceed about 10 [4.43]. It is generally estimated that R_M should be somewhere between 10 and 100 for the maintenance of dynamo action. Here L is a length scale associated with the fluid motion, v is a typical velocity associated with fluid motion, μ, the magnetic susceptibility, and σ, the electrical conductivity. If we take the most generally accepted geophysical values, $L = 10^3$ km and $v = 100$ km/yr, and set $R = 100$, then $\sigma \geqslant 3 \cdot 10^3$ mho/cm. Although this type of dimensional approach is admittedly crude, it provides a rough estimate of the allowable lower limit on the value of the electrical conductivity of the metallic liquid core. Figure 4.18 shows a comparison of the calculated lower limit with the experimentally measured conductivity values for iron and two of its alloys. It can be seen that the 20 % Si-80 % Fe alloy, which provided the best fit to the limits calculated by Birch is approaching the upper limit for stable dynamo action in the core, based on the simple arguments presented here. If this alloy is in the solid state under shock conditions at 1.4 Mbar, then estimates of the conductivity of the liquid alloy [4.44] under core conditions yield a magnetic Reynolds number of 100, which is close to the upper limiting value. Thus, the 20 % Si-80 % Fe alloy is a much less probable candidate for a core material than pure iron or alloys with a much lower silicon content, based on magnetic considerations.

This is an example of how electrical conductivity measurements under shock conditions can be used to provide valuable geophysical data. Continuing studies must be made on materials of geophysical and astrophysical interest. Conductivity may prove to be a powerful tool for diagnosis of melting on the Hugoniot. The determination of the melting curve of iron and its alloys would be of prime interest in defining the temperature profile within the core of the earth.

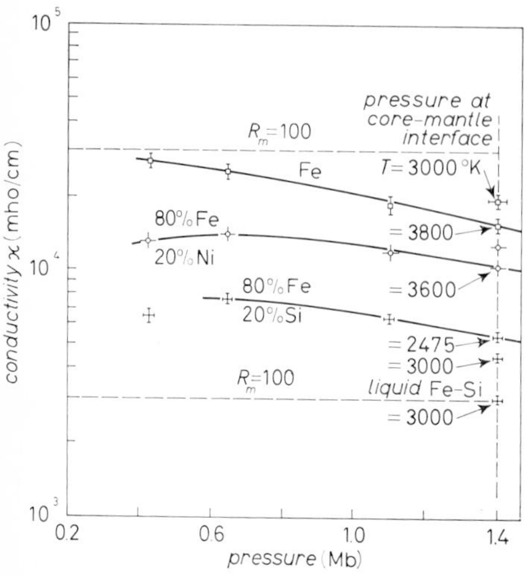

Fig. 4.18. – The electrical conductivity of several core materials at high pressures.

REFERENCES

[4.1] B. J. ALDER and R. H. CHRISTIAN: *Phys. Rev.*, **104**, 550 (1956).

[4.2] B. J. ALDER and R. H. CHRISTIAN: *Discussions Faraday Soc.* **22**, 44 (1956).

[4.3] H. G. DAVID and S. D. HAMANN: *Journ. Chem. Phys.* **28**, 1006 (1958).

[4.4] S. D. HAMANN: *Australian Journ. Chem.*, **11**, 391 (1958).

[4.5] S. D. HAMANN: *Intern. Congr. Pure Appl. Chem.*, Vol. **2** (1959), p. 277.

[4.6] H. G. DAVID and S. D. HAMANN: *Trans. Faraday Soc.*, **55**, 72 (1959).

[4.7] H. G. DAVID and S. D. HAMANN: *Trans. Faraday Soc.*, **56**, 1043 (1960).

[4.8] S. D. HAMANN and M. LINTON: *Trans. Faraday Soc.*, **62**, 2234 (1966).

[4.9] S. JOIGNEAU and J. THOUVENIN: *Compt. Rend.*, **246**, 3422 (1958).

[4.10] J. THOUVENIN and A. RAUCH: *Compt. Rend.*, **5**, 868 (1962).

[4.11] A. A. BRISH, M. S. TARASOV and V. A. TSUKERMAN: *Žurn. Ėksp. Teor. Fiz.*, **37**, 1544 (1959); English translation in *Sov. Phys. JETP*, **10**, 1095 (1960).

[4.12] A. A. BRISH, M. S. TARASOV and V. A. TSUKERMAN: *Žurn. Ėksp. Teor. Fiz.*, **38**, 22 (1960); English translation in *Sov. Phys. JETP*, **11**, 15 (1960).

[4.13] L. V. AL'TSHULER, L. V. KULESHOVA and M. N. PAVLOVSKII: *Žurn. Ėksp.*

Teor. Fiz., **39**, 16 (1960); English translation in *Sov. Phys. JETP*, **12**, 19 (1961).
[4.14] R. A. GRAHAM, O. E. JONES and J. R. HOLLAND: *Journ. Phys. Chem. Solids*, **27**, 1519 (1966).
[4.15] D. G. DORAN and T. J. AHRENS: Stanford Research Inst., Final Report. PGU-4100 (August 1963).
[4.16] T. J. AHRENS: *Journ. Appl. Phys.*, **37**, 2532 (1966).
[4.17] B. HAYES: *IV Symposium on Detonation*, edited by S. J. JACOBS, Vol. **2** (Silver Spring, Md., 1965).
[4.18] P. J. A. FULLER and J. H. PRICE: *Nature*, **193**, 262 (1962).
[4.19] The physical model and approximations on which this relationship is based will be discussed in G. V. MATISSOV: Thesis, Department of Applied Science, University of California, Davis (Livermore) (1971).
[4.20] R. N. KEELER and A. C. MITCHELL: *Solid State Communications*, **2**, 271 (1969).
[4.21] See for example, the work cited by ALDER in *Solids Under Pressure*, edited by W. PAUL and D. M. WARSCHAUER (New York, 1963), p. 411; and the subsequent discussion of these results by S. B. KORMER: *Usp. Fiz. Nauk*, **94**, 641 (1968).
[4.22] A. C. MITCHELL and R. N. KEELER: *Rev. Sci. Instr.*, **39**, 513 (1968).
[4.23] S. D. HAMANN: private communication.
[4.24] J. M. WALSH and M. H. RICE: *Journ. Chem. Phys.*, **26**, 815 (1957).
[4.25] R. S. HAWKE, A. C. MITCHELL and R. N. KEELER: *Rev. Sci. Instr.*, **40**, 632 (1969).
[4.26] A. S. KUSUBOV: unpublished work.
[4.27] K. B. YUSHKO, G. V. KRISHKEVICH and S. B. KORMER: *JETP Lett.*, **7**, 7 (1968).
[4.28] S. B. KORMER: *Usp. Fiz. Nauk*, **94**, 641 (1967); English translation in *Sov. Phys. Usp.*, **11**, 229 (1968).
[4.29] E. A. KRAUT and G. C. KENNEDY: *Phys. Rev. Lett.*, **16**, 608 (1966).
[4.30] M. ROSS and B. J. ALDER: *Phys. Rev. Lett.*, **16**, 1077 (1966).
[4.31] M. ROSS: unpublished work.
[4.32] B. J. ALDER: in *Solids Under Pressure*, edited by W. PAUL and D. WARSCHAUER (New York, 1963), p. 385.
[4.33] S. B. KORMER, M. V. SINITSYN, G. A. KIRILLOV and L. T. POPOVA: *Žurn. Ėksp. Teor. Fiz.*, **49**, 135 (1965); English translation in *Sov. Phys. JETP*, **22**, 97 (1966).
[4.34] R. N. KEELER, M. VAN THIEL and B. J. ALDER: *Physica*, **31**, 1437 (1965).
[4.35] M. ROSS: *Phys. Rev.*, **171**, 777 (1968).
[4.36] R. E. DUFF: in *Properties of Matter under Unusual Conditions*, edited by H. MARK and S. FERNBACH (New York, 1967), p. 73.
[4.37] M. VAN THIEL and B. J. ALDER: *Journ. Chem. Phys.*, **44**, 1056 (1966).
[4.38] A. S. BALCHAN and H. G. DRICKAMER: *Rev. Sci. Instr.*, **32**, 308 (1961).
[4.39] P. W. BRIDGMAN: *Proc. Am. Acad. Arts Sci.*, **81**, 165 (1952).
[4.40] E. B. ROYCE: *Proceedings of IUTAM Symposium on Behaviour of Dense Media Under High Dynamic Pressures* (New York, 1968), p. 419; and unpublished data.
[4.41] A. S. BALCHAN and G. R. COWAN: *Journ. Geophys. Res.*, **71**, 3577 (1966).
[4.42] F. BIRCH: in *Solids Under Pressure*, edited by W. PAUL and D. WARSCHAUER (New York, 1963), p. 137.
[4.43] F. D. STACEY: *Earth and Planets Scientific Letters*, **3**, 204 (1967).
[4.44] A. I. GUBANOV: *Quantum Electron Theory of Amorphous Conductors* (New York, 1965).

V. – Properties of Magnetic Materials Under Shock Compression.

E. B. Royce

5·1. – Introduction.

The passage of a strong shock wave through a biased sample of ferromagnetic material will cause a change in its permeability and magnetization. Three mechanisms have been identified as producing this effect. 1) Iron and iron-silicon alloys exhibit a *first-order phase transition* in the neighborhood of 130 kb from ferromagnetic bcc α-iron to nonmagnetic hcp ε-iron [5.1-5.3], resulting in a shock-induced total demagnetization [5.4-5.6]. 2) Iron-nickel alloys with nickel content greater than 30 % remain in the fcc ferromagnetic state, but the Curie temperature, characterizing the *second-order phase transition* between ordered and disordered states, is lowered under compression [5.7-5.8]. The combination of this Curie temperature shift and shock heating, together with a possible change of the moment in the totally polarized state ($T = 0°$) [5.9, 5.10] will cause a change in the magnetization [5.5, 5.11-5.13]. 3) Because of the nonzero yield strength of normally solid materials, shock-wave compression is not completely isotropic on a microscopic level; it is axial on a macroscopic level [5.14]. The anisotropic component of the compression can make the direction of shock propagation, the longitudinal direction, an easy direction of magnetization through an inverse magnetostriction [5.15]. Since it is usual to bias the magnetization in a transverse direction, *this shock-induced magnetic anisotropy* produces a rotation of the direction of the magnetization and a transverse demagnetization. This has been observed in the ceramics, nickel ferrite [5.6, 5.16], yttrium-iron garnet (YIG) [5.6, 5.17], and manganese-zinc ferrite [5.18].

The first-order transition is, of course, well known from equation-of-state measurements [5.1]. The effect of the second-order magnetic transition has also been seen in equation-of-state measurements [5.19, 5.19*a*].

5·2. – Experimental procedures.

All of the demagnetization experiments are basically similar in that a sample of material is biased to saturation (or nearly to saturation) by a magnetic field transverse to the direction of shock propagation. As the shock wave propagates through the sample, measurements are made of the component of the magnetization along the applied field.

Figure 5.1 shows the experimental geometry used in our laboratory to make such measurements [5.6, 5.16, 5.17]. The high-explosive assembly produces a plane shock wave in the thin sample of the material being studied. The use of a thin sample minimizes the effects of lateral rarefaction waves, which propagate part of the way into the sample from the edges while the shock wave is propagating longitudinally across the narrow dimension of the sample.

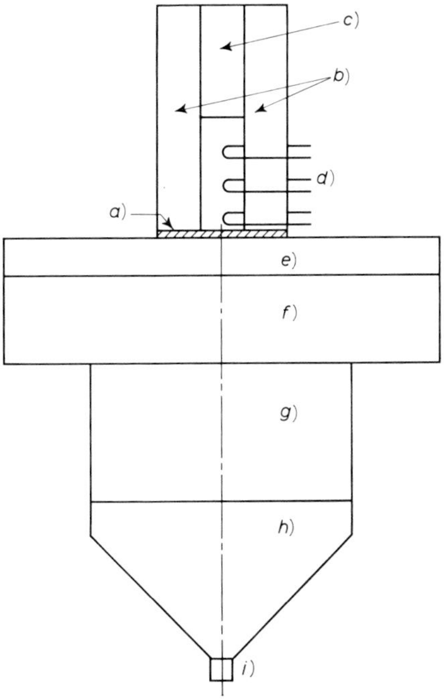

Fig. 5.1. – Assembly for shock-induced demagnetization experiments as carried out at the Lawrence Radiation Laboratory, Livermore (schematic, not to scale): *a*) Sample of material to be studied. For insulators, the sample is solid; for metals it is a sandwich of metal surrounded by alumina ceramic. *b*) Ferrite bars or arms completing a magnetic circuit. *c*) Permanent magnet. *d*) Single-turn pickup coils on printed-circuit boards. The magnetic circuit is normally surrounded by epoxy. *e*) metal base plate; *f*) impedance mismatch shock attenuators or flying plate assembly; *g*) high explosive; *h*) plane-wave lens; *i*) detonator.

In this geometry, most of the sample may be considered to be in a uniformly compressed state after the initial passage of the shock wave through the sample. No correction is made for the partial rarefaction of material near the edges of the sample, but such a correction would be small in any case.

In Fig. 5.1, the element *c* is a permanent magnet and produces the bias field for the sample. The elements *b* are ferrite bars or arms which complete the magnetic circuit between the sample and bias magnet. The shock-induced reduction in the magnetic flux through the sample is detected by the pick-up coils, *d*, wound on one arm of the magnetic circuit. The demagnetization of insulating materials takes place as the shock wave progresses through the sample. The magnetic circuit *cbabc* is interrupted, and the magnetic flux through the sample before compression redistributes itself across the gap between the arms. Signals are generated as this flux cuts the single-turn pick-up coils, and the flux change is determined by integrating these signals. The use of more than one pickup coil and appropriate data correction procedures allows one to determine the true change in the transverse magnetization of the sample to an accuracy of $\pm 5\%$ [5.16].

The demagnetization of insulating materials is completed as soon as the shock wave passes from the sample into the remainder of the magnetic circuit. Thus, the distortion and destruction of the magnetic circuit presents no problem. For metals, on the other hand, the demagnetization rate is determined by eddy-current diffusion in the sample and is longer than the shock-transit time through the sample. To prevent the distortion of the magnetic circuit during this time, the metal samples are sandwiched between thicker plates of aluminum oxide ceramic [5.20, 5.21]. The alumina plates extend beyond the dimensions of the iron sample plate, to minimize the effects of lateral rarefactions in the iron and alumina.

Shock pressure in the thin metal samples may be assumed to be the same as that characterizing the alumina, the final state in the iron being reached by a series of reverberating shock waves and rarefactions. The shock impedance of the alumina is not far from that of iron and its alloys, so these reverberations are small compared to the initial shock wave. Therefore, the final state in the sample may be assumed to be just the state reached in a single shock to the alumina pressure.

Figure 5.2 (taken from ref. [5.13]) shows an arrangement more typical of the work in other laboratories. The sample is in the form of a tape-wound core of the magnetic material to be studied, and the bias field is provided by

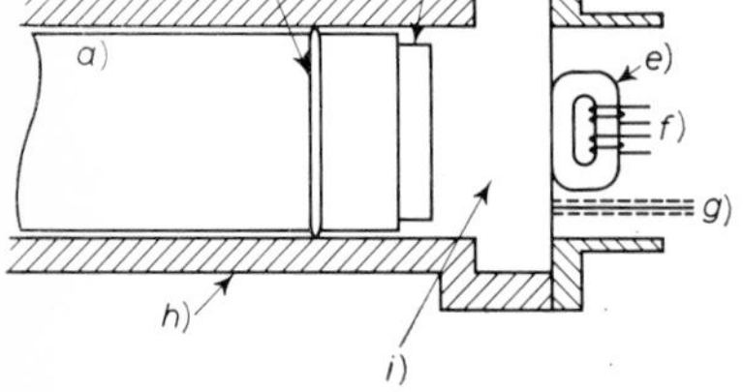

Fig. 5.2. – Assembly for shock-induced demagnetization experiments as carried out at Sandia Laboratories, Albuquerque (schematic, not to scale). Ref. R. C. WAYNE: *Journ. Appl. Phys.*, **40**, 15 (1969). *a*) projectile; *b*) « O » ring seal; *c*) projectile facing; *d*) sample holder; *e*) magnetic core; *f*) magnetizing and detection coils; *g*) coaxial scope trigger pin; *h*) 24.4 m gun barrel; *i*) vacuum.

the magnetizing coils. High explosives can be used to produce the shock wave, instead of the gun shown.

5'3. – First-order transitions.

Figure 5.3 shows the demagnetization of iron measured in our laboratory at different shock pressures. Most of the points shown are averages of the results from several samples of different thickenesses and on different shots. The data show no significant effect of varying the sample thickness. Bias fields between 200 and 300 Oe were employed. The bracket on the right indicates the measured initial magnetization of the sample in these fields.

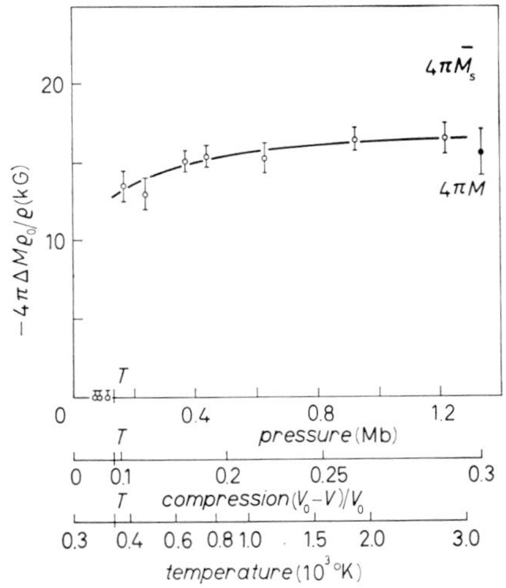

Fig. 5.3. – Shock-induced demagnetization of iron, $-(4\pi\,\Delta M)\,\varrho_0/\varrho = -(4\pi M_0)(\Delta\sigma/\sigma_0)$. Also shown at the right are the values of the measured initial magnetization of the iron samples ($4\pi M$) in the bias fields used ((200 ÷ 300) Oe), and the accepted value of the saturation magnetization ($4\pi M_s$) at 20 °C.

Figure 5.4 shows Graham's measurements [5.5] on the demagnetization of Silectron (grain-oriented 3 wt % Si-97 % Fe), also a bcc structure. These measurements were made on tape-wound cores.

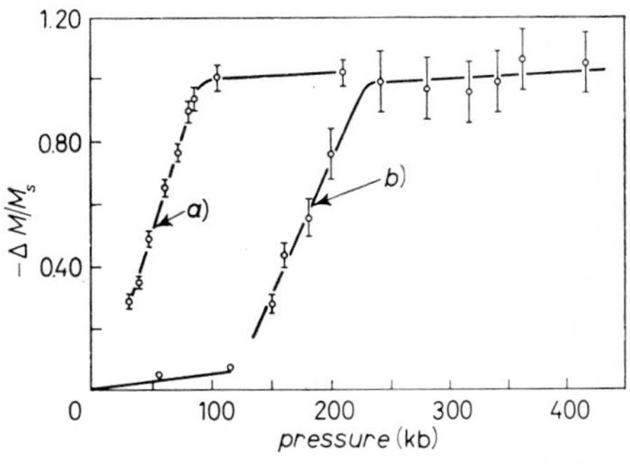

Fig. 5.4. – Shock-induced relative demagnetization of *a*) Invar (36% Ni-64% Fe) and *b*) Silectron (3% Ni-97% Fe). Ref. R. A. GRAHAM: *Journ. Appl. Phys.*, **39**, 437 (1968).

At high pressures, both sets of measurements show a complete shock-induced demagnetization, within experimental uncertainty. As the final pressure is lowered toward the « 130-kbar » transition (130 kbar in iron, 145 kbar in Silectron [5.22], less demagnetization is observed. Pressure-volume data indicate that the transition may not be complete in this region in the shock experiments [5.1], and these magnetic studies indicate the same result. Below

130 kbar, the Silectron experiments show a small demagnetization. No such demagnetization is seen in the iron work within experimental uncertainty. Calculations based on static measurements of the pressure-induced shifts in the Curie temperature (negligible) and magnetization at zero temperature, together with an estimated temperature rise on shock compression, indicate that this demagnetization should not exceed 3% for iron or 6% for Silectron [5.9, 5.10], in agreement with the experimental results.

It can be shown that the decay of the demagnetization pulse after the passage of the shock wave through the iron sample is exponential, if the process is governed by the decay of eddy currents. The time constant of the decay is given by

$$\tau = \varkappa\mu(d/\pi)^2 , \tag{5.1}$$

where $\varkappa$ and μ are the conductivity and permeability in the compressed state, and d is the sample thickness. This effect provides a means of determining the conductivity from the demagnetization measurements. At pressures of 168, 420, and 920 kbar, several experiments were performed in which the sample thickness was varied. The observation of the predicted quadratic variation with thickness verified that the observed decay is, in fact, due to the eddy current decay. The conductivity has been measured at several other pressures, by the magnetic technique, and the results were shown by KEELER in Fig. 4.15 *b*). The data there were reduced to 300 °K by assuming the resistivity on the Hugoniot to be proportional to the temperature. The magnetic measurements of the conductivity are considerably less accurate than the direct measurements of the conductivity, also shown, but the two techniques appear to be in general agreement. Other dynamic [5.23] and static [5.24] measurements are also shown and are in agreement with this work as well.

It is interesting that, while the demagnetization measurements give the conductivity as a by-product, the conductivity measurements also show the demagnetization, as discussed in Sect. **4** [5.25-5.29]. Again the two techniques are in general agreement, except at pressures below 130 kbar. There the conductivity experiments show a small demagnetization, whereas the direct demagnetization measurements do not. The difference may be that the demagnetization measurements are made only in the first 50 ns after the sample is shock compressed, whereas the conductivity work shows a slow demagnetization over a longer time.

In general, all of the shock work on equation of state, conductivity, and magnetic properties shows that the high-pressure state of iron existing in dynamic experiments is the same as that in static experiments.

5·4. – Second-order transitions.

Figure 5.4 also shows measurements of the demagnetization of Invar (36 % Ni-64 % Fe) tape-wound cores performed by GRAHAM [5.5]. The straight-line portion indicates a value $(\mathrm{d}\ln M_s/\mathrm{d}P)$ of $-1.3\cdot10^{-2}\,\text{kbar}^{-1}$, in good agreement with the statically measured value of $-1.1\cdot10^{-2}\,\text{kbar}^{-1}$ [5.9]. It appears that roughly half of the effect comes from a pressure-induced lowering of the Curie temperature and half from a lowering of the zero-temperature magnetization.

WAYNE [5.13] has made detailed studies of the demagnetization of 31.4 % Ni-Fe using both static and shock techniques. In hydrostatic measurements he found $(\mathrm{d}\ln\sigma_s/\mathrm{d}P) = -3.5\cdot10^{-2}\,\text{kbar}^{-1}$ at low pressures, and $-3.7\cdot10^{-2}\,\text{kbar}^{-1}$ at higher pressures. The shock measurements yield values of $-2.8\cdot10^{-2}$ to $-3.3\cdot10^{-2}\,\text{kbar}^{-1}$, depending on how corrections are made for unloading due to lateral rarefaction waves from the edges of the tape-wound cores.

The essential agreement between the hydrostatic measurements of $(\mathrm{d}\ln\sigma_s/\mathrm{d}P)$ and the shock-wave measurements seems to make it clear that the shock-induced demagnetization in the fcc Ni-Fe alloys is due to the pressure-induced shifts in the Curie temperature and in the zero-temperature magnetization.

5·5. – Magnetic anisotropy.

Figure 5.5 shows the shock-induced demagnetization of nickel ferrite [5.16] and demonstrates the induced anisotropy effect. The solid lines indicate the demagnetization predicted on the basis of shock heating and compression, and an estimate of the initial magnetization [5.30]. One calculation was made assuming the Néel temperature to be independent of compression, while the other was made on the assumption that the Néel temperature increases with compression at a rate determined in static measurements [5.31].

The observed, virtually total demagnetization results from the development of an induced magnetic anisotropy, which makes the shock propagation direction an easy direction of magnetization. It was these results on nickel ferrite that first brought the induced-anisotropy effect clearly to light.

Magnetic anisotropy [5.32-5.35] in ferrites arises from the interaction of the nonspherical charge distribution of the transition-metal ions and the nonspherical environment of their lattice sites, together with the well-known spin-orbit interaction coupling the spin angular momentum with the orbital angular momentum and orientation of the charge distribution. In cubic materials, this anisotropy energy is usually written

$$W_{\text{cub anis}} = -\frac{1}{2}K_1\left(\alpha_x^4+\alpha_y^4+\alpha_z^4-\frac{5}{3}\right)+K_2\left(\alpha_x^2\alpha_y^2\alpha_z^2-\frac{1}{105}\right), \tag{5.2}$$

where the α's are direction cosines of the magnetization in a co-ordinate system parallel to the cubic crystalline unit cell axes. The first term in eq. (5.2) is sometimes written in the equivalent form

$$+K_1(\alpha_x^2\alpha_z^2+\alpha_x^2\alpha_y^2+\alpha_y^2\alpha_z^2-\tfrac{1}{5}).$$

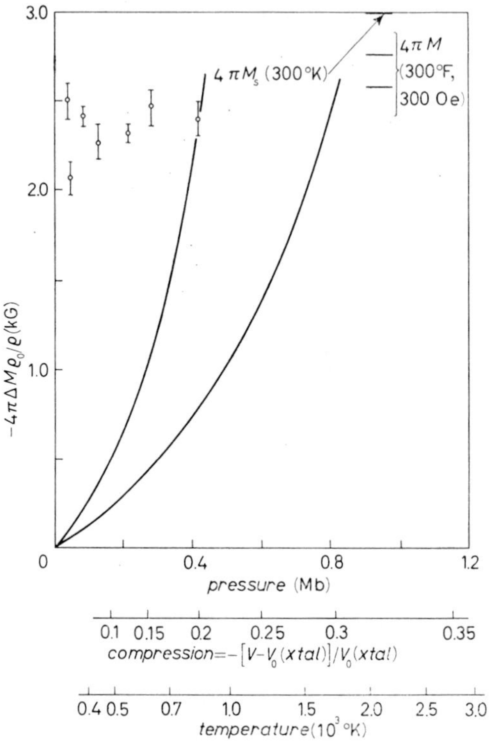

Fig. 5.5. – Shock-induced demagnetization of nickel ferrite, $-(4\pi\Delta M)\varrho_0/\varrho=-(4\pi M_0)\cdot(\Delta\sigma/\sigma_0)$. At the right are indicated the saturation magnetization ($4\pi M_s$) and the estimated magnetization at the bias field employed, 300 Oe. The curves show the theoretical demagnetization predicted from shock compression and heating alone. The upper curve is based on the assumption that the Néel temperature is constant at 590 °C, the lower curve is based on an increase in Néel temperature of 1.16 °C kbar^{-1}.

We shall be concerned with a much larger shock-induced axial anisotropy and shall, henceforth, ignore this cubic anistropy. Its effect on our problem is only in determining the shape of the initial magnetization curve, and we may account for this separately. Since we shall be discussing polycrystalline materials, we would have to average eq. (5.2) over all crystal orientations.

The ordinary magnetoelastic energy is written

$$(5.3)\qquad W_{\text{mag el}} = b_1\left[\alpha_x^2 e_{xx} + \alpha_y^2 e_{yy} + \alpha_z^2 e_{zz} - \tfrac{1}{3}(e_{xx}+e_{yy}+e_{zz})\right] + \\ + b_2(\alpha_x\,\alpha_y\,e_{xy} + \alpha_y\,\alpha_z\,e_{yz} + \alpha_x\,\alpha_z\,e_{xz})\,.$$

To this we must add the elastic energy

$$(5.4)\qquad W_{\text{el}} = \tfrac{1}{2}\,C_{11}(e_{xx}^2 + e_{yy}^2 + e_{zz}^2) + \tfrac{1}{2}\,C_{44}(e_{xy}^2 + e_{yz}^2 + e_{xz}^2) + \\ + C_{12}(e_{yy}\,e_{zz} + e_{xx}\,e_{zz} + e_{xx}\,e_{yy})\,.$$

Minimizing the sum of eqs. (5.3) and (5.4) yields the usual striction coefficients, describing the (100) and (111) axis strictions existing when the saturation magnetization lies along these axes,

$$(5.5)\qquad \lambda_{100} = -\frac{2}{3}\,\frac{b_1}{C_{11}-C_{12}}\,,$$

$$(5.6)\qquad \lambda_{111} = -\frac{1}{3}\,\frac{b_2}{C_{44}}\,.$$

For a sintered ceramic, such as the ferrites used in these experiments, we must then average these expressions over all orientations, with the constraints that the strains are equal in all of the crystallites. The result is

$$(5.7)\qquad \lambda = \frac{2}{5}\left(\frac{C_{11}-C_{12}}{\overline{C}_{11}-\overline{C}_{12}}\right)\lambda_{100} + \frac{3}{5}\left(\frac{C_{44}}{\overline{C}_{44}}\right)\lambda_{111}\,,$$

where

$$(5.8)\qquad \overline{C}_{11} = C_{11} - \frac{2(C_{11}-C_{12}) - 4C_{44}}{5}\,,$$

$$(5.9)\qquad \overline{C}_{44} = \tfrac{3}{5}C_{44} + \tfrac{1}{5}(C_{11}-C_{12})\,,$$

$$(5.10)\qquad \overline{C}_{11} - \overline{C}_{12} = 2\overline{C}_{44}\,.$$

Contrast eq. (5.7) with the more commonly quoted result for a powder

$$(5.11)\qquad \lambda = \tfrac{2}{5}\,\lambda_{100} + \tfrac{3}{5}\,\lambda_{111}\,.$$

The difference is that eq. (5.11) is derived under the assumption that the striction of each crystallite is independent from that of its neighbors.

On the assumption that the strains are equal in all of the crystallites forming our ceramic, and that the properties of the ceramic are isotropic, we may write

$$W_{\text{anis}} = K_1' \sin^2\theta + K_2' \sin^4\theta \,, \tag{5.12}$$

where θ is the angle between the axial anisotropy and the magnetization, and

$$K_1' = -(\tfrac{2}{5}b_1 + \tfrac{3}{5}b_2)e \,. \tag{5.13}$$

The strain has been divided into an isotropic component ($\Delta V/V_0$) and an axial component (e).

In the simple elastic-plastic theory [5.14], $e = -\sigma_{\text{HEL}}/\bar{C}_{11}$, where σ_{HEL} is the longitudinal stress at the Hugoniot elastic limit. Hugoniot elastic limits between 40 and 80 kbar have been measured on samples of nickel ferrite; we take a mean value of 50 kbar for both nickel ferrite and YIG. The published values of the ordinary magnetoelastic coefficients [5.15, 5.36], then yield $K_1' = 1.07 \cdot 10^6$ erg/cm^3 for nickel ferrite and $0.102 \cdot 10^6$ erg/cm^3 for YIG. The inclusion of terms quadratic in the strain [5.36] reduces the value of K_1' to $0.094 \cdot 10^6$ erg/cm^3 for YIG. Measurements of the higher-order coefficients do not appear to have been made for nickel ferrite. These energies must be compared with the energy resulting from the application of the bias field

$$W_{\text{bias}} = -\boldsymbol{M} \cdot \boldsymbol{H} \,. \tag{5.14}$$

This energy is typically $0.05 \cdot 10^6$ erg/cm^3 in the bias fields used. The orientation of the magnetization will be determined by the competition between W_{anis} and W_{bias}, and for nickel ferrite, it is clear that the anisotropy term will be dominant. Thus, the observed transverse demagnetization is accounted for.

Since the applied bias-field energy and anisotropy energies are comparable for YIG, this material offered the possibility of investigating the induced anisotropy in some detail by varying the bias field in a series of shots. Ordinary magnetoelastic theory predicts $K_2' = 0$, in which case the theory predicts the transverse magnetization to be proportional to the bias field, the line «axial anisotropy» in Fig. 5.6. Figure 5.7 shows the experimental transverse magnetization under shock compression *vs.* bias field [5.17]. The values were calculated from the measured change in magnetization and an assumed initial magnetization [5.37]. Although the data are linear in the bias field, the experimental lines do not go through the origin, particularly at the higher pressures! In previous publications [5.6, 5.17], this discrepancy was accounted for by invoking a large value for K_2' at high pressure, even though all of the known higher-order magnetoelastic terms contribute only to K_1'. These include the terms quadratic in the direction cosines of the magnetization and in the strain [5.36], and terms quartic in the direction cosines of the magnetization and linear in the strain [5.38].

The correct explanation of the demagnetization of YIG comes from a consideration of some recent experiments of WAYNE, SAMARA, and LEFEVER [5.39]. They observed a partial demagnetization of biased samples of YIG and nickel ferrite under ***hydrostatic*** compression. The materials were the same as those used in the shock work. They explained their results by noting that the ceramics are slightly porous (2% in YIG, 5% in nickel ferrite), and that even under hydrostatic conditions, nonhydrostatic stresses will be set up in individual crystallites in the vicinity of voids. What they were observing was a «random anisotropy», in which different parts of the sample developed axial anisotropies randomly oriented.

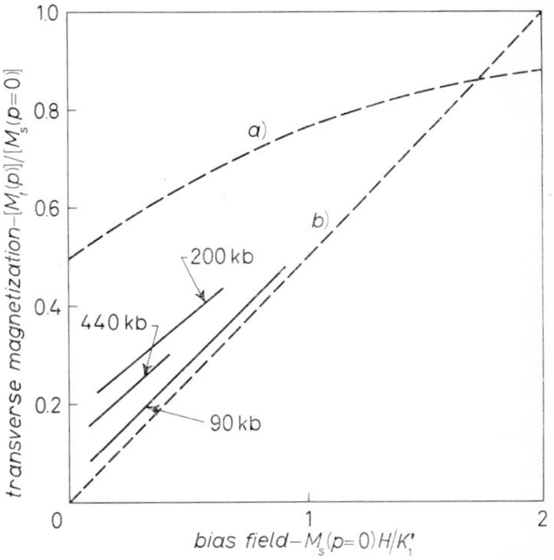

Fig. 5.6. – Transverse magnetization *vs.* bias field for two models of induced anisotropy *a*) random, *b*) axial. Also shown are the experimental data (solid lines)—see text and Fig. 5.7.

It seems clear that this must also apply to the shock work, and on Fig. 5.6 we show the predictions of a random anisotropy model. Note the interesting feature that even in small bias fields, there is only a 50% demagnetization. The solid lines plotted in Fig. 5.6 show the experimental data. Values of $K_1'(P)$ were selected to keep the lines between the theoretical lines and more or less parallel to them; K_2' was assumed to be zero. The experimental results show that there is a mixture of axial and random anisotropy, with the ran-

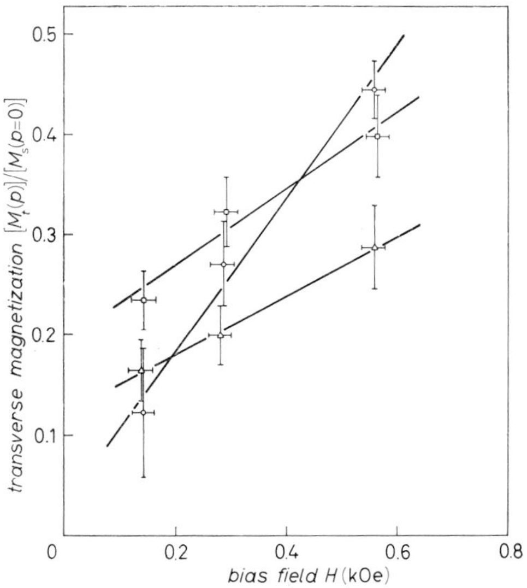

Fig. 5.7. – Transverse magnetization of shock-compressed YIG ($4\pi M_t$) *vs.* bias field. ○ 0.09 Mbar, □ 0.20 Mbar, △ 0.44 Mbar.

dom component growing with pressure. However, simple axial anisotropy remains as the dominant effect.

It should be noted that a pressure-induced increase in Néel temperature [5.40] almost exactly cancels the effect of the temperature rise of the sample due to the shock compression. Thus, we have assumed the saturation magnetization to be constant in constructing the Figures. Actually, there is probably a 5 to 10 % reduction in saturation magnetization for the 440-kbar shots, and the 440-kbar line in Fig. 5.6 could be shifted upward to lie on top of the 200 kbar line.

The values of K_1' used in the data reduction for Fig. 5.6 were $0.09\cdot10^6$, $0.14\cdot10^6$, and $0.21\cdot10^6$ erg/cm³, respectively, for pressures of 90, 200, and 440 kbar. The lower-pressure value is in good agreement with our estimated value of $0.094\cdot10^6$ erg/cm³. However, it appears that either the magnetoelastic coefficients or the yield strength are increasing at higher pressures. Experiments on aluminum oxide ceramics [5.20] show that the yield strength is relatively independent of the pressure. The values of K_1' above then indicate that the magnetoelastic coefficients roughly double for a 17 % volume compression. Such a rapid variation of the magnetoelastic energy is not unreasonable when it is recalled [5.34] that the contributions to this energy from the two magnetic sublattices in ferromagnetic YIG are comparable but opposite in sign. Thus, a rather small change in one of the contributions could have a large effect on the net magnetoelastic energy. Since the Fe-O tetrahedral and octahedral bonds are of different length and character, it seems reasonable to expect these bonds to have different compressibilities. Thus, the surroundings of the magnetic ions on the two sublattices could be compressed by different amounts, where each sublattice is made up of sites of a particular co-ordination.

It appears that the shock-induced anisotropy is completely consistent with statically measured properties.

* * *

I am grateful to R. A. Graham, G. A. Samara, and R. C. Wayne of the Sandia Laboratories for discussing their work with me prior to its pubblication. I have made free use of their results in preparing this Section.

REFERENCES

[5.1] D. Bancroft, E. L. Peterson and S. Minshall: *Journ. Appl. Phys.*, **27**, 557 (1956).

[5.2] J. C. Jamieson and A. W. Lawson: *Journ. Appl. Phys.*, **33**, 776 (1962);

T. Takahashi and W. A. Bassett: *Science*, **145**, 483 (1964); R. L. Clendenen and H. G. Drickamer: *Journ. Phys. Chem. Solids*, **25**, 865 (1964).
[5.3] D. N. Pipkorn, C. K. Edge, P. Debranner, G. de Pasquali, H. G. Drickamer and H. Fraunfelder: *Phys. Rev.*, **135**, A 1604 (1964).
[5.4] R. W. Kulterman, F. W. Nelson and W. B. Benedick: *Journ. Appl. Phys.* **29**, 500 (1958).
[5.5] R. A. Graham: *Journ. Appl. Phys.*, **39**, 437 (1968).
[5.6] E. B. Royce: in *Behaviour of Dense Media Under High Dynamic Pressures* (New York, 1968), p. 419. Note that Fig. 4, 6 and 7 in this reference are either incorrect or were significantly altered by the editor; they should be disregarded.
[5.7] J. M. Leger, C. Susse and B. Vodar: *Solid State Comm.*, **5**, 755 (1967).
[5.8] L. Patrick: *Phys. Rev.*, **93**, 384 (1954).
[5.9] J. S. Kouvel and R. H. Wilson: *Journ. Appl. Phys.*, **32**, 435 (1961).
[5.10] J. S. Kouvel: in *Solids Under Pressure*, edited by W. Paul and D. M. Warschauer, Chap. 10 (New York, 1963).
[5.11] J. E. Besançon, J. L. Champetier, Y. Leclanché, J. Vedel and J. P. Plantevin: in *Megagauss Magnetic Field Generation by Explosives and Related Experiments*, edited by H. Knoepfel and F. Herlach (Brussels, 1966), p. 331.
[5.12] I. G. Clator and M. F. Rose: *Brit. Journ. Appl. Phys.*, **18**, 853 (1967).
[5.13] R. C. Wayne: *Journ. Appl. Phys.*, **40**, 15 (1969).
[5.14] D. S. Wood: *Journ. Appl. Mech.*, **19**, 521 (1952); L. W. Morland: *Phil. Trans. Roy. Soc. London*, A **251**, 341 (1959); G. R. Fowles: *Journ. Appl. Phys.*, **32**, 1475 (1961); J. W. Taylor and M. H. Rice: *Journ. Appl. Phys.*, **34**, 364 (1963); C. D. Lundergan and W. Herrmann: *Journ. Appl. Phys.*, **34**, 2046 (1963).
[5.15] A. B. Smith and R. V. Jones: *Journ. Appl. Phys.*, **34**, 1283 (1963); **37**, 1001 (1966); A. E. Clark, B. DeSavage, W. Coleman, E. R. Callen and H. B. Callen: *Journ. Appl. Phys.*, **34**, 1296 (1963); *Phys. Rev.*, **130**, 1735 (1963); K. Andres and B. Lüthi: *Phys. Chem. Solids*, **24**, 584 (1963).
[5.16] E. B. Royce: *Journ. Appl. Phys.*, **37**, 4066 (1966).
[5.17] J. W. Shaner and E. B. Royce: *Journ. Appl. Phys.*, **39**, 492 (1968).
[5.18] G. E. Seay, R. A. Graham, R. C. Wayne and L. D. Wright: *Bull. Am. Phys. Soc.*, **12**, 1129 (1967).
[5.19] D. R. Curran: *Journ. Appl. Phys.*, **32**, 1811 (1961).
[5.19*a*] R. A. Graham, D. H. Anderson and J. R. Holland: *Journ. Appl. Phys.*, **38**, 223 (1967).
[5.20] T. J. Ahrens, W. H. Gust and E. B. Royce: *Journ. Appl. Phys.*, **39**, 4610 (1968).
[5.21] Data reported by F. Birch: in *Handbook of Physical Constants*, edited by S. P. Clark jr. (New York, 1966), p. 154.
[5.22] E. G. Zukas, C. M. Fowler, F. S. Minshall and J. O'Rourke: *Trans. AIME*, **222**, 746 (1963).
[5.23] P. J. A. Fuller and J. H. Price: *Nature*, **193**, 262 (1962).
[5.24] A. I. Balchan and H. G. Drickamer: *Rev. Sci. Instr.*, **32**, 308 (1961).
[5.25] R. E. Duff, W. H. Gust, E. B. Royce, M. Ross, A. C. Mitchell, R. N. Keeler and W. G. Hoover: in *Behaviour of Dense Media Under High Dynamic Pressures* (New York, 1968), p. 397.
[5.26] A. C. Mitchell and R. N. Keeler: *Bull. Am. Phys. Soc.*, **12**, 1128 (1967).
[5.27] R. N. Keeler and A. C. Mitchell: *Solid State Comm.*, **7**, 271 (1969).

[5.28] J. Y. Wong, R. K. Linde and P. S. De Carli: *Nature*, **219**, 713 (1968).
[5.29] J. Y. Wong: *Journ. Appl. Phys.*, **40**, 1789 (1969).
[5.30] R. Pauthenet: *Ann. de Phys.*, (Ser. 12), **7**, 710 (1952).
[5.31] C. L. Foiles and C. T. Tomizuka: *Journ. Appl. Phys.*, **36**, 3839 (1965).
[5.32] C. Kittel: *Solid State Physics*, Chap. 4 and 15 (New York, 1956).
[5.33] J. Kanamori: in *Magnetism I*, edited by G. T. Rado and H. Suhl, Chap. 4 (New York, 1963).
[5.34] E. Callen and H. B. Callen: *Phys. Rev.*, **139**, A 455 (1965); **130**, 1735 (1963); **129**, 578 (1963).
[5.35] L. C. Bartel: *Journ. Appl. Phys.*, **40**, 661 (1969).
[5.36] D. E. Eastman: *Phys. Rev.*, **148**, 530 (1966).
[5.37] R. Pauthenet: *Ann. de Phys.*, (Ser. 13), **3**, 424 (1956); E. E. Anderson: *Phys. Rev.*, **134**, A 1581 (1964).
[5.38] R. Becker and W. Döring: *Ferromagnetismus*, Chap. III, Sect. **11** (Berlin, 1939).
[5.39] R. C. Wayne, G. A. Samara and R. A. Lefever: *Journ. Appl. Phys.*, **41**, 633 (1970).
[5.40] D. Bloch, F. Chaissé and R. Pauthenet: *Journ. Appl. Phys.*, **37**, 1401 (1966).

VI. – New Experimental Developments in Shock Wave Physics.

R. N. Keeler

In preceding lectures we have described the application of rather elementary experimental techniques to the measurement of material properties. This lecture covers a second generation of shock-wave experiments, in which somewhat more sophisticated techniques are applied to obtain the properties of matter at high pressure. In view of the considerable difficulties encountered in carrying out the experiments described in this lecture, it is obvious that it will be some time before the third generation of experiments, such as cyclotron resonance or microwave Hall measurements, will be possible.

6·1. – X-ray diffraction studies of solids under shock compression.

6·1.1. *Background.* – It was pointed out at the conclusion of one of the preceding lectures that there is a lack of understanding of the disruptive forces within a shock front, and the effect of shock-wave passage on crystalline order. This kind of detailed knowledge of the microscopic state of matter in and behind a shock front will be necessary before it will be possible to apply more sophisticated techniques and analyses to the study of shock-compressed solids. The situation is even more complicated when conditions behind the shock front are such that a polymorphic phase transition can occur. Although it is possible to infer the occurrence of a shock-wave-driven phase transition from discontinuities on the Hugoniot [6.1], breaks on the U_s-U_p curve [6.2], multiple-wave structure [6.3], discontinuities in the electrical conductivity [6.4], or in optical properties [6.5], there is as yet no direct and positive way to identify the new high-pressure phases formed during the very brief time of passage of a shock wave. In certain instances, it has been possible to recover and identify crystallographically a metastable phase formed as a result of the shock wave [6.6]. Unfortunately, the « quenched » phase need not necessarily be the phase stable at high pressure [6.7]. In the absence of a more suitable alternative, attempts are usually made to correlate dynamic observations with static high-pressure X-ray results where the crystallographic form of the high-pressure phase has been identified *in situ*.

6·1.2. *Static studies.* – The requirements for a system with which X-ray diffraction measurements could be carried out are the same as for systems presently used to carry out other measurements in shock-wave experiments. The measurements must be completed in fractions of a microsecond, the sensing element must penetrate the shocked medium with no hydrodynamic disturbance (or at least, the effect of the disturbance must be understood), and the nonexpendable diagnostics must be remote, or protected from the destructive effects of the high-explosive or gun system used to drive the shock wave.

Flash X-ray radiography or densitometry has been used as a diagnostic technique for many years in shock-wave experiments. A high-intensity pulsed X-ray source is directed at a shocked medium. The transmitted X rays are registered on film and give a densitometric « map » of the region traversed by the X rays. The technique has been used to study spallation and general metallic behavior under shock loading [6.8], hypervelocity particles in flight [6.9], the equation of state of shocked dense media [6.10], and phase-transition kinetics [6.11, 6.12].

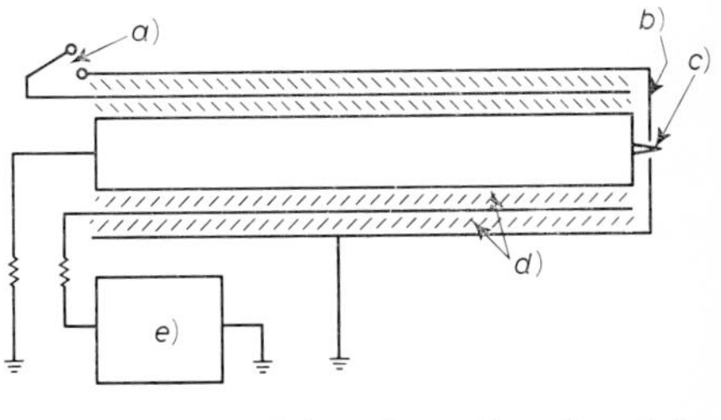

Fig. 6.1. – Blumlein schematic. *a*) switch; *b*) cathode; *c*) anode; *d*) dielectric; *e*) 50 kV power supply.

A high-intensity, pulsed X-ray source in common use at the Lawrence Radiation Laboratory for such experiments is the Blumlein X-ray generator [6.13, 6.14]. This device, shown in Fig. 6.1, consists of three essential components, which are the evacuated anode-cathode gap, the three-conductor coaxial transmission line, and the switch. The anode is conical in shape, is connected to the central conductor, and projects through a sharp-edged, circular opening in the cathode plate. This plate, in turn, is connected to the outer conductor. A floating middle conductor, separated by dielectric from the outer and inner conductors, is negatively charged to (40 ÷ 50) kV by a power supply. These generators were designed to give an impedance of 1 ohm although no experiments have actually been performed to confirm this value. The low-impedance design requires that the switch have an impedance which is small relative to 1 ohm. To date, this switch has been a detonator which breaks down the dielectric between the outer and the middle conductors.

In operation, the Blumlein is charged to operating potential. This situation is shown in Fig. 6.2 *a*). A pulse from a capacitance-discharge unit (CDU) is sent to the detonator which affects the actual switching, Fig. 6.2 *b*). The

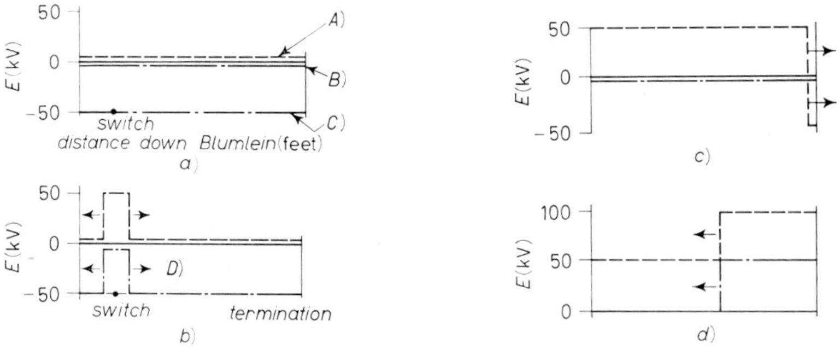

Fig. 6.2. – Discharge of Blumlein. *a*) Blumlein charged: *A*) anode (aluminum tube, grounded), *B*) cathode (outer aluminum layer, grounded), *C*) internal aluminum layer connected to power supply. *b*) Switch closed, voltage removal wave commences on internal aluminum layer; *D*) voltage removal wave on internal aluminum layer. *c*) Instant before reflection. *d*) Several nanoseconds after reflection.

Blumlein circuit is a voltage-doubling circuit so that a peak voltage substantially above the charging potential is reached during the discharge, as shown in Fig. 6.2 *c*), *d*). Electrons undergo field emission from the sharp cathode edge, are accelerated across the gap by the high potential, and lose their energy in the form of heat and X rays upon striking the anode. Peak electron currents probably exceed 10 kA. After a short interval of approximately 30 ns, a plasma forms from the material vaporized at the anode. This shorts out the gap and terminates the X-ray pulse. Although originally designed for radiographic studies, it has been shown that these devices have sufficient intensity to be used as X-ray sources for pulsed X-ray diffraction experiments [6.15].

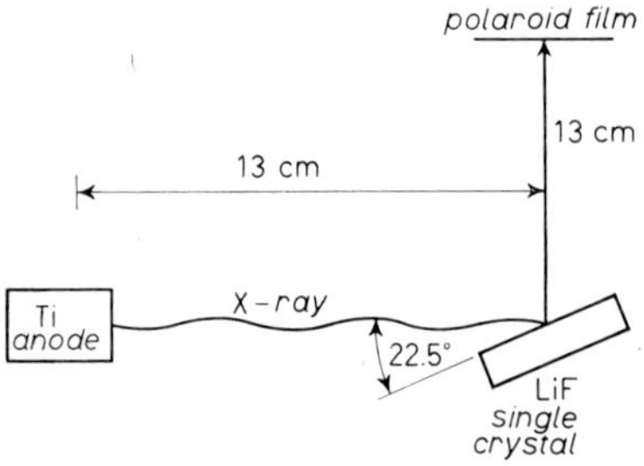

Fig. 6.3. – Bragg reflection experiment.

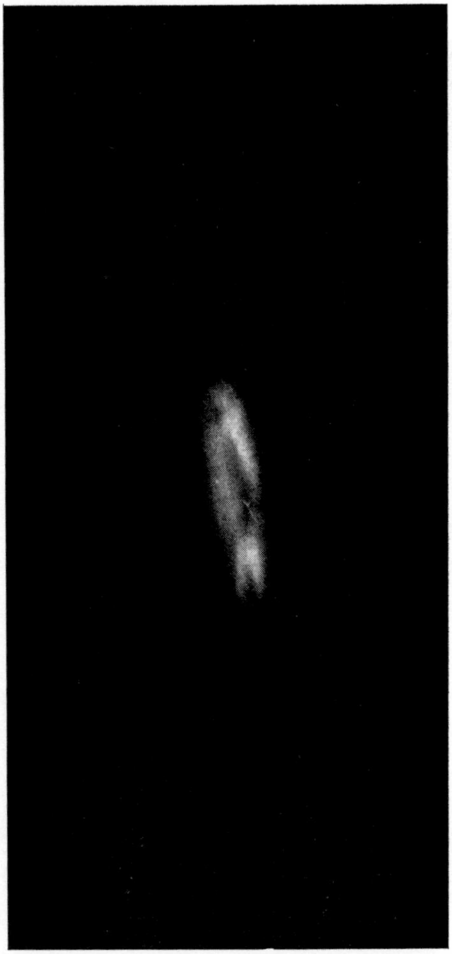

Fig. 6.4. – Bragg reflection, film record.

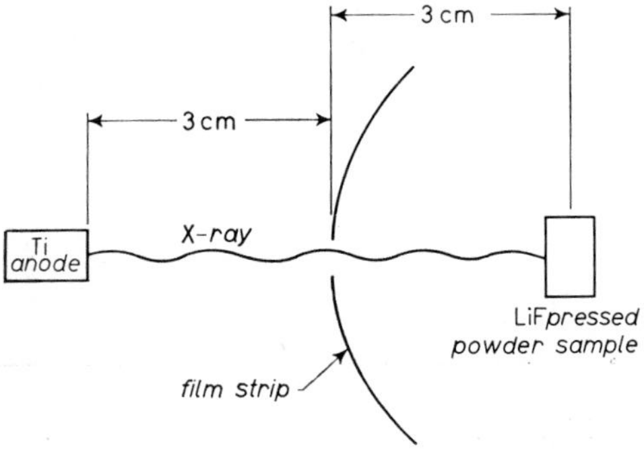

Fig. 6.5. – Debye-Scherrer experiment.

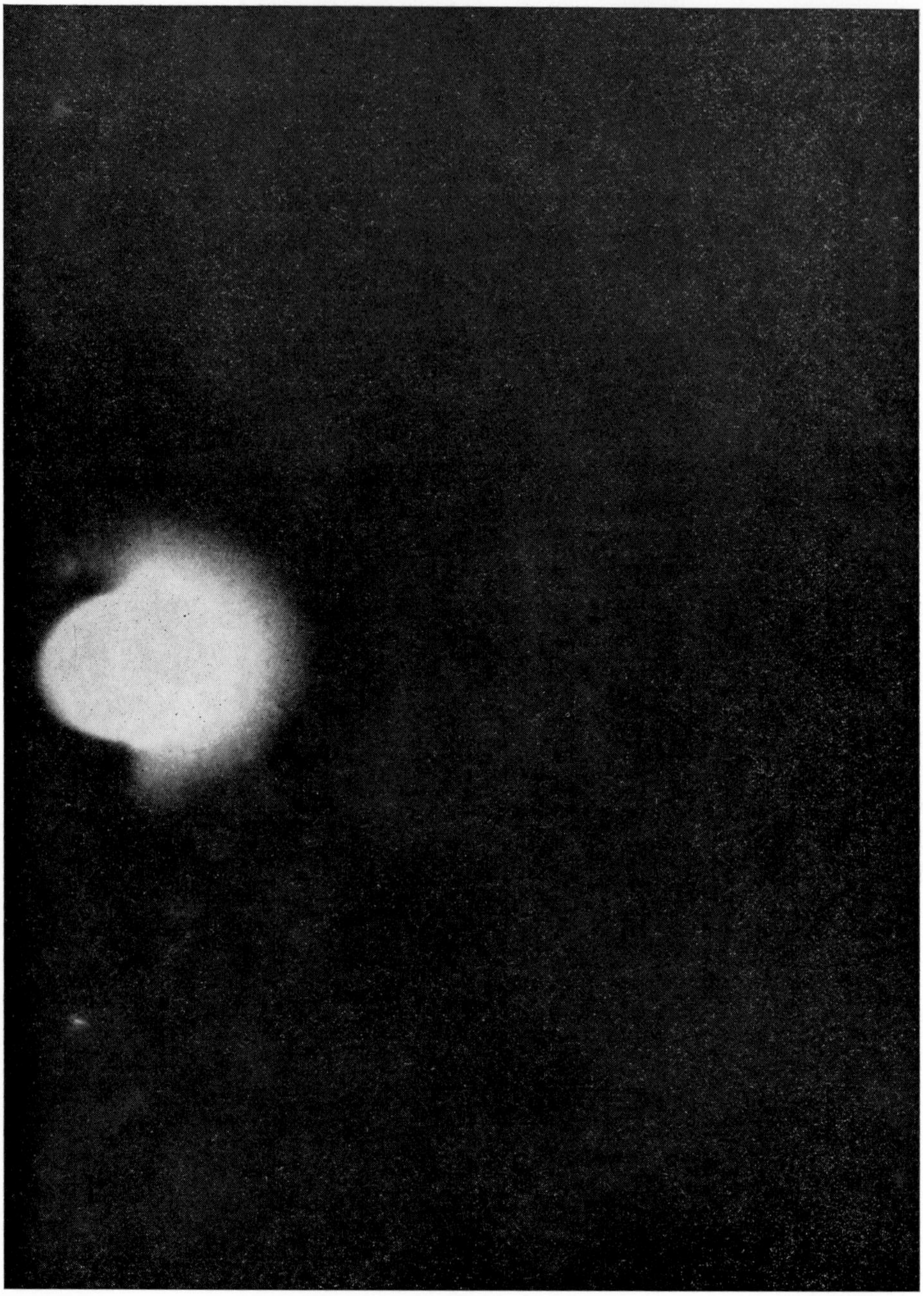

Fig. 6.6. – Recorded powder pattern.

Diffraction from a single crystal was demonstrated with the experimental arrangement shown in Fig. 6.3. A single X-ray pulse from a Blumlein equipped with a copper anode was directed at a single crystal of NaCl. This crystal was set at the angle appropriate for diffraction of 1.54 Å radiation from the 220 reflection. The event was recorded on film and is shown in Fig. 6.4. Even starting with a single crystal sample in a shock-wave experiment, it is doubtful that the material behind the shock wave remains as a single crystal, particularly if a phase transition is involved. It was necessary, therefore, to develop a diffraction technique which could utilize powdered samples.

A Debye-Scherrer experiment in which the sample was a flat pellet of pressed LiF powder was next carried out. The arrangement for this experiment is shown in Fig. 6.5. Results were again successful and are shown in Fig. 6.6. In this photograph, both the 111 and the 200 K_α lines are clearly seen. The significance of these two experiments is that an X-ray source has been found which is suitable for X-ray diffraction studies of either single crystal or powder samples during extremely short time intervals. While it is immediately obvious that the study of materials under shock compression can then be undertaken, it should be stressed that this is but one possible use to which this new tool can be put. Indeed, it should now be possible to perform practically all types of conventional diffraction experiments during this short ($\sim$ 30 ns) time interval.

6·1.3. *Dynamic studies.* – The destructive nature of the techniques used to generate shock waves and the timing requirements pose special problems in carrying out this kind of experiment. One of the primary constraints stems from the requirement that the Blumlein and associated diagnostics be shielded from the detrimental effects of the explosive or other shock-wave generator. It is clear that use of high explosives would lead to almost certain destruction of film or detectors, and severely damage the Blumlein. For this reason, the initial experiments were carried out with a magnetically driven flying plate.

The experimental assembly using the magnetically driven flying plate is shown in Fig. 6.7. The plate accelerator is pulsed with an intense current surge, which flows through accelerator and plate but in opposite directions. The magnetic pressure generated accelerates the plate, which shears off from the accelerator, breaking the circuit. Velocities up to 1.0 mm/μs can be obtained,

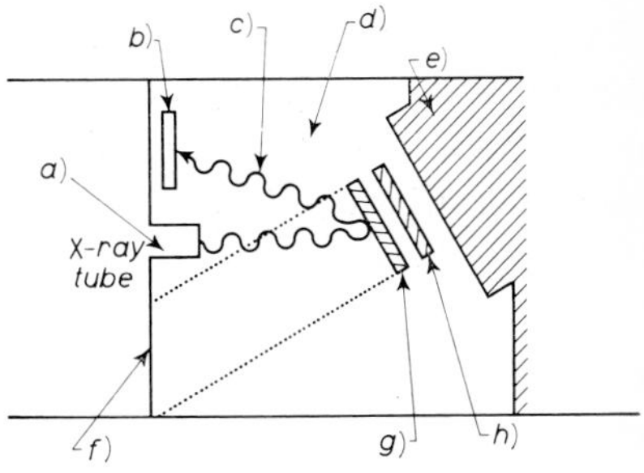

Fig. 6.7. – Dynamic X-ray experimental configuration. *a*) collimator; *b*) detector; *c*) diffracted beam; *d*) vacuum chamber; *e*) flying-plate accelerator; *f*) impact area; *g*) sample; *h*) aluminum plate.

but the velocities achieved in this work were 0.25 mm/μs, to prevent excessive deformation and heating. This technique, of course, is limited to rather low pressures. The advantage of using this technique is that the entire process is carried out in a vacuum, and, except for the area directly in line with the accelerating flying plate, there is essentially no damage. A block diagram of the experiment, set up for use with detectors, is shown in Fig. 6.8.

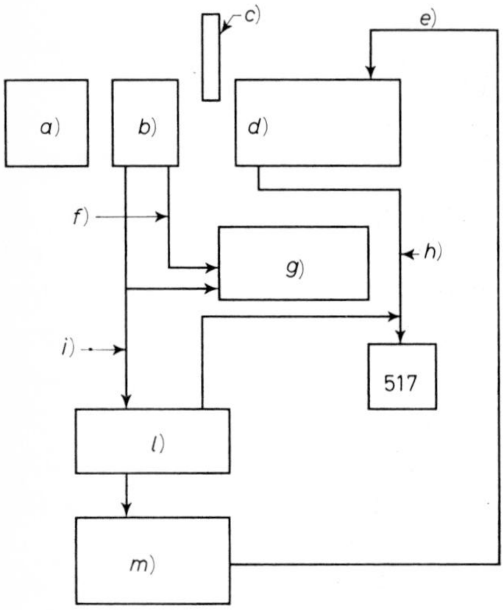

Fig. 6.8. – Block diagram, dynamic X-ray experiment. *a*) flying plate assembly; *b*) sample; *c*) detector; *d*) X-ray tube; *e*) switch; *f*) timing pin; *g*) raster oscilloscope; *h*) X-ray voltage monitor; *i*) trigger pin; *l*) pulse shaper; *m*) capacitive discharge unit.

The two different detector schemes employed in these experiments were X-ray sensitive films and scintillation detectors. The arrangement for both types of experiment is identical with that shown in Fig. 6.7. A total of eight experiments was carried out with pressed LiF in which photographic identification of the shift of the 422 line was attempted. For the film experiments, a normal pressure picture was recorded, in which part of the film was protected from exposure. A second (double) exposure was then carried out with the sample under shock compression. It was expected that the effect of a pressure shift would be especially noticed at the single-double exposure boundary. Seven of these experiments were unsuccessful because of timing problems and equipment malfunction. On the shot where timing was within tolerance, a slight shift of the 422 line was observed at the single-double exposure boundary. Because of inadequate Blumlein intensity, however, this line of experimentation was discontinued, and effort shifted to scintillation detectors. Again, timing difficulties were encountered, but one shot was timed successfully, and showed the expected intensity enhancement at the detector set to record X rays diffracted from the compressed 422 planes, along with the corresponding diminution in intensity at the detector set to record X rays diffracted from uncompressed 422 planes. A more detailed discussion of this work can be found elsewhere [6.16].

6.1.4. *Future work.* – The major difficulty encountered with these experiments was the inability to discharge the Blumlein with sufficient timing accuracy to insure consistent synchronization with the shock-wave arrival at the face surface. Some idea of the precision required can be had when we consider the following relationship, governing the ratio between intensity diffracted from a sample thickness, δ, and the intensity diffracted from the total thickness.

$$(6.1)\qquad \frac{I(\delta)}{I(\infty)} = 1 - \exp\left[-\mu\delta\left(\frac{1}{\cos\gamma} - \frac{1}{\cos\gamma\cos 2\theta - \sin\gamma\sin 2\theta\cos\varphi}\right)\right].$$

Equation (6.1) refers to the geometry of Fig. 6.9, where μ is the linear absorption coefficient. When $\varphi = 90°$ and $\delta = 0.29$ mm, then 95 % of the diffraction comes from the top 0.3 mm of the sample. In terms of a shock-wave experiment, if the Blumlein were fired instantaneously when the shock wave was only 0.9 mm from the free surface, only 5 % of the diffraction pattern would be from the compressed solid behind the shock front. At a shock velocity of 3 mm/μs, the experimental window is only 100 ns wide if it is assumed that at the instant of shock breakout, the disintegration of the free surface eliminates the possibility of obtaining any further signal.

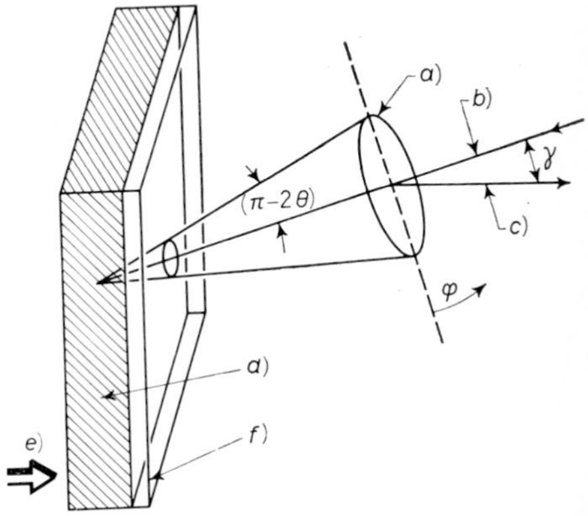

Fig. 6.9. – Geometry of X-ray diffraction at sample surface. *a*) diffraction cone; *b*) X-ray beam; *c*) surface normal; *d*) shock-compressed region; *e*) direction of shock; *f*) unshocked region.

For these series of experiments using detonators as a switch component, a jitter of ~100 ns was encountered in the actual Blumlein switching. An additional complication was met in the long turn-on time of the detonator switch (~3 μs). This necessitated timing the whole event from the instant that the flying plate reached a known point in space but before it impacted on the sample. This is extremely difficult to accomplish without timing uncertainties. Indeed, timing errors introduced by this problem were responsible in large measure for the rather large number of unsuccessful shots. The problems of Blumlein jitter as well as Blumlein turn-on time have recently been solved at the Lawrence Radiation Laboratory by the use of techniques described by GUENTHER [6.17].

6·2. – Stimulated Brillouin scattering as a diagnostic tool in shock-wave physics.

Brillouin scattering is the scattering of light waves from sound waves. In a normal fluid, there are always a number of thermally excited sound waves, whose wavelengths vary from a distance greater than the mean free path of a molecule in the fluid, to the size of the container. When the proper conditions for conservation of momentum and energy hold, light may be scattered off these fluctuations. Since fluctuations are traveling at acoustic velocity for the medium, the light will be Doppler-shifted either up or down, depending on the direction of movement of the sound waves.

In stimulated Brillouin scattering, the light intensity is so intense that the incident and scattered light beating against one another amplifies the acoustic waves from which they are scattered by electrostriction. When the gain of this amplification process exceeds the losses due to light absorption and sound-wave dissipation, a runaway situation occurs and a large fraction of the incident light is scattered from the greatly amplified sound waves. For example, when a giant pulse ruby laser is fired into a liquid, and the threshold power density is exceeded, as much as 30 % of the incoming light can be scattered back directly along its incoming path. The light is monochromatic, coherent, and is Doppler-shifted in frequency by an amount proportional to the velocity of sound in the medium [6.18]

$$\Delta\nu = \nu_0 \frac{2n v_s}{c} \sin\frac{\theta}{2}, \tag{6.2}$$

where $\Delta\nu$ is the difference in frequency between the incident and the scattered light, ν_0 is the frequency of the incident light, n the refractive index of the medium, v_s the velocity of sound in the medium, and θ the scattering angle. $\Delta\nu$ is also the frequency of the sound waves, or acoustic phonons. This formula holds for both ordinary and stimulated scattering. In stimulated scattering, gain is maximum in the back-scattered direction, the scattering angle is 180°, and eq. (6.2) becomes simply

$$\Delta\nu = \nu_0 \frac{2n v_s}{c}. \tag{6.3}$$

The threshold for stimulated Brillouin scattering is given by [6.19]

$$\frac{E_0^2}{8\pi} > \frac{2\varepsilon B}{\varrho(\mathrm{d}\varepsilon/\mathrm{d}p)^2 k_s k}(L_s^{-1}+L^{-1})^2, \tag{6.4}$$

where ε is the dielectric coefficient, ϱ the density, and $B = -v(\partial p/\partial v)_T$ the bulk modulus. k_s and k are the wave numbers of the acoustic and scattered light wave, respectively, and L_s and L are the decay lengths, or inverse absorption coefficients of the sound and light waves, respectively. Typical values for this threshold are power levels of 10^4 MW/cm^2, well within the capability of current laser technology.

In stimulated Brillouin scattering, we have a technique with which it is possible to make accurate ($\sim 0.5\%$) measurements of the velocity of sound in transparent materials in times less than 0.1 μs. The diagnostics for this measurement can be remote, and are recoverable. Such an experiment appears to be ideally suited for use in shock-wave applications, particularly since an accurate measurement of the velocity of sound in a shocked material has never been obtained, although some attempts have been made [6.20, 6.21].

The first stimulated Brillouin scattering experiment in a shock-compressed fluid has been reported previously [6.22]. This initial experiment was carried out with a rather crude laser arrangement. The oscillator laser was actually activated by a trigger laser which was used to provide timing by saturating a bleach filter in the oscillator cavity. This system was not particularly stable, but provided the first demonstration that the application was feasible, plus a semi-quantitative measurement of the velocity of sound in shock-compressed liquid acetone.

The main disadvantage of the system used was its instability. It has been shown that the use of optically active elements in the laser cavity such as Kerr cells or bleach filters is inferior to techniques using passive elements such as Pockels cells and mode selectors. It was also found that in the actual Brillouin scattering experiment, the back-scattered light which was Doppler-shifted by the amount given in eq. (6.3) returned to the amplifier and began the process known as multiple-stimulated Brillouin scattering, or multiple-returning, Doppler-shifted pulses. In the experiment cited, the light was focused behind the shock front into a sharp pressure gradient, which resulted from the use of a small high-explosive system to generate the shock pressures. This spectrally broadened the Brillouin scattered light, which returned to the oscillator amplifier system, and was amplified in several cavity modes, greatly degrading the accuracy obtained in the Fabry-Perot interferogram.

In view of these difficulties, an improved laser system was designed for use with a single-stage gas gun. As discussed in previous lectures, this provides a flat pressure profile behind the shock front, eliminating the troublesome pressure gradient encountered previously. A schematic of this system is shown in Fig. 6.10. This system utilizes a water-cooled (± 0.01 °C) oscillator for increased stability. The oscillator ruby is a ($2\frac{1}{2} \times \frac{1}{4}$)-in. Brewster cut ruby sealed in sapphire windows. The xenon flash lamps are helical to provide for a more uniform optical pumping. The oscillator is switched in by a Pockels cell, and mode selection is accomplished internally by two tilted-plate mode selectors. The oscillator is optically isolated from the returning laser light by a Faraday rotator. The returning Brillouin scattered light is taken out of the optical path by a polarizing prism, and directed to a group of Fabry-Perot interferometers, where spectral comparison is made between the outgoing and returning Brillouin light.

The initial step of operation is the firing of the light gas gun. As the projectile accelerates, it passes a light port, interrupting a light beam and initiating the pumping of the oscillator flash tubes. The Pockels cell is discharged by a signal from a trigger pin set in the sample. The laser discharges, and the light beam is amplified and focused behind the shock front where it stimulates Bril-

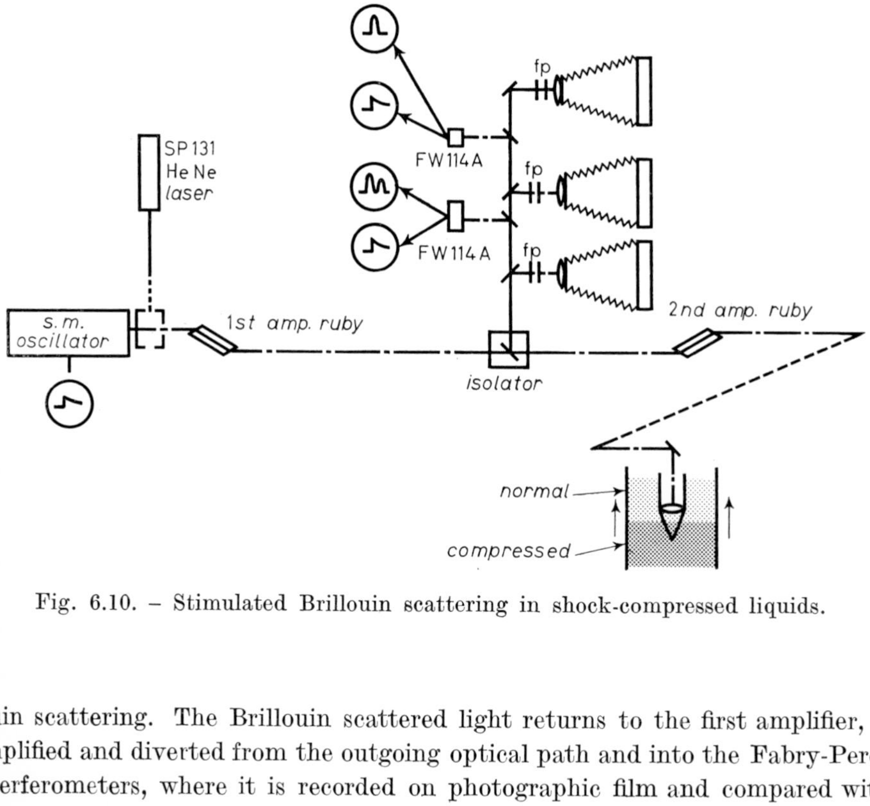

Fig. 6.10. – Stimulated Brillouin scattering in shock-compressed liquids.

louin scattering. The Brillouin scattered light returns to the first amplifier, is amplified and diverted from the outgoing optical path and into the Fabry-Perot interferometers, where it is recorded on photographic film and compared with the outgoing light.

The simple formula, eq. (6.3), does not apply in the case of stimulated Brillouin scattering experiments done in shocked media. The appropriate relationship is

$$\Delta\nu' = \frac{2\nu_0}{c}\left[n_2 v_s + (n_2 - n_1)\, U_s - n_2\, U_p\right], \tag{6.5}$$

with $\Delta\nu'$ the downward frequency shift in the dynamic experiment, and n_1 and n_2 the refractive indices of unshocked and shocked medium, respectively.

The total shift, therefore, is the sum of three shifts: the shift associated with the term $n_2 v_s$, the static Brillouin shift; the shift associated with the term $n_2 U_p$, which reflects the fact that the scattering region behind the shock front is moving in the laboratory frame of reference with velocity U_p, and the third term $(n_2 - n_1) U_s$, which is associated with the movement of an interface between regions of two different refractive indices.

Since Brillouin shifts are around $(2 \div 50)$ GHz and the single-mode laser linewidths are ~ 20 MHz, the anticipated accuracy of these measurements is about 0.5 %.

The measurement of sound velocity along the Hugoniot will permit a far more accurate determination of the equation of state than was available previously.

* * *

This Section was prepared in collaboration with Dr. Q. C. Johnson, Chemistry Department, University of California, Lawrence Radiation Laboratory, Livermore, Cal.

REFERENCES

[6.1] D. Bancroft, E. L. Peterson and S. Minshall: *Journ. Appl. Phys.*, **27**, 291 (1956).

[6.2] See the comprehensive discussion by R. G. McQueen: *Metallurgy at High Pressures and High Temperatures*, edited by K. A. Gschneidner jr., M. T. Hepworth and N. A. D. Parlee (New York, 1964), p. 74.

[6.3] D. B. Larson: *Journ. Appl. Phys.*, **38**, 1541 (1967).

[6.4] P. J. A. Fuller and J. H. Price: *Nature*, **193**, 262 (1962).

[6.5] S. B. Kormer, M. V. Sinitsyn, G. A. Kirillov and V. D. Urlin: *Žurn. Ėksp. Teor. Fiz.*, **48**, 1033 (1965); English translation in *Sov. Phys. JETP*, **21**, 689 (1965).

[6.6] P. S. De Carli and J. C. Jamieson: *Science*, **133**, 1821 (1961).

[6.7] F. P. Bundy and J. S. Kasper: *Science*, **139**, 340 (1963).

[6.8] A. S. Balchan: *Journ. Appl. Phys.*, **34**, 241 (1963).

[6.9] A. H. Jones, W. M. Isbell and C. J. Maiden: *Journ. Appl. Phys.*, **37**, 3493 (1966).

[6.10] J. Dapoigny, D. Kiefer and B. Vodar: *Compt. Rend.*, **238**, 215 (1954).

[6.11] B. R. Breed and D. Venable: *Journ. Appl. Phys.*, **39**, 3222 (1968).

[6.12] R. Schall: *Proceedings of the Third Congress on High Speed Photography* (London, 1957), p. 228.

[6.13] A. D. Blumlein: British patent No. 589127, June 12, 1947.

[6.14] R. A. Fitch and V. T. S. Howell: *Proc. Inst. Elec. Engrs.*, **111**, No. 4, 849 (1964).

[6.15] Q. Johnson, R. N. Keeler and J. W. Lyle: *Nature*, **213**, 1114 (1967).

[6.16] Q. Johnson, A. C. Mitchell, R. N. Keeler and L. Evans: *Trans. Am. Crystallog. Assoc.*, **5**, 133 (1969).
[6.17] A. H. Guenther and J. R. Beths: *Journ. Quant. Electr.*, QE-**31**, 581 (1967).
[6.18] L. Brillouin: *Ann. de Phys.*, **17**, 88 (1922).
[6.19] R. Y. Chiao and C. H. Townes: *Phys. Rev. Lett.*, **12**, 592 (1964).
[6.20] L. V. Al'tshuler, S. B. Kormer, M. I. Brazhnik, L. A. Vladimirov, M. P. Speranskaya and A. I. Funtikov: *Žurn. Ėksp. Teor. Fiz.*, **38**, 1061 (1960); English translation in *Sov. Phys. JETP*, **11**, 766 (1960).
[6.21] T. J. Ahrens and M. H. Ruderman: *Journ. Appl. Phys.*, **37**, 4758 (1966).
[6.22] R. N. Keeler, G. H. Bloom and A. C. Mitchell: *Phys. Rev. Lett.*, **17**, 852 (1966).

Acceleration of Projectiles to Hypervelocities.

J. G. LINHART

Laboratorio Gas Jonizzati
Associazione Euratom-Cnen - Frascati

1. – Introduction.

Let us define the term hypervelocity as a projectile-speed higher than 1 cm/μs. There is no deep physical reason for this threshold—it is motivated by several effects that are related to the speed of this order of magnitude, such as the escape velocity from Earth, meteorite impact speed, speed limit of chemically propulsed projectiles; pressures generated by impact of solid projectiles travelling at 1 cm/μs corresponds to maximum pressure at the centre of Earth. With impact-velocities of the order of 10 cm/μs we enter the domain of plasma physics, whereas a speed or 100 cm/μs corresponds to solar escape velocities and on impact, thermonuclear temperatures may be generated. We shall discuss how projectiles may be accelerated to hypervelocities and what are the limitations of these acceleration mechanisms.

2. – Acceleration methods.

It seems appropriate to distinguish between two different types of projectiles. The first corresponds to the conventional concept of a solid piece of material driven by some set of forces. However, it is also conceivable to start with a dilute mass, such as a plasmoid or a large number of disconnected grains, and condense this mass to a single large-density projectile during, or after acceleration.

Both types of projectile (mass M) can be driven by a suitable force (F). This drive can be either direct, described by

$$M\frac{\mathrm{d}^2x}{\mathrm{d}t^2} = F\,, \tag{1}$$

or indirect, in which case the acting forces cause the driven side of the projectile to explode with a speed v_j and provide thus a rocket-drive, described by

$$\frac{\mathrm{d}}{\mathrm{d}t}\left(M\frac{\mathrm{d}x}{\mathrm{d}t}\right)=\frac{\mathrm{d}M}{\mathrm{d}t}\left(\frac{\mathrm{d}x}{\mathrm{d}t}+v_j\right). \tag{2}$$

In many cases both drives are present. For some driving forces (electrostatic drive, magnetically driven super conductors or ferromagnetic materials) only the direct drive exists. Table I shows the different projectile and driver

TABLE I.

Projectile	A) *single, dense projectile* (nonconductor, conductor, superconductor)	gas propulsion ion beam electron beam photon beam magnetic fields	acceleration by pressure or/and reaction drive of the incident beam or field
		electrostatic fields	
	B) *dust*	electrostatic acceleration and subsequent condensation	
	C) *plasma*	magnetic acceleration and snow-plough accumulation photon-beam drive	

possibilities arranged according to the above-described criteria. Let us describe briefly each of these possibilities and indicate the corresponding limitations.

2·1. *A dense projectile accelerated by a driver gas.* – This is the closest relative of the conventional gun (Fig. 1). In a plane geometry we have for the maximum speed from energy conservation

$$\tfrac{1}{2}\overline{M}v_m^2=\int_0^l F\,\mathrm{d}x\,,\qquad \overline{M}=M+aM_1\,,\qquad a<1$$

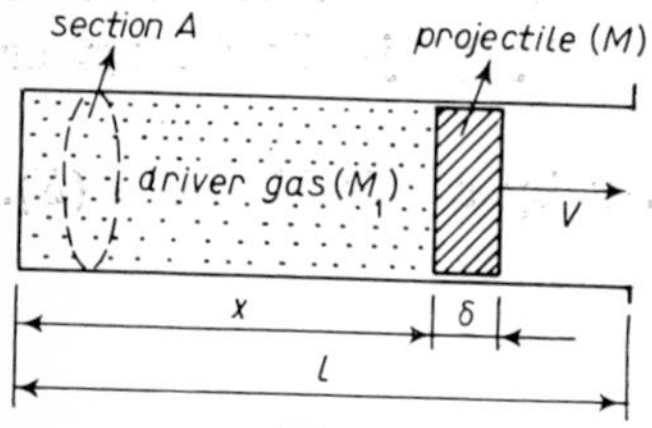

Fig. 1.

and in case of a constant pressure p_1 we get

$$v_m^2=\frac{2p_1 l}{\varrho\cdot\delta+a\varrho_1 l}\,,$$

where $M=A\cdot\delta\cdot\varrho$ and the density of the driver-gas is ϱ_1.

Then

$$v_m = \sqrt{\frac{2p_1}{\varrho_1}} \sqrt{\frac{\varrho_1 l}{\varrho \cdot \delta + a\varrho_1 l}} \,.$$

As $\sqrt{2p_1/\varrho_1}$ is approximately the speed of sound v_s in the driver-gas and $\varrho_1 l/\varrho\delta$ the ratio of the total gass mass M_1 to projectile mass M we can write

$$v_m \sim v_s \sqrt{\frac{M_1}{M + aM_1}} \,. \tag{3}$$

It is clear that the driver-gas can exert little force when $v > v_s$ and, therefore, making $M_1 \gg M$ would not help to increase the speed, it could only get it closer to the relalization of the hypothesis of constant p_1. The optimum ratio of M_1/M has to be determined as a function of the driver gas used and of the application. When we wish to get the maximum speed possible $M_1/M > 1$, when on the contrary an efficient propulsion is required, then $M_1/M < 1$.

The expression for the sound speed, which is related to the limiting speed, is $v_s = \sqrt{\alpha kT/m}$ where m is the mass of the gas molecule (atom, ion) and α depends on the number of degrees of freedom per molecule. Consequently the only hope to increase v_m is to increase the temperature T of the driver gas. Take as example nonionized but dissociated hydrogen ($\alpha = \frac{5}{3}$) and $v_s = 1$ cm/μs. The required temperature is $T = 1.4 \cdot 10^4$ (°K). It is clear that no chemical reaction can directly give such temperatures and, consequently Jules Verne's gun could never shoot a projectile on the moon. One way of getting higher driver-gas temperatures is by rapid conversion of electromagnetic energy into heat, such as occurs in exploding wires [1, 2]. It is conceivable that temperatures of 10^6 (°K) can be obtained in Li-wire explosions [3]. In sucha case a sound speed of the order of a few cm/μs can be obtained. In this case the driver-gas is a high-temperature plasma, presenting a number of stringent requirements, such as 1) the thermal insulation of the plasma from the gun walls 2) short driving period so as not to vaporize completely the projectile, etc. Further limitation is provided by the rate of transfer of electromagnetic energy in the heat energy of the driver-gas. From this point of view the best situation corresponds to a reservoir of magnetic energy B_y situated directly behind the driver-plasma (Fig. 2). The currents i_z induced in the plasma dissipate heat and the driving pressure is then of the order of $B^2/8\pi$. As in such a case much of the driver-plasma derives from the projectile mass itself the acceleration resembles that of a magnetically driven projectile

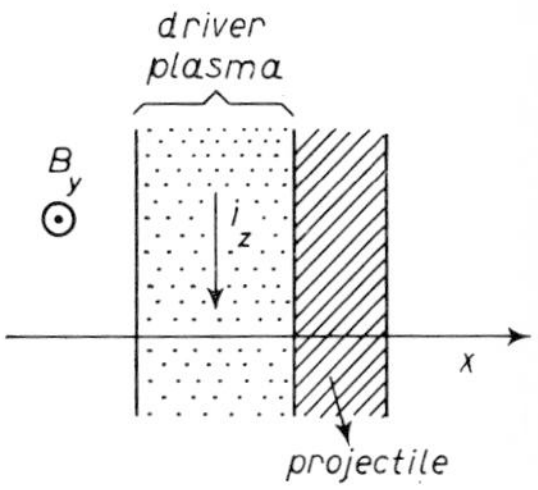

Fig. 2.

which will be treated later. In order to get projectile speeds of the order of 100 cm/μs we would have to use driver temperatures of the order of 10^8 (°K) which are usually associated with nuclear explosion. As at such temperatures the pressure p_r of the radiation modes in thermodynamic equilibrium exceeds considerably the kinetic pressure p_c of the gas, we shall mention it in the Section on radiation-driven projectiles (*e.g.* in hydrogen at solid state density and at 10^8 (°K) the ratio $p_r/p_c \simeq 140$).

So far, we have dealt with the direct drive. The indirect (or rocket) drive allows the projectile to reach $v_m > v_s$ at the expense, as one would expect, of efficiency.

Using eq. (2) and assuming $v_j \sim v_s$ and a constant rocket drive, one can write

$$M \cdot \frac{\mathrm{d}x^2}{\mathrm{d}t^2} = \frac{\mathrm{d}M}{\mathrm{d}t} \cdot v_s ,$$

which once integrated gives

$$v_m = v_s \cdot \ln \frac{M_1}{M_2} , \tag{4}$$

Where M_1, M_2 respectively are the initial and final mass of the projectile.

Let us work out the efficiency η of the propulsion. We have

$$\eta = \tfrac{1}{2} M_2 v_m^2 / W_r ,$$

where W_r is the energy dissipated by the rocket, which is

$$W_r = \tfrac{1}{2} (M_1 - M_2) v_s^2 .$$

Then

$$\eta = \frac{M_2/M_1 (\ln M_1/M_2)^2}{1 - M_2/M_1} . \tag{5}$$

TABLE II.

M_1/M_2	1/2	1/4	1/10	1/100	10^{-x}
η	0.48	0.64	0.59	0.21	$5.3 \cdot x^2 \cdot 10^{-x}$

Numerical values of efficiencies as a function of M_1/M_2 are found in Table II.

It is clear that efficiency does not fall rapidly with increasing M_1/M_2, *e.g* for $M_1/M_2 = 100$ it is still 21% and $v_m = 4.6v_s$. This is the reason for the success of rocket propulsion in astronautics.

In discussing the indirect drive we have assumed that the energy for rocket propulsion was inherent in the driver gas, *e.g.* it could have been one of the known rocket propellents. In such a case we would rely on chemical energy and, therefore, again $v_s < 1$ (cm/μs). If, on the other hand, the rocket energy is transfered to the projectile by dissipating electromagnetic or other energy in it, one can hope to obtain $T > 10^4$ (°K) on the driven face of the projectile and consequently $v_s > 1$ cm/μs. However, this belongs already to following Sections.

There remains one other possibility of increasing the sound speed v_s of the driver fluid, *i.e.* one related to cumulation processes. It is well known that in converging cylindrical and spherical shocks the pressure and temperature reached at the centre of convergence are much higher than the corresponding initial values [4, 5]. If one can use the central volume of such a shocked fluid to drive a projectile (or to produce a jet) one may reach v_s and, therefore, v_m much higher than those corresponding to the outer layers of the fluid [6]. The use of *e.g.* shaped charges is based on this principle. Unfortunately the efficiency of mo t energy-density cumulation processes is very poor and if a large increase of v_s is desired it is generally more convenient to resort to a more noble type of energy, which can be easily concentrated.

2·2. *Ion or electron beam acceleration of a dense projectile.* – In this type of acceleration proposed originally as a direct propulsion [7], an electron or a ion wind drives a projectile charged to such a potential φ_0 that none of the wind-particles can reach it. Thus

$$e\varphi_0 \geqslant \tfrac{1}{2} mv^2 = e\varphi ,$$

where m, v and $e\varphi$ are the mass, speed and energy, respectively, of the wind particles. Neglecting first the self-potential φ_b of the beam, we can treat the problem as a scattering problem of the charged wind-particle on the charged spherical projectile. The latter (radius r_0, density ϱ, charge $Q = r_0 \varphi_0 = E_0 r_0^2$, speed v_0) presents to the wind a large-angle scattering cross-section (the « sail-area » of the projectile)

$$\Sigma = \pi \left[\frac{2Q \cdot e}{m(v - v_0)^2}\right]^2 ;$$

for $v \gg v_0$ we get

$$\Sigma = 4\pi \left(\frac{e}{m}\right)^2 \left(\frac{E_0 r_0^2}{v^2}\right)^2 . \tag{6}$$

The mean pressure of the wind on this sail will be

$$p = m \cdot n(v - v_0)^2$$

and the total force

$$F \sim \pi r_0^2 \left(\frac{\varphi_0}{\varphi}\right)^2 nmv^2 . \tag{7}$$

The length l of an accelerator capable of pushing the projectile to a speed v_0 will be, therefore,

$$l = \frac{1}{2}\frac{Mv_0^2}{F} \sim \frac{2}{3}\frac{\varrho}{nm}\left(\frac{v_0}{v}\right)^2\left(\frac{\varphi}{\varphi^0}\right)^2 . \tag{8}$$

Iti s clear that the large disparity between the density ϱ of the projectile and that of the wind ($\varrho_w = n \cdot m$) will result in a very large l. *E.g.* let $v/v_0 = 10$, $\varphi_0/\varphi = 10$ and $n = 10^{11}$ ions/cm³. Assume that one uses heavy-ion wind, *e.g.* $m = 10^{-22}$ g and $\varrho = \frac{3}{2}$, one gets $l = 10^7$ (cm) $= 10^2$ (km).

It is impossible to try to shorten l by increasing n much above 10^{11}, as one will soon encounter difficulties in formation and propagation of such highly charged beams owing to the self-potential φ_b becoming larger than φ and choking off the electron or ion flow [8]. Thus the pressure $\varrho_w v^2$ developped by the beam will always be very low.

On the other hand it is possible to generate highly focused, powerful electron beams in which positive ions neutralize partially or even completely the electron beam space-charge [9]. It has been demonstrated that MeV electron beams can be thus focused, giving a current-density of almost 0.1 MA/cm². The power-flow in such a beam is, therefore, of the order of 10^{11} W/cm² and the pressure $\varrho_w v^2 \sim 200\, i\gamma$ where $\gamma = (1 - \beta^2)^{-\frac{1}{2}}$ and i is in A/cm²] is of the order of 100 atm. Unfortunately a negatively charged projectile in such a neutralized beam will attract the positive ions and get rapidly discharged. At most one can hope to keep it charged to the potential φ, in which case from eq. (8)

$$l = \frac{2}{3}\frac{\varrho v_0^2}{\varrho_w v^2} \sim \frac{1}{3}\,10^{-2}\,\frac{\varrho v_0^2}{i\gamma} ;$$

e.g. for $\varrho = \frac{3}{2}$, $v_0 = 10^8$ (cm/s), $i = 10^5$ A/cm² and $\gamma \sim 50$ we get $l = 10^7$ (cm), as in the above example of a ion beam. If ion beams of comparable intensities could be produced $\varrho_w v^2$ would increase considerably but so would the problem of keeping the projectile charged.

A better solution would be to consider an indirect propulsion in which one could convert the very large electron-beam power in jet power on the driven side of the projectile.

Let us suppose that the incident power is absorbed in a layer Δ. This layer explodes within a time

$$\tau_\Delta \cong \frac{\Delta}{v_{ex}} .$$

As the mean density of this layer when ejected at v_j will certainly be lower than the projectile density ϱ the explosion of the projectile will propagate itself with a speed v_{ex} lower than v_j. Assuming that $v_{ex} = \frac{1}{10} v_j$ we have for the available acceleration time

$$\tau \sim \frac{\delta_1}{v_j} 10 .$$

Within this time the energy W consumed by the jet is

$$W = P \cdot \tau = \tfrac{1}{2} \delta_1 \varrho v_j^2 .$$

From which the order of power required for a given v_j is

(9) $$P \sim \frac{1}{20} \varrho v_j^3 .$$

As we have seen the rocket drive can give v_m which is at most an order of magnitude higher than v_j and therefore,

(9a) $$P \sim \tfrac{1}{2} 10^{-4} \varrho v_m^3 .$$

Assuming $\varrho \sim 2$ we have the following Table showing the (optimistic) values

TABLE III.

v_m (cm/s)	10^6	10^7	10^8
P (W/cm^2)	10^7	10^{10}	10^{13}

of $P(v_m)$. Examining a little further the case $v_m = 10^7$ we see that the total energy required is

$$W = 10^4 \cdot \delta_1 \text{ (J/cm}^2\text{)} ,$$

delivered during a time $\tau = \delta_1$ (μs). If one is to finish with a macroscopically usable projectile, after a reproducible acceleration procedure δ_1 must be of the order of several millimeters, which points to an accelerator length of the order of 10 cm. Consequently it appears that up to v_m of about 10 cm/μs the rocket drive powered by an incident electron beam could be used, provided the electrons could be absorbed in a suitably thin surface layer of the projectile [10].

2·3. *Acceleration by a photon beam.* – Electroagnetic radiation incident on an electrical conductor may be reflected, absorbed or transmitted. The latter case occurs only with a plasma medium in which the plasma-frequency ω_p is smaller than the frequency of the incident radiation.

Let us first treat the case of an incoherent radiation, *e.g.* one emitted by a black-body, driving a plane projectile. Let us assume first a total reflection. The pressure on the projectile is

$$p = \frac{2}{\pi}\frac{\sigma T^4}{c}\,. \tag{10}$$

The Table IV gives numerical values of $p(T)$ and for comparison values of a magnetic field B capable of exerting the same pressure $p = B^2/8\pi$. In the same Table is plotted the ratio π between $p(T)$ and $p_c(T)$, the kinetic pressure of Z-times ionized plasma at solid-state density and temperature T. Here $p(T) = 7\cdot(Z+1)\,T$ (atm).

TABLE IV.

T (°K)	10^5	10^6	10^7	10^8
p (atm)	0.12	1220	$12.2\cdot10^6$	$122\cdot10^9$
B (MG)	0.0017	0.175	17.5	1750
$(Z+1)\pi$	$1.75\cdot10^{-7}$	$1.75\cdot10^{-4}$	0.175	$1.75\cdot10^2$

The table shows clearly that at temperatures $T > 10^8$ the kinetic pressure of even uranium plasma (in thermodynamic equilibrium with its radiation modes) can be neglected with respect to the radiation pressure and any projectile driven by such a medium can be considered radiation-driven. The speed of a projectile driven over a length l by radiation pressure will be

$$v_m = \frac{1}{2}10^{-7}\sqrt{\frac{l}{\varrho\delta}}\,T^2\,. \tag{11}$$

E.g. a copper projectile driven through its own thickness by black-body radiation at $T = 10^8$ reaches a speed of approximatively $1.7 \cdot 10^8$ cm/s.

It is clear that one cannot hope to obtain a total reflection, a part of the incident radiation will be always absorbed in the projectile, causing a vaporization wave to progress into the projectile and explode it. Consequently the direct drive will be always mixed-up with the indirect one.

The nearest one can get to the above-mentioned radiation pressures, without using nuclear explosions, is in the focused laser beam. The focusing of a monochromatic light-beam is a better energy-concentration process than those related to particle-beam focusing, owing to the absence of repulsion between photons. The ideal maximum for energy density and, therefore, pressure obtainable from a focused laser-beam of power $\dot{W}$ is

$$p = \frac{\dot{W}}{c\lambda^2},$$

where λ is the wavelength of the radiation.

Obviously the lens errors, the divergence of the beam in the laser and other imperfections do not permit a focalization of the beam into a cross-section $S = \lambda^2$. Taking the top laser and lens performances of today we get for $\dot{W} = 10^{13}$ W, $S = 3 \cdot 10^{-4}$ cm² [11] a pressure of about 10^7 atm, equivalent to black-body pressure at $T \sim 10^7$ (°K). The duration of such light pulses could be as short as several picoseconds, the total light energy rarely exceeding 100 J. This indicates that only very small projectiles could be accelerated in this way—at the present time their dimensions would not exceed 100 μm. In such a case the best use of the large focused laser power would be to explode a considerable part of the projectile, heating thus the driven side to a temperature T, corresponding to a jet speed v_j of the same order of magnitude as the desired final speed v_m of the unexploded portion of the projectile. This heating process is, however, connected with until now not well-understood anomalous dissipation processes and, therefore, little can be said about the attainability of such high driver-temperatures [12]. In any case considerations similar to eq. (9) and Table III will apply.

The laser beam could be also applied to acceleration of dense plasmoids provided

$$\omega < \omega_p \,. \tag{12}$$

which can be written as $n_e > 1.13 \cdot 10^{13}\, \lambda^{-2}$ (electrons/cm³) and provided the focus could be made to follow the plasmoid (or the laser beam could be focused over some length into a nearly parallel beam). Even here one should observe a certain type of « vaporization » wave caused by the heating and subsequent expansion of the plasma below the critical density n_e. Once such an expansion occurs the hot plasma ceases pratically to be pushed by the radiation pressure and is lost to the acceleration process. If no disastrous loss of projectile occurs and the hot parts radiate as a black body one finds for a helium plasmoid whose ion density $n_i = 10^{21}$, accelerated to $v_m = \frac{1}{2}c$ (*i.e.* nonrelativistic limit)

$$l \sim \frac{1}{8}\frac{c^2}{\ddot{x}} \sim 2\cdot 10^{21}\frac{\delta}{\dot{W}_F}\,, \tag{13}$$

where $\dot{W}_F = \dot{W}/S$ (W/cm²) is the power flow in the beam focus. Putting $\dot{W}_F = 2\cdot 10^{17}$ and $\delta = 10^{-2}$ (cm) we get $l \sim 100$ (cm).

2'4. *Acceleration by electric fields.* – The simplest proposal for electric propulsion is related to an electrically charged projectile (charge $Q = E_0 r_0^2$, radius r_0) in an electric field E [7]. The length of an accelerator producing a final projectile speed v_m is

$$l \cong 2\cdot 10^5 \varrho (E E_0)^{-1} r_0 v_m^2 \text{ (cm, g/cm}^3\text{, V/cm, cm/s)}\,. \tag{14}$$

E_0 is limited to 10^7 V/cm for negative Q owing to field emission of electrons, whereas in case of positive Q it may be as high as 10^8 V/cm, at which level the electrostatic stress $E_0^2/8\pi \sim 4000$ (atm.) and the projectile will tend to break up into smaller pieces having a lower E_0.

Before we fix the value of E let us distinguish two possible modes of acceleration. The first is effected by an electrostatic accelerator, such as a Van de Graaf.

In this case $l\cdot E = \varphi$, which is usually limited to about 10^7 V and the eq. (14) for positive Q becomes

$$v_m \sim 0.7\cdot 10^5 (\varrho r_0)^{-\frac{1}{2}}\,. \tag{14a}$$

For an Al projectile we get a speed of 0.43 cm/μs for $r_0 = 0.1$ mm and a speed of 100 cm/μs for $r_0 = 0.002$ μm. It is clear that this method is of interest only for speeds of at most a few cm/μs.

The second conceivable acceleration method could be based on linear accelerators in which case it is unlikely that one can use $E > 10^5$ (V/cm). In such a case (for Al)

$$l \sim 5.4\cdot 10^{-8} r_0 v_m^2\,. \tag{14.b}$$

E.g. for $r_0 = 0.1$ mm we get $l = 540$ m for $v_m = 10$ cm/μs showing that the results are not much better than those obtained by electrostatic accelerators.

A rather remote possibility of accelerating a relatively large projectile ($R_0 = 1$ mm) to speeds in excess of 10^7 would be to accelerate a cloud of « dust » to that speed, in which case each « dust » particle could be sufficiently small so as to satisfy eq. (14*a*) and try *after* acceleration to condense the cloud into a single projectile. Consequently the total number N of the dust particles should be

$$N = \left(\frac{R_0}{r_0}\right)^3 .$$

The dimension D of the « dust cloud » should be such as not to produce electrostatic potentials comparable with φ. Therefore,

$$\frac{N \cdot Q}{D} \ll \varphi , \qquad \text{or} \qquad D \gg \frac{R_0^3 E_0}{r_0 \varphi} .$$

In the case of $v_m = 10^8$ cm/s and $R_0 = 1$ mm we get impossibly large values for D.

2.5. *Acceleration by magnetic fields.* – There are two purely direct acceleration mechanisms:

$$\left.\begin{cases}\text{permanent magnet (dipole)}\\ \text{superconductor}\end{cases}\right\} \quad \text{accelerated in a moving nonhomogeneous magnetic field.}$$

In the first case [7], under the most favorable condition (synchronous acceleration) and for $\mathrm{d}B/\mathrm{d}x \sim B/2r_0$ we get for the length

$$l \sim \tfrac{1}{3} \varrho B^{-2} v_m^2 r_0 . \tag{15}$$

For permanent magnets $B < 20$ (kG) and B^2 is, therefore, comparable with the equivalent term, *i.e.* EE_0, in eq. (14).

In the second case [13, 14] the projectile is pushed by the Lorentz force, whose density $j \wedge B$ has an upper limit of about $\frac{1}{2} \cdot 10^{11}$ (A/cm² G) for V_3Ga. The total length of the accelerator is, therefore,

$$l \sim 0.5 \cdot 10^{-11} v_m^2 , \tag{16}$$

giving about 50 km for $v_m = 10^8$ (cm/s). Experiments have shown that a V_3Ga superconducting ring can survive the bumpy ride as superconductor [15]. The length is still impracticably long and one must wait for superconductors whose $j \wedge B$ is at least an order of magnitude better than the best known today.

In ordinary conductors and for sufficiently short time $j \wedge B$ can be several orders of magnitude larger than that mentioned above—unfortunately the current circulating in the projectile generates Joule's heat which explodes the projectile. Let us consider the simplest analysis of a plane projectile pushed by a magnetic field, assuming a uniform current distribution throughout the thickness δ_0 of the projectile (ref. [16]).

The equation of motion is

$$\varrho \cdot \delta \cdot \frac{\mathrm{d}^2 x}{\mathrm{d}t^2} = \frac{B^2}{8\pi} .$$

The magnetic energy dissipated is converted in heat energy according to

$$\varrho \frac{\mathrm{d}(CT)}{\mathrm{d}t} = \varrho_e j^2 ,$$

where

$$j \cdot \delta = \frac{B}{4\pi} \qquad \text{and} \qquad \varrho_e \sim \varrho_0 \frac{T}{T_0} .$$

Assuming C and δ to be constant up to the boiling point we get from equations

$$\frac{\dot{T}}{T} = \frac{\varrho_0}{2\pi C T_0 \delta} \ddot{x} ,$$

integrating from T_1 to the boiling temperature T_b we get for the limiting speed for a direct drive

$$v_{\text{lim}} = \frac{2\pi C T_0}{\varrho_0} \delta \ln \frac{T_b}{T_1} \tag{17}$$

above which the projectile explodes. For metals such as Cu, Al, Li and assuming $T_1 = 273$ (°K) we get

$$v_{\text{lim}} \sim 10^7 \delta \text{ (cm/s, cm)} , \tag{17a}$$

which does not depend on the magnitude of B and acceleration time (for more precise values see [17]. This shows that for projectiles whose dimension is of the order of one millimeter the direct drive can give speeds of the order of one cm/μs in the most favorable circumstances.

This limit is, however, not final and a considerable improvement can be made by letting most of the projectile explode. In such a case it is important to show that the propulsion efficiency does not suffer catastrophically even if 99% of the projectile explodes. Let us combine eqs. (1) and (2) as follows:

$$\varrho\delta \frac{d^2x}{dt^2} = \frac{B^2}{8\pi} + v_j\dot{\delta} \,. \tag{18}$$

For constant B, $\dot{\delta}$ and v_j we have

$$v_m = (v_j + v_a^2/2\dot{\delta}) \ln \frac{\delta_0}{\delta} \,, \tag{19}$$

where $v_a = B/\sqrt{4\pi\varrho}$, the speed of sound in that part of projectile in which B has diffused. The efficiency of the propulsion is

$$\eta = \frac{\frac{1}{2}\varrho\delta v_m^2}{(B^2/8\pi)[l + (\delta_0 - \delta)]} \,, \tag{20}$$

where we assume that the energy $B^2/8\pi \cdot l$ can be recuperated and

$$l = \int_0^t v\,dt = \left(\frac{v_j}{\dot{\delta}} + \frac{v_a^2}{2\dot{\delta}^2}\right)\left(\delta_0 - \delta - \delta \ln \frac{\delta_0}{\delta}\right) \,, \tag{20a}$$

giving

$$\eta = \eta^+ \frac{(v_j/v_a + v_a/2\dot{\delta})^2}{1 + (1/\dot{\delta})(v_j + v_a^2/2\dot{\delta})\left(1 - \left(\delta/(\delta_0 - \delta)\right) \ln (\delta_0/\delta)\right)} \,, \tag{21}$$

where η^+ is the efficiency of the jet-drive alone (eq. 5). When $v_a \gg v_j$ (and, therefore, $v_a \gg \dot{\delta}$)

$$\eta = \frac{1}{2} \frac{\eta^+}{1 - \left(\delta/(\delta_0 - \delta)\right) \ln (\delta_0/\delta)} \tag{21a}$$

and in the important case of $\delta/\delta_0 \to 0$ we get $\eta \to \frac{1}{2}\eta^+$.

The values of η can, therefore, be read directly from Table II. *E.g.* in case of $\delta/\delta_0 = 1/100$ we still get $\eta \sim 10\%$. Let us suppose that we wish to obtain v_m considerably larger than the v_{lim} of eq. (17). Then

$$10^7\delta_0 \ll (v_j + v_a^2/2\dot{\delta}) \ln \frac{\delta_0}{\delta} \,; \tag{22}$$

if the jet drive is not important we have

$$v_a^2 \gg 2 \cdot 10^7 \frac{\delta_0 \dot{\delta}}{\ln (\delta_0/\delta)} \,,$$

or

$$B > 1.6 \cdot 10^4 \sqrt{\varrho \delta_0 \dot{\delta} / \ln \frac{\delta_0}{\delta}} \,. \tag{23}$$

Take as an example: $\varrho = 2.7$, $\delta_0 = 0.1$, $\dot{\delta} = 10^6$, $\delta_0/\delta = 100$, then $B > 3.8$ (MG), showing the importance of using MG fields. Let us consider the length l of such an accelerator. Equation (20a) gives for $v_a^2 \gg v_s \dot{\delta}$

$$l \simeq \frac{1}{2}\, \delta_0 \left(\frac{v_a}{\dot{\delta}}\right)^2 \left[1 - \frac{\delta}{\delta_0}\left(1 + \ln \frac{\delta_0}{\delta}\right)\right] \tag{24}$$

and for $\delta \ll \delta_0$ we have

$$l \sim \frac{1}{2}\, \delta_0 \left(\frac{v_a}{\dot{\delta}}\right)^2 . \tag{24a}$$

It is not clear what relation there can be between the explosion speed $\dot{\delta}$ and the Alfvén speed v_a; it is not excluded that the ratio may be independent of B for very strong fields. In any case

$$l < l^* ,$$

where l^* is the length required for acceleration of nonvaporizing projectiles to v_m and obviously

$$l^* = \left(\frac{v_m}{v_a}\right)^2 \delta_0 = \frac{1}{4}\left(\frac{v_a}{\dot{\delta}}\right)^2 \delta_0 \left(\ln \frac{\delta_0}{\delta}\right)^2$$

and, therefore,

$$l = \frac{2l^*}{(\ln \delta_0/\delta)^2} \,. \tag{25}$$

The shortening factor $s = \frac{1}{2} (\ln \delta_0/\delta)^2$ and for $\delta_0/\delta = 100$ it is equal to about 10. Let us work out l^* for $v_m = 10^8$ (cm/s). We get for Al

$$l^* = 5.83 \cdot 10^{16} \frac{\delta_0}{B^2} ,$$

thus for $B = 10^7$ (G), $\delta_0 = 0.2$ we get $L^* = 117$ (cm) and considering $s \sim 10$

$$l \sim 10 \text{ (cm)} .$$

The greatest incertitude of this mechanism is our ignorance of $\dot{\delta} = f(B)$ and both pleasant or unpleasant surprises are possible. One of the pleasant ones would be the creation of a magnetized plasma heat shield on the driven side of the projectile, delaying the explosion. Unpleasant could be an unstable explosion wave, similar to the saw-effect encountered with MG fields [18].

Talking about plasma heat-shields brings us to the possibility of driving a relatively small-density plasma ($n > 10^{18}$ ions/cm^3) by a strong magnetic field to hypervelocities and then trying to condense it into a projectile of almost solid density. It is known [19] that the acceleration of plasma to $(2\div5)\cdot10^7$ cm/s does not represent great difficulties and it is likely that in MG magnetic fields speeds of several times 10^8 cm/s could be achieved. The problem is how to effect the condensation. It has been shown, that this can be resolved to a large extent using heavy-element plasmas, such as argon and neon (ref. [20]). The plasma of these elements is accelerated and condensed by an almost pure snow-plough mechanism [21], *i.e.* a magnetically driven shock-front. Generally a large $\varrho_1/\varrho_0 = (\gamma+1)/(\gamma-1)$ across a strong shock is connected with many degrees of freedom α (*i.e.* $\gamma \to 1$ as $\alpha \to \infty$). In high-Z plasmas this is associated with multiple ionization and heavy radiation losses.

The snow-plough model in plane geometry is described for perfect sweeping by

$$\frac{\mathrm{d}}{\mathrm{d}t}\left(\varrho_0 x\,\frac{\mathrm{d}x}{\mathrm{d}t}\right) = \frac{B^2}{8\pi}, \tag{26}$$

which gives

$$x\,\frac{\mathrm{d}x}{\mathrm{d}t} = \int_0^t \frac{B^2}{8\pi\varrho_0}\,\mathrm{d}t\,.$$

For a constant magnetic field

$$\dot{x} = \mathrm{const} = \frac{B}{\sqrt{8\pi\varrho_0}} = \frac{v_a}{\sqrt{2}}\,. \tag{27}$$

The efficiency of this process is

$$\eta = \frac{\frac{1}{2}\varrho_0 x\dot{x}^2}{\alpha x B^2/8\pi}, \tag{28}$$

where $\alpha = 1$ if the energy $x(B^2/8\pi)$ accumulated in the magnetic field is not dissipated, otherwise $\alpha = 2$. In the first case $\eta = 50\,\%$ in the second $\eta = 25\,\%$.

The thickness of the projectile is composed of a layer in which B has diffused, of a field-free layer and of a precursor layer. During the acceleration phase the B-diffusion dominates and gives the projectile a thickness

$$\delta_B \sim \frac{10^6}{T^{\frac{3}{4}}} \sqrt{t}\,. \tag{29}$$

The mean density $\bar{n}$ is, therefore,

$$\bar{n} \sim \frac{10^5}{\sqrt{A}} B T^{\frac{3}{4}} \sqrt{n_0 x} \qquad (A = \text{atomic mass n.})$$

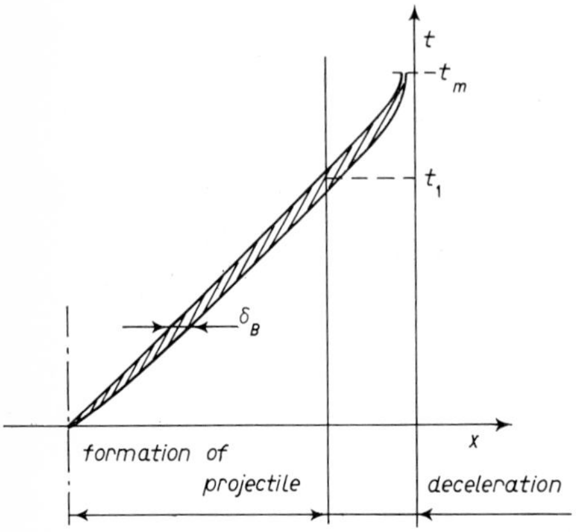

Fig. 3.

and the field diffusion becomes less important as acceleration proceeds. Although the snow-ploughing already represents an important condensation process, a much more important accumulation is expected to occur during the plasma-projectile impact on a target. The formation of the projectile and its condensation are represented in Fig. 3. It is expected that the final thickness is given by the Haley expression (which is really equivalent to pressure balance)

$$(Z+1)kT = \delta_{\mathrm{H}} \frac{M_v}{\tau_{\mathrm{imp}}} \sim \delta_{\mathrm{H}} M \frac{v^2}{x_{\mathrm{imp}}}$$

and, therefore

$$\delta_{\mathrm{H}}/x_{\mathrm{imp}} \sim \frac{(Z+1)kT}{Mv^2}\,.$$

For reasonably thin liners [22]

$$\delta_{\mathrm{H}}/x_{\mathrm{imp}} \ll 1$$

and the density at t_m could be, therefore, an order of magnitude higher than that at $t = t_1$.

REFERENCES

[1] *Exploding Wires*, edited by G. Chace and H. K. Moore (New York, 1964).

[2] J. L. Bohn *et al.*: *Exploding Wires*, edited by G. Chace and H. K. Moore (New York, 1964), p. 339.

[3] J. Katzenstein and M. Sydor: *Journ. Appl. Phys.*, **33**, 718 (1962).
[4] G. Guderley: *Zeit. Luftfahrtforsch.*, **19**, 302 (1942).
[5] A. Kantrowitz and R. W. Perry: *Journ. Appl. Phys.*, **22**, 878 (1951).
[6] R. F. Flagg and I. I. Glass: *Phys. Fluids*, **11**, 2282 (1968).
[7] E. R. Harrison: *Macron accelerators*, rep. NURL/M/60 of Rutherford High-Energy Lab. (1964).
[8] J. R. Pierce: *Theory and Design of Electron Beams* (Amsterdam, 1954).
[9] S. E. Graybill and S. V. Nablo: *Appl. Phys. Lett.*, **8**, 18 (1966).
[10] F. Winterberg: *Phys. Rev.*, **174**, 212 (1968).
[11] J. L. Hughes: *Appl. Opt.*, **6**, 1411 (1967).
[12] J. L. Bobin, F. Delobeau, G. De Giovanni, C. Fauquignon and F. Floux: *Nucl. Fus.*, **9**, p. 115 (1969).
[13] Ch. Maisonnier: *Nuovo Cimento*, **42** B, 332 (1966).
[14] F. Winterberg: *Nucl. Fusion*, **6**, 152 (1966).
[15] M. Sauzade, *Etude Préliminaire d'un accélérateur à macroparticules supraconductrices* (priv. com.).
[16] G. Lehner, J. G. Linhart and Ch. Maisonnier: Rep. Labor. Gas Ionizzati RTI/FI (64)2, Nov. 1964.
[17] E. C. Cnare: *Journ. Appl. Phys.*, **37**, 3812 (1966).
[18] H. P. Furth, M. A. Levine and W. Waniek: *Rev. Sci. Instr.*, **28**, 949 (1957).
[19] J. Marshall: *Phys. Fluids*, **3**, 134 (1960).
[20] M. Haegi and J. G. Linhart: *Proc. III Europ. Conf. on Plasma Physics* (Utrecht, 1969).
[21] J. G. Linhart: Report Lab. Gas Ionizzati 67/12, June 1967.
[22] J. P. Somon: Report Lab. Gas Ionizzati, LGI 66/11 Nov. 1966.

Cumulation of Electromagnetic Energy.

H. Knoepfel

Laboratori Gas Ionizzati (Associazione EURATOM-CNEN) - Frascati

1. – Introduction.

The generation of high energy densities, as discussed at this Summer Course, depends primarily on the availability of convenient energy sources, which typically must deliver energies in the range of megajoules in a time of the order of microsecond (Tables I and II). As energy source we actually understand the whole energy system, including the storage, the transfer, eventual transformation stages, and the concentration in space and time of the primary energy. In most experiments on high energy density the required final form is thermal energy. The electromagnetic energy is then only a most convenient intermediate energy form used to store, to transfer and to concentrate the required energy.

If we limit our attention to the generation of electromagnetic power pulses we see from Table II that various energy sources exist, which store the necessary energy and transform and deliver it within times varying from submicroseconds to multiseconds. Among the systems used most is the capacitor bank, because of its flexibility, high power rate and high-energy transfer-efficiency (see, *e.g.*, [19]). A particularly interesting modification of such an electrostatic-energy storage system is discussed in [20]. Also interesting is the inductive storage system, but due to its intrinsic technical difficulties (*e.g.* large superconducting storage inductors, opening switches, etc.) it is not yet commonly used. Mechanical generators, as well as chemical batteries, are interesting when the required pulse duration is longer than about 0.01 s. Of particular interest are explosive-driven generators based on the principle of magnetic flux compression.

For the cumulation of electromagnetic energy it is required that the energy delivered by these storage systems is concentrated in space. When the wave nature of electromagnetic fields can be neglected, *i.e.* within the validity of the quasi-stationary electromagnetic theory, the transfer of magnetic energy

TABLE I. – *Terrestrial energy sources for the generation of high energy density.*

Primary source	Final form	Energy density (MJ/cm^3)	Energy per atom involved (eV; element)		Total involved energy (MJ)	Ref.
Chemical explosives	—	$8\cdot10^{-3}$	1	(H, N, O)	100	—
	Metallic jets	1	1700	(Cu)	10^{-3}	[25]
	Metallic plates	0.8	60	(Fe)	3	[7]
	1 MOe	$4\cdot10^{-3}$	0.3	(Cu)	5	[1]
	5 MOe	0.1	8	(Cu)	1	[1]
	25 MOe	2.5	200	(Cu)	1	[8]
Capacitor	—	10^{-8}	—		5	—
	Exploding wires	0.05	4	(Cu)	10^{-3}	[9]
	Plasma focus	0.01	1000	(D)	10^{-4}	[10]
Optical storage	—	10^{-6}	—		10^{-3}	[11]
	Focused laser beam	0.4	100	(D, T)	10^{-5}	[12]
Nuclear explosive (D, T)	—	10^{4}	$9\cdot10^{6}$	(D, T)	10^{11}	—

TABLE II. – *Energy sources for pulsed current generators* (*).

Primary energy source	Density of stored energy ($J\cdot cm^{-3}$)	Possible stored energy (MJ)	Produced electrical power			
			Energy (MJ)	Pulse duration (ms)	System efficiency (%)	Location
Chemical explosive	10 000	100	2	0.1	2	Frascati
Capacitor	$10^{-2}\div10^{-1}$	10	3	0.01	80	Garching
Rotating machine	$100\div500$	100	2	100	30	Orsay
Inductor	10	100	—	—	—	—
Lead-cell battery	500	1000	0.5	1000	0.5	Frascati

(*) Typical, good figures, not necessarily maximum values.

is bound to conductors of electric currents. The concentration of magnetic energies can then be obtained by simply forcing the current through a relatively small conductor; this is the case when a capacitor bank is discharged into relatively small single-turn coils [15]. It is then convenient to express the concentration in terms of the magnetic field (although the energy density goes with the field squared). It may be instructive to remind that the energy density in a field of 1.5 MOe corresponds to that released chemically by a detonating explosive.

The local concentration of electromagnetic energy also demands that the stored energy is discharged into the concentrating system at a high power rate (typically 10^{12} W). This requirement depends on the energy losses in the system, which (as we shall see in detail later on) drastically increase at high energy densities; thus, as the electromagnetic energy density increases, the power influx must increase rapidly to compensate at least for the losses. The shortening of the characteristic discharge time of a given energy storage system, as required by the power condition, is in principle possible for many systems, but always difficult to realize ([21], Chs. 6 and 9). In general it requires a transformation of the stored energy, such that the final load remains decoupled from the primary energy source. As an example we can mention the magnetically driven flux compression system, a cylindrical system similar to the explosively driven one that will be described later on. A capacitor bank discharges through a cylindrical shell forcing it to implode coaxially (as in the explosive experiment), thereby converting the stored electrostatic energy into kinetic energy. If a magnetic flux is trapped initially in the bore, it will be compressed by the imploding shell and the kinetic energy will, towards compression end, rapidly be transformed into magnetic one, as described in detail later on for the explosively driven generator. It should be noted that the characteristic time of this last transformation (the time constant that matters) can be an order of magnitude smaller than the discharge time of the capacitor bank.

We shall limit our discussion in this lecture to only one example of electromagnetic energy concentrators, the explosively driven ultrahigh magnetic field generator. For one reason, this system has generated among the largest electromagnetic energy densities on terrestrial scale; then, it allows one to describe fairly well typical difficulties and limitations encountered when cumulating electromagnetic energy. We shall discuss first the principle of cylindrical flux compression, then analyse the limitations introduced by taking the electrical conductivity and the compressibility into account. One Section is dedicated to the instabilities of the conductor surface, whereas the last Section describes more specifically the explosively driven generator.

2. – Cylindrical flux compression.

The principle is shown in Fig. 1. Let us consider here the simple case where an ideal (*i.e.* incompressible, infinitely conducting and very long) cylindrical shell with radii r_1, r_2 compresses an initially trapped magnetic field H_0.

From the general continuity equation we deduce for the axisymmetric and and incompressible fluid the condition

(1) $$\frac{\partial(rv)}{\partial r} = 0 ,$$

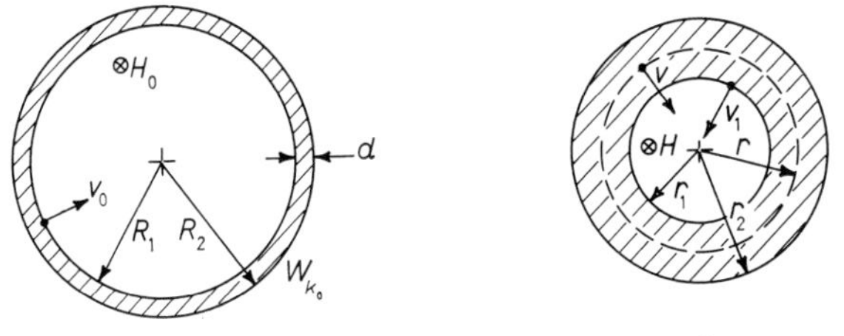

Fig. 1. – Imploding ideal shell: *a*) initial and *b*) intermediate position.

i.e. we have, for example,

(2) $$r_1 v_1 = rv ,$$

or

(3) $$r_2^2 - r_1^2 = R_2^2 - R_1^2 = R_0^2 ,$$

where R_0 is the (outer) radius of the fully collapsed shell ($r_1 \to 0$).

For the kinetic energy of the shell (per unit of axial length) comprised between the radii r_1 and r_2 we calculate with eq. (2)

(4) $$W_K(r_2, t) = \int_{r_1}^{r_2} 2\pi r \varrho \frac{v^2}{2}\, dr = \pi r_1^2 \varrho v_1^2 \ln \frac{r_2}{r_1} .$$

As a consequence of the collapse of the shell from R_1 to r_1, work is done against the magnetic pressure

(5) $$p_1 = \frac{\mu_0 H^2}{2} = \frac{\mu_0 H_0^2}{2} \left(\frac{R_1}{r_1}\right)^4$$

(here we have used the flux conservation: $Hr_1^2 = H_0 R_1^2$). Consequently the potential magnetic energy increases by

$$W_M - W_{M_0} = \int_{R_1}^{r_1} 2\pi r_1 p_1 \, dr_1 - W_{M_0} = \pi R_1^2 \frac{\mu_0 H_0^2}{2} \left(\frac{R_1^2}{r_1^2} - 1\right) ,$$

where

$$W_{M_0} = \pi R_1^2 \frac{\mu_0 H_0^2}{2} \tag{6}$$

is the initial magnetic energy. The relation between the implosion velocity v_1 and the inner radius r_1, in dependence of the initial magnetic field H_0, follows from the conservation of energy

$$W_{K_0} + W_{M_0} = W_K + W_M ; \tag{7}$$

W_{K_0} is the initial kinetic energy characterized by the radii R_1, R_2 and the velocity $v_1 \to v_0$. For the case where the linear thickness d is initially very small

$$R_0^2 \simeq 2dR_1 , \qquad d = R_2 - R_1 ,$$

and when

$$W_{K_0} \gg W_{M_0}$$

we obtain from eq. (7) the approximation

$$\left(\frac{v_1}{v_0}\right)^2 \simeq \frac{1 - (r_t/r_1)^2}{(r_1/R_0)^2 \ln[1 + (R_0/r_1)^2]} . \tag{8}$$

The radius r_t, defined as

$$\left(\frac{r_t}{R_1}\right)^2 = \frac{W_{M_0}}{W_{K_0} + W_{M_0}} , \tag{9}$$

is called the *turn-around radius*, since it is the smallest radius attained by the shell ($v_1 \to 0$). Correspondingly we define the turn-around (maximum) field

$$\frac{H_t}{H_0} = \left(\frac{R_1}{r_t}\right)^2 . \tag{10}$$

From these results we see that within the validity of this simple model the maximum field H_t increases with increasing initial kinetic energy W_{K_0} and with decreasing trapped flux W_{M_0}/H_0. We also see that the velocity of the inner boundary gradually increases and at $r_1 \to 0$ would diverge as $(r_1^2 |\ln r_1|)^{-\frac{1}{2}}$. Thus this behaviour also means that during implosion of an incompressible shell the (constant) kinetic energy concentrates more and more towards the inner boundary.

If, in addition, the liner also remains thin in the vicinity of the turn-around radius, so that

$$\frac{R_0^2}{r_1^2} \ll 1 ,$$

eq. (8) reduces to

$$\left(\frac{v_1}{v_0}\right)^2 \simeq 1 - \left(\frac{r_t}{r_1}\right)^2 ,$$

from which we calculate for later use the deceleration

$$\left|\frac{\mathrm{d}v_1}{\mathrm{d}t}\right| \simeq \frac{v_0^2 r_t^2}{r_1^3} . \tag{11}$$

3. – Magnetic-flux losses.

For a reasonable understanding of flux compression systems, particularly as far as their maximum performance is concerned, one must at least include the *compressibility* and the *finite electrical conductivity* of the imploding shell. The proper description of a flux compression experiment can be done only within the framework of *magnetohydrodynamic theory* (MHD). A detailed treatment of such problems, which involve a large numerical effort (see, *e.g.*, [3]), lies, however, outside the scope of this lecture. In the following we shall proceed only to a discussion of the most important physical consequences that derive from considering a real conductor.

If finite electrical conductivity of the shell is taken into account, a fraction $(1-\lambda)$ of the initially trapped magnetic flux $\pi R_1^2 \mu_0 H_0$ will diffuse into the conductor during implosion and be lost for further compression. For the field amplification as a function of radius we can write, therefore,

$$\frac{H}{H_0} = \lambda \left(\frac{R_1}{r_1}\right)^2 . \tag{12}$$

Differentiation of this equation yelds

$$\frac{1}{H}\frac{\mathrm{d}H}{\mathrm{d}t} = v_1 \left(\frac{2}{r_1} - \frac{1}{\lambda}\left|\frac{\mathrm{d}\lambda}{\mathrm{d}t}\right|\right) .$$

This expression shows that the field maximum can be obtained at a finite radius, which eventually may be larger than the turnaround radius r_t.

It may be instructive to give a simple, if only qualitative, physical meaning to the maximum field amplification, when finite electrical conductivity of the shell is taken into account. The characteristic diffusion time of a magnetic flux out of a cylindrical cavity with radius r is ([21], Ch. 3)

$$\tau_{\text{diff}} \simeq r^2 \sigma \mu_0 ,$$

whereas the remaining compression time, at constant implosion velocity v_0, is

$$\tau_{\text{compr}} \simeq \frac{r}{v_0} .$$

The two time values become equal at the radius

$$r_{md} = \frac{1}{v_0 \sigma \mu_0} . \tag{13}$$

Consequently, in the first part of the implosion (from R_1 to r_{md}) we may neglect flux losses, but thereafter flux is rapidly lost. Hence, we can say that the maximum magnetic field is attained at about r_{md}, *i.e.*

$$\frac{H_{md}}{H_0} = \alpha \left(\frac{R_1}{r_{md}}\right)^2 , \tag{14}$$

where α is a numerical parameter, yet unknown. The problem, however, is: what value of the electrical conductivity do we have to take in eq. (13)? In fact, one knows that the conductivity of the most interesting metals for this application follows the temperature-dependence

$$\sigma = \frac{\sigma_0}{1 + \beta c_v \theta} , \tag{15}$$

where θ is the temperature in degrees Celsius, c_v the volume specific heat, β a temperature coefficient and σ_0 the conductivity at 0 °C. On the other hand, according to a general result of the magnetic diffusion theory [21], the surface temperature of the conductor is approximately given by the equipartition law

$$c_v \theta \simeq \frac{\mu_0 H^2}{2} . \tag{16}$$

By combining the two expressions we obtain at high temperatures

$$\sigma \simeq \sigma_0 \left(\frac{h_c}{H}\right)^2 ,$$

where

(17) $$h_c = \sqrt{\frac{2}{\mu_0 \beta}}$$

is a critical magnetic field above which the heating due to eddy currents becomes an important effect in the field diffusion (non-linear diffusion). Since the flux is lost mainly toward the end of compression, when the conductivity is lowest

$$\sigma = \sigma_{\min} \simeq \sigma_0 \frac{h_c^2}{H_{md}^2} ,$$

we can extend eqs. (13), (14) into

$$\frac{H_{md}}{H_0} = \alpha R_M^2 \left(\frac{h_c}{H_{md}}\right)^4 ,$$

where

$$R_M = R_1 v_0 \sigma_0 \mu_0$$

is defined as the *magnetic Reynolds number.* It is instructive to rewrite this expression in the form

(18) $$\frac{H_{md}}{h_c} \simeq (\alpha)^{\frac{1}{5}} \left[\frac{\sigma_0^2 \mu_0^2}{h_c} v_0 (H_0 R_1^2)\right]^{\frac{1}{5}} .$$

(A numerical calculation based on an incompressible shell imploding with constant velocity gives $\alpha \simeq 1.1$, [13].) We then see clearly that the maximum attainable fields, obtained by the implosion of an incompressible shell with variable conductivity, can be influenced, but only weakly, by the initial velocity v_0 of the initial trapped flux $(\pi R_1^2 \cdot \mu_0 H_0)$.

For *example,* for a copper conductor [$\sigma_0 = 63.3 \cdot 10^6 \ (\Omega \cdot \mathrm{m})^{-1}$; $h_c = 430$ kOe] and the very good initial values $H_0 = 100$ kOe, $R_1 = 5$ cm, $v_0 = 0.5$ cm/μs (*i.e.* $R_M = 20\,000$), we obtain from eq. (18) $H_{md} = 18$ MOe and from eq. (14) $r_{md} = 0.1$ cm.

4. – Influence of metal compressibility.

By taking *the compressibility* of the shell into account we make a further important step towards reality. For the purpose of separating this problem, for the time being, from magnetic field diffusion, we will assume that the shell is an ideal conductor ($\sigma \to \infty$). The problem then becomes purely hydrodynamical and the magnetic field compression enters only through the boundary condition (5). It is clear that these equations can be solved only numerically. In the following we shall first outline the most important physical effects and give then some results obtained by numerical computation.

As the magnetic pressure at the shell surface increases during compression, the inner layers of the shell are gradually compressed. When the magnetic field is of the order of megaoersted and its rise time in the microsecond range, the pressure pulse may eventually pile up in front and form a shock wave characterized by its front velocity v_s and flow velocity u. We can say that the magnetic pressure

$$p_H = \frac{\mu H^2}{2} \tag{19}$$

drives a shock wave into the conductor thereby compressing the free surface at a velocity u into the conductor. This problem is analogous to the classical shock wave problem, where a piston is driven with constant velocity u into the compressible fluid. Provided p and u remain the same in the whole shocked region (*i.e.* between the free boundary and the shock front), the solution is given by the Hugoniot relation (see, *e.g.*, [21], Ch. 10)

$$p = \varrho_0 v_s u\,, \tag{20}$$

together with the empirical velocity law

$$v_s = c_0 + Su\,, \tag{21}$$

where S is a material constant, c_0 is the sound velocity and ϱ_0 the mass density of the unshocked fluid. In our problem, the velocity of the free surface u is thus determined by the equation

$$u(c_0 + Su)\varrho_0 = \frac{\mu}{2} H^2 \tag{22}$$

(u and v_s are measured with respect to the unshocked fluid).

When the pressure p_H (hence the magnetic field H) increases with time, the pressure distribution in the fluid, as well as the flow velocity, become time- and space-dependent. Nevertheless, various numerical results show (see, *e.g.*, [21]) that eq. (22) provides a useful, though rough approximation for the surface velocity u.

Applying these results to our imploding shell (Fig. 2), we see that maximum field is attained when

$$u = v_0 , \tag{23}$$

where v_0 is the (constant) implosion velocity of the shell.

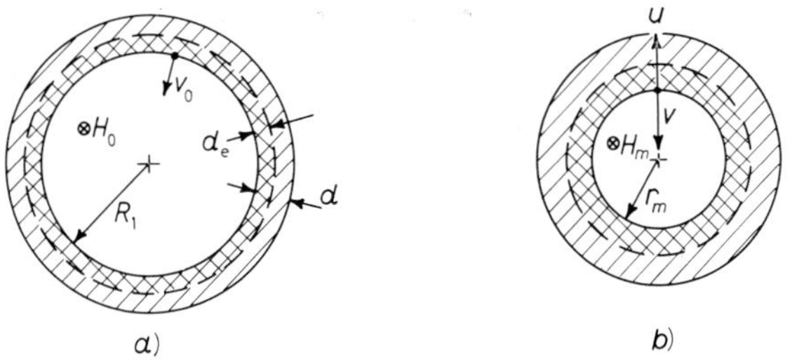

Fig. 2. – Implosion of a compressible shell.

We obtain the *ideal « maximum » magnetic field* H_{m0} by equating the magnetic pressure and shock pressure (eq. (20))

$$\frac{\mu_0 H_{m0}^2}{2} = \varrho_0 v_0^2 \left(\frac{c_0}{v_0} + S\right) . \tag{24}$$

Since c_0/v_0 and S are both of the order of unity, if follows that the energy density of the maximum attainable fields is about twice the kinetic-energy density of the liner.

Due to the finite velocity of the shock wave, only a fraction of the initial kinetic energy is, therefore, transformed into magnetic energy of the compressed field. This effect can be expressed by the *effective thickness* d_e which we define as the thickness of an inner shell of the liner at the initial position whose kinetic energy is transformed totally into the magnetic energy of the final field. According to this definition we have (for $d_e \ll R_1$)

$$2\pi d_e R_1 \varrho_0 \frac{v_0^2}{2} = \frac{\mu_0 H_{m0}^2}{2} \pi r_{m0}^2 ,$$

where r_{m0} is the radius corresponding to maximum field. Hence, by using eq. (24) and the flux conservation relation ($H_{m0} \cdot r_{m0}^2 = H_0 \cdot R_1^2$) we obtain

$$d_e = \frac{H_0 R_1}{v_0} \sqrt{\frac{\mu_0}{2\varrho_0} \left(\frac{c_0}{v_0} + S\right)} . \tag{25}$$

In an imploding cylindrical generator the final field can be larger than H_{m0}, because the velocity of the imploding shell increases as approximately shown by eq. (8). The results of a numerical calculation [13] based on the hydrodynamical problem and an appropriate equation of state for copper can be approximated by the simple expression

$$\frac{H_{mc}}{H_{m0}} \simeq \sqrt[3]{\frac{d}{d_e}}, \quad \left(1 < \frac{d}{d_e} \leqslant 30\right). \tag{26}$$

Considering the same example mentioned in connection with the diffusion problem, we now obtain by dynamic limitation for a copper conductor of thickness $d=0.3$ cm: $d_e=0.1$ cm, $H_{m0}=11.3$ MOe, $H_{mc}=16.5$ MOe, $r_{mc}=0.4$ cm. Thus we obtain about the same maximum field as by considering the magnetic diffusion limit (eq. (18)). Here, however, the limit is more stringent than for magnetic diffusion alone. In fact, we dispose of only one «free» parameter, v_0, whereas in eq. (18) we have, in addition, the flux $(\mu_0 H_0 \pi R_1^2)$.

5. – Flux compression by real conductors.

A realistic evaluation of the compression dynamics requires that field diffusion and the compressibility of the conductor be simultaneously taken into

TABLE III. – *Theoretical results* (*).

Conductivity law	Max. field H_m (MOe)	Min. radius r_m (cm)	Rel. magn. energy in πr_m^2 (%)	Rel. magn. energy tot (%)
Kidder's numerical calculations				
a) conductivity $\sigma=\infty$;	13.1	0.32	44	44
b) conductivity of metal as in eq. (15) (nonconducting metal vapor)	11.4	0.26	22	30
c) same as in *b*), but temperature coefficient β increased by a factor of 3	10.5	0.21	12	26
Our approximations				
d) compressible shell, $\sigma=\infty$; see eq. (26)	13.1	0.32	44	44
e) conductivity as in eq. (15), incompressible shell with constant implosion velocity v_0; see eqs. (14), (18)	13.9	0.08	—	—

(*) Initial values: Copper shell, $R_1=3.81$ cm, $R_2=4.13$ cm, $v_0=0.37$ cm/μs, $H_0=90$ kOe ($W_{H_0}=1.47$ kJ/cm, $W_{K_0}=480$ kJ/cm).

account. This general problem, which includes, therefore, the complete set of MHD equations, has been solved through a numerical calculation by KIDDER [3]. Without going into details (which, in fact, turn out to be not so important), we have used Table III to summarize Kidder's most significant results and compare them with the data obtained from the approximated expressions given in the previous Sections.

From this comparison and from the previous discussions on the most significant effects playing a role near turnaround, we can draw the following conclusion.

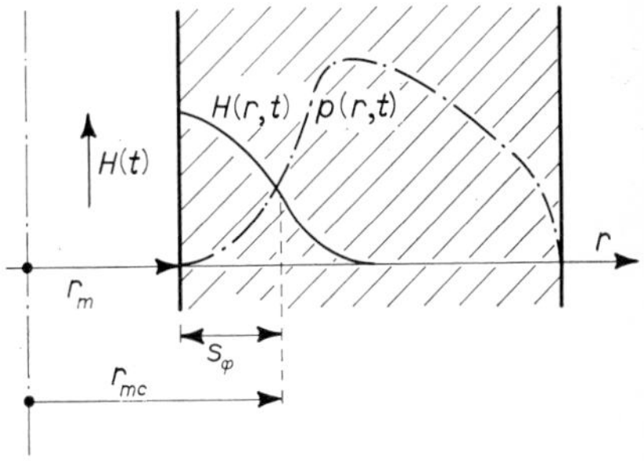

Fig. 3. – Cross-section through a real flux compressing shell at maximum field.

a) *Maximum field* H_m is determined principally by the implosion dynamics of the compressible piston, *i.e.* $H_m \simeq H_{mc}$. In fact, if a flux skin depth s_φ is taken into account (Fig. 3), the conclusions we have drawn with respect to the cylindrical compressible piston of radius r_{mc} remain valid; they are simply extended to an inner surface of an approximated radius $r_m + s_\varphi$, to which the magnetic pressure is substantially applied.

b) *Minimum radius* r_m is determined principally by the conductivity of the piston. If H_m is fixed by pt. *a*), this radius follows naturally from flux conservation: $H_m(r_m + s_\varphi)^2 = H_0 R_1^2$.

These conclusions apply to a generator giving typically $H_m = (5 \div 15)$ MOe and $r_m = (0.3 \div 1)$ cm. In other, extreme conditions, other effects may predominate. For example, if the final radius tends to become of the same order of the diffusion radius r_{md}, or smaller (*i.e.* of the order of millimeters), it is clear that the maximum field is fixed by diffusion losses alone, *i.e.* $H_m \simeq H_{md}$.

6. – Metal-field interface.

Ultrahigh-field generators are subjected, in addition to enhanced field diffusion and shock compression, to other effects that can, in practice, limit their performance. Among the most disturbing we mention the boiling-off of the shell surface layer and the instability of the metal-field interface.

As the temperature of the flux compressing conductor predicted by the equipartition law (16) easily exceeds the melting and vaporization points at normal pressures (at 10 MOe the surface temperature for copper amounts according to eq. (16) to about 10^5 °K), the surface layer will undergo phase changes and a certain amount of it will be projected ahead of the piston as vapor. The metallic fluid is exposed, in addition, to all sorts of magnetohydrodynamic instabilities. For all practical cases of interest to us, the *Rayleigh-Taylor instability* is by far the most important. In flux compression systems the effect of such instabilities is to penetrate into the volume where the ultra-high fields are generated and used and thus seriously limiting the performance of the generators.

Let us recall the most simple case of the Rayleigh-Taylor instability. Consider an ideal fluid of density ϱ, whose surface is displaced as

$$\eta(t=0) = \eta_0 \cos kx \,,$$

where k is the wave number (see Fig. 4). According to linear perturbation theory, the displacement develops in time as

$$\eta = \eta_0 \cosh \omega t \cdot \cos kx \,,$$

with ω given (in this simple case) by the dispersion relation

$$\omega^2 = \mp gk \;;$$

here the plus sign applies when the fluid is supported against gravity by the magnetic field (unstable case), and the minus sign applies for the opposite, stable case.

This result can be applied to study the stability of a moving sheet, provided we replace the (vector of) gravity with the inertia, *i.e.* with the opposite of the acceleration of the fluid (Fig. 5).

Therefore, when the imploding cylindrical shell is being slowed down towards compression end by the growing magnetic pressure, azimuthal perturbations

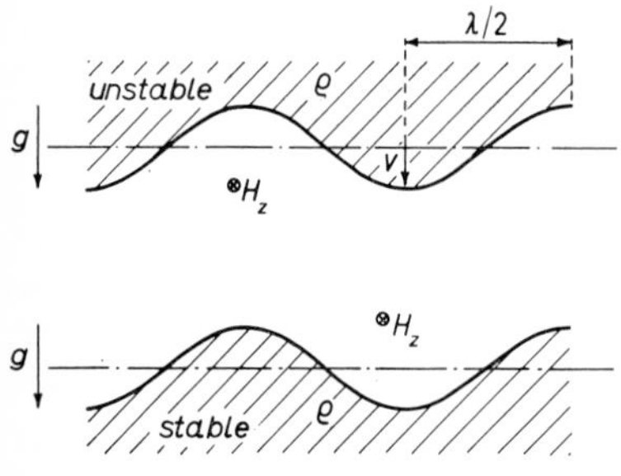

Fig. 4. – Rayleigh-Taylor hydromagnetic instability.

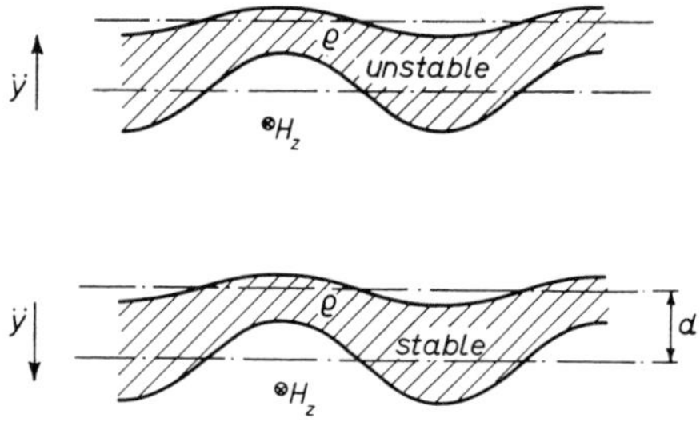

Fig. 5. – Rayleigh-Taylor instability of an accelerated sheet.

will grow at an approximated growth rate of

$$\omega^2 \simeq |\ddot{r}|k ,$$

where for k we must take into account that the possible wave lengths are limited by the condition $n\lambda = 2\pi r_1$; $n = 1, 2, \ldots$. Near the turnaround radius r_t, the deceleration can become very large (typical value: $10^{10}\ \text{m}\cdot\text{s}^{-2}$), since the magnetic pressure increases as $1/r^4$ (eq. (5)). By using the approximation (11) (valid for thin shells) at $r_1 = r_t$, we obtain the maximum growth time

$$\frac{1}{\omega} \simeq \frac{1}{\sqrt{n}} \frac{r_t}{v_0}, \tag{27}$$

where v_0 is the initial implosion velocity. We see that for short wavelengths ($n > 1$) this time becomes shorter than the characteristic implosion time r_t/v_0. (Time it takes the shell to implode from $2r_t$ to r_t, during which the field increases by a factor of 4 and the magnetic pressure by a factor of 16). When reaching the turn-around point, Rayleigh-Taylor instabilities can thus already be worrysome. Immediately thereafter, however, in a time of less than $1/\omega$, the instability spikes, which continue their growth, will penetrate into the useful field volume and destroy any probe or sample in it.

Some doubts about these results may arise because they are based on small amplitude perturbations, *i.e.* linearization of the problem. In fact, as the amplitude Δr of the harmonic perturbation becomes larger than about half the wave length, nonlinear effects become recognizable by changing the shape of the perturbation toward sharper spikes.

Nevertheless, the estimates made previously are confirmed by a detailed study of Rayleigh-Taylor instabilities in ultrahigh-field generators made by SOMON [22]. He used a particular numerical method to solve the full, nonlinear, time and two space co-ordinate-dependent fluid equation for a compressible, nonviscous fluid. This study shows that the instabilities can reduce the maximum field, and, in any case, they exclude any well-defined and « clean » turn-around condition.

7. – Explosively driven generators.

The working principle of an imploding ultrahigh field generator is based on the three following processes:

a) Generation of an initial magnetic flux in the bore of the cylindrical shell (the « liner »).

b) Implosion of the liner by means of an outer energy source (*e.g.*, explosives).

c) Compression of the trapped magnetic flux by the imploding liner, *i.e.* transformation of (a part of) the kinetic energy into magnetic energy.

Technically, the most difficult problem is the generation of a cylindrically imploding detonation wave, since in addition to the problem of simultaneously detonating a large surface (Fig. 6), one must also require that the variations

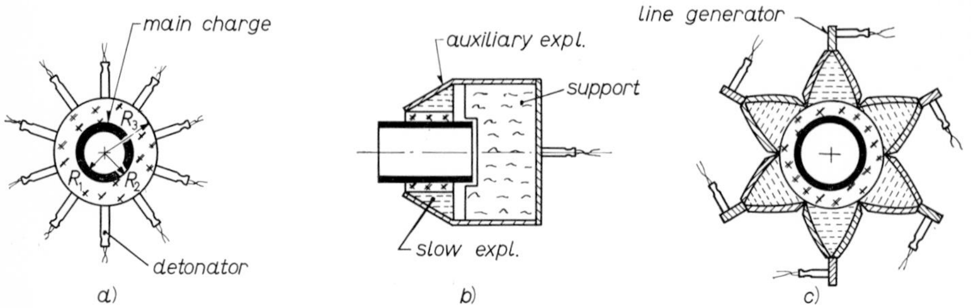

Fig. 6. – Various detonation methods to obtain a cylindrically converging detonation wave.

Δr on the mean detonation radius remain within $\Delta r \leqslant 0.1$ cm. This condition is, in fact, obtained if one considers the small final radii in high-performance generators and if one requires that maximum fields be produced around the predetermined generator axis, where the field probes or other experimental gadgets are placed. In addition, initial perturbations of the liner are also required to be as small as possible by stability considerations. On the other hand, it is obvious that the constancy of the detonation radius along the axis is not so important. In fact, the flux compression by a convex, but otherwise coaxial, liner does not limit the maximum field that can be produced locally on the axis. Thus, in many cases, it is sufficient to solve the detonation problem in the relatively simple way shown by Fig. 6 *a*) or 9.

Here we are mainly interested in the kinetic energy imparted to the imploding liner by the detonation pressure. The simple arrangement consisting of an explosive driving a plane liner has been studied extensively, also theoretically [23]. Our cylindrical geometry is more complex, the main differences with respect to the plane case being:

the finite distance over which the liner can be accelerated, *i.e.* from R_2 to about R_0, see eq. (3);

the relative pressure increase due to the convergence of the explosive products during implosion, and the decreasing surface ($2\pi r_2$) on which the explosive products can act.

The transfer efficiencies and the liner velocities obtained by extending the calculation methods and the results of the plane case [23] to cylindrical geometry are summarized in Fig. 7. We find, for example, that maximum transfer efficiencies can attain values of up to 25%. To obtain these efficiencies it is required, of course, that end losses be negligible. In practice this condition is fulfilled if the length of the charge is about equal to its outer radius R_3, or for shorter charges, if heavy tampers are applied at the ends.

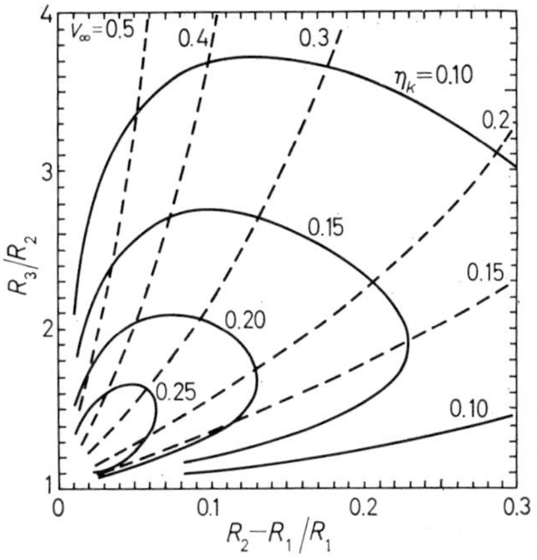

Fig. 7. – Transfer efficiency η_k (kinetic energy over chemical energy, per unit length) and center-of-mass velocity v_∞ of the imploding liner at compression end ($r_{2m} \simeq R_0$). The curves are obtained from an approximated calculation made by SOMON [13], in which the results of the plane case [23] have been transformed to describe the case of a thin, incompressible cylindrical copper liner.

The injection of a sufficiently large magnetic flux into the bore of the cylindrical shell represents a serious technical and experimental difficulty. In fact, points *a*) and *c*) (mentioned earlier in this Section) pose, in principle, two opposite conditions. In the beginning it is required that magnetic flux be easily introduced from the outside into the liner's bore; but, then, during implosion, no flux should diffuse out of the compression volume.

Two methods are currently used to cope with these requirements. In one, the liner is *slotted* longitudinally; since the conductor is now open, it is no problem to build up an inner flux. When the detonation wave hits the liner, the slot is violently forced to close and the flux trapping is perfect. The disadvantage of this method lies in the implosion dynamics. In fact, the slot represents a mass discontinuity, giving rise to irregularities and even jets, which seriously disturb the implosion symmetry. In the second method one uses a continuous, seamless liner. The diffusion through it of a magnetic field generated by outer coils is obtained by choosing the parameters (resistivity, liner thickness, time constants, etc.) as required by the conditions for magnetic-field diffusion (see [21], Ch. 4). The explosive-driven implosion is then sup-

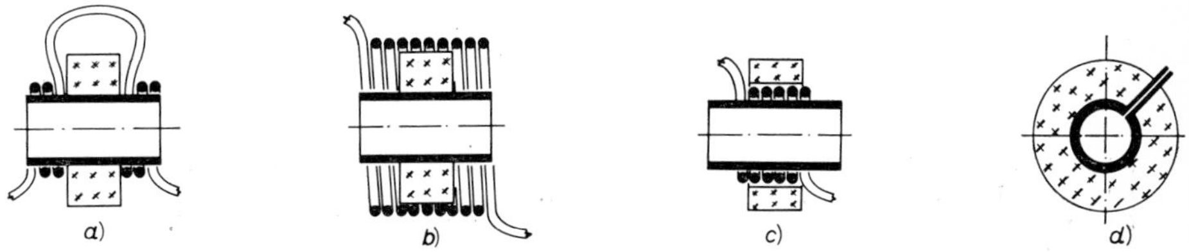

Fig. 8. – Various arrangements concerning the coil-explosive system.

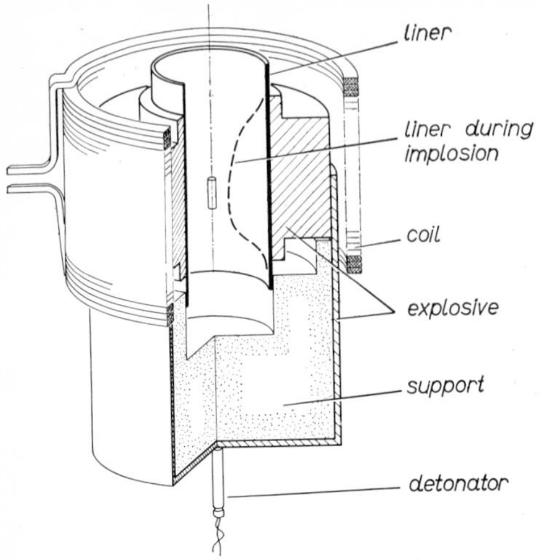

Fig. 9. – Charge type 45 as used by the Frascati group for the generation of fields in the $(5 \div 6)$ MOe range (liner diameter is $2R_1 = 7.7$ cm).

posed to be sufficiently rapid to avoid any flux leakage to the outside space. This condition is certainly fulfilled toward the compression end, when the liner has anyway thickened appreciably. A major practical difficulty arises because the coil generating the initial flux should actually be placed where the explosive is, since just beneath the explosive the liner is propelled and flux compression takes place (Fig. 8). The solution shown in Fig. 8 *c*) is the best, but clearly it can disturb the propulsion of the liner. Thus various other solutions have been adopted in practice, all having their own merits and disadvantages

TABLE IV – *Expl*

Location		Ref.	Explosive			
			Type	Ignition syst. (Fig. 6)	Length (cm)	*R* (c
Los Alamos	1960	[24]	comp. *B*	Fig. *a*); 22 det.	7.6	[illegible]
	1966	[4]	comp. *B*	Fig. *a*); 25 det.	7.6	[illegible]
Moscow (*e*)	1966	[7, 26, 1]	—	Fig. *a*) or *d*)	—	3
	1966	[26]	—	Fig. *a*) or *d*)	—	~3
Foulness	1967	[27]	comp. *B*	Fig. *a*); 2 rings of 60 det.	10	[illegible]
Limeil	1967	[29, 5]	—	Fig. *d*); 4 lenses	5	[illegible]
Frascati	1967	[30, 2]	CP8	Fig. 9	8	

(*a*) Brass (BR) and copper liners are slotted, stainless steel liners (SS) are seamless.
(*b*) The coil was, however, placed inside the liner.
(*c*) An explosively driven current generator was used to produce the initial field.
(*d*) Have been produced once or twice.
(*e*) Most of the indications are estimated.

(see, *e.g.*, Fig. 9). The use of a capacitor bank to energize the coils has many advantages, although it is by no means the only solution. The main advantage is the relatively short rise-time of the current, which often makes it possible for the coil to be contained by inertial forces only. This effect can greatly simplify the coil construction and also reduce its cost.

The experimental situation in ultrahigh field generation may be said to be still at its beginning, although about 10 years have elapsed since FOWLER *et al.* first reported their work. The results, as shown in Table IV, are widely spread and not systematic. Although SAKHAROV *et al.* claim to have produced fields in excess of 25 MOe, it seems safe to say that up to 1969 reproducible and applicable magnetic fields, are somewhere between 6 and 10 MOe, probably much nearer to the former figure than to the latter. In any case, fields that have been applied to physical measurements other than purely inductive probes (notably the Zeeman and Faraday effects [31] and temperature-dependent field diffusion [18]) lie below the 5 MOe mark. One can offer two main explanations for this situation:

a) The experimental means required to generate reproducible fields in the (6÷10) MOe region and above are sophisticated and expensive. Consequently, the reproducibility of these experiments is very low.

ated ultrahigh fields.

Liner			Initial field			Max. performance	
$_2$ m)	R_1 (cm)	Metal ([a])	Coil type (Fig. 8)	Bank en. W_{CB} (kJ)	H_0 (kOe)	H_m (MOe)	r_m (cm)
81	3.49	BR	Fig. *c*) ([b])	30	90	14 ([d])	— 0.25
4	5.24	SS	Fig. *a*)	300	30	5 —6	~ 0.13
—	—	SS (+20 μ Cu)	Fig. *c*)	(C.B.)	—	25 ([d])	~ 0.2
—	15	Cu	Fig. *c*)	([c])	~ 60	5	~ 1.5
3	5	Cu	Fig. *c*)	—	80	3.5÷5	—
25	4.1	SS	Fig. *a*)	200	50	5 ÷7	~ 0.25
95	3.85	SS	Fig. 9	200	65	5 ÷6	~ 0.25

b) Often the most advanced techniques, and therefore the corresponding experimental results as well, are classified for military reasons. Consequently, the exchange of information is extremely difficult and a critical appreciation of the results virtually impossible.

In Fig. 9 the explosive generator used by the Frascati group in the 5 to 6 MOe range is shown. A typical implosion sequence and a field record are shown in Fig. 10 and 11.

The generation of a field of 10 MOe in a final diameter of 1 cm (a dimension which is necessary for a reasonable reproducibility of the results) should be possible, but requires a formidable experimental and technological effort.

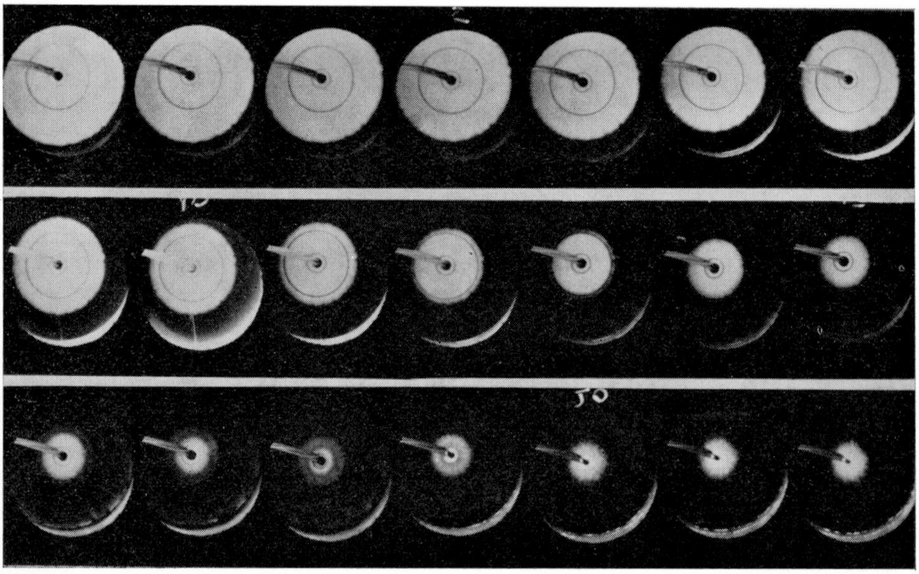

Fig. 10. – Framing camera sequence showing the implosion of a stainless steel liner of the seamless type, with $R_1 = 3.65$ cm, $R_2 = 3.75$ cm. The compression device is of the type shown in Fig. 9 but with less explosive ($R_3 = 4.9$ cm). Time interval between pictures is 1 μs; diameter of the ring marker is 3 cm.

For example, the explosive charge will have an outer diameter of about 50 cm and the energy source feeding the coil for the generation of the initial field must store an energy of at least 500 kJ (see [21], Ch. 9).

Finally, the generation of still higher fields (at the 20 MOe level and higher) is determined mainly by the hydrodynamic conditions established previously. In practice, it requires implosion velocities in excess of 1 cm/μs at half initial radius, *i.e.* values that can hardly be obtained with classical high explosives.

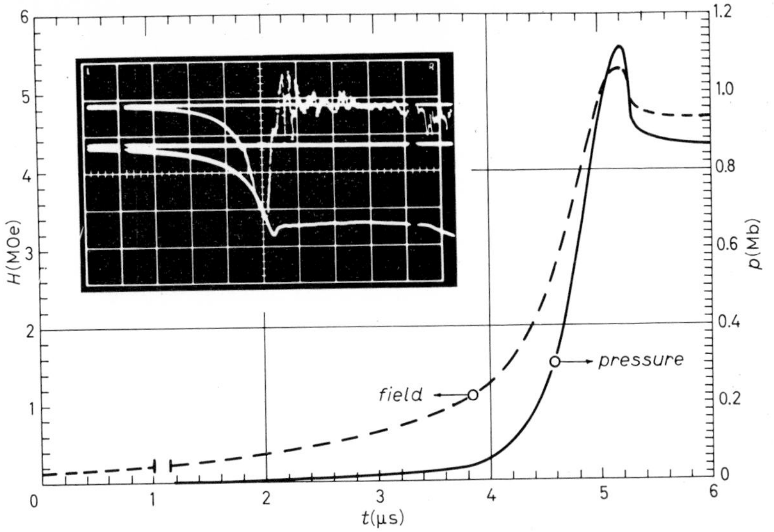

Fig. 11. – The inserted oscillogram reproduces a typical magnetic field trace and the corresponding dH/dt signal obtained with a generator as shown in Fig. 9. The field curve is redrawn in its last phase, together with the correspondign $\mu_0 H^2/2$ pressure curve. Initial field was 62 kOe, total implosion time 11.5 μs.

Unless new, more powerful chemical explosives are developed, the only solutions that remain are to resort to magnetically driven shells (which can attain velocities of around 2 cm/μs), or use nuclear explosives, or apply a combination of both.

REFERENCES

[1] H. Knoepfel and F. Herlach, Editors: *Megagauss Magnetic Field Generation by Explosives and Related Experiments*, EUR 2750 e (Euratom, Brussels 1966).

[2] F. Herlach and H. Knoepfel, Editors: *Megagauss Magnetic Field Generation by Explosives and Related Experiments*, EUR 2750 e (Euratom, Brussels, 1966), p. 147.

[3] R. E. Kidder, Editor: *Megagauss Magnetic Field Generation by Explosives and Related Experiments*, EUR 2750 e (Euratom, Brussels, 1966), p. 37.

[4] R. S. Caird, W. B. Garn, F. B. Thomson and C. M. Fowler, Editors: *Megagauss Magnetic Field Generation by Explosives and Related Experiments*, EUR 2750 e (Euratom, Brussels, 1966), p. 101.

[5] A. Brin, J. E. Besançon, J. Champetier, J. P. Plantevin and J. Vedel, Editors: *Megagauss Magnetic Field Generation by Explosives and Related Experiments*, EUR 2750 e (Euratom, Brussels, 1966), p. 21.

[6] F. Herlach, H. Knoepfel, R. Luppi and J. E. van Montfoort, Editors: *Megagauss Magnetic Field Generation by Explosives and Related Experiments*, EUR 2750 e (Euratom, Brussels, 1966), p. 471.
[7] L. V. Al'tshuler: *Sov. Phys. Usp.*, **8**, 52 (1965).
[8] A. D. Sakharov: *Sov. Phys. Usp.*, **9**, 294 (1966).
[9] W. G. Chace and H. K. Moore, Editors: *Proceedings of the Exploding Wires Conference III* (New York, 1964).
[10] J. W. Mather: *Phys. Fluids*, **8**, 366 (1965).
[11] A. M. Levine: *Lasers*, vol. **1** (New York, 1966).
[12] N. G. Basov, S. D. Zaharov, P. G. Kriukov, U. V. Senatski and S. V. Tsekalin: *Experiments for the observation of neutrons produced by focusing a high-power laser onto a lithium-deuterite target*, in *Proc. Int. Quantum Electronics Conf.* (Miami, 1968).
[13] J. P. Somon: EURATOM Rep. EUR 4197 f (1968).
[14] G. Lehner: in *Springer Tracts in Modern Physics*, vol. **47** (Berlin, 1968), p. 67.
[15] J. W. Shearer: *Journ. Appl. Phys.*, **40**, 4490 (1969).
[16] F. Herlach and H. Knoepfel: *Rev. Sci. Instr.*, **36**, 1088 (1965).
[17] H. Knoepfel, H. Kroegler, R. Luppi and J. E. van Montfoort: *Rev. Sci. Instr.*, **40**, 60 (1969).
[18] H. Knoepfel and R. Luppi: *Exploding Wires*, vol. **4** (New York, 1968), p. 233.
[19] R. A. Gross: This volume, p. 245.
[20] F. Winterberg: This volume, p. 370.
[21] H. Knoepfel: *Pulsed High Magnetic Fields* (Amsterdam, 1970).
[22] J. P. Somon: *Journ. Fluid. Mech.*, **38**, 769 (1969).
[23] A. K. Aziz, H. Hurwitz and H. M. Sternberg: *Phys. Fluids*, **5**, 380 (1961).
[24] C. M. Fowler, W. B. Garn and R. S. Caird: *Journ. Appl. Phys.*, **31**, 588 (1960).
[25] F. H. Harlow and W. E. Pracht: *Phys. Fluids*, **9**, 1951 (1966).
[26] A. D. Sakharov, R. Z. Lyudaev, E. N. Smirnov, Yu. I. Plyushchev, A. I. Pavlowskii, V. K. Chernyshev, E. A. Feoktistova, E. I. Zharinov and Yu. A. Zysin: *Sov. Phys. Dokl.*, **10**, 1045 (1966).
[27] C. S. Speight: A.W.R.E. Report n. 0-71/67 (1967).
[28] *Les champs magnétiques intenses*, (Paris, 1967).
[29] J. E. Besançon and J. Vedel: *Les champs magnétique intenses*, (Paris, 1967), p. 345.
[30] H. Knoepfel: in *Physics of Solids in Intense Magnetic Fields* (New York, 1969), p. 467.
31] W. B. Garn, R. S. Caird, D. B. Thomson and C. M. Fowler: *Rev. Sci. Instr.*, **37**, 762 (1966).

Cumulation Processes. Self-Similar Solutions in Gas Dynamics.

J. P. Somon

Laboratori Gas Ionizzati (Associazione EURATOM-CNEN) - Frascati

Introduction.

High pressures generated statically in the laboratory are limited to the tenth of kilobars range by technical restrictions such as the weakness of the material used for compression. Much larger pressures of the order of megabars and much higher energy densities are produced in a dynamical way. The basic scheme is the following single-shock process: a plane piston, driven by some external energy source (explosive, electromagnetic energy, ...), moves impulsively at a constant velocity v_0 in a medium at rest of density ϱ_0 and pressure P_0. A uniform shock wave is generated behind which the pressure P_1 reaches high values. If $P_1 \gg P_0$, which is the case of strong shock waves in gases and (except for the weakest) of most shock waves in solids, the increase of specific internal energy through the shock, $\Delta e = e_1 - e_0$, is equal to the specific kinetic energy of the piston $v_P^2/2$; if moreover, the increase in density is appreciable, the pressure $P_1 \approx v_P^2/\Delta\tau$ ($\tau = 1/\varrho$) is of the order of $\varrho_0 v_P^2$. Achievable energy densities and pressures are obviously limited by the characteristics of available energy sources. High explosives allowed to accelerate metallic plates to the velocity of $1.4 \cdot 10^6$ cm/s and the impact with solid targets generated pressures of 10 Mb [4]. It is interesting not only to reach high pressures but also to cover a large region of the thermodynamical plane (P, τ). Unfortunately the initial uncompressed state of the medium cannot be chosen arbitrarily. For solids it is restricted to a small region around the normal state (p_N, τ_N) and a single-shock process only allows to create compressed states which are situated within a narrow zone along the Hugoniot curve passing through (p_N, τ_N).

It is possible to extend the pressure range or more generally the thermodynamical domain reached by means of this single-shock process. The first part of these lectures is intended to give a brief review of some principles allowing such an extension and to provide an introduction to the study of cumulation processes. In cumulation processes it appears to be possible, at least in ideal conditions, to concentrate the energy provided by the external source in order to generate infinite pressures or temperatures within a vanishingly small volume. Cumulation is due to unsteady motions for which the solution of fluid equations can generally be obtained only through numerical calculation. Nevertheless, dimensional considerations show that the solution of cumulation problems for perfect gases can be obtained in many cases of interest; several examples of the so-called self-similar solution are treated in the second part. Clearly a perfect cumulation is not possible and physical imitations are briefly outlined.

1. – Some processes to reach higher energy densities.

1·1. *Interaction of shock waves. Converging shocks.* – Let a medium be shocked twice and brought from initial state (0) to state (1) by a first shock wave, from state (1) to state (2) by a second shock wave. Hugoniot adiabatics $\mathscr{H}_1$ and $\mathscr{H}_2$ relating these states, isotherms $\mathscr{I}$ and adiabatics $\mathscr{A}$ are shown in the thermo-dynamical co-ordinates P, $\tau = 1/\varrho$ of Fig. 1. A state of given pressure $P_2 = P_{2'} = P$ may be reached through a single-shock or a double-shock process. The diagram shows immediately that a two-step process leads to higher densities and lower temperatures than a single step process. A multi-step process could lead to very high densities. It is thus expected that the successive interaction of shock waves increases the pressure and energy density of a medium in an intermediate way between an adiabatic compression and a single-shock compression.

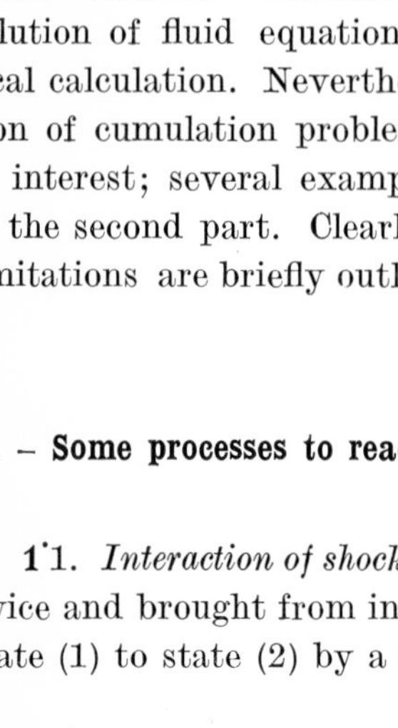

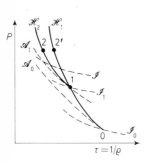

Fig. 1. – One-step and two-step shock compression processes. Pressure *vs.* specific volume.

Some figurative examples of shock wave collisions are given. The discussion is limited to the collision of shock waves of equal strengths. In fact the equivalent problem of the collision of a shock wave against an infinitely rigid wall will be considered.

1·1.1. Head-on collisions. A given uniform plane shock wave (S_1) strikes an infinitely rigid wall and a uniform shock wave (S_2) is reflected (Fig. 2). All quantities are determined by the condition of zero velocity $v_2 = 0$ between the wall and the reflected shock. They are easily calculated for a perfect gas with a polytropic coefficient γ and the state behind the reflected shock is given by [1-2]

$$(1)\quad \begin{cases} \dfrac{\varrho_2}{\varrho_1} = \dfrac{K(\varrho_1/\varrho_0)-1}{(\varrho_1/\varrho_0)+K-2}\,, \\[2ex] \dfrac{P_2}{P_1} = \dfrac{(K+2)(P_1/P_0)-1}{P_1/P_0+K}\,, \end{cases}$$

where

$$K = \frac{\gamma+1}{\gamma-1}\,.$$

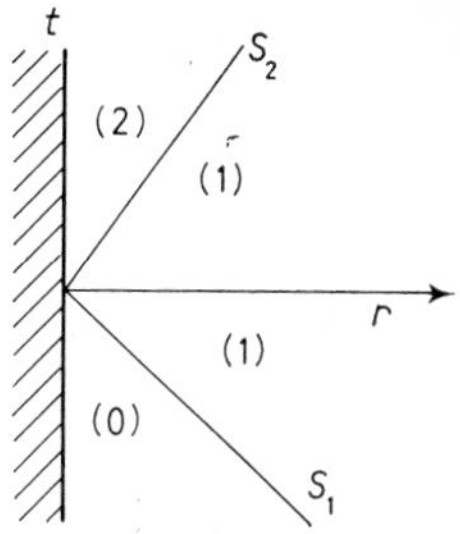

Fig. 2. – Plane shock reflection on an infinitely rigid wall.

For a weak incident shock or small overpressure $P_1 - P_0 \ll P_0$, the ratio of overpressures $(P_2 - P_1)/(P_1 - P_0)$ is close to 2 (acoustic approximation). For very strong incident shocks ($\varrho_1/\varrho_0 \to K$, $P_1/P_0 \to \infty$), the overpressure is large and $P_2/P_1 \to K+2$; because $\varrho_2/\varrho_1 \to (K+1)/2 < K$, the reflected shock can never be strong.

The effect of the reflection of a very strong shock is thus to achieve pressures much higher than those which are reached behind the incident shock ($P_2 \approx 6P_1$, for a monatomic gas with $\gamma = \frac{5}{3}$, $K = 4$). The overpressure is due to comparable density and temperature increase of the medium ($\varrho_2 \approx 2.5\,\varrho_1 \approx 10\,\varrho_0$; $T_2 = 2.4\,T_1$). Such a pressure $P_{2'} = P_2$ would have been obtained through a single shock but with a much greater temperature $T_{2'} = 2.5\,T_2$ and smaller density $\varrho_{2'} = 0.4\,\varrho_2$. A head-on collision of strong shock waves is then a two-step process which provides a mean to reach high pressures and densities without increasing the temperature too much.

1.1.2. Oblique collisions. Plane shock waves collide now obliquely with an angle 2α. The phenomenon is rather complex and is only briefly outlined. It is complicated by the existence of two different regimes depending on the value of α compared to a critical angle α_{cr} [1-4].

When $\alpha < \alpha_{cr}$, the reflection is said to be regular and an oblique plane shock is reflected (Fig. 3*a*). The condition that the velocity of state (2) is parallel to the

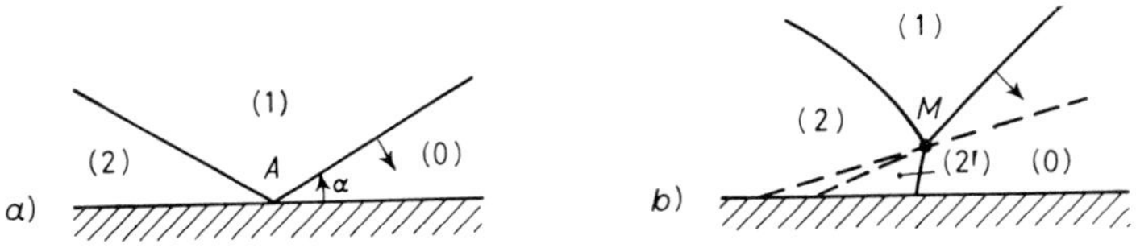

Fig. 3. – Oblique reflection of a plane shock: *a*) regular, *b*) Mach.

wall together with the Hugoniot relations through oblique shock waves dedetermine completely all quantities but solutions become imaginary for $\alpha > \alpha_{cr}$. The angle α_{cr} depends on the strength of the incident shock wave [3].

When $\alpha > \alpha_{cr}$, the reflection is irregular (Mach reflection) and experiments show the occurrence of a 3-shock waves configuration in which the triple point M moves along a straight line (Fig. 3 b)). States (2) and (2′), respectivly reached through a two-step and a one-step process, are separated by a slip-line ($P_2 = P_{2'}$, $\varrho_2 > \varrho_{2'}$, parallel but different velocities $V_2 > V_{2'}$).

The need for two different configurations may be understood considering the perturbations which can be created on the wall immediately behind the point A in the region (2) of a regular reflection. If the angle α is small enough, the phase velocity of point A is high and a sonic perturbation generated in zone (2) cannot react on the motion of point A. Increasing the angle α, these perturbations start to influence the propagation of the reflected shock and finally catch up the point A. As a consequence point A becomes detached from the wall at large angles α. The case of very weak shock waves illustrates also interesting features of the oblique reflection. For a regular reflection, the acoustical approximation shows that $P_2 - P_0 \approx 2(P_1 - P_0)$ whatever is the angle α. Nevertheless for $\alpha = 90°$, the shock merely slips along the wall and $P_2 - P_0 = P_1 - P_0$. This proves the necessity to introduce a discontinuity. Careful considerations show that a Mach reflection occurs at $\alpha_{cr} \approx 90°$, creating an overpressure peak $P_2 - P_0 \approx 3(P_1 - P_0)$ which smooths the discontinuity.

As an example of oblique collision, Fig. 4 shows the function $P_2(\alpha)$ relative to the collision of relatively weak shock waves in aluminium ($P_1 = 0.33$ Mb). The head-on collision produces a pressure of 0.82 Mb ($P_2/P_1 = 2.5$) but a much higher pressure of 2 Mb is reached for an angle $\alpha \geqslant \alpha_{cr}$ ($P_{2'}/P_1 \approx 6$). The accessible pressure range can thus be extended using the Mach reflection phenomena. Experimental arrangements are theoretically simple (Fig. 5)

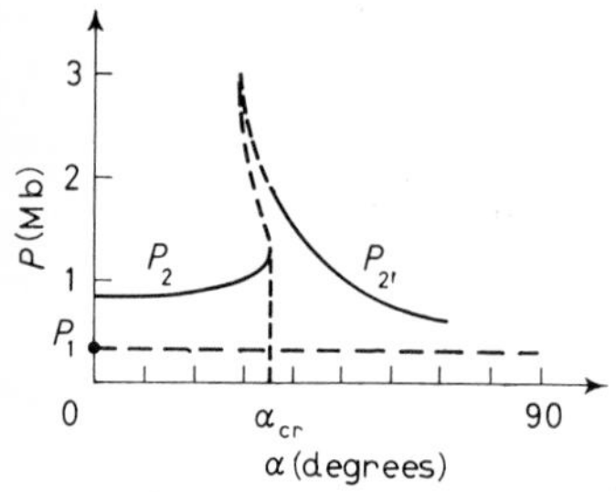

Fig. 4. – Overpressure due to oblique reflections of shocks in aluminium [4].

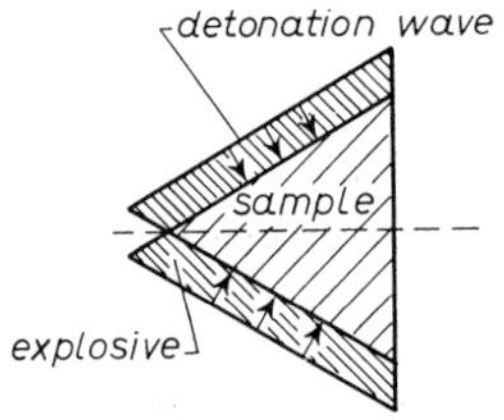

Fig. 5. – Experimental arrangement for Mach-reflection observations.

but large overpressures are only obtained for angles very close to the critical angle at which the regime of Mach reflection is somewhat unstable and not as yet well known. Similar but conical Mach arrangements have also been experimented and produced a pressure of 18 Mb in copper samples [5].

A remarkable property of oblique collision and especially of Mach reflection at the vicinity of α_{cr} is that it may produce pressures greater than the pressure obtained in a head-on collision. However this property is not general and depends on the medium and on the strength of the shock wave. In other cases the curve $P_2(\alpha)$ may have a behaviour very different from the curve of Fig. 4 [1, 22].

1·1.3. Converging shocks. Multiple interaction of oblique shock waves can be considered as, for example, plane shock fronts regularly tangent to a circular cylinder and moving towards the axis (Fig. 6). At the limit of an infinite number of shock fronts one gets the simple picture of a cylindrical converging shock for which very high pressures may be expected.

Cylindrical and spherical imploding shock waves are studied in Subsect. 2·3.2 for a perfect gas and a cumulation process is found to occur when the shock reaches the axis or center. Pressure and temperature become then infinite while density tends to a finite value in a way which does not much depend on how the wave was created. Cumulation may be explained by the fact that the shock wave encounters a vanishing mass of fluid when its radius tends to zero.

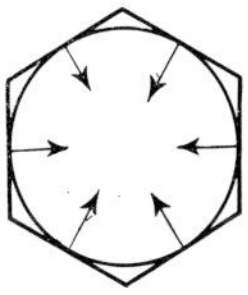

Fig. 6. – Converging shock waves.

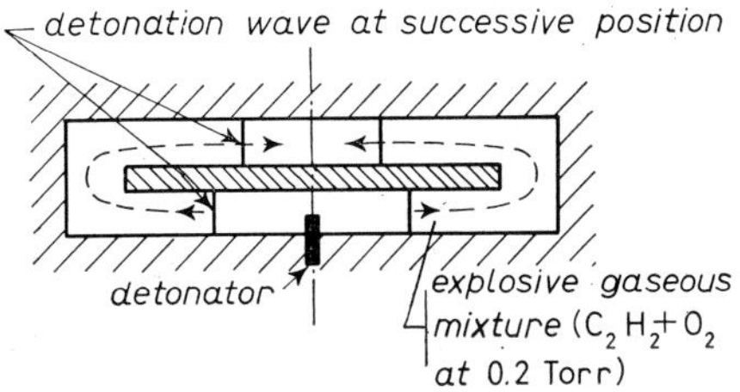

Fig. 7. – Experimental arrangement for a converging cylindrical detonation wave [26].

Converging shock waves have been produced experimentally [26, 27, 31] (review in [2]) but many results connected with weapons research have probably not been published [32]. Converging detonation-waves have also been considered. They reduce to converging shock waves when the cumulation becomes important or the chemical energy negligible compared to the shock-wave energy. An experimental cylindrical arrangement is shown in Fig. 7. It allowed to verify that the motion is self-similar near the axis and to reach pressures as large as 18 times the Chapman-Jouguet pressure [26].

1·2. *Collision of moving media.*

1·2.1. Plane geometry. Let us first consider the head-on collision at velocity v_0 of an uniform semi-infinite medium against an infinitely rigid wall.

A receeding uniform shock wave is created through which the incoming kinetic energy is transformed into internal energy. Velocities are known on both sides of the front and Hugoniot conditions consequently determine the density and pressure reached against the wall. For a metallic medium, due to the well-known linear relation between the velocity behind the shock and the shock velocity, the pressure on the wall turns out to be

$$P_M = \varrho_0 v_0^2 \left(\alpha + \frac{1}{M}\right), \qquad \alpha \approx 1.5, \qquad M = \frac{|v_0|}{c_0}, \tag{2}$$

while for a polytropic gas with a high Mach number M

$$P_M = \frac{\gamma + 1}{2} \varrho_0 v_0^2, \qquad M \gg 1 \tag{3}$$

In both cases the wall pressure becomes of the order of the kinetic energy density $\varrho_0 v_0^2$ for moderate or high shock strengths.

Figure 8 shows a more complicated scheme in which a compressible medium (1) has been inserted between the wall and the semi infinite piston (0).

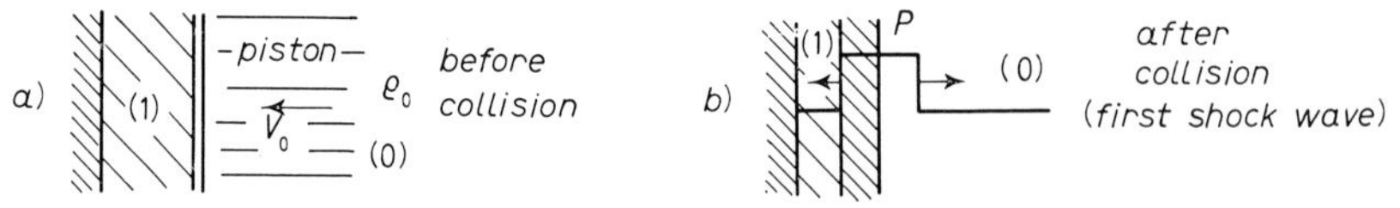

Fig. 8. – Plane compression of a medium (1) between a piston (0) and an infinitely rigid wall.

Asymptotically, when $t \to \infty$, medium (1) represents only a perturbation of the preceding case and the pressure exerted on the wall depends only on the piston. The asymptotic pressure P_M is progressively build up. At the instant of collision a first shock wave is transmitted in medium (1) and a shock wave reflected in medium (0). Let us assume that conditions are such that the pressure P_f reached behind this first shock is small compared to P_M. In medium (1) the first shock wave is reflected back and forth between the wall and the piston. Even if this first shock is strong, the shocks which follow are weak and their effect is roughly equivalent to an adiabatic compression of medium (1). Their successive reflections on the interface (0)-(1) transmit in the piston weaker and weaker shock waves which overtake the leading shock. Such reflections occur until the asymptotic pressure P_M is reached. The final state of medium (1) being obtained through one shock and an adiabatic compression, the final density can be very high. Obviously the asymptotic pressure P_M is also reached for a finite piston whose thickness d is smaller than some characteristic thickness d_c. If $d < d_c$, the asymptotical shock wave should be created outside of the piston; a maximum pressure $P_m < P_M$ is obtained when the piston stops before rebouncing and is simply calculated writing that the initial kinetic energy of the piston is equal to the internal energy of medium (1) at the pressure P_m. The same energy condition applied for $P_m = P_M$ determines the characteristic thickness d_c [8].

The oblique collision of plane media is beyond the scope of these lectures and is not studied. As for oblique collisions, two regimes are possible i) when the collision angle α is greater than some critical angle α^* a high-speed jet is created (hollow-charge effect) ii) when $\alpha < \alpha^*$ the regime becomes jet-free [9].

1'2.2. Cylindrical and spherical geometries. Many examples of converging motions may be found which depend on the initial conditions. Let us consider only extreme cases, according to the values of the initial Mach number M:

$M = \infty$ (or $c_0 = 0$). The medium may be considered as formed of independent particles converging towards the axis or the center with a radial velocity. Time $t = 0$ is the instant of collision and initial conditions are the distributions $\varrho_0(r)$ and $v_0(r)$. The collision creates a diverging shock wave behind which high pressures are reached. The simplest case $\varrho_0 = c^t$, $v_0 = c^t$, is studied in Subsection **2'3.1**. Such a converging motion is useful in plasma physics to simulate a hollow pinch [10].

$M = 0$ (or $c_0 = \infty$). The medium is incompressible and the initial conditions cannot be chosen arbitrarily. The most famous example is the Rayleigh problem or the collapse of a spherical bubble in an incompressible liquid [11, 15]. Small bubbles are created in a liquid, due to the vapor of the liquid or to undissolved gases. The liquid motion moves them in regions where the pressure is higher than in the region of formation and the bubbles collapse. Initially take a spherical empty cavity of radius $R = R_0$ into the liquid at rest. The motion starts under the action of the surrounding pressure P_0. The continuity equation $\nabla \cdot \boldsymbol{v} = 0$ gives the velocity distribution.

$$v = \dot{R}\left(\frac{R}{r}\right)^2, \quad \text{or} \quad \frac{v}{\dot{R}} = \frac{1}{\eta^2} \quad \text{with } \eta = \frac{r}{R}. \tag{4}$$

The pressure is obtained substituting (4) into the equation of motion (13) and integrating with respect to r from R to ∞:

$$P = P_0 + \varrho\,\frac{R\ddot{R} + 2\dot{R}^2}{\eta} - \varrho\,\frac{\dot{R}^2}{2\eta^4}. \tag{5}$$

From the condition $P=0$ at the bubble boundary, $\eta=1$, one gets easily the motion equation.

$$\dot{R}^2 = \frac{2P_0}{3\varrho}\left[\left(\frac{R_0}{R}\right)^3 - 1\right]. \tag{6}$$

When the bubble radius R tends to zero, eqs. (6) and (5) tend towards the limiting form $(\eta \ll (R_0/R)^3)$

$$\dot{R}^2 \approx \frac{2P_0R_0^3}{3\varrho}\frac{1}{R^3}, \qquad P \approx \frac{\varrho\dot{R}^2}{2}\left(\frac{1}{\eta}-\frac{1}{\eta^4}\right). \tag{7}$$

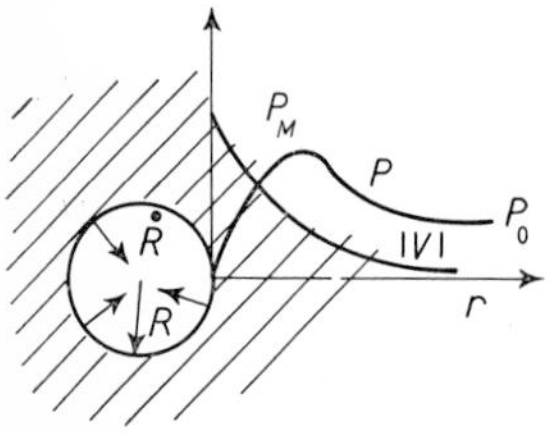

Fig. 9. – Imploding spherical bubble.

A cumulation process takes place in which $\dot{R}\to\infty$. Indeed a maximum of pressure P_M (Fig. 9) exists inside the liquid for $\eta=\eta_M\to 4^{\frac{1}{3}}$ and the shell $(1\leqslant\eta\leqslant\eta_M)$ of evanescent mass submitted to this external pressure acquires an infinite velocity.

Notice that for a small radius the motion has forgotten a part of the initial conditions R_0, P_0 and retains only the combination $W_0 \sim P_0R_0^3$ which is proportional to the initial energy necessary to create the bubble. The motion of the boundary will be $R\sim(W_0/\varrho)^{\frac{1}{5}}(-t)^{\frac{2}{5}}$, taking for $t=0$ the time of collapse. When $R\to 0$ one gets

$$v \sim -\frac{R^{\frac{1}{2}}}{r^2}, \qquad P\sim\frac{1}{R^2r}-\frac{R}{r^4}.$$

At collapse time the velocity of the medium tends to zero at any finite radius $r\neq 0$ and the total initial kinetic energy is thus entirely concentrated at the origin within a region of vanishing volume. The pressure becomes infinite at any finite radius which shows that the incompressible assumption is not valid anymore. Indeed the pressure remains finite when the compressibility is taken into account (Subsect. 2˙3.3).

The case of a collapsing incompressible cylindrical or spherical shell in vacuum is also of interest. Initially the shell has a converging motion. Writing the mass conservation, the kinetic energy conservation and the velocity law ($v_{cyl}=\dot{R}R/r$, $v_{sph}=\dot{R}R^2/r^2$) one gets immediately the motion law of the inner radius R. When $R\to 0$

$$\dot{R}_{cyl}\sim-[R^2|\log R|]^{-\frac{1}{2}}, \qquad \dot{R}_{sph}\sim-[R]^{-\frac{3}{2}} \tag{8}$$

and a cumulation of velocity is obtained exactly as in the preceding example. The only difference is in the pressure which becomes still infinite at the immediate vicinity of the inner wall but remains now finite inside the shell when $R \to 0$. Formulae (8) show that the cylindrical cumulation of velocity is much less efficient than the spherical one, a character which remains when the compressibility is taken into account.

The cylindrical collapse of metallic shells has some application in magnetic flux compression experiments [12, 23, 33]. (Figure 10). Such shells, driven by explosives, are used to compress an axial magnetic field which is flux-conservative ($BR^2 = \varphi = C^t$ for an infinitely conducting shell) and exerts an increasing pressure $P = B^2/8\pi \sim \varphi^2/R^4$ on the inner-shell boundary. The pressure reaches a maximum value P_m and repells the shell. For an incompressible shell, the initial kinetic energy $W_K(0)$ is equal to the magnetic energy at maximum pressure $(B_m^2/8\pi)\cdot\pi R_m^2 \sim B_m\varphi$. A cumulation of the field $B_m \sim \varphi^{-1}$ in an area $A \sim \varphi^2$ is thus obtained for a vanishing flux $\varphi \to 0$. Actually the shell compressibility limits the cumulation. According to Subsect. 1˙2.1 the pressure P_m is given by eq. (2) in plane geometry and for a thick enough piston; the weak concentration obtained in the cylindrical geometry does not allow pressures much greater than (2). Fields of the order of 10 MG have been experimentally produced and it should be theoretically possible to generate fields of about 30 MG [12].

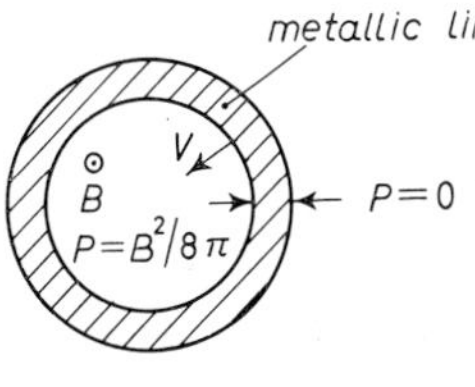

Fig. 10. – Cylindrical compression of an axial magnetic field by a metallic liner.

1˙3. *Cumulation in a medium of variable density.* – Elementary mechanics show that a small rigid body of mass m at rest acquires a velocity $v/v_0 = 2/(1 + m/M) \in (1, 2)$ when stricken by another body of mass $M/m > 1$ and at velocity v_0. Thus an increase in velocity can be expected from successive collisions between bodies of decreasing mass. At the limit of an infinite number of bodies and collisions, a body of vanishing mass acquires an infinite velocity. Some attempts have been made to apply this principle using collisions of metallic plates but practical difficulties are considerable [13]. The propagation of a shock wave through a system which consists in layers of alternate light and heavy media and successive decreasing thicknesses has also been investigated and a cumulative process has been found [14].

A practical example is given by astrophisics. Near the edge of a star the density decreases to zero and due to gravity and radiative conduction effects, follow approximately a law $\varrho_0 \sim x^\delta$ with $\delta \approx 3$. When a shock wave, created at the center of a star, reaches the surface, a cumulation effect takes place. This cumulation is similar to the cumulation obtained with imploding shock waves, in the sense that the energy is imparted to a vanishing mass of material. Nevertheless an important difference will be found in Subsect. 2˙3.4, which is due to the cause of the decrease in the mass (volume or density decrease). As a consequence of the density decrease, the pressure and the energy density tend to zero as $x \to 0$ and only the specific energy (or temperature) increases indefinitely as in the case of imploding shocks. The velocity imparted to a vanishingly small mass becomes infinite, which could explain the origin of cosmic rays and of interstellar particles of very high energies [15]. Indeed very strong shock waves are created by gravitational instability at the center of an exploding supernova.

1'4. *Sonic cumulation in the cylindrical geometry.* – In the cylindrical geometry, the wave equation relative to a quantity G is written as

$$\frac{\partial^2 G}{\partial t^2} - \frac{c^2}{r}\frac{\partial}{\partial r}\left(r\frac{\partial G}{\partial r}\right) = 0\,, \tag{9}$$

where $c = c^t$ is the propagation velocity.

Equation (9) has the unexpected property of admitting the solution [16, 17]

$$G = r^{-\frac{1}{2}} f(\tau)\,, \qquad \tau = ct/r\,,$$

in which τ is negative for converging motions, positive for diverging motions and $f(\tau) \to \infty$ for $\tau = +1$. The radius r may be taken as the radius of a discontinuity imploding at velocity c ($\tau = -1$). Such a discontinuity becomes infinite as $r^{-\frac{1}{2}}$ as it approaches the axis. It remains infinite along the line $r = ct$ ($\tau = +1$) after its reflection and a propagating cumulation has thus been obtained outside of the axis. This particular property is intrinsec to the cylindrical geometry and does not occur in the plane or spherical geometry.

For real media, eq. (9) represents the acoustical approximation which is not valid near the axis where G reaches high values. An example of strict validity of eq. (9) is found for a pure electromagnetic shock wave (invariance of the light velocity c). Assume a uniform magnetic field B_0 is limited by a perfect plane conductor which is suddenly moved at uniform velocity v. An electromagnetic discontinuity will propagate at light velocity c, behind which the magnetic field is uniform and equal to B_1. Due to the magnetic flux conservation, $B_1 = B_0/(1 - v/c)$. In cylindrical geometry the light shock wave is created in the same way by a suddenly imploding wall and Maxwell's equations show that B satisfies eq. (9). Consequently, the described cumulation occurs for the magnetic field [17]. Obviously the wall cannot be infinitely conducting and the diffusion of the field inside it leads to a smoothing of the shock wave and of the cumulation.

1'5. *Magnetohydrodynamic cumulation near a zero-field line.* – This example shows that a cumulation process may also be bidimensional. A circular cylinder of perfectly conducting and incompressible liquid is supposed to be initially immersed into a quadrupolar steady external magnetic field $\boldsymbol{B}_e = B_0 \nabla xy$ (Fig. 11). In the absence of currents and velocities the liquid is an equilibrium. This equilibrium is unstable as it can be seen by introducing a linear velocity perturbation.

$$v_x = Ux\,, \qquad v_y = Vy\,. \tag{10}$$

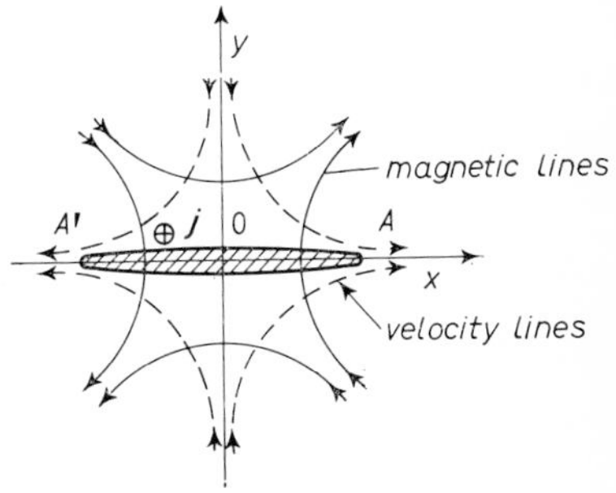

Fig. 11. – Magnetohydrodynamic cumulation near a zero (axial) field line.

It turns out that the MHD equations of the motion, which do not depend on the axis variable z, can be exactly solved with the initial conditions (10). The motion induces an axial current density j_z and, due to the pinch effect (or to the Lorentz force $\boldsymbol{j}\times\boldsymbol{B}$), the circular section is deformed into ellipses of axes $(0x, 0y)$. After a finite time, comparable to the ratio of the initial cylinder radius to the Alfvén velocity $V_A = B_0/(4\pi\varrho_0)^{\frac{1}{2}}$ a cumulation takes place and the elliptical cross-section is stretched along the x-axis (or y-axis depending on the values of U and V). The point A tends to infinity while the fluid kinetic energy and the current density increase without limit [18].

The MHD equations are again exactly solved when the fluid is supposed to be a perfect gas with a finite electrical conductivity. At the finite cumulation time, the fluid initially contained within a circular cross-section has been squeezed inside an ellipse of zero volume or a segment AA' of the x (or y) axis. Density, velocity V_y, current density j_z become infinite inside the limiting segment, as well as the internal, kinetic or magnetic energies. The cumulation requires an energy source supplied by the magnetic field. The assumed external field B_e can be created by four conductors disposed symmetrically with currents J_e of alternative directions. In the fluid the induced current J_i is becoming infinite at cumulation and infinite currents J_e will be necessary to maintain the assumed quadrupolar field B_e. An interesting consequence of the cumulation is also that J/n, *i.e.* the electron velocity, tends also to infinity which provides a mean to accelerate particles to high velocities [19].

2. – Self-similar solutions.

2'1. *Dimensional considerations.* – The solution of a fluid cumulation problem is unsteady and the resolution of the motion equations leads to complicated numerical calculations. Nevertheless in some cases of great interest the resolution of partial differential equations reduces to the resolution of ordinary differential equations. Such cases are predicted by the dimensional analysis. The subject of dimensional methods is quite vast. Essential notions necessary to our purpose will only be reported and more details could be found in the specialized literature [20].

2˙1.1. Transformation groups. A well-defined problem of physics, involving n quantitites $G_1, \ldots, G_n$, time t and space co-ordinates x_K is composed of the following elements:

The indefinite equations (E) which establish differential or partial differential relations between the quantities G_i, t and x_K.

The initial and boundary conditions, plus some functional relations (Equation of state ...) (C) which are necessary and sufficient to solve the problem.

The solution (S) which gives G_i as function of x_K, t for the conditions (C).

Let us introduce a family or group of transformations T_E which acts on the quantities G_i, x_K, t and has the property to let (E) formally invariant (Galilean group, change of scales ...). In the same family some transformations T_C will be found which let (C) invariant. Hence the solution (S) must remain invariant under the transformations T_S which let both (E) and (C) invariant. In the general case $T_S \equiv T_C \equiv T_E$ and the problem cannot be simplified. But when the conditions (C) are simple enough, the transformations T_C are reduced compared to T_E and the solution (S) loses its generality, becoming easier to find.

The scale transformations

$$G_i = \lambda_i \bar{G}_i\,, \qquad x_K = \lambda_K \bar{x}_K\,, \qquad t = \lambda_0 \bar{t} \tag{11}$$

are the simplest and the most important transformations; it is well known that the mathematical expression of the physical laws must not depend on the choice of the units. The form of the solution relative to a given problem will be directly obtained by introducing (11) in (E) and (C), and expressing the invariance conditions. This determines relations between the λ and consequently the combinations of the G_i, x_K and t from which depends the solution.

The substitution of (11) in (E) leads immediately to the usual notion of dimensions. In the case of one-dimensional fluid dynamics this substitution reduces to 3 the number of independent λ and it is equivalent to consider the 3 fundamental units M, L, T instead of these λ.

2˙1.2. One-dimensional unsteady motion of a perfect gas. Equations (E) are

$$\frac{\partial \varrho}{\partial t} + \frac{\partial \varrho v}{\partial r} + (\nu - 1)\frac{\varrho v}{r} = 0\,, \tag{12}$$

$$\frac{\partial v}{\partial r} + v\frac{\partial v}{\partial r} + \frac{1}{\varrho}\frac{\partial p}{\partial r} = 0\,, \tag{13}$$

$$\left(\frac{\partial}{\partial t} + v\frac{\partial}{\partial r}\right)(p/\varrho^{\gamma}) = 0\,, \tag{14}$$

where $\nu = 1, 2, 3$ in the plane, cylindrical and spherical geometry respectively. Transformation (11) applied to (12)-(14) gives the dimensions of the quantities $v = LT^{-1}$ etc.

In the general case the conditions (C) introduce 3 independent dimensional parameters a, b, c and a number of nondimensional quantities $\mu_1, \ldots, \mu_l$ (such as the ratio of some initial radii etc ...). Characteristic time, length and density (which contains the mass), t_0, r_0, ϱ_{00} are obtained by combining the parameters a, b, c. Hence, the solution of (12), expressed under an invariant adimensional form will be:

$$v = \frac{r}{t} V\left(\frac{r}{r_0}, \frac{t}{t_0}; \mu_1 \ldots \mu_l\right),$$

$$p = \varrho_{00}\frac{r^2}{t^2} \Pi\left(\frac{r}{r_0}, \frac{t}{t_0}; \ldots\right),$$

$$\varrho = \varrho_{00} H\left(\frac{r}{r_0}, \frac{t}{t_0}; \ldots\right).$$

In special cases which are here of interest, the conditions (C) introduce only 2 independent dimensional parameters, one of which contains necessarily the mass (the medium density). These parameters may be written.

$$A = LT^{-\alpha}\,, \qquad B = \varrho_{00} L^m\,.$$

It will thus be impossible to build independent characteristic time and length with A and B.

As a consequence, the functions V, Π and H depend on r and t only through the nondimensional variable

$$\xi = r/At^{\alpha}$$

and their substitution into the motion equations (12)-(14) leads to a system of ordinary differential equations. These functions of a single variable describe completely the motion and the profiles $G_i(\xi)$ give the distribution of the quantities G_i at any time by means of appropriate changes of scales in r and G_i. For that reason these solutions are called « self-similar ».

Equation (14) is only valid for a perfect gas. It could have been replaced by the energy conservation equation completed by a general equation of state $e = e(\varrho, p)$. Such an equation of state depends on the interaction forces between the particles (which do not appear for a perfect gas) and involves dimensional parameters. These parameters can be hidden or appear into the equations (for instance the density and sound velocity at zero temperature for the case of a solid). The number of dimensional parameters becomes anyhow too elevated and a self-similar solution exists no more. This will generally also be the case for gas dynamics problems which involve phenomena related to cross-sections (nuclear reactions, etc...).

For incompressible fluids, (14) is replaced by $\varrho = \varrho_0$ and problems become trivial. Nevertheless the dimensional analysis is still useful. For example in the problem of spherical bubble collapse (Subsect. **1**˙2.2), if one assumes that the particular initial conditions have been forgotten and replaced by the energy W_0 necessary to create the bubble, the two-dimensional parameters become $W_0 \sim \varrho_0 L^5 T^{-2}$ and $B = \varrho_0$. One gets directly the bubble motion $R \sim \sim (W_0/\varrho_0)^{\frac{1}{5}}(-t)^{\frac{2}{5}}$ in agreement with (7).

2˙2. *Some properties of self-similar gas dynamics.*

2˙2.1. The equations. The procedure of SEDOV [20] will be used and involves some slight change of notations. The two independent dimensional parameters are chosen as:

$$a = ML^K T^S\,, \qquad A = LT^{-\alpha}\,. \tag{15}$$

The sound velocity c is considered instead of the pressure $p = \varrho c^2/\gamma$ and the self-similar solution is written

$$v = \frac{r}{t}\,V(\xi)\,, \qquad c^2 = \frac{r^2}{t^2}\,Z(\xi)\,, \qquad \varrho = \frac{a}{r^{K+3}t^S}\,G(\xi) \tag{16}$$

with

$$\xi = r/At^\alpha\,.$$

Substituting (16) into the equations of motion (12)-(14) one gets after some transformations

$$\frac{\mathrm{d}Z}{\mathrm{d}V} = \frac{Z}{V-\alpha}\cdot \frac{[2(V-1)+\nu(\gamma-1)V](V-\alpha)^2-(\gamma-1)V(V-1)(V-\alpha)-[2(V-1)+\beta(\gamma-1)]Z}{V(V-1)(V-\alpha)+(\beta-\nu V)Z}\,, \tag{17}$$

$$\frac{\mathrm{d}\log\xi}{\mathrm{d}V} = \frac{Z-(V-\alpha)^2}{V(V-1)(V-\alpha)+(\beta-\nu V)Z}\,, \tag{18}$$

$$\frac{\mathrm{d}\log G}{\mathrm{d}\log\xi} = \frac{1}{V-\alpha}\left\{[s+(k-\nu+3)V]-\frac{V(V-1)(V-\alpha)+(\beta-\nu V)Z}{Z-(V-\alpha)^2}\right\}, \tag{19}$$

with

$$\beta = [s + 2 + \alpha(k + 1)]/\gamma \,.$$

These equations have several remarkable properties. First the quantities ξ and G appear only through their differential $\mathrm{d}\log\xi$ and $\mathrm{d}\log G$. It is a consequence of the dimensional structure of the eqs. (16). The quantities ξ and G involve the dimensional parameters A and a which can be chosen arbitrarily and must thus not appear into the equations of motion. The logarithmic dependence eliminates A and a from eqs. (17)-(19). Second, eq. (17) involves only the quantities V and Z. Indeed it has been obtained by eliminating $\mathrm{d}\log\xi$ and $\mathrm{d}\log G$ from the three equations directly derived when substituting (16) into (12)-(14).

Solutions will be found, solving eq. (17) in Z and V. Substituting $Z(V)$ into (18) and (19) one gets $V(\xi)$ and $G(\xi)$ through two quadratures. Quadratures are not always necessary and integrals can be found, due to the conservative properties of the fluid equations. The order of the system is then lowered; SEDOV proves that [20]:

The mass-conservation eq. (12) leads to a mass integral $f_1(\xi, G, V, M) = 0$. $M(\xi)$ is defined by $M = r^{K+3} t^S \mathscr{M}/ar^\nu$ where $\mathscr{M}$ is the mass contained between a fixed surface and a surface $\xi = c^t$.

The entropy invariance gives a function $f_2(\xi, G, Z, M) = 0$. The « adiabatic integral » $f_3(\xi, V, Z, G) = 0$ is obtained eliminating M between f_1 and f_2.

If a constant with the dimensions $ML^{\nu-1}T^{-2}$ of an energy, an energy per unit length or unit surface in respectively the spherical, cylindrical or plane case can be found from the parameters a and A (*i.e.* if $s + 2 + \alpha(\nu - 1 + k) = 0$), an integral $f_4(\xi, V, Z) = 0$ exists and represents the energy conservation.

If a constant with the dimension $ML^{-1}T^{-1}$ of a momentum per unit area can be formed with a and A, an integral $f_5(G, V, Z) = 0$ relative to the momentum conservation is obtained.

Integrals are not too much complicated and are often useful. For instance the problem of an intense spherical explosion is solved in a closed analytical form, using the adiabatic and energy integrals (the parameters are $\varrho_0 = ML^{-3}$ and $W_0 = ML^2T^{-2}$).

2·2.2. *Properties of eq. (17) in the (Z, V)-plane.* Equation (17) is obviously the basis of the problem. Pressures $P = ZG/\gamma$ are zero on the V-axis, velocities are zero on the Z-axis. By definition, Z is positive. The integral curves of (17) have generally a complicated behaviour (Fig. 12 gives an example) which mainly depends on the singular points, *i.e.* the points obtained by annulating both numerator and denominator in the right-hand side of (17). One of these curves is determined by the conditions of the problem and represents the solution. It will be further clearer that two types of solution exist [15]:

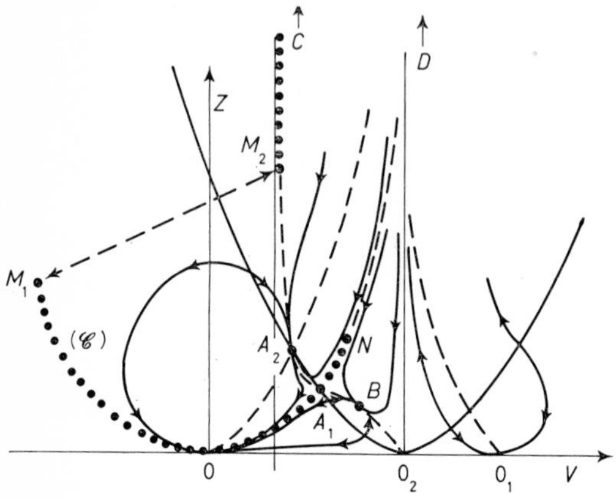

Fig. 12. – Self-similar integrals in the (Z, V)-plane for a converging shock wave. The solution is the curve $NA_1OM_1M_2C$. Arrows indicate increasing ξ.

Solutions of the first kind for which the initial and boundary conditions are sufficient to determine the power coefficient α and the solution in the (Z, V)-plane (example of Subsect. 2˙3.1).

Solutions of the second kind for which the conditions are not sufficient. The parameter A is not known and the coefficient α as well as the solution $Z(V)$ are *a priori* undetermined. This is generally the case in cumulation problems in which a part of the initial conditions has been forgotten when the cumulation occurs. The solutions are nevertheless determined by expressing that the solution $Z(V)$ must be singular and moreover must pass through some particular points related to boundary conditions.

The solution is not necessarily a continuous curve in the (Z, V)-plane. A shock wave may happen and introduce a discontinuity.

Shock discontinuities. Let a shock occur at a radius $r_s(t)$. Dimensional considerations show that $r_s/At^\alpha = \xi_s = c^t$. The shock velocity is thus $\dot{r}_s = \alpha r_s/t$. The states (1) and (2) on both sides of the shock are fixed points in the (Z, V)-plane. Writing the Hugoniot conditions in reduced co-ordinates one gets

$$(20)\quad \begin{cases} V_2 - \alpha = (V_1 - \alpha)\left[1 + \dfrac{2}{\gamma+1}\,\dfrac{Z_1 - (V_1-\alpha)^2}{(V_1-\alpha)^2}\right], \\[2ex] Z_2 = \left(\dfrac{\gamma-1}{\gamma+1}\right)^2 \dfrac{1}{(V_1-\alpha)^2}\left[(V_1-\alpha)^2 + \dfrac{2Z_1}{\gamma-1}\right]\left[\dfrac{2\gamma}{\gamma-1}(V_1-\alpha)^2 - Z_1\right], \end{cases}$$

and also

$$G_1(V_1 - \alpha) = G_2(V_2 - \alpha)\,.$$

Relations (20) establish the correspondance between the states (1) and (2) or a topological transformation the (Z, V) plane. Several parabolas are of interest (Fig. 13).

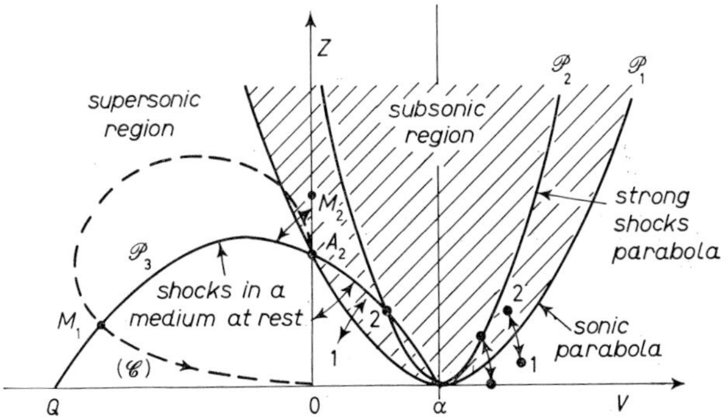

Fig. 13. – Relevant curves in the (Z, V)-plane. $(\mathscr{C})$ is relative to a converging motion integral.

First, on the parabola $(\mathscr{P}_1)$

(21) $$Z = (V - \alpha)^2 \, (\mathscr{P}_1)$$

the points transform into themselves. Therefore a point on this parabola represents a weak discontinuity, *i.e.* a characteristic. As shown by writing (21) in terms of dimensional quantity the characteristic is $\mathrm{d}r/\mathrm{d}t = v + c$ if $V < \alpha$ and $\mathrm{d}r/\mathrm{d}t = v - c$ if $V > \alpha$. It is also seen that velocities relative to the shock are subsonic above $(\mathscr{P}_1)$ and supersonic under $(\mathscr{P}_1)$. Finally $(\mathscr{P}_1)$ has the essential property that it cannot be crossed by a curve $Z(V)$ which has a physical meaning, unless at a singular point determined by the annulation of the denominator of (18); indeed ξ has an extremum on $(\mathscr{P}_1)$ (eq. (18)) and physical quantities become multivalued at the crossing of $(\mathscr{P}_1)$ except at a singular point.

Second the parabola $(\mathscr{P}_2)$

(22) $$Z_2 = \frac{2\gamma}{\gamma - 1} (V_2 - \alpha)^2 \, (\mathscr{P}_2)$$

corresponds to the limiting case in which the state (1) is on the V-axis ($Z_1 = 0$ or $c_1 = 0$), *i.e.* to infinitely strong shock waves. As a consequence of the properties of $(\mathscr{P}_1)$ and $(\mathscr{P}_2)$ a state (1) ahead of a shock will belong to the zone under $(\mathscr{P}_1)$ and transform into a state (2) behing the shock which is situated between $(\mathscr{P}_1)$ and $(\mathscr{P}_2)$.

Third, when one state, (1) or (2), is at rest *i.e.* belongs to the Z-axis, the other state lies on the parabola $(\mathscr{P}_3)$

$$Z = -\alpha(V-\alpha)\left(1+\frac{\gamma-1}{2\alpha}V\right) \quad (\mathscr{P}_3)\,. \tag{23}$$

Singular points. They are obtained by writing that the numerator and the denominator of the right-hand side of (17) are both zero. Only some of them are useful in a given problem. Their properties depend on the values of the parameters $\alpha, k, s, \ldots$, which makes the configuration of the integrals $Z(V)$ rather complex (see Fig. 12 [21]).

Point 0 $(Z=0,\ V=0)$ is a node. Equations (17)-(18) give the first order development

$$Z \sim V^2 \sim \xi^{-2/\alpha}, \qquad G \sim \xi^{-S/\alpha}\,. \tag{24}$$

Point 0 is reached for $\xi \to \infty$ (if $\alpha > 0$). At finite time it represents the state at infinite $r \to \infty$; it represents as well the state at every finite distance $r \neq 0$ when $t \to 0$. A singular curve $Z = (\alpha^2/\beta)\,V$ starts also from 0.

Point O_1 $(Z=0,\ V=1)$ is a node, nonsingular integrals being tangent to the V-axis. Point O_2 $(Z=0,\ V=\alpha)$ is a saddle point.

Points A_1 and A_2 are on the parabola $(\mathscr{P}_1)$ and are determined by

$$(\nu-1)\,V^2 - (\beta+\alpha\nu-1)\,V + \alpha\beta = 0\,. \tag{25}$$

They are real or imaginary and their nature has to be checked in each case. The value of ξ is finite at these points.

Point B $\big(V = 2/[2+(\gamma-1)\,\nu]\,,\ Z = V(V-1)(V-\alpha)/(\nu V-\beta)\big)$ is a node if situated above $(\mathscr{P}_1)$ in which case $\xi \to \infty$. It is a saddle point if located below $(\mathscr{P}_1)$.

Points C $(V = \beta/\nu,\ Z=\infty)$, D $(V=\alpha,\ Z=\infty)$ and E $(V=\infty,\ Z=\infty)$ are singular points at infinity. Point C is a node and point D is a saddle point if $\alpha\gamma\nu < 2(1-\alpha)$. The situation is reversed if the inequality holds in the opposite sense; $\xi \to 0$ at C when C is a saddle point, $\xi \to c^t$ at D when D is a node.

2·3. *Self-similar cumulation.* – The solutions of unsteady motions relative to some of the cumulation processes described in Sect. **4** are now determined. The case of converging motions is also investigated; it is not a cumulation problem but it gives an example of a self-similar solution of the first kind and shows clearly the concentrative effect of the geometry. Converging shocks are treated with some details in order to illustrate a typical and important case of self-similar solutions of the second kind; only the main results are given for the other examples of such solutions.

2˙3.1. Converging motions The gas is initially in a uniform state with a density ϱ_0, a sound velocity c_0 and a velocity v_0 directed towards the plane, axis or center of symmetry, depending on the geometry. It enters in collision with itself at time $t = 0$ (Fig. 14). The dimensional parameters are

$$a = \varrho_0 = ML^{-3}, \qquad A = |v_0| = LT^{-1}.$$

The considerations of Subsect. 2˙2. apply with $k = -3$, $s = 0$, $\alpha = 1$, $\beta = 0$. Solutions depend also on the dimensionless parameter $M = |v_0|/c_0$. They are self-similar and the self-similar variable is $\xi = r/|v_0|t$. The singular points of interest are O and A_2 ($V = 0$, $Z = \alpha^2$; eq. (25)).

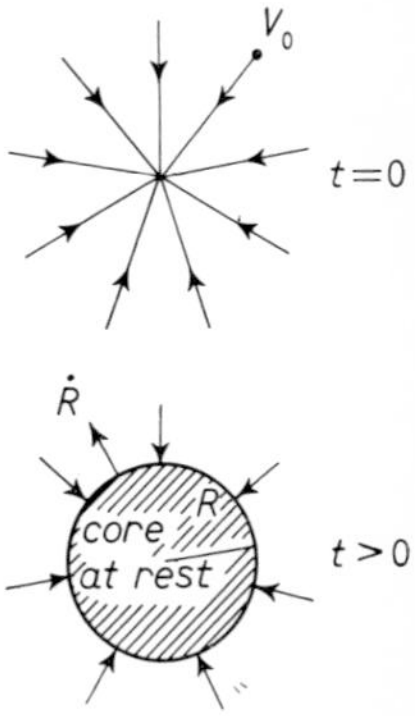

Fig. 14. – Converging motion.

Consider the (Z, V)-plane (Fig. 13). Point O is relative to the fluid state at infinity; the initial conditions determine the asymptotic behaviour $Z = V^2/M^2$ eq. (24) and thus the integral curve $(\mathscr{C})$. The image point (Z, V) starts from O and follows the curve until it reaches the parabola $(\mathscr{P}_1)$ on which ξ has a minimum value, *i.e.* on which $Z(\xi)$, $V(\xi)$ become multiform. A continuous solution is therefore impossible. A shock has to be introduced at the intersection M_1, of $(\mathscr{C})$ and $(\mathscr{P}_3)$. Indeed according to the property of $(\mathscr{P}_3)$ this shock transforms M_1 in a point M_2 of the Z-axis. In physical space the gas is first compressed adiabatically in its motion from infinity until it meets a shock which brings it to rest. The shock velocity is $\dot{R} = |v_0|\xi_1$.

The motion equations (17)-(19) have to be numerically solved. For very high Mach number, $M \gg 1$, a simple approximation is easily found. In this case the integral curve $(\mathscr{C})$ is close to the very flat parabola $Z = V^2/M^2$ and point M_1 is in the vicinity of point Q. A straightforward first-order approximation gives [10]

$$(26) \qquad \frac{\varrho}{\varrho_0} = \left(\frac{P}{P_0}\right)^{1/\gamma} \approx \left(1 + \frac{1}{\xi}\right)^{\nu-1}, \qquad v \approx v_0$$

in the adiabatic region of the motion and

$$(27) \qquad \begin{cases} \dfrac{\varrho_2}{\varrho_1} \approx \left(\dfrac{\gamma+1}{\gamma-1}\right)^{\nu}, \qquad P_2 \approx \left(\dfrac{\gamma+1}{\gamma-1}\right)^{\nu-1} \dfrac{\gamma+1}{2}\, \varrho_0 v_0^2, \\[2ex] T_2 \sim c_2^2 = \dfrac{\gamma(\gamma-1)}{2}\, v_0^2 \end{cases}$$

in the uniform core at rest behind the shock front.

The physical meaning of this solution is simple. The condition $M \gg 1$

shows that thermal effects are negligible in the adiabatic region where the particles are quite free. The quantity $1+1/\xi=(r+|v_0|t)/r$ is equal to the ratio of the radius r_0 where the particle is located initially to the actual radius r. Equation (26) writes as the mass-conservation law $\varrho r^{\nu-1}=\varrho_0 r_0^{\nu-1}$. The Hugoniot condition for a strong shock wave determines successively (27). The next term of the approximation would be of order $1/M^2$.

The effect of the geometry on the state inside the core is clearly seen on eqs. (27), taking $\nu=1, 2, 3$. Passing through successive geometries one gets each time large density and pressure amplifications equal both to the ratio $(\gamma+1)/(\gamma-1)$. The temperature remains independent of the geometry.

The problem could be extended to the case of an initial nonuniform density $\varrho_0=ar^{-\delta}$ ($\delta>0$). The injection of fast particles at constant flux corresponds to $\delta=\nu-1$, $M\gg 1$. The similarity coefficient is still $\alpha=1$, determined by dimensional analysis. The solution starts as before from point O of the (Z, V)-plane but the shock point M_2 has now to be determined on the singular curve passing through the singular point C ($V=\delta/\nu$, $Z=\infty$) which represents the origin $r=0$ (see also the next Section). The curve M_2C gives the nonuniform state of the core.

2·3.2. Converging shock waves. Let us consider a cylindrical or spherical shock wave of radius $R(t)$ which propagates towards the axis or the center of symmetry into a gas at rest of uniform density and zero pressure. Such a shock wave has been created by some cylindrical or spherical piston which has imparted the energy to gas. When the shock enters into a region of radius small enough compared to the piston radius, the collapse occurs within a time so small that, due to the finiteness of the mean propagation velocity of perturbations, it cannot be influenced by the gas motion at large radii. The motion becomes thus autonomous and has forgotten a great part of the initial conditions.

In the limiting autonomous regime the only evident parameter is the gas density ϱ_0. Initial conditions related to the piston are of no use because they have been partly forgotten in some unknown way. Remain the position R and the velocity $\dot{R}$ of the shock which depend on time and cannot be retained as characteristic length or time. It will be assumed that, for dimensional reasons, $\dot{R}=\alpha R/t$ ($t<0$). The motion is self-similar with $k=-3$, $s=0$, $\beta=2(1-\alpha)/\gamma$. The parameter α is not determined directly by the dimensional analysis and the solution is of the second kind.

Initial conditions must enter in some way. As it will be proved later, α depends only on the polytropic coefficient γ. Therefore initial conditions are contained entirely into the parameter A which defines $\xi=Ar/(-t)^\alpha$. One can take $A=1$ and, on the shock front, $\xi=\xi_s=1$ without any loss of generality. During the imploding phase $t<0$. The instant of collapse is $t=0$.

The problem of the imploding shock wave can be solved numerically, for a given piston motion. The motion can be observed experimentally [26]. A theoretical study of a symmetrically perturbed self-similar motion can also be made [28]. In all cases it has been found that the non self-similar flow tends asymptotically to a self-similar regime when $R \to 0$.

The integral curve in the (Z, V)*-plane.* – The boundary conditions on the shock and at infinity give the end points of the integral curve $(\mathscr{C})$. The shock is strong and moves in a medium at rest. The end point N on the shock will therefore be at the intersection of the parabolas $(\mathscr{P}_2)$ and $(\mathscr{P}_3)$. On the other hand the quantities $v = (r/t)V$ and $c^2 = (r^2/t^2)Z$ must remain finite in all space $r \neq 0$ at the instant of collapse, as well as at infinity at any time. Thus $V(\infty) = Z(\infty) = 0$ and the origin O is the end point of $(\mathscr{C})$ for $\xi \to \infty$.

The points O and N are on opposite sides of the sonic parabola $(\mathscr{P}_1)$. (Figure 12). As shown in Subsect. 2'2.2 the solutions are physically meaningful only if $(\mathscr{C})$ crosses $(\mathscr{P}_1)$ at a singular point A_1 or A_2 given by (25). It turns out that the singular point can only be A_1 $(V_{A_1} > V_{A_2})$ [6, 22]. For a given α exists just one curve, a singular integral, which passes by the singular points O and A_1. The requirement that N lies also on this singular integral determines α. In practice eq. (17)-(19) are solved numerically with trial values of α until the condition is fulfilled. Point A_1 has to be real which limits the possible values of α. Expressing that the singular points A_1 and A_2 (or A_1 and B) coincide, one gets a trial value of α which turns out to be close to the right value [22].

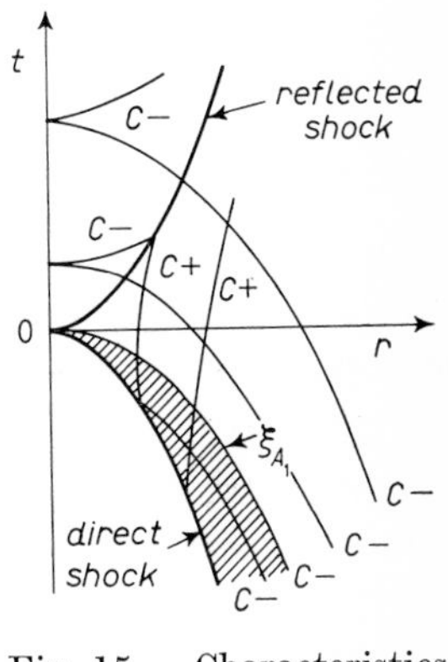

Fig. 15. – Characteristics of a converging shock motion.

As shown before a point lying on $(\mathscr{P}_1)$, such as A_1, represents a characteristic of the fluid motion. Actually $r/(-t)^\alpha = \xi_{A_1}$ is the equation of a C_- characteristic $dr/dt = v - c$ which meets the shock front only at the time of collapse. Characteristics are drawn on Fig. 15. The C_- characteristics which are above the ξ_{A_1}-line never overtake the shock front. Consequently the ξ_{A_1}-line bounds regions of influence. The state of the motion in the shaded region of the (r, t)-plane can in no way be affected by the motion in the unshaded region, before the collapse.

The question arises of what happens after the collapse of the imploding shock. A reflected shock wave is expected which moves into the region of incoming unsteady flow. Before the collapse, $t < 0$, $\xi = r/(-t)^\alpha$; the self-similar solution is extended to $t > 0$ with the same similarity exponent α, taking now $\xi = r/t^\alpha$. At $t = 0_+$ the state of infinity is the same than at $t = 0_-$ and the integral curve starts from point 0 with the same asymptotic behaviour than before but in the $V<0$ direction. As was the case for the imploding motion of Subsect. 2·3.2., a shock discontinuity has to be introduced along the integral curve at some point M_1. After a jump M_1M_2 through the reflected shock the integral curve must proceed and reach the representative point of the center where $\xi=0$. One finds that this point is the singular point C $(Z=\infty, V=\beta/\nu)$ reached along the singular integral passing through A_1, A_2, B and O_2. The two parts of the integral curve ahead and behind the shock are thus known. Only two points M_1 and M_2 on these two parts correspond to the shock transformation (20) which completely determine the solution.

Numerical results. The similarity exponent α has been calculated in a number of cases summarized on the following Table [6, 7, 21, 22, 25, 34) (*).

γ		1	1.2	1.4	5/3	3
$\nu=2$	α	1	0.861	0.835	0.816	0.775
	$2((1-\alpha)/\alpha)$	0	0.332	0.394	0.452	0.581
$\nu=3$	α	1	0.757	0.717	0.688	0.638
	$2((1-\alpha)/\alpha)$	0	0.641	0.788	0.90	1.13

The distributions $v(\xi)$, $\varrho(\xi)$, $P(\xi)$ and $w(\xi)=(\varrho v^2/2)+p/(\gamma-1)$ are given in Fig. 16 for the cylindrical case $\gamma=\frac{5}{3}$ [21].

The converging shock front moves according to the law $R=(-t)^\alpha$ $(A=1)$. The velocity, density and pressure behind the front are

$$(28)\qquad \begin{cases} \varrho_N=\dfrac{\gamma+1}{\gamma-1}\,\varrho_0\,; \qquad v_N=\dfrac{2}{\gamma+1}\,\dot{R}\sim R^{-(1-\alpha)/\alpha}\,, \\[2ex] p_N=\dfrac{2}{\gamma+1}\,\varrho_0\dot{R}^2\sim R^{-2(1-\alpha)/\alpha}\,. \end{cases}$$

Note, from the Table, that the value of the pressure exponent $-2(1-\alpha)/\alpha$ relative to the spherical case is approximately twice the value relative to the cylindrical case (the acoustical approximation should give $p_{\text{sph}}\sim r^{-1}$ and $p_{\text{cyl}}\sim r^{-\frac{1}{2}}$). At any time, Fig. 16 *a*) gives the normalized distribution $v(r)$,

(*) The approximation of Chester-Chisnell-Whitham, based on the linearized equations of the motion characteristics gives an analytical expression $\alpha(\gamma)$ which turns out to be in very good agreement with the tabulated values [35].

$\varrho(r)$, $p(r)$ and $w(r)$ behind the imploding shock; the density increases, the velocity and the energy decrease monotonically with the radius while the pressure has a maximum before dropping; the asymptotic behaviour is

$$\varrho \sim c^t , \qquad p \sim v^2 \sim r^{-2(1-\alpha)/\alpha} , \tag{29}$$

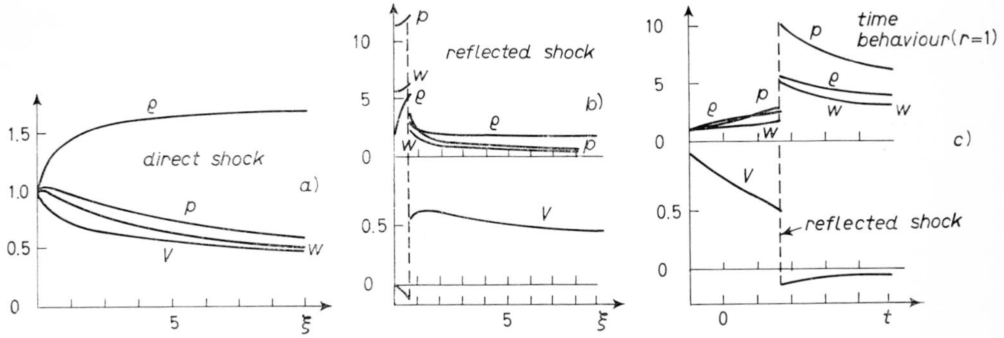

Fig. 16. – Self-similar distributions of density ϱ, pressure p, energy density w and velocity v for a converging cylindrical shock wave ($\gamma = 5/3$ [21]). Quantities are normalized to the values reached on the converging shock front (eq. (28)).

as $r \to \infty$. When $t \to 0$, at the collapse time, the pressure, velocity and temperature on the shock front tend to infinity and the density remains constant and equal to its strong shock value (eq. (2)) while, at any finite radius, the distributions tend to their asymptotic values (29). Figure 16 *c*) also shows that at a given radius, *e.g.* $r = 1$, $\varrho(t)$, $p(t)$ and $w(t)$ increase monotonically with time between the passage of the converging shock ($t = -1$) and that of the reflected shock; in the meanwhile the temperature $T(t) \sim p/\varrho$ remains approximately constant.

The energy inside the whole fluid is infinite (integration with distributions (29)) which proves that the self-similar solution cannot be valid for indefinitely large radii. On the other hand the energy contained inside the autonomous self-similar shaded region of Fig. 15 tends to zero as $R^{\nu+2-(2/\alpha)}$ at the collapse time: the energy density increases indefinitely but inside a region of vanishing volume. Finally the energy contained inside a region of radius r increases with time but remains finite.

The reflected shock front follows the law $R = \xi_R t^\alpha$ ($\xi_R < 1$). Figures 16 *b*) and *c*) give the fluid behaviour after the reflection. At a given time, the pressure and energy density are quite uniform between the axis (or center) and the shock front; as the density vanishes on the axis (or center) the temperature there tends to infinity. The reflected shock is weak. Nevertheless it contributes to increase the cumulation. Consider the values of density, pressure and temperature reached at a given radius r at the passages of the converging (index N) and reflected (index R) shock fronts. For $\gamma = \frac{5}{3}$ they are related by $\varrho_R \approx 8.1\varrho_N \approx 32.5\varrho_0$, $P_R/P_N \approx 12.7$ and $T_R/T_N \approx 1.57$ in the spherical geometry; by $\varrho_R \approx 5.7\varrho_N \approx 22.8\varrho_0$, $P_R/P_N \approx 10.3$, $T_R/T_N \approx 1.81$ in the cylindrical geometry; and, from Subsect. 1˙1.1, by $\varrho_R = 2.5\varrho_N = 10\varrho_0$, $P/P_N = 6$ and $T/T_N = 2.4$ in the plane geometry. Therefore the reflection increases sensibly the cumulation with a marked influence of the geometry on the density and pressure gain (compare with the results of Subsect. 2˙3.1).

2˙3.3. Collapse of compressible bubbles. The problem has been defined in Subsect. 1˙2.2. The motion is considered for very small bubble radii when the initial conditions have been « forgotten ». The medium is assumed to be isentropic with an equation of state [15, 30].

$$p = B[(\varrho/\varrho_0)^\gamma - 1]\,, \qquad \gamma = 7\,, \qquad B = c^t \text{ (water)}\,.$$

For the very high cumulation pressures, the equation of state reduces to the form $p = B(\varrho/\varrho_0)^\gamma$ and the equations of motion (12)-(14) are therefore still valid. The solution depends only on the dimensional parameter $B\varrho_0^{-\gamma}$ and is consequently self-similar of the second kind as for converging shocks. The radius of the bubble is $R = A(-t)^\alpha$.

The motion is determined in the same way as in the preceding Subsect. and has analogous properties. Numerical resolution gives $\alpha = 0.55$ for $\gamma = 7$. The energy of the whole fluid is infinite and zones of influence limited by a C_- characteristic are found. At the instant of collapse, when $R = 0$ the spatial distributions ($r \neq 0$) are $v \sim c \sim r^{-(1-\alpha)/\alpha}$ as for converging shocks but the limiting density and pressure $\varrho \sim r^{-2(1-\alpha)/\alpha(\gamma-1)}$, $p \sim r^{-2(1-\alpha)\gamma/\alpha(\gamma-1)}$ are different. Indeed the cumulation process is isentropic and the increase in pressure is related to an increase in density. After the collapse a diverging shock wave propagates outwards [30].

2˙3.4. Plane shock wave in a medium of variable density. – Let us come back on the example given in Subsect. 1˙3 of the emergence of a shock wave at the surface of a star with a density $\varrho_0 = ax^\delta$ [15, 20, 36]. The solution is again self-similar of the second kind with the single dimensional parameter a. The shock front moves according to the law $x = A(-t)^\alpha$ and emerges at the surface $x = 0$ at time $t = 0$.

The solution is obtained with the preceding method. The exponent α is determined by the condition that the integral curve $Z(V)$ passes through a singular point A_2 which corresponds to a characteristic C_- limiting a zone of influence. The following values have been found [29, 36]

γ	δ			
	0.5	1	2	3.25
1.2	0.920	0.855	0.752	—
1.4	0.905	0.831	0.717	—
5/3	0.877	0.817	0.696	0.59

As $\alpha < 1$, the shock is accelerated at the vicinity of the surface. At the instant of emergence the temperature $T \sim \dot{x}^2$ tends to infinity as $x^{-2(1-\alpha)/\alpha}$. But, as a difference with converging shock waves, the pressure $p \sim \varrho \dot{x}^2 \sim x^{\delta-2(1-\alpha)/\alpha}$ vanishes due to the decrease in density. Figure 17 shows spatial distributions of

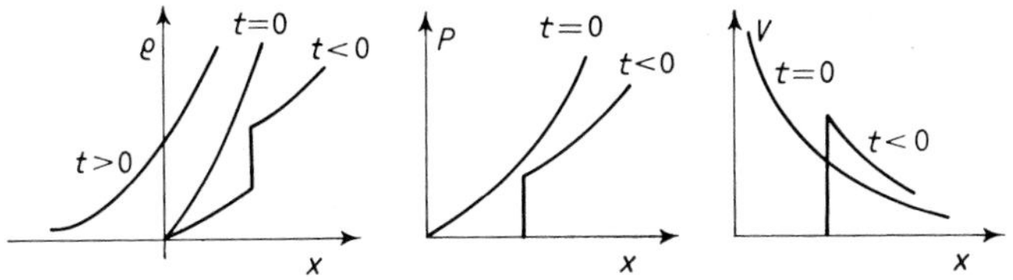

Fig. 17. – Distributions of density ϱ, pressure p and velocity v when a shock moves in a gas of variable density.

density, pressure and velocity. At the time $t = 0$, the distributions are $v \sim \sim x^{-(1-\alpha)/\alpha}$, $T \sim v^2$, $\varrho \sim x^\delta$ and $p \sim x^{\delta-2(1-\alpha)/\alpha}$. And continuous flow in vacuum follows the shock wave emergence, calculated in [36].

2'4. *Limitations in cumulation processes.* – Obviously it will not be possible to reach infinite pressures or temperatures in the cumulation processes described before. Dissipative phenomena such as viscosity, heat conduction, radiation, have to be taken into account. They generally do not alter much the main fluid motion and become important only at the vicinity of the cumulation, introducing some limitating mechanism. When shock waves are concerned, the self-similar solution will be valid as long as the shock radius is greater than the shock thickness associated with these dissipative phenomenas; for smaller radii the cumulation is described no more by fluid equations. Moreover, at high temperatures and densities dissociation and ionization modify the equation of state. The limits of validity of the cumulation have to be checked in every case.

Let us consider with more details another important cause for which cumulative processes cannot be perfect, namely the departure from the symmetry condition. A plane shock wave is known to have a stable form for a perfect gas (more generally for a regular concave Hugoniot curve $p_H(\tau)$ [24]). In other words, if the plane shock wave is perturbed and assumes a sinusoidal shape of amplitude η, the perturbation tends to zero exponentially with increasing time. In the cylindrical or spherical geometry, the stability is no more related to the displacement η but to the aspect ratio η/r_s where r_s is the radius of the unperturbed shock front. (Fig. 18). It follows from the stability of a plane shock that a diverging shock will be stable ($\eta/r_s \to 0$) but that a converging shock could be unstable.

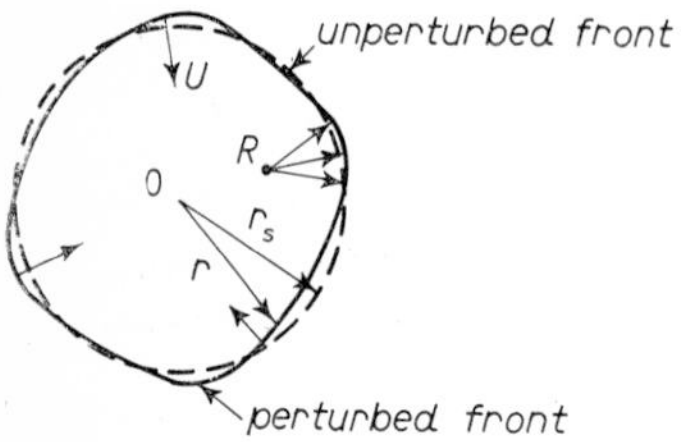

Fig. 18. – The instability of a cylindrical converging shock.

The study of the converging shock stability is rather complex [25] and a very simple approximate theory is given. The approximation lies on the assumption that the shock wave shape can only be influenced by the fluid perturbations which are immediately situated behind the shock front. Indeed a perturbation which, at time t, is not located at the immediate vicinity of the shock overtakes the shock front, following a C_- characteristic (Fig. 15), at a later instant when the shock strength has considerably increased. This perturbation has then become negligible; if it was too far from the shock front it does not even overtake it. If one considers a perturbed cylindrical shock front which has a local curvature radius R, it can therefore be assumed that, this shock will roughly be locally self-similar. It follows that

$$\frac{\ddot{R}}{\dot{R}^2} = \frac{\alpha - 1}{\alpha}\frac{1}{R}. \tag{29}$$

The left-hand side is the ratio $\dot{U}/U$ where U is the normal component of the perturbed shock velocity. The right-hand side is proportional to the local curvature of the perturbed shock. These two quantities, kinetic and geometrical, are easily calculated, assuming that the unperturbed shock of radius $r_s \sim (-t)^\alpha$ has been perturbed into a shock of radius r given by

$$r = r_s[1 + \zeta(r_s, \theta)]. \tag{30}$$

Evaluating both sides of (29) with the help of (30) one gets in the linear approximation:

$$\alpha r_s^2 \frac{\partial^2 \zeta}{\partial r_s^2} + (1+\alpha) r_s \frac{\partial \zeta}{\partial r_s} = (1-\alpha)\frac{\partial^2 \zeta}{\partial \theta^2}. \tag{31}$$

Substituting into (31) the harmonic perturbation $\zeta = r_s^{p/\alpha} \exp[in\theta]$ one obtains the dispersion relation

$$p^2 + p + \alpha(1-\alpha)n^2 = 0\,, \qquad \text{or} \qquad p = -\tfrac{1}{2} \pm i\,[n^2\alpha(1-\alpha) - \tfrac{1}{4}]^{\frac{1}{2}}\,.$$

As the real part of p is negative, the perturbation $(r - r_s)/r_s$ is amplified and the shock is unstable in shape. The distorsion can also be given by the ratio of the two unperturbed and perturbed curvature radii which turns out to be $r_s/R \sim (n^2-1)\,r_s^{p/\alpha}$ and tends to infinity as r_s tends to zero. Nevertheless, the radius of the perturbed shock $r \sim r_s\zeta \sim r_s^{1+p}$ tends to zero as $r_s \to 0$ (with some oscillations due to the imaginary part of p) and the shock converges towards a point.

Another example is given by the implosion of cylindrical shells used for magnetic cumulation (Subsect. **1**.2.2) which turns out to be dynamically unstable [37]. High performances of cylindrical or spherical experimental arrangements will therefore require a very accurate initial symmetry.

REFERENCES

[1] R. Courant and K. O. Friedrichs: *Supersonic Flow and Shock Waves* (New York, 1948).
[2] S. D. Hamann: *Advances in High Pressure Research*, vol. **1** (New York, 1966).
[3] W. Bleakney and A. H. Taub: *Rev. Mod. Phys.*, **21**, 4, 584 (1949).
[4] L. V. Al'tchuler: *Uspekhi*, **8**, 1, 52 (1965).
[5] J. Leygonie and J. Cl. Bergon: *Proc. of IUTAM Symposium on High Dynamics Pressures* (Paris, New York, 1968).
[6] G. Guderly: *Luftfahrtforschung*, **19**, 302 (1942).
[7] A. W. Aikin: Metropolitan-Vickers Report No. 5090 (1956).
[8] J. P. Somon: Laboratori Gas Ionizzati, Frascati, Internal Report LGI No. 66/9 (1966).
[9] F. H. Harlow and W. E. Pracht: *Phys. Fluids*, **9**, 10, 1951 (1966).
[10] J. P. Somon: Laboratori Gas Ionizzati, Frascati, Internal Report LGI, No. 66/11 (1966).
[11] L. Rayleigh: *Phil. Mag.*, **34** (6-th series), 94 (1917).
[12] J. P. Somon: *Sur l'obtention de champs magnétiques intenses au moyen d'une implosion cylindrique*, Euratom, Bruxelles, EUR 4197 (1968).
[13] A. S. Balchan and G. R. Cowan: *Rev. Sci. Instr.*, **35**, 8, 937 (1964).
[14] E. I. Zababakhin: *Sov. Phys. JETP*, **22**, 2, 446 (1966).
[15] Y. B. Zel'dovich and Y. P. Raiser: *Physics of Shock Waves and High Temperature Hydrodynamic Phenomena* (New York, 1967).
[16] Y. B. Zel'dovich: *Sov. Phys. JETP*, **6**, 537 (1958).
[17] E. I. Zababakhin: *Sov. Phys. JETP*, **6**, 2, 345 (1958).
[18] S. Chapman and P. C. Kendall: *Proc. Phys. Soc.*, A **273**, 435 (1963).
[19] S. I. Syrovatskii: *Sov. Phys. JETP*, **27**, 5, 763 (1968).

[20] L. I. Sedov: *Similarity and Dimensional Methods in Mechanics* (London, 1959).
[21] J. P. Somon, H. Knoepfel and J. G. Linhart: *Nucl. Fusion Suppl.*, part 2, 717 (1962).
[22] K. P. Stanyukovich: *Unsteady Motion of Continuous Media* (New York, 1960).
[23] A. D. Sakharov, R. Z. Lyudaev, E. N. Smirnov, Yu. I. Plyushchev, A. I. Pavlovskii, V. K. Chernyshev, E. A. Feoktistova, E. I. Zharinov and Yu. A. Zysin: *Dokl. Akad. Nauk*, **10**, 11, 1045 (1966).
[24] J. J. Erpenbeck: *Phys. Fluids*, **5**, 10, 1181 (1962).
[25] D. S. Butler: A.R.D.E. Report No. 18/56 (1956).
[26] J. H. Lee and B. H. K. Lee: *Phys. Fluids*, **8**, 12, 2148 (1965).
[27] R. W. Perry and A. Kantrowitz: *Journ. Appl. Phys.*, **22**, 7, 878 (1951).
[28] B. H. K. Lee: *A.I.A.A. Journ.*, **5**, 11, 1997 (1967).
[29] G. M. Gandel'man and D. A. Frank-Kamenetskii: *Dokl. Akad. Nauk*, **1**, 223 (1956).
[30] C. Hunter: *Journ. Fluid Mechan.*, **8**, 241 (1960).
[31] R. W. Flagg and I. I. Glass: *Phys. Fluids*, **11**, 10, 2282 (1968).
[32] D. Hawkins: Report No. LAMS 2532 (1961).
[33] H. Knoepfel: This volume p. 168.
[34] R. L. Welsh: *Journ. Fluid Mechan.*, **29**, 1 (1961).
[35] W. Chester: *Advances in Applied Mechanics*, vol. **6** (New York, 1960).
[36] A. Sakurai: *Comm. Pure Appl. Math.*, **13**, 353 (1960).
[37] J. P. Somon: *Journ. Fluid Mechan.*, **38**, 769 (1969).

Vaporization-Wave Transitions.

F. D. BENNETT

U.S. Army Ballistic Research Laboratories,
Aberdeen Proving Ground, Aberdeen, Md.

1. – Introduction.

We outline here some theoretical and experimental aspects of the dynamical process by which an impulsively heated fluid makes the transition from liquid to vapor. Our interest is in the initial stages of the expansion as it occurs in the vaporization of superheated metals.

For polytropic, nonreacting fluid the head of an expansion wave proceeds into the compressed region at the local velocity of sound and the ensuing expansion is a simple wave [1]. When the fluid can be condensed, the first expansion will also be limited by the speed of sound in the liquid; but the vaporization which follows may take place either from the bounding, external surface or by cavitation with vapor-bubble formation. Since formation of internal cavities is limited by inertia and nonequilibrium effects, and may require a considerable time interval for bubble formation, we shall consider only experimental situations where vaporization takes place from the exposed surface. In this case the evaporation will be limited by some maximum speed with which the head of the wave travels. Because a phase change occurs and because the resulting damp vapor is more compressible than the liquid, one would expect the limiting wave speed to be much slower than that of sound in the homogeneous liquid. In other words, the rate at which a uniform vaporization phenomenon should occur will be much slower than the rate at which a purely liquid expansion is propagated.

2. – Wave hypothesis.

We wish to test the hypothesis that the limiting upper velocity of a dynamic vaporization process will be a sound speed characteristic of the two-phase region of the fluid. In particular the chosen speed will be that of the coexistence side of the liquidus line which is the low volume boundary of the two-phase region, represented as the locus $V_3(T)$ in Fig. 1.

For a substance characterized by an equation of state $P=P(V, T)$, where $V=1/\varrho$ is specific volume and P, ϱ, T refer to pressure, density and temperature respectively, one can readily show [2, 3] that the speed, c, of a small disturbance is given by $c^2=(\partial P/\partial \varrho)_S$, or equivalently by

$$c^2=(\partial P/\partial \varrho)_T+ + (T/\varrho^2 C_V)(\partial P/\partial T)^2_\varrho \,, \tag{1}$$

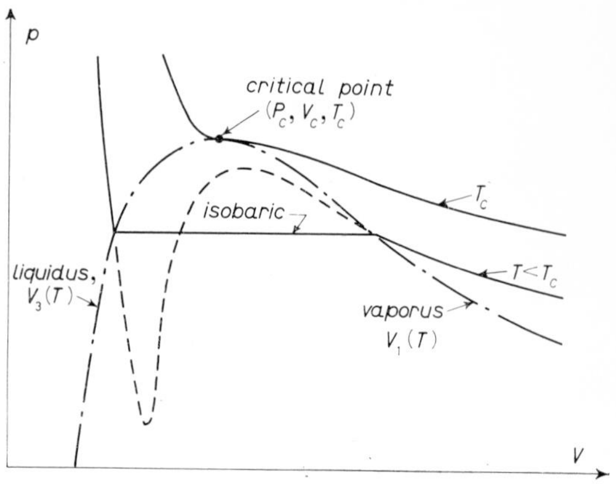

Fig. 1. – Sketch of two-phase region and isotherms. P, pressure, V, specific volume.

where C_V is the specific heat at constant volume. We note that within the two-phase region the number of free variables is reduced to one and $P=P(T)$, the vapor pressure; thus, only the second term of eq. (1) remains and

$$c=(T/C_V\varrho^2)^{\frac{1}{2}}(\mathrm{d}P/\mathrm{d}T)\,. \tag{2}$$

On the boundary given by the liquidus line two values of sound speed occur, *viz.*, those for the liquid and liquid plus damp vapor. Sound speed is double valued on the loci where slope discontinuities occur in the adiabats.

We refine the wave hypothesis slightly, to state: dynamic-vaporization phenomena in superheated liquids move into the undisturbed fluid no more rapidly than the liquid-vapor sound speed characteristic of the liquidus line. A more detailed development of the thermodynamical model will be given in Sect. 6 below.

While our main experience with the vaporization waves comes from experiments with metals, one would expect the hypothesis to apply equally well to a broader class of liquids including elements, compounds, solutions and mixtures. Necessary conditions for any liquid to exhibit vaporization wave phenomena can be tentatively stated as follows. Such a substance should have 1) a piecewise continuous equation of state relating pressure, specific volume and temperature, 2) a change in phase with a large increase in specific volume, 3) a latent heat of vaporization large compared with the specific energy at the boiling point; and, for practical reasons, 4) a temperature high enough so that specific energy exceeds a certain threshold.

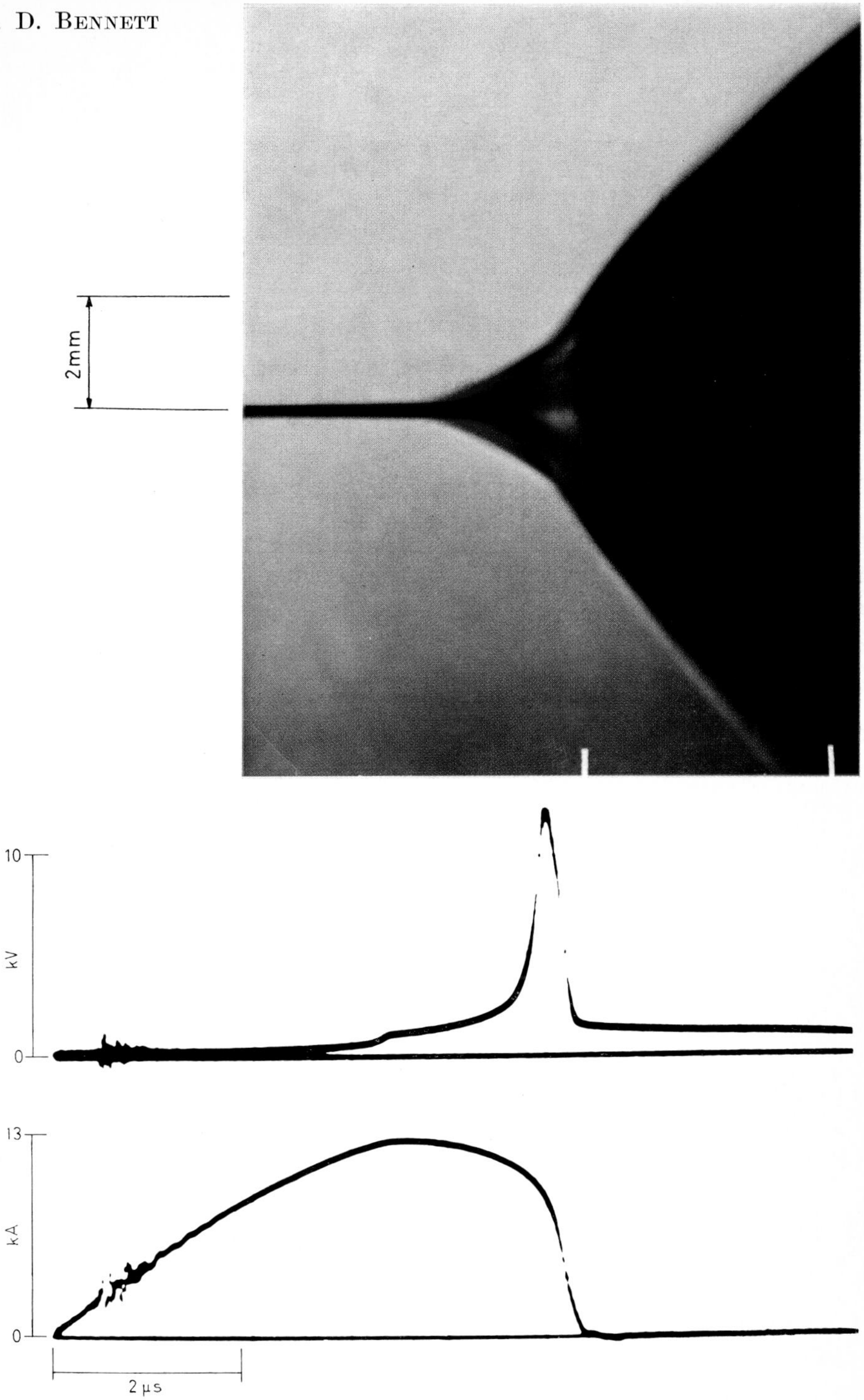

Fig. 2. – Streak photograph of expanding copper wire correlated in time with current and voltage. $d = 0.0254$ cm, $V_0 = 3$ kV, $C = 32$ μF.

Wavelike vaporization phenomena are encountered in the steam shock tube. Here, water contained at temperatures and pressures higher than its normal boiling point is suddenly released, by breaking a diaphragm, into a driven section containing air or other gas at lower pressure. Experiments of this kind have been performed by BROWN [4] and by TERNER [5]. BROWN treats the expansion wave traveling into the driver by assuming a two-stage process. In the first stage the water overexpands to a lower-pressure, metastable state. In the second it undergoes an exponential relaxation with increase of specific volume to its final equilibrium state. The connection between metastable and equilibrium states is found by using certain additional assumptions and the conservation laws of fluid mechanics in their « jump » form. While there is some arbitrariness about choosing the metastable state, Brown's experimental data appear to agree somewhat better with the results of the two-stage theory than with those of an equivalent equilibrium expansion.

3. – Exploding-wire experiments.

From the point of view of historical order, the vaporization wave hypothesis was first developed [3] to account for some of the peculiar phenomena observed when wires are exploded by a condenser discharge. Somewhat later [6] it was proposed as a more general concept applying to the explosive expansion of superheated liquids in general.

In an exploding-wire experiment a long, thin, conducting cylinder is impulsively heated by an energetic condenser discharge. Depending on the choice of the conductor and the initial conditions a variety of complex phenomena occur in the ensuing expansion. We shall not describe the entire process here but for a more complete discussion refer the reader to a recent review [7]. What will interest us are the early stages of the explosion as depicted in Fig. 2. There one sees a streak camera record of the expansion of a cylinder of copper correlated in time with simultaneous oscilloscope traces of the current passing through the wire and the voltage measured across it. The small ramp in the voltage curve represents melting of the wire. Shortly thereafter the streak photograph shows a haze of vapor expanding about the more-dense, interior of the wire. During this process the current decays rapidly to zero in the same interval that voltage goes through a peak, which in this case is about four times as high as the initial condenser voltage. For the example chosen, the wire is so well matched to the condenser that nearly the entire initial energy is deposited during the first pulse. The physical evidence for a vaporization wave is contained in the interval of wedge-shaped expansion terminated by the zero of the current pulse.

An interpretation of the exploding-wire data can be obtained as follows.

To first approximation we suppose the wire material to exist in two states only, *viz.* 1) fully conducting metal not yet affected by the expansion wave, and 2) expanded, nonconducting, wet vapor. Thus, for a vaporization wave propagating with velocity $v(t)$ inward from the wire periphery, the radius of the conducting core diminishes with time as $r = r_0\left[1-\int_{t_0}^{t}(v(\xi)/r_0)\,d\xi\right]$, where t_0 is the time chosen to represent the start of the expansion. In the example of Fig. 2, t_0 would be a little more than 4 μs. With this model, conduction will cease when $r = 0$. Thus, the model provides an interpretation of the current shut-off; and, in fact, initiates an explanation of dwell.

4. – Constant-velocity vaporization waves.

Since wire resistance, R, varies inversely with area of cross-section we can write

$$R = R_0(T)\left[1-\int_{t_0}^{t}(v(\xi)/r_0)\,d\xi\right]^{-2}. \tag{3}$$

If, for simplicity, we neglect temperature effects and assume the wave speed to be constant, then, with $\tau = r_0/v$ and $s\tau = t - t_0$, eq. (3) becomes

$$R = R_0/(1-s)^2 . \tag{4}$$

As Fig. 2 shows, the expansion takes place in a short time interval near the maximum of current. When current is at its maximum the charge on the condenser is zero and the energy supply is stored in the magnetic field. If we assume condenser voltage to be negligible during the current decay, the circuit equation becomes $L(di/dt) + Ri = 0$; and, with the previous work, can be written

$$di/ds + ai/(1-s)^2 = 0 , \tag{5}$$

where $a = (r_0/v)(R_0/L)$ is the ratio of the time constant of the wave to that of the circuit. Integrating eq. (5) yields

$$i = i_0 \exp[-as/(1-s)] , \tag{6}$$

and with the definitions for voltage, V_R and power, P_R,

$$V_R = R_0 i_0 (1-s)^{-2} \exp[-as/(1-s)] , \tag{7}$$

and

$$P_R = R_0 i_0^2 (1-s)^{-2} \exp\left[-2as/(1-s)\right] . \tag{8}$$

Differentiation shows that V_R and P_R have maxima at $1-a/2$ and $1-a$ respectively. Furthermore, these maxima have their minimum values, as functions of a, at $a=2$ and $a=1$ respectively. A variety of different pulse shapes can be interpreted [8] with this elementary calculation. Broadly speaking, if the time for e-fold decay of current is small compared with the time of wave passage, *i.e.* $a>2$, the current and voltage decay curves have exponential shapes without peaks. Conversely, if $a<1$ and the wave is comparatively fast, then the current decay is quite flat at first and steepens rapidly as the wave approaches the center. Voltage passes through a peak whose height increases as a gets smaller. The present treatment can be generalized to any portion of the current cycle providing only that the vaporization-wave relaxation time is much smaller than a quarter period of the undamped circuit so that the condenser voltage can reasonably be taken constant. The analysis of the constant-speed vaporization wave shows clearly the interrelation between circuit and wave speed parameters; and shows, furthermore, the possibility of a unified discussion of a wide variety of pulse shapes. It fails to provide an accurate method of calculating actual wave speeds.

5. – Wave speeds from experiments.

If one abandons the program, initiated with the discussion of constant-speed waves, of attempting to provide a wave-speed theory of current and voltage pulses, a considerable shift in point of view becomes possible. We regard the i-V_R curves as the raw data, the physical evidence of the vaporization wave, and attempt to deduce wave speeds from them. Several additional assumptions must be made in order to provide a rational reduction scheme. We outline it here; the details may be found elsewhere [3, 9].

If we neglect all work and heat terms except ohmic heating, conservation of energy allows us to equate rise in specific energy of the current-carrying core with the electrical energy deposited. From the corrected voltage V_R and current i curves one can calculate resistance $R=V_R/i$, power $P_R=V_R i$ and energy $E=\int V_R i \, \mathrm{d}t$. We can write

$$\mathrm{d}e = \begin{cases} i^2 R \, \mathrm{d}t/m\,, & (9) \\ V_R^2 \, \mathrm{d}t/Rm\,, & (10) \end{cases}$$

where e, m are respectively the specific energy, and the mass heated during interval dt. The quantity Rm is an invariant under the vaporization wave. The factor $(r/r_0)^2$ cancels out of the product; however, temperature variation remains and must be accounted for.

To this end we further assume that resistance is linear with specific energy above the melting point. This assumption can be expressed by writing

$$R = R_0[1 + \beta(e - e_0)](r_0/r)^2 , \tag{11}$$

where β is a constant to be determined from the experiment. If specific heat s is constant in the interval, then $\beta = \alpha/s$, $e = sT$ and eq. (11) contains the usual statement that resistance increases linearly with temperature. Using eqs. (10) and (11) and integrating, we find

$$(e - e_0) + (\beta/2)(e - e_0)^2 = (1/m_0 R_0)\int_{t_0}^{t} V_R^2 \mathrm{d}t . \tag{12}$$

To determine time t_0 when the vaporization wave begins, one plots scaled resistance, or resistivity, against apparent specific energy E/m_0. Before vaporization, the radius is taken to be $r = r_0$ and the constant β in eq. (11) can be determined from the linear portion of the resistivity data above melt. After vaporization commences, R rises steeply away from the linear law and this law is extrapolated to account for the effects of temperature variation. The point at which R departs from the line determines R_0 and e_0. Since R is already known as a function of time t_0 is then determined. True (e, t) values may be obtained by integrating V_R^2 as in eq. (12); and (e, R) values by comparison with the (R, t) data. With (e, R) data eq. (11) may be solved for $(r/r_0)^2$. The resulting points are fitted to an interpolation curve and differentiated numerically to obtain wave speed. The initial radius r_0 is taken nominally to be that of the cold wire. A more elaborate reduction method has been devised [9] to correct for thermal expansion of the liquid cylinder; but, because thermodynamic data for metals at high temperatures are lacking, has not yet been used. The correction if it could be made would tend to increase the values of wave speed deduced from measurement by a variable factor which increases approximately from 1 at room temperature to 1.7 at the critical temperature.

The assumption of a linear dependence of resistivity on specific energy is crucial to the reduction. Without it no account can be taken of the effects of temperature rise, and these are known to be too large to neglect. Certain metals, *e.g.*, Fe, Ni, W show anomalous decrease of resistance above the melting point in an exploding wire experiment. For this reason no wave speeds have yet been obtained for these metals or for others like them. If, as suspected, the cause lies in early voltage breakdown and associated current conduction in paths external to the wire there is some hope that a dense ambient fluid may prevent the charge leakage and enable wave-speed measurements to be made.

Wave speeds for Cu have been deduced from electrical measurements as described above and compared with theoretical values calculated from eq. (2). Values of liquid density, specific heat and vapor pressure were obtained from tabulated values as described in [6]. Owing to the scarcity of measured values of these quantities at the desired specific energies, various approximations, *e.g.* constancy of liquid density and specific heat, had to be made. Nevertheless, agreement between measured and calculated wave speeds for Cu [3], in the temperature range from the boiling point to the critical temperature, is quite satisfactory and a considerable encouragement to further work. Accordingly, similar measurements were made on metals Al and Pb. The experimental values may be seen in Fig. 3; the theoretical curve shown there comes from a thermodynamical model and assumed equation of state, see Sect. 6.

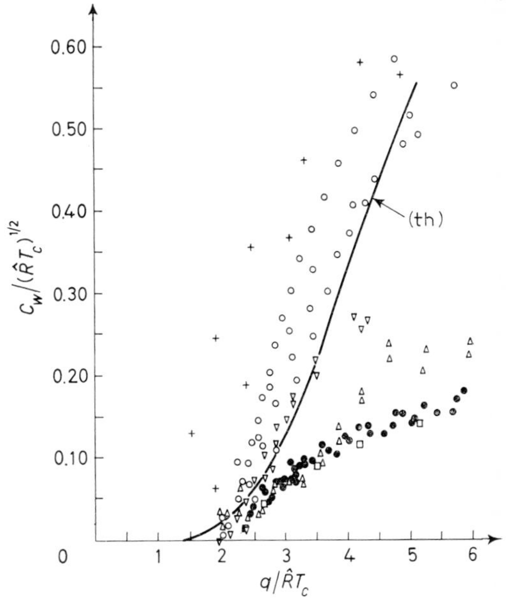

Fig. 3. – Wave speeds for several metals. □ Ag; ▽ Al; △ Au; • Cu; + Hg; ○ Pb.

When the tabulated data were used to calculate theoretical values [6] for wave speeds in Pb, poor agreement with experiment was obtained and the absence of reliable high-temperature vapor pressure data for this element was thought to be responsible. Since, generally speaking, Hg is the only metal for which measurements exist of pressure, density and conductivity up to the critical temperature, there is not much hope that expected values of vaporization wave speed can be calculated for other metals with sufficient precision to provide a correct trend of the function for comparison with experimental values.

6. – Thermodynamic model.

Accordingly, we develop here a simplified fluid model to represent the transient behavior of a material heated rapidly from the solid state up through its critical temperature. Our motivation is to supply a simple yet reasonably accurate basis for describing the state and dynamical behavior of superheated liquids. In order to make any further detailed flow calculations an equation of state is needed; accordingly, the goal will be to find an approximate equation of state which is capable of representing what is known about liquid metals and capable of filling in the large gaps where no data are available. Agreement will be sought with vaporization wave speeds, for example, even though detailed internal checks with other quantities are impossible. Our model is based on the following hypotheses: 1) the temperature rise and transitions take place under conditions of local thermodynamic equilibrium, 2) a form of Van der Waals equation applies, 3) realistic specific heats can be assigned the liquid and vapor states and 4) the liquid expansion takes place along the liquidus line, $V_3(T)$ of Fig. 1.

The first assumption is one mainly of simplification although a plausible argument can be given [3] that vaporization from the free surface should occur nearly at equilibrium conditions.

Van der Waals' equation is representative of a class of state equations which are reasonably accurate, though not precise, and yet embody the main features of most condensable gases, *viz.*, a hard core repulsion and a weak, long-range attractive force. With the modifications to be discussed below Van der Waals' equation allows different and realistic specific heats to be assigned the liquid and vapor phases, a matter which improves later agreement with experiment.

Choice of the liquidus line as the thermodynamic path is somewhat arbitrary although it appears to be reasonable, as may be seen from the following argument. For most physical examples, the isotherms and adiabats in the liquid phase are very steep compared to those of either the coexistence or vapor regions, except very near the critical point. Slope magnitudes of the adiabats are proportional to sound speed squared; consequently the sound speed in the liquids is much higher than in either of the other regions. We may then argue that heating moves the thermodynamic state along the saturated liquid line. Any compressive tendency to drive the expanding system into the all-liquid state will be rapidly counteracted by liquid thermal expansion, which is dominated by the fastest speed of sound. The liquid expansion lowers the pressure to that of the liquidus boundary where any further expansion must be accompanied, under the assumed equilibrium conditions, by partial vaporization, which is governed by the slower vaporizing wave speed. For interior parts of the system where inertial confinement prevails, premature expansion via incipient cavitation will cause local pressure rise which will force the fluid element back toward the liquidus line. This process of excursion about the liquidus continues during the electrical heating pulse until the fluid particle evaporates from the surface by expansion through the vaporization wave.

In what follows we take the thermodynamic system of interest to be a uniform, molten cylinder of metal to which heat is being added and from whose surface partial vaporization occurs. The speed of the leading edge of the vaporization wave is assumed to proceed with the wave speed of the saturated wet vapor at absolute temperature T, as given by eq. (2) evaluated on the liquidus line.

For unit mass of fluid, pressure P and specific volume V are related by the Van der Waals equation in the form

$$P_W = \hat{R}T/(V - b) - a/V^2, \tag{13}$$

where the constants a, b are given in terms of values at the critical point by

$$a = 3P_C V_C^2, \qquad b = V_C/3, \qquad \hat{R}T_C = (8/3)P_C V_C. \tag{14}$$

Inequalities $T > 0$, $V \geqslant b$ are understood to hold. When $T \leqslant T_C$ the Van der Waals equation of state describes a two-phase region where both liquid and vapor phases coexist, as in Fig. 1. In the coexistence region, vapor pressure P_A is not derived directly from eq. (13) which has the form of a cubic, but rather from a generalization of the Maxwell equal-area rule which replaces the two loops by a horizontal line cutting equal areas from the loops. KAHL [10] has proven the remarkable result that the equal-area condition is not necessary, but defines only one of a continuum of possible choices. For $T < T_C$ he writes

$$\int_{V_3(T)}^{V_1(T)} P_W \, \mathrm{d}V = P_A(T)[V_1(T) - V_3(T)] + \varphi(T), \tag{15}$$

where

$$\varphi(T) = \int_T^{T_C} [C_V(\tau, \text{liq}) - C_V(\tau, \text{vap})](1 - T/\tau)\, \mathrm{d}\tau. \tag{16}$$

Specific heat at constant volume C_V is a function not only of temperature but also of the state of the fluid, whether liquid or vapor. As KAHL shows, the features of equilibrium thermodynamics are preserved, the main one being the stationary property of the Gibbs function as the system passes from liquid to vapor on an isotherm. If C_V is the same for both liquid and vapor, then $\varphi(T) = 0$ and eq. (15) gives the Maxwell result. If the two specific heats are not the same, then $\varphi(T) \neq 0$, the areas are not equal and some further, useful flexibility is available in making a realistic choice of the two specific heat functions.

To evaluate eq. (2) for the vaporization sound speed on the liquidus line, P_A and $V_3 = 1/\varrho_3$ can be found numerically from eqs. (15) and (16). The specific heat for the coexistence state, C_{VA}, must also be found formally and then numerically. To do this one finds the internal energy E_A and differentiates, using $C_{VA} = (\partial E_A/\partial T)_V$. E_A may be determined from the general relation

$$(\partial E/\partial V)_T = T(\partial P/\partial T)_V - P\,, \tag{17}$$

and noting from eq. (15) that $M(T) = T(\partial P_A/\partial T)_V - P_A$ is independent of V. Direct integration of eq. (17) gives

$$E_A = E_W + M(T)[V - V_3(T)]\,, \tag{18}$$

where $E_W = E_W(V_3, T)$ is the internal energy on the liquidus given by the Van der Waals function. One finds

$$C_{VA} = (\partial E_W/\partial T)_V + (\mathrm{d}M/\mathrm{d}T)(V - V_3) - M(\mathrm{d}V_3/\mathrm{d}T)\,, \tag{19}$$

and

$$(\partial E_W/\partial T)_V = (\partial E_W/\partial V_3)_T(\mathrm{d}V_3/\mathrm{d}T) + (\partial E_W/\partial T)_V\,; \tag{20}$$

finally, on the saturated liquid line

$$C_{VA} = [(\partial E_W/\partial V_3)_T - M](\mathrm{d}V_3/\mathrm{d}T) + (\partial E_W/\partial T)_W\,. \tag{21}$$

The last term is $C_V(T, \text{liq})$ and may be taken from experiment or represented approximately in some chosen functional form. Van der Waals' equation gives no further information. Putting P_W in eq. (17) gives $\partial E_W/\partial V_3 = a/V_3^2$. With these results, eqs. (15), (16), (21) and Van der Waals' equation, the vaporization wave speed can be numerically evaluated on $V_3(T)$. In the liquid the more general definition of eq. (1) must be used.

While the equation of state is expressed in terms of absolute temperature, the independent variable one obtains more easily from experiment is the added specific heat content Δq. One can make the connection between q and T by integrating $\mathrm{d}q = \mathrm{d}E + p\,\mathrm{d}V$ along $V_3(T)$ from the melting temperature T_M to T using computed values of E_W (V_3, T) and $P_A(T)$. For any $T \leqslant T_C$ a correspondence is thus given between $\Delta q(T)$ and T. One then writes $q(T) = q(T_M) + \Delta q(T)$ where $q(T_M) = q_M$ is the heat content of the liquid at melt, assumed known from other sources. Clearly the scaled variable $\bar{q} = q/\hat{R}T_C$ depends upon the specific heats of the liquid and vapor phases through eqs. (15) and (16), as well as on $q_M/\hat{R}T_C$. This latter quantity has a value near 0.60 for a number of metals. Both scaled wave speed $\bar{c} = C_W/(\hat{R}T_C)^{\frac{1}{2}}$ and scaled heat content are functionals of the specific heats for liquid and vapor. When one can approximate both of the specific heat functions by constants, the parametric dependence of the wave speed curve on the choices of specific heats can readily be shown as in [9].

7. – Comparison of the thermodynamic model with experiment.

Wave speed data for six metals, Au, Ag, Cu, Al, Pb, Hg, are displayed in Fig. 3. While a somewhat similar plot appears in [9] where cross-hatched areas are used to represent the more abundant data, here the individual data points are included for a range of variables which includes most of the two-phase field. Dimensionless wave speeds $\bar{c}$ are plotted against dimensionless specific heat content $\bar{q}$. Because measured, thermal expansion data for liquid metals are lacking, no correction could be made for the effects of expansion prior to the passage of the vaporization wave. Likewise, no measured values of critical point data are available for any metal except Hg so the best recent estimates were used [11-13]. The scaling constants are summarized in Table I.

TABLE I.

Metal	T_M (°K)	T_C (°K)	$\hat{R}T_C$ (kJ/g)	$(\hat{R}T_C)^{\frac{1}{2}}$ (m/s)
Al	933	8650	2.67	1633
Cu	1356	8500	1.09	1043
Ag	1234	7460	0.575	758
Au	1336	9500	0.401	630
Hg	234	1733	0.0718	268
Pb	601	5400	0.217	465

The theoretical curve is computed from a modified Van der Waals model, in dimensionless form, with critical compressibility $\frac{3}{8}$, and with $C_V(\text{liq}) = \frac{5}{2}\hat{R}$ and $C_V(\text{vap}) = \frac{3}{2}\hat{R}$. This choice of specific heats, made possible by relinquishing Maxwell's equal area rule, is a better approximation to the known specific heat data than assumption of identical specific heats for liquid and vapor. In its scaled form $\bar{c}$ is then a function of dimensionless variables only.

The scale of reduced specific heat content, $\bar{q}$, for the theoretical curve is that computed from the Van der Waals function, with an added 0.1 unit to adjust the $\bar{q}_M$ value so that it represents, reasonably well, the corresponding values for the six metals which cluster fairly closely to the nominal, empirical value of 0.60.

As may be seen in Fig. 3 the data and the theoretical curve agree quite well on the point of inception of vaporization wave speed, and on the early rise of the curve. The data points for the noble metals fall below the theoretical curve and go out beyond the critical point limit toward what appears to be a horizontal asymptote. Handbook wave speeds for Cu [3] actually agree better with the plotted data than does the Van der Waals curve. Some recent unpublished data on Cu taken in our laboratory, reduced by a finer-grained method of reduction, adhere more closely to the Van der Waals curve up to an ordinate of 0.15 before bending away. If the correction for thermal expansion could be applied, the wave speed values would be increased by at least $(20 \div 70)\,\%$ depending on $\bar{q}$, the larger increases applying near critical. Such a correction would somewhat improve the agreement for Ag, Cu, Au and Al; but would move Pb and Hg points further away from the theoretical curve.

Some special problems were encountered with the Hg wires which had to be frozen in an acetone dry-ice mixture, kept in a cold box and transferred rapidly to the test cell for explosion. Refinements in this process might result in better reproducibility of the mercury data.

8. – Deviations from the model.

If we confine our attention to phenomena below critical, two deviations from the assumed model appear to be of primary importance.

The theoretical wave speed applies to the head of an expansion wave traveling into the molten fluid. Of necessity the front across which the conductivity drops to zero must come somewhere in the region of reduced density behind the head of the wave. It may occur across a narrow segment of the expansion fan and thus be representable by a sharp front possibly resembling a « conductivity shock wave », but its speed of propagation will be less than that of the sound wave which first travels into the interior. Because we know little about the change of conductivity in the expansion wave we have no way of estimating the decrement in wave speeds to be expected. The noble metals and Al fall below the theoretical curve in the expected way but Pb and Hg do not.

The X-ray pictures of Fansler and Shear [14] demonstrate that for wires $2 \div 5$ times larger and for somewhat slower rates of energy addition the expanding cylinder of molten fluid is not uniform but displays the transverse variations in density called striations. How these striations are related to the expansion wave is not known at present although some speculative hypotheses can be advanced. Their presence in these cases appears to rule out the hypothesis that the vaporization wave always proceeds uniformly through a cylinder of uniform density.

The effect of unvaporized portions of the wire remaining in striations appears to move all data points toward higher values of $\bar{q}$ by variable amounts. There is presently no method of estimating this effect. Experiments now under way are directed toward determining the relationship between number of striations and rate of energy addition during vaporization-wave passage. The present X-ray techniques are unable to resolve any striations in the smaller wires and faster explosions typical of much of the data plotted in Fig. 3; however, their presence in the larger, slower cases is a cautionary warning that density variations somehow play an important part in the heating and expansion of superheated metals.

REFERENCES

[1] R. Courant and K. O. Friedrichs: *Supersonic Flow and Shock Waves* (New York, 1948), p. 92.

[2] C. F. Curtiss, C. A. Boyd and H. B. Palmer: *Journ. Chem. Phys.*, **19**, 801 (1951).

[3] F. D. Bennett, G. D. Kahl and E. H. Wedemeyer: in *Exploding Wires*, vol. **3**, edited by W. G. Chace and H. K. Moore (New York, 1964), p. 80.

[4] E. A. Brown jr.: *Explosive Decompression of Water*, Ph. D. Thesis, Northwestern University, Evanston, Ill. (1959).

[5] T. E. Terner: *Ind. Eng. Chem., Process Design Dev.*, **1**, 84 (1962).

[6] F. D. Bennett: *Phys. Fluids*, **8**, 1425 (1965).

[7] F. D. Bennett: *High-temperature exploding wires*, in *Progress in High-Temperature Physics and Chemistry*, vol. **2**, edited by C. A. Rouse (London, 1968), p. 1.

[8] F. D. Bennett: *High-temperature exploding wires*, in *Progress in High-Temperature Physics and Chemistry*, vol. **2**, edited by C. A. Rouse (London, 1968), p. 29.

[9] F. D. Bennett and G. D. Kahl: in *Exploding Wires*, vol. **4**, edited by W. G. Chace and H. K. Moore (New York, 1968).

[10] G. D. Kahl: *Phys. Rev.*, **155**, 78 (1967).

[11] A. v. Grosse: *Journ. Inorg. Nucl. Chem.*, **22**, 23 (1961).

[12] J. A. Cahill and A. D. Kirshenbaum: *Journ. Phys. Chem.*, **66**, 1050 (1962).

[13] A. v. Grosse: *Rev. Hautes Temp. et Réfract.*, **3** (2), 115 (1966),

[14] K. S. Fansler and D. D. Shear: *Exploding Wires*, vol. **4**, edited by W. G. Chace and H. K. Moore (New York, 1968), p. 185.

Detonation Physics.

R. Schall

Institut Franco-Allemand - St. Louis

1. – Introduction.

High explosives are materials which can undergo exothermic reaction, the ignition of which is transmitted from the reacting to the adjacent layer by mechanical impulse and propagates therefore as a pressure wave. That this « detonation » process can take place, requires that the reaction comes to an end before release waves, starting from free surfaces and travelling at sonic speed, lower the pressure level. For explosive bodies of limited size (the order of 1 cm) and with sound velocity of about 5 mm/μs, the reaction time must therefore be shorter than 1 μs. This is possible only at very high pressures where pressure waves steepen up to shocks. Detonation can consequently be understood as a coupling of a shock with a chemical reaction. The shock provokes the reaction and the reaction strengthens the shock until a balance between energy transfer and dissipation is reached. In a geometry which does not vary with propagation, steady conditions are established. Herefrom stems the characteristic feature of a constant detonation velocity, *e.g.* in a cylindrical charge.

The one-dimensional treatment of this steady-wave phenomenon is the domain of hydrodynamic detonation theory, the most outstanding result of which consists in the finding that the detonation velocity D is independent of the reaction rate. The knowledge of the total energy release and of the equation of state of the reaction products is sufficient to determine D. The success of this theory has not contributed to our knowledge of reaction properties of high explosives.

It became, however, clear very soon that the one-dimensional theory is unable to describe several detonation phenomena, as *e.g.* the dependence of D on the charge diameter. This « diameter effect » can be adequately described by one characteristic parameter, the ratio of the reaction zone length to the charge diameter. More information about the reaction process is needed to treat unsteady detonation waves, in particular converging waves, which are of special interest because they concentrate locally the chemical energy released in a greater volume. The speed of energy transport from the reacting layer to the shock front is here the predominant factor. The shock induced in an inert material reflects the reaction-zone structure only in a thin layer close to the interface. At appreciable depth, the shock strength is determined by the conditions of the produced gases. The projection of a metal plate is the result of multiple shocking. In this way an appreciable amount of the energy content of the detonating gas can be transferred to the liner metal. The conical collapse of such a liner in a « shaped charge » gives rise to the formation of a metal jet, which can be much faster than the impacting liner elements. The conditions for occurrence of jetting limit, however, the energy densities which can be produced by this technique.

2. – Plane detonation waves.

2·1. *The Chapman-Jouguet detonation.* – A one-dimensional detonation process propagating at constant velocity D becomes steady in a co-ordinate system moving at the same speed. In this system (Fig. 1) the explosive enters at speed D into the wave front 0, reacts and leaves the reaction zone at 1 at a flow speed lowered by W, the particle velocity of the reacted gas. For a strictly one-dimensional flow, mass and momentum conservation give

(1) $$\varrho_0 = \varrho_1(D - W)\,,$$

(2) $$\varrho_0 D^2 + p_0 = \varrho_1(D - W)^2 + p_1\,,$$

$$\varrho = \frac{1}{v} = \text{density}\,, \qquad p = \text{pressure}\,.$$

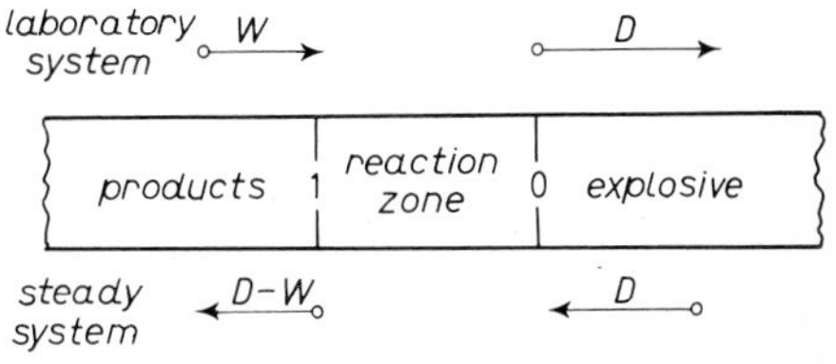

Fig. 1.

Neglecting transport effects (heat conductivity and friction) the energy equation reads:

(3) $$E_0 + \tfrac{1}{2}D^2 + p_0 v_0 = Q + E_1 + \tfrac{1}{2}(D - W)^2 + p_1 v_1\,;$$

$$Q = \text{reaction heat}, \qquad E_0, E_1 = \text{specific internal energy}.$$

A further relation between these quantities is given by the equation of state of the products

(4) $$p = p(v_1, E_1)\,.$$

Before we start showing how D can be determined, we remark that with $p_1 \gg p_0$ and $E_1 \gg E_0$, (1) + (2) gives

$$p_1 = \varrho_0 DW , \tag{5}$$

and (1) + (2) + (3)

$$Q + E_1 = \tfrac{1}{2} p_1 (v_0 - v_1) . \tag{6}$$

From (1) and (2) we get

$$-\varrho_0^2 D^2 = \frac{p_1 - p_0}{v_1 - v_0} .$$

In a (p-v) diagram (Fig. 2) D is therefore determined by the slope of the line (the « Rayleigh line ») linking the initial state and the state of terminated reaction. Two cases have to be considered:

a) The Rayleigh line is steeper than the tangent and intersects the state curve of the products in 2 points.

b) The Rayleigh line is tangent to the state curve in the point $v_1 = c_{\text{C-J}}$, $p_1 = p_{\text{C-J}}$.

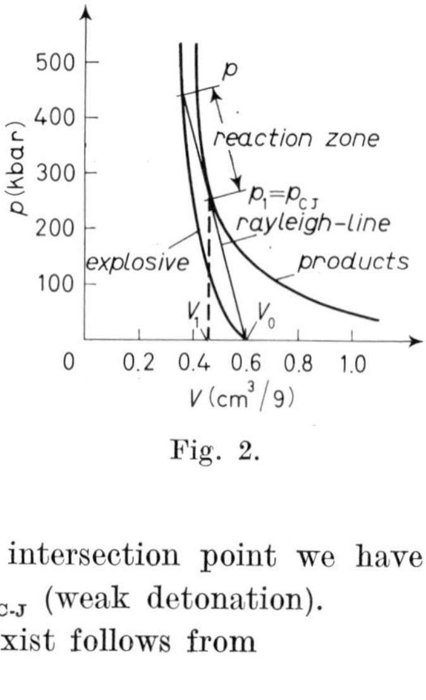

Fig. 2.

Case *b*) represents the C-J-condition and it is

$$D = W + a , \tag{7}$$

a being the sound velocity in the detonation products.

In case *a*) we have $D > D_{\text{C-J}}$; at the upper intersection point we have $p_1 > p_{\text{C-J}}$ (strong detonation), at the lower $p_1 < p_{\text{C-J}}$ (weak detonation).

That unsustained strong detonations cannot exist follows from

$$D - W < a .$$

This means that release waves, which arise as a natural consequence of the compression in the shock front, can catch the shock front and lower the detonation pressure down to a point where the C-J condition is fulfilled. Weak detonations are eliminated by the following argument. It is reasonable to suppose that intermediate states during the reaction lie between the Hugoniot curve of the intact explosive and that of the produced gas. For a weak detonation, the Rayleigh line runs from its intersections with these 2 curves—it is along this line that the states in the reaction zone vary—through a domain, to which correspond no possible reaction states.

The C-J theory therefore supposes that in a steady-state detonation the Rayleigh line is tangent to the product's Hugoniot curve.

The conservation equations (1) to (3), the equation of state (4), and the C-J condition form a system which determines D as well as the states in the reacted explosive (C-J states). Table I gives results for a perfect gas and for a gas obeying an Abel equation. Typical data for a detonating gas and for a condensed explosive show pressures of different orders of magnitude. While for a perfect gas D does not depend on ϱ_0, D increases with ϱ_0 for a condensed explosive. This increase is found experimentally to be roughly linear. Table II

TABLE I.

a) Detonating gas	*b*) Condensed explosive
$p/\varrho = R$	$p(1-\alpha\varrho)/\varrho = R$
$\varrho_0 = 10^{-3}\ \text{g}\cdot\text{cm}^{-3}$	$\varrho_0 = 1.7\ \text{g}\cdot\text{cm}^{-3}$
$Q = 1400\ \text{cal g}^{-1}$	$Q = 1400\ \text{cal g}^{-1}$
$\gamma = 1.4$	$\gamma = 1.25$
$\bar{c}_v = 0.33\ \text{cal g}^{-1}\cdot\text{Grad}^{-1}$	$\bar{c}_v = 0.33\ \text{cal g}^{-1}\cdot\text{Grad}^{-1}$
$\varrho/\varrho_0 = (\gamma+1)/\gamma = 1.71$	$\alpha = 0.4\ \text{cm}^3\ \text{g}^{-1}$
$p = 2\varrho_0 Q(k-1) = 50\ \text{atm}$	$\varrho/\varrho_0 = (\gamma+1)/\gamma = 1.17$
$T = (Q/\bar{c}_v)(2\gamma/(\gamma+1)) = 5200\ °\text{K}$	$p = 2\varrho_0 Q((\gamma-1)/(1-\alpha\varrho_0)) = 170000\ \text{atm}$
$D = [2Q(\gamma^2-1)]^{\frac{1}{2}} = 3400\ \text{ms}^{-1}$	$T = (1/(1-\alpha\varrho_0))[2Q(\gamma^2-1)]^{\frac{1}{2}} = 8000\ \text{ms}^{-1}$
$W = D(1/(\gamma+1) = 1400\ \text{ms}^{-1}$	$T = (Q/\bar{c}_v)(2\gamma/(\gamma+1)) = 5000\ °\text{K}$
	$W = D((1-\alpha\varrho_0)/(\gamma+1)) = 1200\ \text{ms}^{-1}$

TABLE II. – *Detonation velocities of some solid explosives* $D = D_1 + k(\varrho_0 - 1)$.

	D_1 (m/s)	k (m/s/g/cm³)	ϱ_{max} (g/cm³)	ϱ_{cryst} (g/cm³)
TNT	5060	3187	1.63	1.654
PETN	5550	3950	1.73	1.77
RDX	6080	3590	1.77	1.82
HMXd	6090	3590	1.85	1.90
Comp B (RDX/TNT 65/35)	5779	3127	1.72	—
RDX/TNT/Al 45/30/25	3950	4200	1.77	—

	D (m/s)	ϱ_0 (g/cm³)		
Baronal ($Ba(NO_3)_2$/TNT/Al 50/35/15)	5450	2.30		
Baratol ($Ba(NO_3)_2$/TNT 65/35)	5580	2.35		

gives experimental values for D as well as the density coefficients. Such experimental data for D are used to check the equation of state. A very simple description for the shock behaviour of high-energy explosives is given by a perfect gas equation with $\gamma \sim 3$. The Hugoniot curve for the often used Comp B (RDX-TNT 65/35, $\varrho_0 = 1.714$) is perfectly matched by $p \cdot v^{2.77} = \text{const}$ in a pressure range between $5 \cdot 10^4$ and $5 \cdot 10^5$ bar. It must, however, be emphasized that such an equation is meaningful only so far as the p-v relation is concerned; it would be quite incorrect to use it for temperature calculations. The highest detonation speed listed in Table II is that for crystalline HMX ($\varrho_0 = 1.90$) for which we find $D = 9.321$ m/s.

With $\gamma = 3$ we then get

$$p = \varrho_0 \frac{D^2}{4} = 412\,000 \text{ bar} .$$

A more detailed description of the hydrodynamic theory can be found in many text books as *e.g.* [1-3].

2·2. *Criticism of the C-J theory.* – For many years the validity of the C-J-theory was generally accepted. It is only in recent times, that more precise measurements have left some doubts as to whether or not experiments are in full agreement with the C-J assumption [4]. It seems that the observed phenomena are more correctly described in terms of a weak detonation, which means that the flow is supersonic at the end of the reaction zone, the detonation velocity is slightly higher, and pressure and density are slightly lower than predicted by the C-J theory. This is true not only for detonation in condensed media, but also for gas detonations with relatively low energy densities.

The strongest argument for this stems from radiographic density measurements. Product densities are found to be 10% lower in gaseous detonations [5], and about 2% lower in condensed detonations than computed from the C-J theory.

Another indication is the following. With the assumptions of the C-J theory, detonation pressures can be calculated without any knowledge of the equation of state; only the derivatives $(\partial D/\partial \varrho_0)_{E_0}$ and $(\partial D/\partial E_0)_{\varrho_0}$ need be known (ϱ_0 being the initial density of the explosive and E_0 the initial specific energy) [6]. It has been found that pressures calculated in this way do not agree with those measured from the shock strength induced in an adjacent metal plate, and also that $(\partial D/\partial \varrho_0)_{E_0}$ depends on the mechanical state of the high explosive, which cannot be understood on the basis of the hydrodynamic theory. As a matter of fact for liquid TNT, $(\partial D/\partial \varrho_0)_{E_0}$ was measured to be (4240 ± 509) $\text{ms}^{-1}/\text{gcm}^{-3}$ while this derivative for solid TNT at the same initial density is (3187 ± 27) $\text{ms}^{-1}/\text{gcm}^{-3}$. This indicates that the processes of detonation in the liquid and in the solid are different [7].

The most plausible explanation for the discrepancy between theory and experiment is that the flow is not strictly one-dimensional. There is experimental evidence that in very many cases one-dimensional linear gaseous detonations are not strictly stable, but show turbulences and velocity variations. The general character of this feature has been emphasized by ERPENBECK [8] who studied analytically the stability of plane detonation waves; his results suggest that, for any material releasing sufficient heat to be classed as high explosives, the one-dimensional steady solution is unstable and therefore does not exist in reality. Erpenbeck's considerations give no information on the speed of growth and the degree to which disturbances develop. They only support the idea that the energy density in the reaction zone is, at least in a microscale, not homogeneous, but varies locally to considerable extent.

Such an assumption would explain why pressure waves jumping from hot spot to hot spot propagate at a higher speed than in a strictly laminar flow with uniform energy distribution. It would also explain why the detonation process in a granular explosive, where disturbances are enhanced through the ignition process, is different from that in an homogeneous explosive. It gives finally, a possible explanation of the very high electrical conductivity of detonation gas [9] and the abnormously high escape velocities which are observed in a gas of low density in contact with detonating high explosives; these are 3 to 5 times higher than the hydrodynamic theory predicts [10].

To conclude, it can be said that the one-dimensional classical detonation theory gives a reasonable overall description of a plane progressing wave, but that it fails to describe the fine structure in the reaction zone, of which little is known today. It can, however, be guessed that in the reaction zone local energy concentrations occur which are essentially higher than the average value which the classical theory predicts.

3. – Detonation with curved front.

3·1. *Steady waves.* – A cylindrical charge with limited diameter detonates at constant rate with a curved front. The steady shape of the wave front is established after a path of a few charge diameters through the interference of release waves entering into the reaction zone from the free surface. These delay the wave more at the surface than in the core. The result is, that, compared with the plane case, a lesser portion of the reaction energy is available to support the leading shock. Detonation velocity is therefore lower than D_i, the plane-wave velocity. The cylindrical charge can, however, be treated by a modification of the ideal hydrodynamic theory. To obtain representative results it is sufficient to know the length of the reaction zone $l = \tau(D - W)$ or the reaction time τ, which may, on the other hand, depend on the velocity D, of the shock. For a bare charge of diameter $2R$, Eyring and others have found a relation [11]

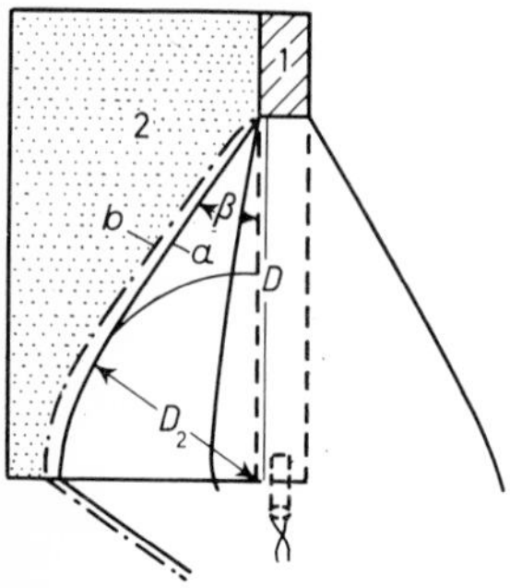

Fig. 3.

$$\frac{D}{D_i}=1-\frac{l}{2R}.$$

For rates D not much smaller than D_i a linear decrease with $1/R$ has been confirmed experimentally (Fig. 3) and Eyring's relation has generally been used to calculate l or τ from $D(1/R)$ plots. There is some evidence that reaction times so determined are too great, but in the absence of more reliable methods, the slope of the $D(1/R)$ curve is generally used as quantitative measure of the reaction zone length.

3·2. *Diverging waves.* – Point initiation—the formation of a spherical detonation wave by energy release in a restricted area—creates a condition wherein the wave speed varies rapidly with time. The distinction between a subsonic and a supersonic region is meaningless. The crucial problem is whether or not pressure waves induced by the energy release can catch and strengthen the leading shock. The analytical treatment of this question must be based on knowledge of the speed of energy transport behind the shock front. Data are scarce in the case of granular explosives, since temperatures are far from being homogenous. In the interstices between the grains, much higher temperatures (« hot spots ») are induced than in the solid explosive [12, 13].

It is only in the case of detonation rates close to D_i that more general results are available. Berger [14] has pointed out that the expanding wave is related to the cylindrical charge as treated in the last paragraph. Curvature of wave front and divergent flow in the reaction zone are indeed characteristic features of both. The relation

$$1-\left(\frac{D}{D_i}\right)^2=\gamma^2(1-2k)(C^2-1)$$

is a general description, where $k=1-\frac{1}{2}(\varrho_{\text{C-J}}/\varrho)$ measures the density behaviour in the reaction zone and C is the ratio of the entrance and exit cross-sections for streamlines entering and leaving the reaction zone. The formula contains the afore-mentioned Eyring relation as an approximation. It furthermore describes, with $C=(r/(r+W\tau))^n$, the cylindrically ($n=1$) and spherically ($n=2$) expanding wave, τ being the reaction time.

3·3. *Converging detonation.* – Converging waves are of particular interest for the physics of extreme energy states. Guderley has shown that an imploding shock wave reaches infinite strength at the center of symmetry, when —as justified for plane waves—radiation, friction and thermal conductivity are neglected. The same holds, of course, for imploding detonation waves, these being reaction-supported shocks. ZEL'DOVICH [15] has treated the imploding detonation under simplifying assumptions ($\gamma = 3$). He finds a slow increase of the detonation rate with r/r_0, r_0 being the initial wave radius. For a cylindrically converging wave, the detonation pressure grows at $r = r_0/10$ by a factor 1.93 and at $r = r_0/50$ by a factor of 3.8.

As long as the wave can be considered as not far from steady, the Berger formula gives an adequate description of the phenomenon with $C = (r/(r - W\tau))^n$. In a special case experiments are in better agreement with this formula than with Zel'dovich's results [16]. Some serious doubts have been expressed about the stability of imploding detonation waves, so far published results show, however, that converging detonations can be used in practice to obtain local energy concentrations.

4. – Wave shaping.

In order to produce detonation waves of special geometry, several methods for shaping detonation fronts have been used. We consider at first techniques to obtain linear or plane waves. They can, however, also be used to form waves of any shape. One way to do this makes use of high explosives having different detonation rates (explosive lenses) (*). If the interface is normal to the propagation in the fast explosive (D_1), a Mach wave is produced in the slow explosive (D_2) the front of which forms, with the interface, an angle β given by $\sin\beta = D_2/D_1$ (Fig. 4) (**). In a conical device (« plane wave generator ») this effect is commonly used to produce waves of plane geometry.

Another way is to place in the path of the detonation front holes or inert bodies around which the wave has to pass. An equidistant arrangement of holes in the explosive assures, *e.g.*, a wave of good linearity (Fig. 5).

A third technique makes use of impacting metal plates to produce instantaneous ignition. Figure 6 shows a device in which a point-initiated slab produces, by the synchronised impact of liner, a linear ignition of an explosive plate which then initiates an explosive block simultaneously at its plane surface.

(*) The combination of plastic bonded HMX ($D_1 \sim 9000$ m/s) and Baratol ($D_2 \sim 5500$ m/s) is frequently used for this purpose.

(**) Close to the interface this is only true if the 2 explosives are matched, *i.e.* if D equals the shock velocity which the explosive 1 produces in the explosive 2. Otherwise D_2 is only established after a transition zone, the length of which depends on the geometry of the device.

These and other wave-shaping techniques (*) can obviously be used to initiate waves with converging fronts. The explosive-lens technique, *e.g.*, can be applied to form cylindrically or spherically collapsing waves. To this end the interface

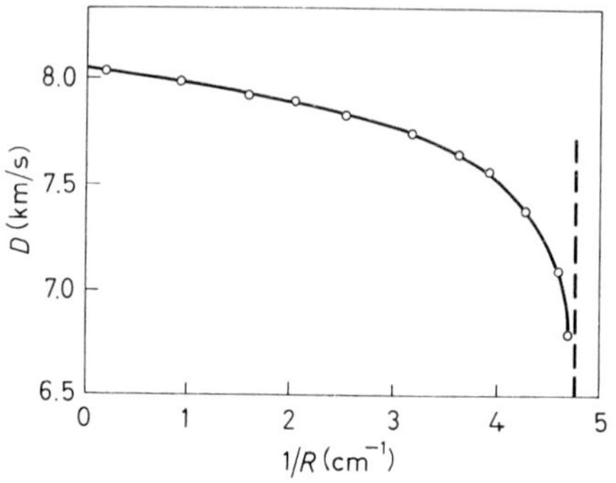

Fig. 4.

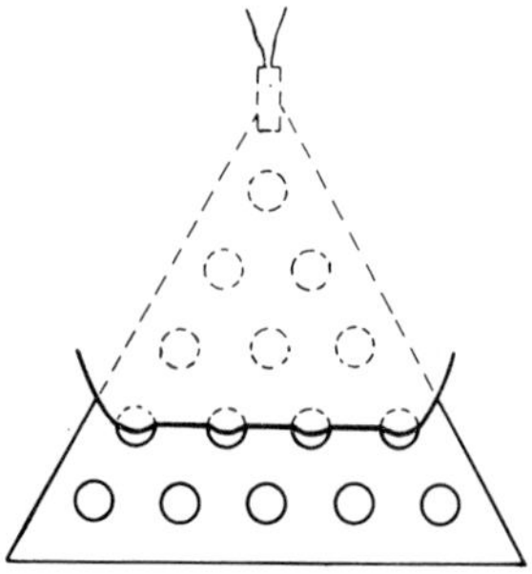

Fig. 5.

between the fast and the slow explosive has to be given (Fig. 7) the form of a logarithmic spiral. In the case of inert inclusions, the appropriate device for the same purpose is sketched in Fig. 8.

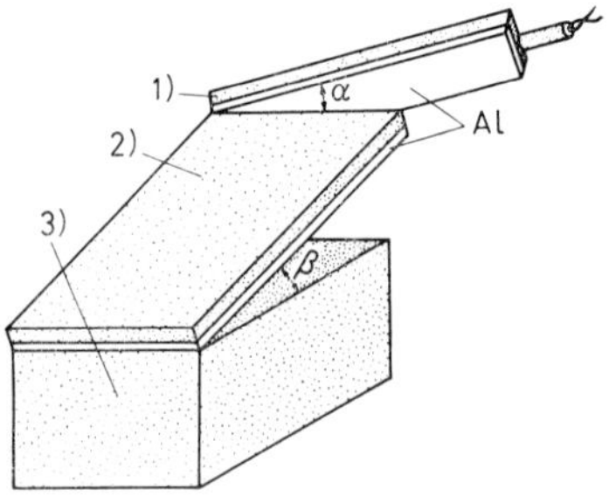

Fig. 6.

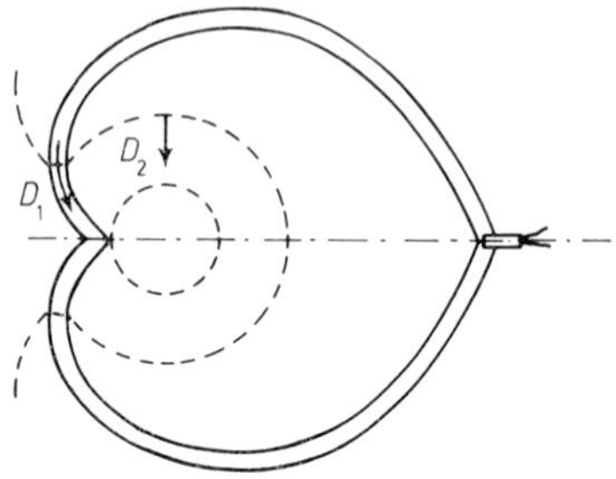

Fig. 7.

Figure 9 gives finally an example of the impact method applied to obtain a conical implosion [18, 19]. It was used to study conically collapsing waves

(*) More recently, DÉFOURNEAUX [17] has suggested as a space-saving method the application of stratified materials consisting of alternate layers of high explosives and inert materials.

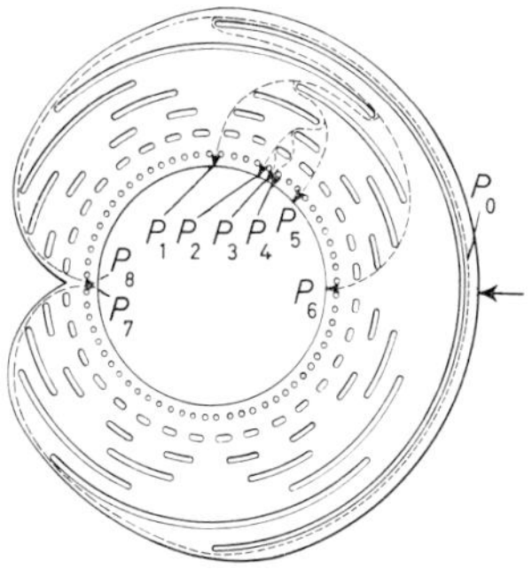

Fig. 8.

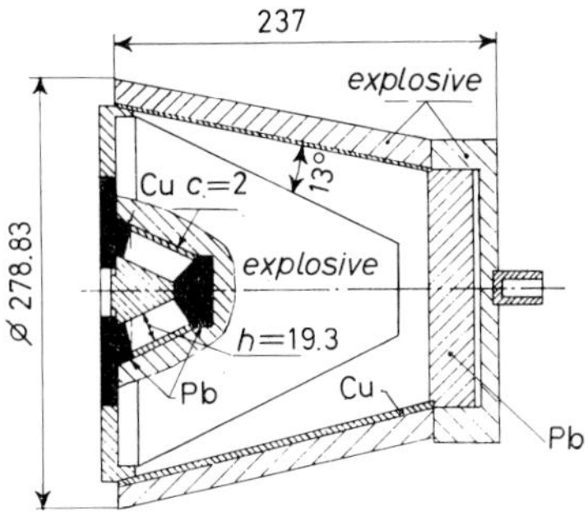

Fig. 9.

in metals but can, of course, also be used to detonate an explosive in converging conical geometry. The devices of Fig. 7 and 8 can also be utilized to initiate converging waves in inert materials.

5. – Detonation-induced shocks.

The detonation wave induces in an adjacent medium a shock wave, the intensity of which is determined by the equilibrium pressure between the expanding detonation products and the induced shock. We consider at first the case of a plane detonation wave propagating in a direction parallel to the interface. In order to compute the pressure equilibrium, we have to know how the velocity of the products increases by adiabatic expansion. This velocity increase is given by the Riemann integral

$$u = \int_{p_0}^{p_1} \frac{\mathrm{d}p}{a\varrho},$$

to be taken from the equilibrium pressure p_0 to the initial value p_1. Here a and ϱ are the sound velocity and density of the expanding gas. For a perfect gas we obtain

$$u = \frac{2a_1}{\gamma - 1}\left[1 - \left(\frac{p_0}{p_1}\right)^{(\gamma-1)/2\gamma}\right].$$

It is convenient to consider the expansion in a p-u-plot. This has been done in Fig. 10 where, according to earlier results, $\gamma = 3$ has been chosen. In this case we have then $p_1 = p_{\text{C-J}}$ and $a_1 = \frac{3}{4}D$. For $p_0 = 0$ (expansion in a vacuum) we get $u_{\max} = a_1$, the sonic velocity at the C-J point. Curve I gives the pressure decrease as a function of the velocity gained.

The motion imparted to the inert particles induces a shock, the pressure of which increases with growing particle velocity W. The possible pressure-velocity states which can be produced for a typical medium of high compressibility (*e.g.* water) are represented by curve II), those for a material of low compressibility (metals) by curve IV). Intersections with the adiabat I) give the equilibrium pressures, *i.e.* the shock strength induced in the respective materials. In the case of a detonation wave impacting with its front parallel to the interface the same procedure applies. It has only to be taken into account that there is at $p_1 = p_{\text{C-J}}$ already a flow velocity $W_{\text{C-J}} = \frac{1}{4}D$. Expansion into the vacuum then equals D. The adiabat III) intersects the shock curve II) at a higher pressure p_2. There is no intersection between the adiabat and curve IV), since the particle velocity at equilibrium is smaller than $W_{\text{C-J}}$. The products are stopped at the interface and a reflected shock in the gas raises the pressure to a value $p_4 > p_{\text{C-J}}$. This pressure p_4 results from the equilibrium between the shock adiabat of the gas (dotted line) and the shock adiabat of the material. If strictly incompressible materials were available, reflected pressures of more than double $p_{\text{C-J}}$ could be obtained.

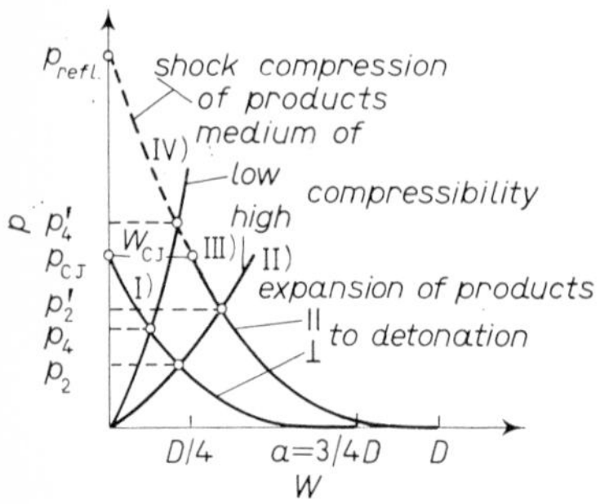

Fig. 10.

6. – High-explosive ballistics.

We consider a metallic plate in contact with a detonating explosive. When the induced shock reaches the free surface, the metal particles are accelerated from a value W behind the shock to

$$W + \int_0^p \frac{\mathrm{d}p}{a\varrho},$$

where the Riemann integral is extended from 0 to the shock pressure p. It has been shown that for any equation of state which describes realistically the shock compressibility of metals, $\int \mathrm{d}p/a\varrho$ is very close to W, so that the free surface moves at a speed which is very precisely double the material velocity behind the shock. When the release wave returns to the explosive metal interface and the gas pressure is high enough, a second shock is produced in the metal; in this way the plate accelerates stepwise by a sequence of shocks. As long as the energy dissipation in the plate is negligible, the final velocity is, however, the same as obtained for an incompressible plate, *i.e.*

$$u = \frac{1}{\varrho_m d}\int p\,\mathrm{d}t\,,$$

$p(t)$ being the pressure at the moving interface and $\varrho_m d$ the mass per unit surface of the plate.

The metal velocities which can be obtained by explosives can be estimated from the « detonation head model » developed by Cook [1]. In this model, release waves are assumed to bring the C-J pressure suddenly down to zero and the gas in the remaining high-pressure region is supposed to act elastically on the metal. In a cylindrical charge the remaining head ($h = \phi$, $\varrho_{\text{C-J}} = \frac{4}{3}\varrho_0$) has a mass

$$M_{\text{H}} = \varrho_{\text{C-J}}\frac{\pi}{3}\cdot\frac{\phi^2}{4}\,h = \frac{\pi}{9}\,\varrho_0\,\phi^3\,.$$

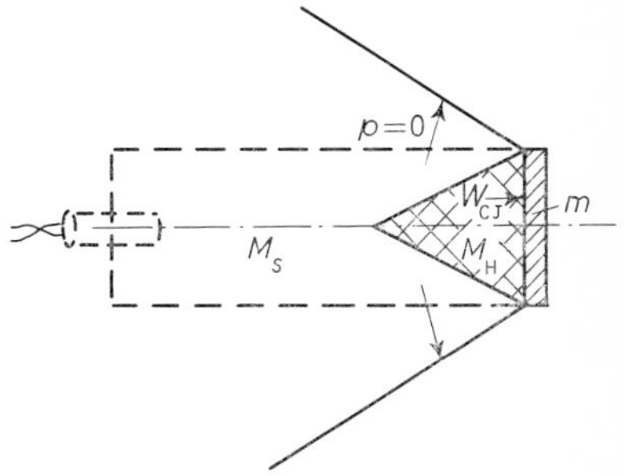

Fig. 11.

Momentum and energy conservation then give for the velocity of the metal projectile

$$v = \frac{2\,W_{\text{C-J}}}{1 + m/M_{\text{H}}}\,.$$

In reality, the upper limit $v = 2W_{\text{C-J}}$ which results from this formula for $m \to 0$ is too low, as we have seen that detonation products escape into vacuum at $D = 4W_{\text{C-J}}$. In practice, however, for $m \sim M_{\text{H}}$, the results are in good agreement with experiments which show that projectile velocities of the order of 5 km/s can be obtained using explosives of high energy content.

7. – Determination of shock Hugoniot curves.

To determine the pressure-density behaviour of condensed materials at very high pressures, different methods have been developed. They all make use of the basic relations which link, for a plane shock, shock velocity S, material velocity W, initial density ϱ_0 and the density ϱ_1 behind the shock:

(8) $$p = \varrho_0 S W\,;$$

(9) $$\frac{\varrho_1}{\varrho_0} = \frac{S}{S - W}\,.$$

a) The free surface method [19, 20]. One measures shock speed S (optically or by electronically using pins) and the free surface velocity, which is supposed to be $2W$. Pressure and density are then found from (8) and (9).

b) Flash radiography [21]. Photometric evaluation of X-ray photographs taken at sufficiently short exposure time allows the density ratio ϱ_1/ϱ_0 to be determined. This value together with S, gives p from (9) and (8).

c) The projectile impact method [22]. One measures the velocity v_i of a plane projectile impacting parallel to the surface and the velocity S of the shock in the target. When projectile and target are of the same material, we have $W = \frac{1}{2}\cdot v_i$ and p follows from (8), ϱ_1 from (9). For non-identical materials the equation of state of the projectile material must be known in order to calculate the established equilibrium pressure and interface velocity. Explosive and light gas gun-driven projectiles are generally used for this technique.

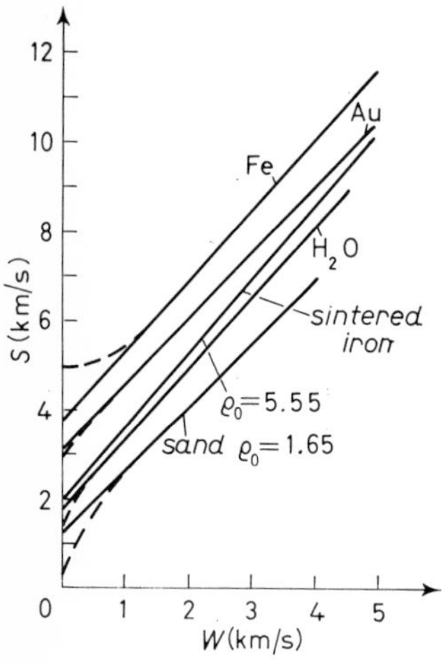

Fig. 12.

It is convenient to represent results obtained from these thecniques in the form $S(W)$. It turns out that for many condensed materials (be they liquid, solid, porous or granular) a linear relation.

$$S = U_0 + \lambda W ,$$

holds with $\lambda \sim 1.6$ in a rather large range, where U_0 is often different from the sound velocity a (Fig. 12).

8. – Shaped charges.

Shaped charges are high-explosive bodies of axial symmetry and a cavity often having conical shape. Ignited at the opposite end, they produce a stronger effect on a target close to the cavity than a charge without a cavity (Fig. 13). The shaped charge effect is explained as an energy concentration in the charge axis. Penetration is particularly high with «lined» charges where the cavity is covered with a metal layer (*e.g.* steel or copper). It reaches a maximum (which can be as high as 8 times the charge diameter in a mild steel target) when the charge is fired at a stand-off of several charge diameter S from the target.

The basic phenomenon is the formation of a metallic jet of very high velocity through concentric collapse of the liner on the charge axis. Theoretically it can be easily treated on the basis of incompressible hydrodynamics. The result is that a portion $\sin^2(\beta/2)$ of the liner mass goes into the jet and that this jet has a velocity $V_j = V_0 \cos(\alpha/2)/\sin(\beta/2)$, where V_0 is the velocity at which the collapse occurs, β the half collapse angle, and α the half cone angle of the cavity [23]. The very high jet velocities which one expects from this formula

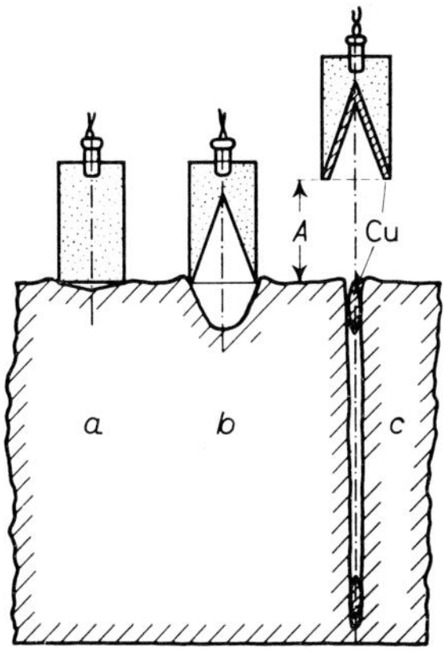

Fig. 13.

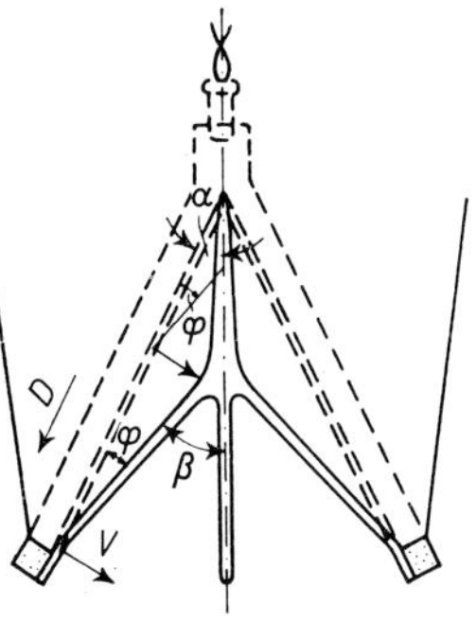

Fig. 14.

for $\beta \to 0$ are in reality not observed. In nature, jetting occurs only if the collapse is subsonic, that means that the collapse point travels at a speed which is lower than the sonic speed, a, of the liner material. Otherwise the collision is jetless [24]. At small β, V_j is roughly twice the velocity of the collapse point; the maximum V_j is close to $2a$. This means that velocities as high as 10 km/s can be reached. Not all liner materials give coherent jets. Brittle materials (like tungsten, titanium, sintered metals, etc.) form particle jets of dustlike consistency. Ductile materials (like Al, Cu, Ag, Au) form coherent jets with a density only slightly lower than that of the compact metal and corresponding to a temperature close to the melting point. The highest concentration of kinetic energy which can be expected from a shaped charge device is not much greater than that which can be obtained with a plane-plate set-up. High velocities can, however, be obtained at a higher efficiency, since the primary energy transfer from the explosive to the liner can take place under the most efficient conditions. With two shaped charges fired one against the other, this kinetic energy can be transformed to pressure energy for a very short time.

REFERENCES

[1] M. A. Cook: *The Science of High Explosives* (New York, 1958).

[2] J. Berger and J. Viard: *Physique des explosifs solides* (Paris, 1962).

[3] Y. B. Zel'dowich and A. S. Kompaneets: *Theory of Denotation* (New York, 1960).

[4] W. Y. Davis and W. Fickett: *Denotation Theory and Experiment*, in *Behaviour of Dense Media Under High Dynamic Pressures* (Paris, 1968).

[5] R. E. Duff, H. T. Knight and J. P. Rink: *Phys. Fluids*, 1, 393 (1958).

[6] N. Manson: *Compt. Rend.*, **246**, 2860 (1958).

[7] W. C. Davis, B. G. Craig and J. B. Ramsay: *Phys. Fluids*, **8**, 2169 (1965).

[8] J. J. Erpenbeck: *Phys. Fluids*, **9**, 1293 (1966).

[9] R. Schall and K. Vollrath: *Sur la conductibilité électrique provoquée par les ondes de détonation*, in *Les ondes de détonation* (Paris, 1962).

[10] C. H. Johansson and S. Ljungberg: *Ark. f. Fys.*, **6**, 269 (1953).

[11] H. Eyring, R. E. Powell, G. A. Duffey and R. B. Parlin: *Chem. Rev.*, **45**, 69 (1949).

[12] *A discussion on the initation and growth of explosion in solids*, *Proc. Roy. Soc.*, A **246**, 146 (1958).

[13] Ch. L. Mader: *Phys. Fluids*, **6**, 375 (1963).

[14] J. Berger: *Recherche d'une théorie intrinsèque de la détonation*, in *Les ondes de détonation* (Paris, 1962).

[15] Y. B. Zel'dowich: *Žurn. Ėksp. Teor. Fiz.*, **36**, 782 (1959) (English translation, *Sov. Phys. JETP*, **36**, 550 (1959)).

[16] A. Cachin: *Les ondes de détonation* (Paris, 1962).

[17] M. Défourneaux: in *Behaviour of Dense Media Under High Dynamic Pressures* (Paris, 1968).

[18] J. Leygonie and J. Cl. Bergon: in *Behaviour of Dense Media Under High Dynamic Pressures* (Paris, 1968).

[19] M. H. Rice and R. G. McQueen: *Compression of solids by strong shock waves* in *Solid State Phys.*, vol. **6** (New York, 1958), p. 1.

[20] R. G. McQueen and S. P. Marsh: *Journ. Appl. Phys.*, **31**, 1253 (1960).

[21] R. Schall and G. Thomer: *Zeits. angew. Phys.*, **3**, 31 (1951).

[22] L. V. Altshuler, A. A. Bakanova and R. F. Trunin: *Žurn. Ėksp. Teor. Fiz.*, **42**, 91 (1962) (English translation, *Sov. Phys. JETP*, **15**, 65 (1962)).

[23] E. M. Pugh, R. J. Eichelberger and N. J. Rostoker: *Journ. Appl. Phys.* **23**, 532 (1932).

[24] J. M. Walsh, R. G. Schreffler and F. J. Willig: *Journ. Appl. Phys.*, **24**, 349 (1953).

The Physics of Strong Shock Waves in Gases.

ROBERT A. GROSS
Columbia University - New York, N. Y.

1. – Introduction.

What happens when energy is concentrated in a small portion of a material medium? Unless the material is constrained, this portion will expand rapidly and do work on the surrounding portions of the medium so that the energy concentration diminishes. At the same time the energy will also be spreading because of conduction, diffusion and radiation. This rapid expansion produces disturbances in the surrounding medium and a shock wave is formed. The expanding medium acts like a « piston » and the shock wave is at the front of the disturbance. The shock wave heats, compresses, and accelerates the medium through which it propagates. By suitable geometrical arrangements, shock waves may be focused to produce unusual concentrations of energy, at least for brief intervals of time. Laboratory temperatures of the order of 10^7 °K have been produced by strong shocks in gaseous plasma, while pressure of 10^7 atm have been created in shocked solids. In nature, for example, shock waves accompany lightning, meteorites entering the atmosphere, sparks, and supernova explosions. In Table I is shown a comparison of some interesting states of matter that occur in nature and states of matter which have been achieved by employing shock waves in the laboratory.

There is an extensive body of knowledge that has been developed about shock waves, which comes principally from aerodynamic studies of high-speed flight and ballistics. The results are based mainly on fluid mechanics. Recently, phenomena associated with very strong shock waves, such as the creation of hot dense plasmas, shock wave interaction with electromagnetic fields, radiation from strong shock waves, and shocks moving at relativistic speeds, have been studied. The fluid-dynamic aspects of shock waves are summarized

(*) This research was supported by the Air Force Office of Scientific Research under Contract AF49(638)-1634.

TABLE I.

	Earth's center	Sun's corona	Sun's center	White dwarf	Controlled fusion
P (atm)	10^6	10^{-9}	10^{11}	10^{18}	20
T (°K)	10^4	10^6	10^7	10^7	10^8

Laboratory achievements (1969)		
shocks in solids	$T\sim 5\cdot 10^4$ °K ,	$p\sim 10^7$ atm
shocks in plasma	$T\sim (10^6\simeq 10^7)$ °K ,	$p\sim 10$ atm

concisely in Sect. **2**. Ionizing shocks and shocks in plasmas are discussed in Sect. **3**, radiative shocks in Sect. **4**, and relativistic shocks in Sect. **5**. The theoretical structure of a thermonuclear detonation wave and the present status of laboratory experimental research with strong-shock waves in gases are discussed in Sect. **6**.

2. – Fluid dynamic shock wave theory.

In this Section are summarized the basic characteristics of shock waves which may be deduced from fluid mechanics. The results are applicable to any compressible medium so long as the shock wave does not cause significant changes in the energy states of the atoms or molecules. Thus, this fluid dynamic theory concerns low-energy shock waves. There are many good texts and monographs on this aspect of shock wave theory, for example [1-4].

Consider the simple problem of a piston, initially at rest, which is accelerated impulsively to a constant velocity, into a stationary gas. In Fig. 1 are shown the piston at time t, an x-t

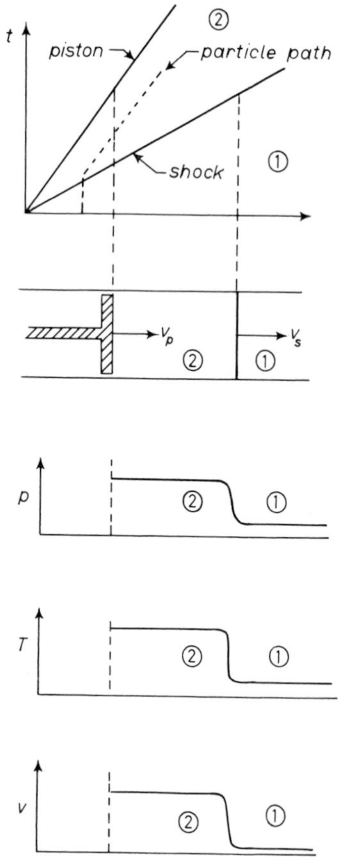

Fig. 1. – The effect of a piston moving at constant speed V_p into a stationary gas. p, T, x and t are pressure temperature, distance and time respectively. v is the fluid speed in the laboratory reference frame.

diagram, and some physical parameters, for this case. The gas is at rest until the shock wave passes, at which time the gas is suddenly accelerated to the speed of the piston V_p. The subscripts 1 and 2 denote the pre- and post-shock states respectively. The shock speed is V_s, and p, T, v, and ϱ represent the pressure, temperature, velocity, and density of the fluid. The shock wave is the very abrupt region (but actually of finite width) of change between states 1 and 2.

It is important to realize that a shock wave is formed regardless of the history of how the piston is accelerated to speed V_p. So long as a small-amplitude disturbance propagates with a higher speed as the medium is compressed, larger amplitude waves will steepen into shock waves. For example, in an ideal gas the speed of sound increases upon adiabatic compression. The steepening process continues until viscous and heat transfer effects (transport properties are diffusive in nature) counter-balance the steepening effect. On the other hand, expansion waves spread out and flatten. These effects are schematically shown in Fig. 2, where $(\varrho - \varrho_0)/\varrho_0$ is plotted for several different times (*). A compression ($\varrho > \varrho_0$) and rarefaction ($\varrho < \varrho_0$) wave is shown at time t_0. At time t_1 the wave has steepened and at t_2 a shock has formed whose steep shape continues to propagate for $t > t_2$. This shock formation occurs in all compressible media whose wave speed increases upon compression.

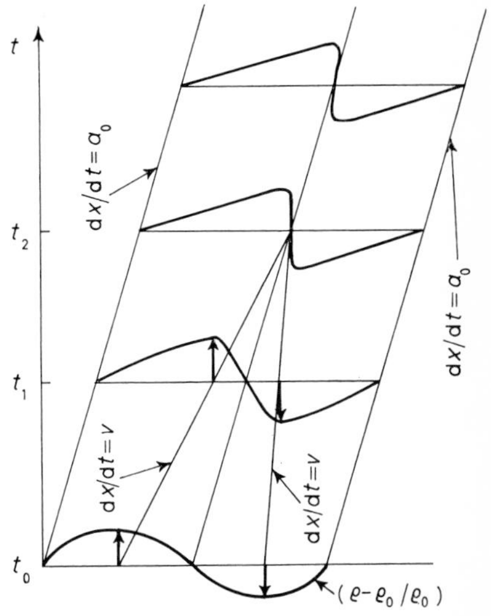

Fig. 2. – The transformation of a disturbance into a shock wave. The left portion of the wave is compressive and steepens; the right portion is a rarefaction and broadens.

2·1. *Shock jump relations.* – The relationship between the pre-shock and post-shock gas states is quite simple, since in fluid mechanics the only small disturbance speed is the sound speed a. If the pre-shock state is known, then the post-shock state depends only on one dynamic variable and the specific-heat ratio γ. The dynamic variable is often chosen to be the Mach number, $M = u/a$. Fluid speeds will be denoted by u when measured relative to the shock wave and v when measured in the laboratory reference frame, the rela-

(*) Subscript 0, corresponds to the undisturbed state.

tionship between them being,

$$u = v - V_s .$$

Any other dynamic variable, such as the pressure ratio, can be chosen, but for convenience we use the Mach number. The principle of conservation of mass momentum and energy, together with the ideal-gas equation of state, when applied across a steady-velocity shock wave, yield a system of algebraic equations which are called the *shock jump equations*. They can be solved, giving the post-shock state in terms of the pre-shock state and the Mach number. For convenience these solutions are listed here:

$$\text{(1)} \qquad \frac{p_2}{p_1} = 1 + \frac{2\gamma}{\gamma+1}(M_1^2 - 1),$$

$$\text{(2)} \qquad \frac{T_2}{T_1} = 1 + \frac{2(\gamma-1)}{(\gamma+1)^2}\frac{\gamma M_1^2+1}{M_1^2}(M_1^2-1),$$

$$\text{(3)} \qquad \frac{\varrho_2}{\varrho_1} = \frac{u_1}{u_2} = \frac{(\gamma+1)M_1^2}{(\gamma-1)M_1^2+2},$$

$$\text{(4)} \qquad M_2^2 = \frac{1+((\gamma-1)/2)M_1^2}{\gamma M_1^2 - (\gamma-1)/2}.$$

It is important to recognize that these jump relations do *not* depend upon the shock wave structure, nor upon the detailed dissipative and collisional processes occurring therein. All shocks must be compressive ($\varrho_2/\varrho_1 > 1$, $T_2/T_1 > 1$ $p_2/p_1 > 1$) and all shocks must be supersonic relative to the medium ahead ($M_1 > 1$) and subsonic behind ($M_2 < 1$). This can be seen by examining the increase of entropy s across a shock wave. Thus,

$$\text{(5)} \qquad \frac{s_2 - s_1}{R} = \ln\left[1 + \frac{2\gamma}{\gamma+1}(M_1^2-1)\right]^{1/(\gamma-1)}\left[\frac{(\gamma+1)M_1^2}{(\gamma-1)M_1^2+2}\right]^{(-\gamma)/(\gamma-1)},$$

where R is the gas constant.

For weak shocks ($M \approx 1$) the entropy increase is

$$\text{(6)} \qquad \frac{s_2 - s_1}{R} \approx \frac{2\gamma}{(\gamma+1)^2}\frac{(M_1^2-1)^3}{3} + \dots,$$

and hence weak shocks are nearly isentropic. In the limit $M \to 1$, the wave is an acoustic wave and is isentropic.

It is sometimes convenient to relate the pre- and post-shock thermodynamic properties by eliminating the velocities from the conservation equations. The resulting equations are called the Rankine-Hugoniot shock jump relations and they are:

$$\frac{\varrho_2}{\varrho_1}=\frac{1+((\gamma+1)/(\gamma-1))(p_2/p_1)}{(\gamma+1)/(\gamma-1)+p_2/p_1}, \tag{7}$$

$$\frac{T_2}{T_1}=\frac{p_2}{p_1}\frac{((\gamma+1)/(\gamma-1))+p_2/p_1}{1+((\gamma+1)/(\gamma-1))(p_2/p_1)}. \tag{8}$$

2·2. *Strong-shock theory.* – The shock jump conditions simplify in the *strong-shock limit*, $M_1\to\infty$. In this limit eqs. (4) and (5) reduce to

$$\frac{\varrho_2}{\varrho_1}=\frac{\gamma+1}{\gamma-1}, \tag{9}$$

$$M_2^2=\frac{\gamma-1}{2\gamma}. \tag{10}$$

For $\gamma=\frac{5}{3}$, the strong-shock density ratio is 4, while the post-shock Mach number M_2 (relative to the shock wave) is 0.45. It will be shown later that these simple results are altered when radiation is included in the analysis. The pressure and temperature ratios increase without limit as M_1 increases, and therefore shock waves can be used to create high pressures and high temperatures.

In the limit of very high Mach number equipartition of energy occurs between the kinetic and internal energy of the gas. This is readily proved as follows:

For simplicity, assume that the shock wave propagates into a stationary gas; *i.e.* $v_1=0$. The shocked gas has a laboratory Mach number,

$$M_{2\,\mathrm{lab}}=\frac{v_2}{a_2}=\frac{u_1-u_2}{a_2}=M_2\left(\frac{\varrho_2}{\varrho_1}-1\right), \tag{11}$$

where the equation of conservation of mass ($\varrho u=$ constant) has been used. Combining the strong-shock limit results (eqs. (9), (10)) with eq. (11), we obtain,

$$M_{2\,\mathrm{lab}}\to\left(\frac{2}{\gamma(\gamma-1)}\right)^{\frac{1}{2}}\ \text{as}\ M_1\to\infty, \tag{12}$$

which, for $\gamma=\frac{5}{3}$, has a value of $M_{2\mathrm{lab}}=1.35$. That is in the laboratory reference frame the shock-heated gas has been accelerated to a very high speed but only a slightly supersonic Mach number. Using the fact that, for an ideal gas $a^2=\gamma RT$ and the internal energy $e=c_v T=RT/(\gamma-1)$, eq. (12)

yields the result

$$\frac{v_2^2}{2} = \frac{RT_2}{\gamma - 1} = e_2 \,. \tag{13}$$

Thus, for strong-shock waves, the energy that has been transmitted to the shocked medium has been redistributed such that half the energy is in macroscopic kinetic energy $(v_2^2/2)$ and the other half is in internal energy, $(RT_2/(\gamma-1))$. This result is quite general, it does not depend on the form of the internal energy function, and can be shown to be valid for relativistic shock waves.

2·3. *Reflected shocks.* – It is possible to convert the macroscopic kinetic energy of the shock-accelerated gas into increased internal energy by shock reflection from a solid wall. When a shock wave strikes a flat stationary wall, another shock is reflected back into the oncoming gas with just that Mach number to bring the new post-shock gas (traditionally designated by subscript 5) to rest. The relationships between the post-reflected shock gas and the initial undisturbed gas can be shown to be [5]

$$\frac{p_5}{p_1} = \left[\frac{2\gamma M_1^2 - (\gamma-1)}{\gamma+1}\right] \cdot \left[\frac{(3\gamma-1)\,M_1^2 - 2\,(\gamma-1)}{(\gamma-1)\,M_1^2 + 2}\right], \tag{14}$$

$$\frac{T_5}{T_1} = \frac{[2\,(\gamma-1)\,M_1^2 + (3-\gamma)][(3\gamma-1)\,M_1^2 - 2(\gamma-1)]}{(\gamma+1)^2\,M_1^2}, \tag{15}$$

and the ratio of the speeds of the incident (v_i) and reflected shocks (v_r) is,

$$\frac{v_r}{v_i} = \frac{2 + \big(2/(\gamma-1)\big)(p_1/p_2)}{(\gamma+1)/(\gamma-1) - p_1/p_2}\,. \tag{16}$$

In the limit of a very strong incident shock $(M_1 \to \infty)$ we get [5],

$$\frac{p_5}{p_2} = 2 + \frac{\gamma+1}{\gamma-1} = 6\,,$$

$$\frac{T_5}{T_2} = \frac{p_5}{p_2} \cdot \left[\frac{(\gamma+1)/(\gamma-1) + p_5/p_2}{1 + \big((\gamma+1)/(\gamma-1)\big)(p_5/p_2)}\right] = 2.40\,,$$

$$\frac{\varrho_5}{\varrho_2} = \left(\frac{p_5}{p_1}\right)\left(\frac{T_2}{T_5}\right) = 2.50\,,$$

$$\frac{v_r}{v_i} = \frac{2\,(\gamma-1)}{\gamma+1} = 0.50\,,$$

where the numerical values shown represent values for a gas with $\gamma = \frac{5}{3}$. It is interesting to note that by stopping the incident gas the temperature has been increased by a factor of 2.40 and the reflected shock speed is one half the incident shock speed. The pressure, temperature and density ratios of state 5 to state 2 for reflected shocks are all limited to finite values. As the temperature of the shock-heated gas increases, the gas dissociates and ionizes, and radiation effects become significant. These effects alter the simple results of this Section and will be discussed later.

2·4. *Shock thickness.* – All the afore-mentioned results apply to a shock wave in which the changes in the fluid dynamic variables remain essentially constant for a time period of the order of the flow time through the shock-wave structure itself. Therefore it is necessary to know the thickness of the shock structure. The shock thickness l is usually defined as

$$l = \frac{(u_2 - u_1)}{(\mathrm{d}u/\mathrm{d}x)_{\max}} .$$

For a given shock speed, this thickness depends only upon the atomic nature of the gas. The time for inertial steepening of the velocity profile in a shock is of the order $(l/\Delta u)$, where l is the shock thickness and Δu is the change in velocity across the shock. The time for smoothing by viscous diffusive action only is of the order of (l^2/ν) where ν is the kinematic viscosity. Equating the times for these two opposing effects provides a crude estimate of shock thickness, namely,

$$l \sim \frac{\nu}{\Delta u} \sim \frac{\nu}{u} . \tag{17}$$

This result can also be obtained by simply equating the fluid viscous stress $\nu\varrho(\mathrm{d}u/\mathrm{d}x)$ to the momentum flux, ϱu^2. From kinetic theory $\nu \approx 0.5\lambda\bar{c}$, where λ is the particle mean free path and $\bar{c}$ is the mean thermal speed of the particles $[\bar{c} = (8kT/\pi m)^{\frac{1}{2}}]$. But $\bar{c} \approx a$ and therefore,

$$l \sim \frac{\lambda a}{2\,\Delta u} = \left(\frac{\lambda}{2}\right) \frac{1}{M(\varrho_1/\varrho_2 - 1)} . \tag{18}$$

Thus this crude estimate indicates that the thickness of a shock wave is proportional to the mean free path in the gas and is a function of the Mach number of the shock. More refined analyses may be found in [4] and particular details can be found in the works of GILBARG and PAOLUCCI [6], and MOTT-SMITH [7]. In particular, the later work uses the Boltzmann equation to study the shock-wave structure, and finds

$$l = \frac{4\lambda_1}{B_0(M)}, \tag{19}$$

where λ_1 is the pre-shock mean free path and $B_0(M)$ is a tabulated function of the Mach number. For moderate-speed shocks ($M \approx 5$), $B_0 \approx 2$ and therefore $l \approx 2\lambda$. Recent numerical computations by CHAHINE and NARASHIMA [8] of shock-wave structure are particularly detailed and give good agreement with experimental measurements of real shock-wave profiles. At higher Mach numbers the gas temperature becomes sufficiently high that dissociation and ionization effects occur, and the shock-wave structure becomes more complex. We return to this matter in later Sections as additional physical phenomena affect the shock structure and thickness. The major conclusion at low Mach numbers is that the shock thickness is of the order of a mean free path.

2'5. *Shock power.* – The rate at which energy is being absorbed by the shock-heated and accelerated gas can be evaluated. For a shock wave traveling at the speed V_s into a stationary gas, the power per unit area, ψ is,

$$\psi = p_2 v_2, \tag{20}$$

or

$$\psi = \left[1 + \frac{2\gamma}{\gamma+1}(M_1^2 - 1)\right] p_1 u_1 \left(1 - \frac{\varrho_1}{\varrho_2}\right),$$

which in the strong-shock limit can be reduced to simply

$$\psi \approx \frac{4\varrho_1 V_s^3}{(\gamma+1)^2}, \tag{21}$$

This is the rate at which an expanding high-energy density region does work on its surroundings and is exactly equal to the rate at which energy per unit volume is being added to the surrounding medium. For example, a shock wave propagating at a speed of 10^7 cm/s into room temperature hydrogen at 0.10 Torr pressure is doing work at the rate of about 10^6 W/cm^2. The speed at which a shock wave is driven depends therefore upon the rate at which energy per unit area can be delivered. Some examples of energy storage systems and their delivery rates are shown in Table II. There are many other aspects of the fluid dynamics of shock waves, such as oblique (2- or 3-dimensional) shock wave interactions, shock diffraction, etc. but it is not our purpose to discuss these matters here and the interested reader is referred to references [1-4].

TABLE II. – *Typical energy storage and release rates.*

1. *Chemical energy.* (For example TNT or nitroglycerine)
   energy density $\sim 5 \times 10^3$ J/g
   detonation wave speed $\sim 10^6$ cm/s
   wave front energy release rate $\sim 5 \cdot 10^9$ W/cm^2

2. *Capacitor bank.* (Columbia University, R. A. GROSS)
   energy stored $\sim 5 \cdot 10^5$ J
   half-cycle time $\sim 6 \cdot 10^{-6}$ s
   energy release rate $\sim 0.8 \cdot 10^{11}$ W

3. *Fast Blumlein system.* (Naval Research Lab., A. KOLB)
   energy stored ~ 560 J
   energy delivery time $\sim 50 \cdot 10^{-9}$ s
   energy release rate $\sim 1.1 \cdot 10^{10}$ W

4. *High-power laser.* (U.S.S.R., N. G. BASOV)
   energy in pulse ~ 10 J
   pulse width $\sim 10^{-11}$ s
   energy release rate $\sim 10^{12}$ W

As a means of comparison, the total electrical generating capacity in the United States in 1966 was $1.4 \cdot 10^{11}$ W, and Consolidated Edison of New York had an installed power of $7 \cdot 10^9$ W in 1968.

2'6. *Blast waves.* – The equations discussed so far do not describe the *shape and locus* of a shock wave as it evolves. The temporal shape of the shock-wave envelope depends upon initial and boundary conditions as well as the characteristics of the media through which it propagates. For example, in the simple one-dimensional piston problem, if the medium is uniform and homogeneous and the power input to the piston is constant, the shock speed is constant and greater than the piston speed by the ratio $(\varrho_2/(\varrho_2 - \varrho_1))$. There is one particularly important case which has been studied in detail and that is where the energy is all deposited instantaneously at time $t = 0$. The shock-wave thus is described as a *blast wave*. The pioneering studies of this phenomena were carried out by TAYLOR [9] and SEDOV [10].

Following the Taylor-Sedov theory we convert eqs. (2)-(4) into the laboratory reference frame for the case where the initial medium is stationary ($v_1 = 0$). The shock jump equations are:

$$v_2 = \frac{2}{\gamma+1} V_s \left[1 - \left(\frac{a_1}{V_s}\right)\right], \tag{22}$$

$$\varrho_2 = \frac{\gamma+1}{\gamma-1} \varrho_1 \left[1 + \frac{2}{\gamma-1} \left(\frac{a_1}{V_s}\right)^2\right], \tag{23}$$

$$p_2 = \frac{2}{\gamma+1} \varrho_1 V_s^2 \left[1 - \frac{\gamma-1}{2\gamma} \left(\frac{a_1}{V_s}\right)^2\right]. \tag{24}$$

In the strong-shock limit, *i.e.* $a_1/V_s \to 0$, the jump conditions simply reduce to

$$v_2 = \frac{2}{\gamma+1} V_s \,, \tag{22a}$$

$$\varrho_2 = \left(\frac{\gamma+1}{\gamma-1}\right) \varrho_1 \,, \tag{23a}$$

$$p_2 = \frac{2}{\gamma+1} \varrho_1 V_s^2 \,. \tag{24a}$$

By employing these strong-shock relations and assuming that the flow field behind the blast wave is isentropic, SEDOV showed by dimensional analysis that the motions and flow fields of all blast waves are mathematically self-similar. Although a blast wave is unsteady, its velocity decreasing with time, its motion and the flow field behind it can be readily evaluated. The position of the blast wave from the origin r_2, can be determined as a function of time t, initial density ϱ_1, and total impulsive input energy $E_0 = AE$. A is a constant determined by the geometry and the specific heat ratio γ and has been tabulated by SEDOV. The Taylor-Sedov blast-wave solutions give the position r_2 and speed of the shock front, V_s as:

spherical wave:

$$r_2 = \left(\frac{E}{\varrho_1}\right)^{\frac{1}{5}} t^{\frac{2}{5}} \,, \qquad V_s = \frac{2}{5}\left(\frac{E}{\varrho_1}\right)^{\frac{1}{5}} t^{-\frac{3}{5}} = \frac{2}{5}\left(\frac{E}{\varrho_1}\right)^{\frac{1}{2}} r_2^{-\frac{3}{2}} \,; \tag{25a}$$

cylindrical wave:

$$r_2 = \left(\frac{E}{\varrho_1}\right)^{\frac{1}{4}} t^{\frac{1}{2}} \,, \qquad V_s = \frac{1}{2}\left(\frac{E}{\varrho_1}\right)^{\frac{1}{4}} t^{-\frac{1}{2}} = \frac{1}{2}\left(\frac{E}{\varrho_1}\right)^{\frac{1}{2}} r_2^{-\frac{1}{2}} \,; \tag{25b}$$

plane wave:

$$r_2 = \left(\frac{E}{\varrho_1}\right)^{\frac{1}{3}} t^{\frac{2}{3}} \,, \qquad V_s = \frac{2}{3}\left(\frac{E}{\varrho_1}\right)^{\frac{1}{3}} t^{-\frac{1}{3}} = \frac{2}{3}\left(\frac{E}{\varrho_1}\right)^{\frac{1}{2}} r_2^{\frac{1}{2}} \,. \tag{25c}$$

Given the initial energy input E_0, together with ϱ_1 and γ, the wave speed and position can be determined as a function of time. The validity of these similarity solutions has been well verified by observation of spherical blast waves emanating from explosions and are frequently used to estimate the yield of nuclear explosions [9, 10]. The problem of blast waves in a nonuniform environment (*e.g.* an exponential atmosphere) has been studied by several authors and the reader is referred to LAUMBACH and PROBSTEIN [11] for results and further bibliography.

The temperature behind a blast wave can be determined by employing an equation of state. For the ideal gas, $p_2 = \varrho_2 R T_2$,

$$T_2 = \frac{2}{R} \frac{(\gamma - 1)}{(\gamma + 1)^2} V_s^2 , \tag{26}$$

where the temperature T_2 (r_2) decreases as the wave moves away from the origin as

$$T_2 \propto r_2^{-\nu}$$

$\nu = 1, 2, 3$, for plane, cylindrical and spherical waves respectively.

The velocity, density and pressure just behind the blast wave are,

$$v_2 = \frac{4}{(\nu + 2)(\gamma + 1)} \left(\frac{E}{\varrho_1}\right)^{\frac{1}{2}} r_2^{-\nu/2} , \tag{27}$$

$$\varrho_2 = \left(\frac{\gamma + 1}{\gamma - 1}\right) \varrho_1 , \tag{28}$$

$$p_2 = \frac{8E}{(\gamma + 2)^2 (\gamma + 1)} r_2^{-\nu} . \tag{29}$$

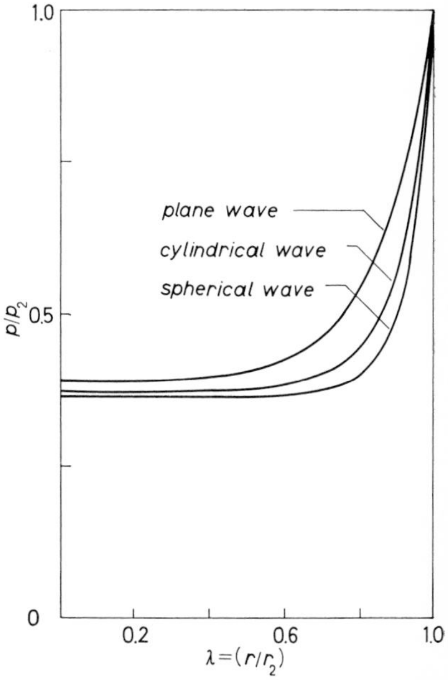

Fig. 3. – The pressure field behind a blast wave from SEDOV [10].

The state immediately behind the blast wave can be determined as a function of E_0, γ, p_1, and t (or r_2), by using eqs. (25) through (29). The flow field behind the blast wave can be evaluated by recalling that the expansion is isentropic. An example of the pressure field behind a blast wave is shown in Fig. 3.

3. – Ionizing and plasma shock waves.

When the shock-heated medium has an energy such that $\exp[-E_i/kT_2]$ is of the order of 0.1 or larger, a significant number of atoms become ionized and a plasma is formed. E_i is the ionization energy of the atom and k is Boltzmann's constant. E_i is usually expressed in eV, and typically, $E_i \sim 10$ eV; it is convenient to note that 1 eV corresponds to 11600 °K. Shock speeds greater than about Mach 20 in room temperature gas create some ionization although exact values depend upon the atomic composition of the medium and upon the pressure. For example, in hydrogen, initially at one atmosphere pressure, ionization begins at $M \sim 25$, and the post-shock gas is completely ionized at $M \sim 75$. The effects of molecular dissociation begin at even lower temperature, and for hydrogen, ($E_D = 4.48$ eV) dissociation begins at $M \sim 5$.

The effects of ionization and a plasma post-shock state are discussed in this Section. A comprehensive review of strong-ionizing shock wave theory and experiments was published by GROSS in 1965 [12] and the role of shock waves in plasma physics has been extensively described by CHU and GROSS [13].

3·1. *Ionizing shock jump relations.* – It is convenient to classify shock waves in terms of their pre- and post-shock electrical conductivity σ. There are three different situations and they are:

$\sigma_1 = \sigma_2 = 0$ fluid dynamic shocks,

$\sigma_1,\ \sigma_2 > 0$ plasma (or MHD) shocks,

$\sigma_1 = 0,\ \sigma_2 > 0$ ionizing shocks.

Fluid dynamic shock theory was discussed in Sect. **2**. Plasma shocks have no change in chemical composition between pre- and post-shock gas, and the shock jump relations developed in Sect. **2** are valid for plasma shocks so long as no magnetic fields are present. However, significant differences may occur in the plasma shock wave structure and this will be discussed later.

The consequences of a magnetic field on plasma (MHD) shock jump conditions are numerous and have been extensively treated in ref. [13]. In brief, plasma shock characteristics are determined by three parameters rather than the single parameter Mach number for fluid dynamic shock theory. These three parameters depend upon the acoustic speed a, the Alfvén speed $b = (B^2/4\pi\varrho)$, where B is the magnetic field, and the angle between the vector $\boldsymbol{B}$ and $\boldsymbol{n}$, the direction of the wave normal. For plasma shocks one must specify the acoustic Mach number, the Alfvén Mach number $M_A = v/b$, and the angle between $\boldsymbol{B}$ and $\boldsymbol{n}$ to uniquely determine the post-shock state. The solutions of the plasma shock jump equations are catagorized in terms of these two Mach numbers and are called: fast or super-Alfvénic shocks and slow or sub-Alfvénic shocks. There are many combinations of these parameters, all of which have been examined to see if shocks can exist. Suffice it to say that the presence of a magnetic field in plasma shocks greatly enriches the variety of shock waves that can be obtained and these waves and their properties are summarized in ref. [13].

Ionizing shock waves are somewhat intermediate between fluid dynamic shocks and plasma shocks. Since the post-shock gas is a plasma and can interact with a magnetic field, these waves exhibit the rich variety of solutions characterized by plasma shocks. In addition, because the pre-shock gas is a dielectric and can support an electric field, ionizing shocks have some special properties. However, these electric-field effects do not appear to be very important at very high shock speeds and will not be discussed here. The interested reader should consult refs. [12] or [13] for further details. The conservation of energy equation for ionizing shocks includes terms which account for the fact that energy is being employed to strip electrons from atoms. For a given ionizing shock, beside the fluid dynamic variables p, ϱ, u, and T, there are now additional unknowns representing the mole fractions for the post-shock chemical composition. The system of shock jump equations may be completed by assuming chemical equilibrium in the post-shock state. For example, if the gas is just singly ionized, the Saha equation closes the system

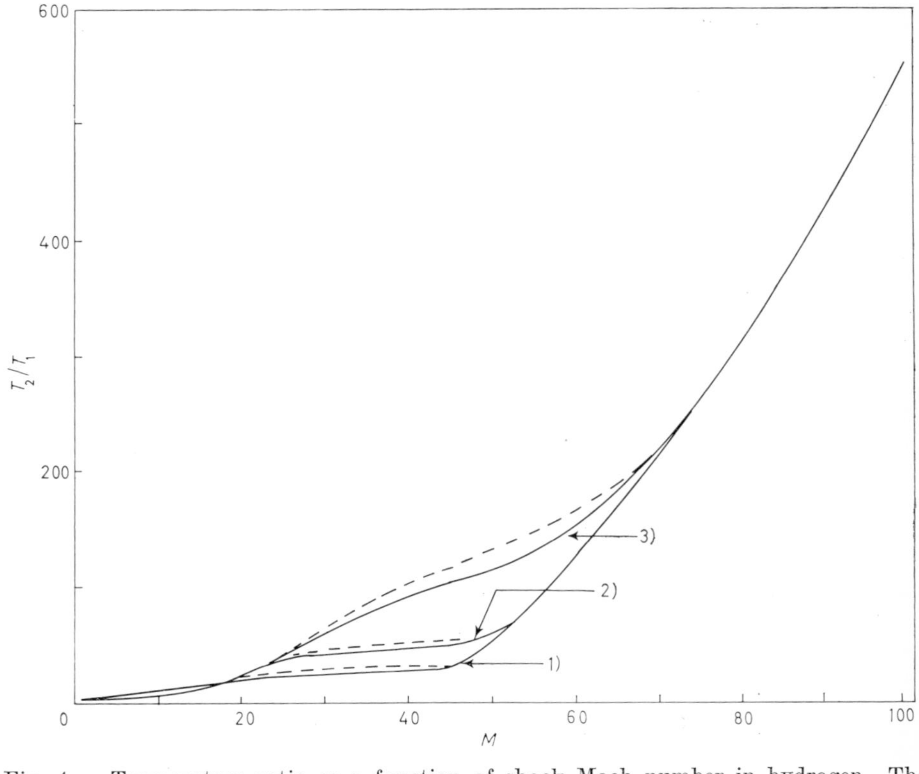

Fig. 4. – Temperature ratio as a function of shock Mach number in hydrogen. The multivalued regions correspond to different degrees of ionization at various pressures. $T_1 = 300$ °K. ——— Saha equilibrium; – – – optically thin. 1) $p_1 = 10^{-5}$ Torr, 2) $p_1 = 10^{-1}$ Torr, 3) $p_1 = 1$ atm.

of equations. Chemical equilibrium involves both the temperature and the pressure as well as the energy levels of the atomic species. Consequently, there are no simple general solutions and numerical calculations are carried out to evaluate specific shock cases. For example, in Fig. 4 and 5 are shown the

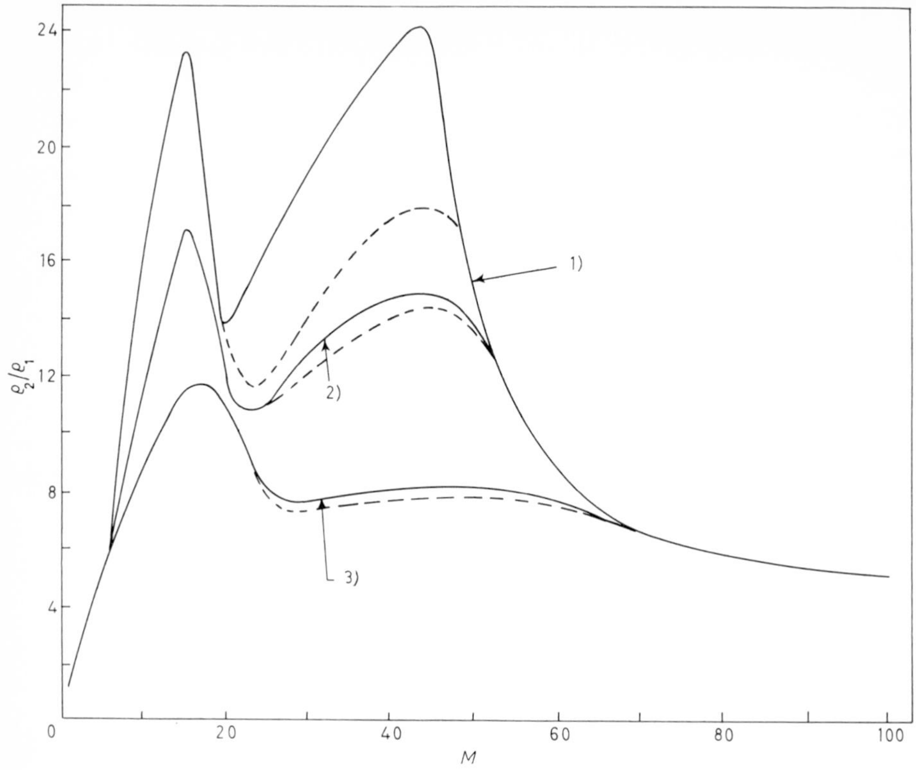

Fig. 5. – The density ratio as a function of shock Mach number in hydrogen. ——— Saha equilibrium; – – – optically thin. $T_1 = 300$ °K. 1) $p_1 = 10^{-5}$ Torr, 2) $p_1 = 10^{-1}$ Torr, 3) $p_1 = 1$ atm.

solutions for the temperature and pressure ratios behind ionizing shock waves in hydrogen.

The very-strong shock limits discussed earlier are still valid for ionizing shocks because in the strong-shock limit E_i/kT_2 and $E_i/\varrho u_1^2$ both tend to zero.

3˙2. *Ionizing and plasma shock wave thickness.* – The characteristic length for the thickness of a fluid mechanical shock l, was previously shown to be of the order of ν/u. In a similar manner there are other characteristic lengths based upon other transport properties, *e.g.* heat conductivity $\varkappa$. We distinguish between these lengths by subscripts denoting the transport property; ν, $\varkappa$, σ. These length scales are:

$$l_\nu = \nu/u\,,$$

$$l_\varkappa = \varkappa/c_p u\,,$$

$$l_\sigma = \mu\sigma/u\,,$$

where μ is the permeability of the medium; the Prandtl number P is the ratio $l_\nu/l_\varkappa$ and in ordinary gases $P \approx 1$. The ratio l_ν/l_σ is called the magnetic Prandtl number. In a plasma the viscosity is controlled by the ions whereas the thermal conduction is dominated by the electrons. As a result, in a plasma, $l_\varkappa/l_\nu \gg 1$ by a factor proportional to $(m_i/m_e)^{\frac{1}{2}}$ where $m_{i,e}$ are the ion and electron masses respectively.

These different length scales in a plasma lead to the concept of a shock within a shock. The plasma shock will have a density jump (ion shock) of

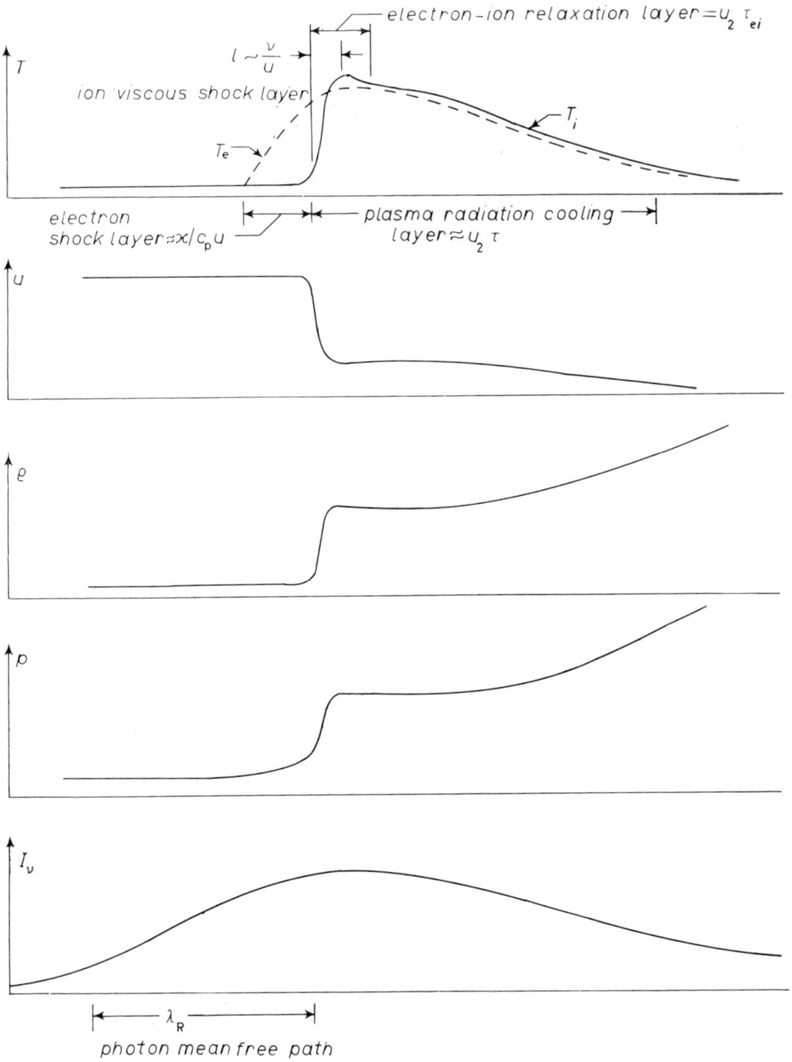

Fig. 6. – The structure of a strong plasma shock wave (schematic) τ_{ei} is the electron-ion relaxation time, τ is the characteristic time for cooling by radiation (see Sect. **4**) and I_ν is the radiative intensity of frequency ν.

width of $O(l_\nu)$ contained within a broader electron shock of width $(l_\varkappa)$. The temperature of the electrons is nearly raised to the post-shock temperature ahead of the ion shock. This pre-heating of the electrons is accomplished primarily by electron thermal conduction from the hot post-shock region. A schematic of the structure of a plasma shock wave is shown in Fig. 6.

For a very strong plama shock TIDMAN [14] analysed the plasma shock structure using the Fokker-Planck equation in the same spirit as did MOTT-SMITH for ordinary gas shocks, and found the thickness of a strong plasma shock to be

$$l \approx 1.3 \frac{\nu_2}{V_s} \approx \frac{\lambda_2}{4}, \tag{30}$$

where λ_2 is the Coulomb mean free path in the post-shock plasma. Tidman's result indicates that a strong plasma shock has a width of the order of the

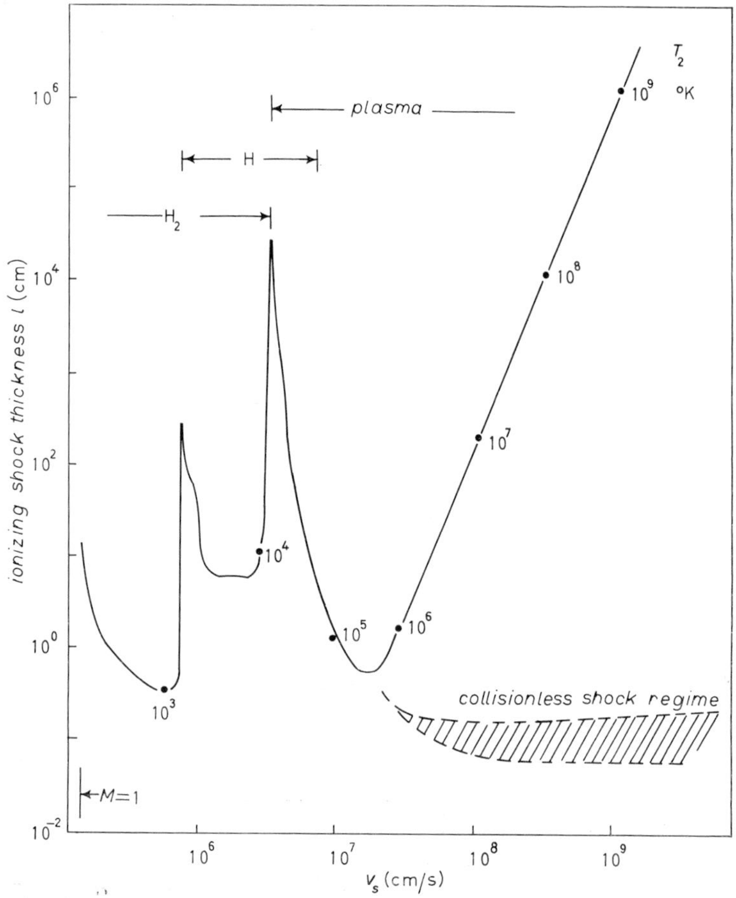

Fig. 7. – The thickness of a ionizing shock wave in hydrogen (adapted from ref. [17], $p_1 = 0.10$ Torr, $T_1 = 300$ °K).

post-shock mean free path. The structure of a plasma shock, showing the influence of the different length scales, was studied by JAFFRIN and PROBSTEIN [15].

If a magnetic field is present, the plasma-shock width is influenced by the value of the electron and ion gyro-radius. The structure of magnetohydrodynamic shocks has been studied by many authors, *e.g.* MARSHALL [16]. A counterpart of electron pre-heating is found in strong, inviscid (*i.e.* $l_\nu \ll l_\sigma$) MHD shocks, which have an imbededd viscous shock preceded by a magnetic compression whose width is of the order l_σ.

The width and structure of an ionizing shock wave is controlled by ionization rate processes. An estimate of the thickness of a hydrogen shock over a wide range of speeds has been given by GROSS [17] and is shown in Fig. 7. A recent detailed study of ionizing shock structure in argon has been made by CHUBB [18].

3·3. *Blast waves with ionization.* – There is not much published literature on blast-wave theory where the flow field behind the wave is ionized. The degree of ionization in the flow field can be readily estimated by employing a multicomponent equation of state,

$$p = \sum_j n_j kT \tag{31}$$

and the Saha equation,

$$\frac{n_i n_e}{n_0} = \frac{(2\pi mkT)^{\frac{3}{2}}}{h^3} \frac{2B_1}{B_0} \exp\left[-\frac{E_i}{kT}\right],$$

where n is the particle number density, subscripts i, e, o, refer to ions, electrons and neutrals, h is Plancks constant, and B_1 and B_0 are the partition functions for the ionized and neutral atoms respectively. Assuming Taylor-Sedov similarity GROSS [19] solved the system of blast-wave equations including ionization. For complete ionization behind the blast wave, the equations reduce exactly to the single-fluid Taylor-Sedov solution.

3·4. *Collisionless shock waves.* – The notion of a shock wave without collisions may seem like a contradiction in view of the previous discussions of shock thickness, in all of which the concept of collisional lengths was central to the physics of shock structure. However in a plasma a characteristic feature is collective phenomena and the concept of a distinct two-body collision loses its sharpness. Strictly speaking no plasma is collisionless. By a collisionless shock is therefore usually meant a physically observed sharp transition zone whose thickness is substantially smaller than all the collisional mean free paths in the plasma.

In recent years, collisionless shock waves in plasma have become a topic of intensive study and interest. This interest stems partly from possible applicabilty of rapid plasma heating to thermonuclear temperatures, partly to the insight they give in the nature of plasma collective effects, and finally because they have been found in nature, an example being the Earth's bow shock which is formed by the solar wind and the earth's magnetosphere.

Although collisionless shock wave research is still evolving, some insight is now available to understand their structure, at least phenomenologically. The most widely studied case is one in which a magnetic field lies in the plane of the shock, the so-called transverse shock. The small-amplitude signal speed in the direction perpendicular to the magnetic field (and in the direction of shock motion) in a plasma is $(a^2+b^2)^{\frac{1}{2}}$ where a and b are the acoustic and Alfvén speeds respectively. The appropriate Mach number is,

$$M_c = \frac{u}{(a^2+b^2)^{\frac{1}{2}}} = \frac{(u/b)}{\left((\gamma\beta/2)+1\right)^{\frac{1}{2}}}, \tag{32}$$

which is sometimes written in the latter form displaying the role of β, the ratio of particle to magnetic pressure $(\beta = nkT/(B^2/8\pi))$.

The two basic scale lengths that enter into collisionless shock theories are c/ω_{pe} and R_{Li}, the collisionless-skin depth and the ion thermal gyro-radius. c is the speed of light, ω_{pe} is the electron plasma frequency defined as $\omega_{pe} = (4\pi n_e^2/m_e)^{\frac{1}{2}}$, e being the electron charge and m_e its mass. The ion thermal gyro-radius R_{Li} is defined as

$$R_{Li} = \frac{c(m_i kT)^{\frac{1}{2}}}{eB}.$$

There are laminar and turbulent collisionless shocks. In the latter, apparently the electrons and ions streaming towards the shock front undergo a plasma collective instability which acts like an anomalous resistivity or viscosity. The type of instability depends upon the value of the Mach number and the orientation of the magnetic field and is not yet a settled question. For transverse collisionless shocks at $M_c \leqslant 3$, the instability is believed to be the two-stream instability. Analyses based upon this instability lead to a shock thickness of the order,

$$l \sim 10 \frac{c}{\omega_{pe}} \quad \text{for } M_c < 3. \tag{33}$$

There are experiments which indicate that this expression does give a proper estimate of low Mach number transverse collisionless shocks.

For transverse collisionless shocks where $M_c > 3$ the structure changes and the precise nature of the instabilty is being debated. Attempts to estimate the thickness of such shocks have led to the length scale

$$l \sim \frac{c}{\omega_{pi}}, \tag{34}$$

where $\omega_{pi} = (4\pi n_i^2/m_i)^{\frac{1}{2}}$ is the ion plasma frequency and m_i is the ion mass. If an ion gyro-radius R_{LA} is defined which is based upon the Alfvén speed, then,

$$R_{LA} = \left(\frac{B^2}{4\pi\varrho}\right)\frac{eB}{m_i c} = \frac{c}{\omega_{pi}} \tag{35}$$

and since $R_{Li}/R_{LA} = (\beta/2)^{\frac{1}{2}}$ it follows that,

$$\frac{c}{\omega_{pi}} = R_{Li}\left(\frac{2}{\beta}\right)^{\frac{1}{2}}. \tag{36}$$

In this way the role of β can be seen and in particular, since

$$\frac{c}{\omega_{pe}} = \left(\frac{m_e}{m_i}\right)^{\frac{1}{2}}\frac{c}{\omega_{pi}} = R_{Li}\frac{2m_e/m_i}{\beta}. \tag{37}$$

Therefore, if $\beta < 2m_e/m_i$, then $c/\omega_{pe} > R_{Li}$ and the collisionless-skin depth is the important length. For $\beta \sim 1$, $R_{Li} \gg c/\omega_{pe}$ and the ion thermal gyro-radius becomes the important length.

There are also electrostatic collisionless shocks which contain no magnetic field. The entire scope of collisionless-shock theory is still evolving. The interested reader may obtain further details and references in the article by CHU and GROSS [13] or in the Proceedings of the Second IAEA Conference on Plasma Physics and Controlled Thermonuclear Research [20].

It is possible that such collisionless shock phenomena will substantilly alter the higher-temperature part of initially collisional shocks such as shown in Fig. 7. Rather than having a shock whose thickness increases with shock speed, collisionless collective effects may occur in the high-temperature tail and thereby substantially shorten the shock transition region.

4. – Radiative shock waves.

When the post-shock temperature is high enough to generate excited atomic states or ionize the medium, radiation becomes an important contributor to shock phenomena. It is often convenient to consider the photons thus released as an additional component of the medium with their own characteristic speed c, the speed of light, and their characteristic path length λ_ν, the mean free path of a photon of frequency ν and energy $h\nu$. The existence of still another signal speed and characteristic length, adds further to the richness of phenomena associated wih radiative shocks. The photons interact with the atoms of the media by means of absorption and scattering. In this Section we summarize the equations of radiative gas dynamics, examine the criteria for optically thick and thin limits, and discuss shock jump equations, shock stucture, and cooling of the post-shock plasma.

4·1. *Radiative shock equations.* – The equation of radiative transfer is well known from astrophysical studies [21]. Radiation has pressure, momentum and energy associated with it and therefore these effects must be included in the mathematical description of a moving medium. The equations of radiative transfer in moving gases are discussed in ref. [4], [22] and [26]. There is a rapidly growing literature concerning radiative shock waves, and many papers are listed in ref. [12] and [13]. We concentrate here upon those radiative effects of most interest in very high temperature strong-shock waves.

There are three types of descriptions of the interaction of radiation with matter. They are: 1) photon-atom interactions as described by quantum mechanics; 2) electromagnetic wave propagation through continuous media as described by Maxwell's equations; and 3) the equations of radiative transfer. If the radiant energy flux and its transformations are of paramount interest, the later approach as developed in astrophysics, is commonly employed. For convenience, the equations of radiative gas dynamics which describe the motion of a fluid including the effects of radiative transfer, are listed below in a rather general form:

mass conservation,

$$\frac{\partial \varrho}{\partial t} + \frac{\partial \varrho v_i}{\partial x_i} = 0\,; \tag{38}$$

momentum conservation,

$$\frac{\partial}{\partial t}\left(\varrho v_i + \frac{F_i}{c^2}\right) + \frac{\partial}{\partial x_j}(\varrho v_i v_j + p_{ij} + p^R_{ij}) = 0\,; \tag{39}$$

energy conservation,

$$\frac{\partial}{\partial t}\left[\varrho\left(e + \frac{v^2}{2}\right) + e^R\right] + \frac{\partial}{\partial x_j}\left\{v_j\left(e + \frac{v^2}{2}\right) + v_i(p_{ij} + p^R_{ij}) + F_j - q_j\right\} = 0\,; \tag{40}$$

and the radiative transfer equation,

$$\frac{1}{\varrho k_\nu c}\frac{\partial I_\nu}{\partial t} = \frac{1}{\varrho k_\nu} l_i \frac{\partial I_\nu}{\partial x_i} = -I_\nu + \left(\frac{j_\nu}{k_\nu}\right). \tag{41}$$

The superscript R refers to radiative quantities defined below. The new quantity introduced by radiative transfer is the specific intensity I_ν defined by

$$\mathrm{d}E_\nu = I_\nu \cos\theta \, \mathrm{d}\tau \, \mathrm{d}\omega \, \mathrm{d}\nu \, \mathrm{d}t \,, \tag{42}$$

where $\mathrm{d}E_\nu$ is the quantity of radiant energy crossing an element of area $\mathrm{d}\sigma$, whose normal makes an angle θ with the direction of the radiation flowing through the solid angle $\mathrm{d}\omega$ in the frequency interval $\mathrm{d}\nu$ during the time $\mathrm{d}t$. The fluid pressure tensor p_{ij} and thermal heat-conduction vector q_j are related to the fluid motion and temperature field by,

$$p_{ij} = p\delta_{ij} + \mu \left[\left(\frac{\partial v_i}{\partial x_j} + \frac{\partial v_j}{\partial x_i} \right) - \frac{2}{3} \frac{\partial v_k}{\partial x_k} \delta_{ij} \right] \tag{43}$$

and

$$q_j = -\varkappa \frac{\partial T}{\partial x_j} \,. \tag{44}$$

The radiation energy density e^R, radiation flux vector F_j and radiation pressure p_{ij}^R are defined in terms of the intensity by,

$$e^R \equiv \frac{1}{c} \int_0^\infty \mathrm{d}\nu \int_0^{4\pi} I_\nu \, \mathrm{d}\omega \,, \tag{45}$$

$$F_j \equiv \int_0^\infty \mathrm{d}\nu \int_0^{4\pi} l_j I_\nu \, \mathrm{d}\omega \,, \tag{46}$$

$$p_{ij}^R \equiv \frac{1}{c} \int_0^\infty \mathrm{d}\nu \int_0^{4\pi} l_i l_j I_\nu \, \mathrm{d}\omega \,, \tag{47}$$

where the direction of radiation is indicated by the unit vector $\hat{l}$ with direction cos l_i.

The emission coefficient is j_ν and the absorption coefficient is k_ν. Stimulated emission can be treated as negative absorption and is included in the definition of k_ν. The above set of equations, together with initial and boundary conditions, are complete and if the material properties (μ, $\varkappa$, j_ν, k_ν) are known, they describe completely the motion of a radiative gas. They do *not* assume radiative equilibrium, and are quite general. However, they are a nonlinear set of integro-differential equations whose solution, even for the most simple physical cases, is difficult.

The distance a photon of frequency ν travels depends upon its energy, the atomic nature of the medium through which it propagates, and whether it is absorbed or scattered. The cross-sections for photon absorption and scattering are discussed in ref. [4]. In an analogous way that particle paths may be reduced to a single number by the definition of mean free path, for photons one defines (by averaging over frequency) a Rosseland mean free path or a Planck mean free path. The Rosseland mean free path is applicable if the photons are nearly in equilibrum with the matter and it weights the high-energy photons ($h\nu > kT$) more heavily [4]. On the other hand the Planck mean free path is used where the photons travel over distances large compared to the particle mean free path and hence the photon gas is not in equilibrium with matter.

For example, for bremsstrahlung (free-free radiation) the Rosseland mean free path is

$$\lambda_R = 4.8 \cdot 10^{24} \frac{T_e^{\frac{7}{2}}}{Z^2 n_e n_i} \text{ (cm)} , \tag{48}$$

where Z is the atomic number, T_e the electron temperature in °K, and the particle number density is in cm^{-3}. For a fully ionized plasma undergoing Coulomb scattering, the ratio of the bremsstrahlung photon mean free path to the particle mean free path is,

$$\frac{\lambda_R}{\lambda} \approx 6 \cdot 10^{19} \frac{T_e^{\frac{3}{2}} Z^2 \ln\Lambda}{n_e} , \tag{49}$$

where $\ln\Lambda$ is sometimes called the plasma parameter and its value is tabulated by SPITZER [23]. It is a slowly varying function whose order of magnitude is 10. For most laboratory plasmas $\lambda_R/\lambda \gg 1$ so such radiation is rarely in thermodynamic equilibrium with the matter.

If L is a characteristic physical length and if $\lambda_\nu \gg L$, then the medium is called optically thin to radiation of frequency ν. On the other hand if $\lambda_\nu \ll L$, the medium is characterized as optically thick. The physics of these two situations is distinctly different. Another way of viewing this distinction is by comparing the time for radiative cooling τ to the mean time of photon flight λ_ν/c. The time τ for a stationary, radiatively cooling plasma undergoing free-free emission is [24]:

$$\tau = 2.3 \cdot 10^{-16} \frac{p^{\frac{1}{2}}}{\varrho^{\frac{3}{2}}} . \tag{50}$$

If one assumes that the gas can transfer all of its energy to the radiation field, then the time for the pressure to reach zero is just 3τ. If $\tau \ll \lambda_\nu/c$, phenomena which occur over this smaller time scale can be characterized as optically thin. For times of the order λ_ν/c, radiation is reabsorbed by the medium at a distance λ_ν from where it was emitted and radiative smoothing takes place. If $c\tau/\lambda_\nu \ll 1$, transient phenomena may be treated as optically thin regardless of the physical size of the system.

For free-free radiating hydrogen [24],

$$\frac{c\tau}{\lambda_\nu} \approx 5.8 \cdot 10^{-3} \frac{n}{T^3}, \tag{51}$$

where n is in cm^{-3} and T is in °K. Equation (36) is, apart from its numerical constant, true for all perfect gases in local thermodynamic equilibrium, regardless of the mechanisms of emission and absorption.

The ratio of gas dynamic pressure to radiation pressure is,

$$\frac{p}{p^R} = \frac{nkT}{\frac{1}{3}a_R T^4} = 5.4 \cdot 10^{-2} \frac{n}{T^3}, \tag{52}$$

where a_R is the Stefan-Boltzmann constant; hence we find that

$$\frac{c\tau}{\lambda_\nu} \approx 1.1 \cdot 10^{-1} \frac{p}{p^R} \tag{53}$$

and in the limit of $c\tau/\lambda_\nu \gg 1$, the radiation pressure is negligible compared with the gas pressure.

For the *optically thin case*, the nonrelativistic equations of radiative gas dynamics reduce to that of an ordinary fluid, except for an additional term in the energy equation to represent the energy loss by radiation. The equations of conservation of mass and momentum do not contain radiative terms. The energy equation for an optically thin gas undergoing one-dimensional motion is

$$\frac{D\varrho}{Dt}\left(e + \frac{u^2}{2}\right) + \frac{\partial pu}{\partial x} + 4\pi\varrho\varepsilon = 0, \tag{54}$$

where e is the internal energy of the gas, and $4\pi\varrho\varepsilon$ is the total radiative energy emitted per unit volume per unit time.

For the *optically thick case in thermodynamic equilibrium*, the equations of radiative gas dynamics contain a radiative pressure term p^R, and a radiative energy density term e^R where

$$e^R = 3p^R = a_R T^4.$$

For the case where the gas is not in equilibrium with the radiation or is neither optically thick or thin, it is necessary to resort to the full hierarchy of the radiative gas dynamic equations, whose solution presupposes a knowledge of all the radiative absorption and emission cross-sections of the medium. Such knowledge and the corresponding solutions are rare.

4·2. *Radiative shock jump equations.* – In the optically thin limit, the shock jump conditions have been studied by KOCH and GROSS [24] who show that, for a given Mach number, the post-shock state differs from the nonradiative case in the partially ionized regime. They compare steady-state solutions using Saha equilibrium with the optically thin case where the post-shock is constant in time because of a balance between collisional and radiative transitions. They also compute the shock jump conditions for the case which is optically thick for Lyman radiation and optically thin at all other frequencies. Examples of these radiative post-shock states are shown in Fig. 4 and 5. If the optical and collisional rates do not balance, then there is no steady-state solution and a study of these rates must be made to compute the time evolution of the shocked medium.

The optically thick, thermodynamic equilibrium, shock jump equations have been studied by KOCH [25]. The resulting set of algebraic equations were solved and, for a given shock speed, the radiative equilibrium shock has a lower post-shock temperature than the nonradiative shock. The physical reason for this is that in the radiative case the medium consists of a mixture of photons and particles, and for a given shock speed there are more degrees of freedom to share the energy.

The strong-shock radiative equilibrium density ratio has the value of 7 rather than its gas dynamic value of 4. The plasma consisting of an equilibrium mixture of electrons, protons and photons behaves as a gas with a specific heat ratio $\frac{4}{3}\leqslant\gamma\leqslant\frac{5}{3}$; a pure photon gas has $\gamma=\frac{4}{3}$ while a monatomic particle gas has $\gamma=\frac{5}{3}$. Another interesting result for the radiative equilibrium gas is the speed of propagation of a small disturbance (sound) which is

$$a=(\gamma^1 RT)^{\frac{1}{2}}\,, \tag{55}$$

where

$$\gamma^1 = 1+\frac{(1+\frac{4}{3}\varepsilon)^2}{(\gamma-1)^{-1}+4\varepsilon}$$

and

$$\varepsilon=\frac{a_R T^4}{nkT}=55.6\,\frac{T^3}{n}\,.$$

Here ε can be recognized as the ratio of the equilibrium radiation energy density to the internal energy density of the gas. It can be shown that the radiative equilibrium post-shock speed is always subsonic and there is always an increase in entropy, in correspondence with fluid dynamic shock theory. The structure of radiative equilibrium shock waves has been solved numerically by KOCH [25]. Such waves are usually very thick because $l \sim \lambda_R(c/V_s)$ and the photon mean free paths are normally large (see Fig. 6).

Intermediate cases, where $(c\tau/\lambda_\nu) \sim 1$, or shocks where there is no equilibrium, are very complex. Several studies are given in ref. [13].

The temperature decay of optically thin media heated by a strong shock has been studied [24]. For strong shocks the density and pressure both increase in the gas flowing behind the shock while the temperature decreases (see Fig. 6). For a bremsstrahlung radiating, optically thin, shock heated. plasma, the time rate of change of the temperature immediately behind the shock front, as measured in the laboratory reference frame, in the strong-shock limit is [24],

$$\frac{dT}{dt} \approx -1 \cdot 10^{-15} n_1 v_1 \; [^\circ\text{K s}^{-1}] \, , \tag{56}$$

where t is time, n is in cm^{-3} and v_1 is in cm/s.

The radiation from blast waves in the optically thin case is rather straightforward and has been estimated in ref. [19].

5. – Relativistic shock waves.

The theory of relativity indicates that shock speeds cannot increase without limit but shoud be bounded by the speed of light c. The theoretical foundations of relativistic shock waves has been studied by ECKART [27], TAUB [28], LANDAU and LIFSHITZ [29] and others (see ref. [12] for a more complete bibliography). In the progression to higher shock speeds the first manifestation of relativity occurs in the electron component of the post-shock gas. For a hydrogen plasma, a significant fraction of the electrons are relativistic at $T_2 \sim 5 \cdot 10^9$ °K (a shock speed of about $2 \cdot 10^9$ cm/s). Consequently, the equation of state of the gas must be suitably altered to account for these relativistic electrons, and this has been done by SYNGE [30]. At still higher speeds ($V_s \sim 10^{10}$ cm/s) the post-shock ions begin to become relativistic, the plasma density is affected, and the shock front speed finally becomes asymptotic to the speed of light. If the radiation associated with the hot post-shock plasma is in equilibrium with the matter (optically thick case) then pair production is significant when $kT_2 \approx m_e c^2 \approx 0.5 \cdot 10^5$ eV $\approx 6 \cdot 10^9$ °K. In such a circumstance, particle number density is not conserved and the energy density of positrons must be taken ino account.

5·1. *Relativistic shock jump relations.* – The equations of motion of a fluid interacting with radiation in a relativistically correct form were given by THOMAS [34] and can be expressed by setting the four-divergence of the stress-energy tensor to zero. The stress-energy tensor can be expressed as the sum of two terms

$$T_{ij} = T^m_{ij} + T^R_{ij}\,, \qquad i, j = 0, 1, 2, 3, \tag{57}$$

where the superscripts m and R refer to matter and radiation. If the fluid is ideal and we employ the Minkowskian co-ordinates $x_0 = ict$, $x_1 = x$, $x_2 = y$, $x_3 = z$, then in the rest frame of the fluid,

$$T^m_{ij} = \begin{pmatrix} -e_m & 0 & 0 & 0 \\ 0 & p & 0 & 0 \\ 0 & 0 & p & 0 \\ 0 & 0 & 0 & p \end{pmatrix}, \tag{58}$$

$$T^R_{ij} = \begin{pmatrix} -\iint I_\nu \,d\nu \,d\omega & i\iint I_\nu \hat{l} \,d\nu \,d\omega \\ i\iint I_\nu \hat{l} \,d\nu \,d\omega & \iint I_\nu \hat{l}\hat{l} \,d\nu \,d\omega \end{pmatrix}. \tag{59}$$

For a problem, such as a shock wave, which depends upon only one spatial dimension and time, the radiation stress-energy tensor reduces to,

$$T^R_{ij} = c \begin{pmatrix} e^R & \dfrac{iF}{c} \\ \dfrac{iF}{c} & p^R \end{pmatrix}. \tag{60}$$

and the radiative transfer equation in its relativistic correct form is,

$$\frac{1}{c}\frac{dI_\nu}{\partial t} + \mu \frac{\partial I_\nu}{\partial x} = -\varrho_0 k_{\nu_0} \gamma (1-\mu\beta) I_\nu + \varrho_0 \varepsilon_{\nu_0} \frac{1-\beta^2}{(1-\mu\beta)^2}\,, \tag{61}$$

where $\mu = \cos\theta$, $\beta = v/c$, $\gamma = (1-\beta^2)^{\frac{1}{2}}$ and the subscript 0 refers to the fluid rest frame. These equations and their consequences for shock wave problems are discussed by KOCH and GROSS [24].

The relativistic shock jump relations, where the radiation is in equilibrium with the matter, has been studied by MASANI *et al.* [31], for conditions of interest in astrophysics, and GROSS [12] studied this problem in terms of the radiation parameter ε.

SYNGE [30] showed that, in the proper reference frame for a single species gas,

$$e + p = \varrho\, c^2 G(\alpha) \quad \text{and} \quad p = \frac{\varrho c^2}{\alpha}, \quad \text{where } \alpha = \frac{mc^2}{kT}.$$

Here e is the energy per unit volume, p the pressure of a nonradiating relativistc monatomic gas, ϱ the proper density, c the speed of light, and α is a reciprocal dimensionless temperature. $G(\alpha)$ is a function which can be approximated by

$$G(\alpha) = \begin{cases} 1 + \dfrac{5}{2\alpha}, & \alpha \geqslant \dfrac{3}{2}, \\ \dfrac{4}{\alpha}, & \alpha < \dfrac{3}{2}. \end{cases}$$

For ionized hydrogen

$$(62a) \qquad e + p = \varrho c^2 Q(\alpha) = \varrho c^2 \left[G(\alpha) + \frac{m_e}{m_i} G\left(\frac{m_e \alpha}{m_i}\right)\right].$$

If equilibrium radiation is assumed, then

$$(62b) \qquad e + p = \varrho c^2 Q(\alpha) + \frac{4}{3} a_R T^4,$$

$$(63b) \qquad p = \frac{2\varrho c^2}{\alpha} + \frac{a_R T^4}{3},$$

or

$$(62c) \qquad e + p = \varrho c^2 \left[Q + \frac{4}{3}\left(\frac{\varepsilon}{\alpha}\right)\right],$$

$$(63c) \qquad p = \left(\frac{\varrho c^2}{\alpha}\right)\left(2 + \frac{\varepsilon}{3}\right)$$

and

$$\frac{\varepsilon}{\alpha} = \frac{a_R T^4}{\varrho c^2}$$

The relativistic shock jump equations representing conservation of mass momentum and energy are (following ref. [28]).

$$(64) \qquad n_1 \beta_1 \gamma_1 = n_2 \beta_2 \gamma_2,$$

$$(65) \qquad \beta_1 \gamma_1^2 (e_1 + p_1) = \beta_2 \gamma_2^2 (e_2 + p_2),$$

$$(66) \qquad (e_1 + p_1)\beta_1^2 \gamma_1^2 + p_1 = (e_2 + p_2)\beta_2^2 \gamma_2^2 + p_2,$$

where $\beta = u/c$, $\gamma = (1 - \beta^2)^{-\frac{1}{2}}$ and $n = n_i = n_e$.

The set of eqs. (44), (45) and (46) have been studied and if the initial gas state is cold then it is appropriate to assume $\alpha_1 \gg \frac{3}{2}$, $p_1 \ll e_1$, $\varepsilon_1 \ll 1$.

If the post-shock state is such that terms of the $O(\alpha_2)^{-1}$ can be ignored, the density ratio reduces to

$$\frac{\varrho_2}{\varrho_1} = \left(2 + \frac{\varepsilon_2}{3}\right)^{-1} \cdot \left\{8 + \frac{\varepsilon_2}{3}\left[7 + \frac{4}{3}\left(\frac{\varepsilon_2}{\alpha_2}\right)\right]\right\}. \tag{67}$$

If $\varepsilon_2 = 0$ (nonradiative case), we recover $\varrho_2/\varrho_1 = 4$, the same resulta s forstrong shocks in a monatomic gas. If $\varepsilon_2 \gg 1$, then

$$\frac{\varrho_2}{\varrho_1} = 7 + \frac{4\varepsilon_2}{3\alpha_2} \tag{68}$$

and if ε_2/α_2 can be ignored, $\varrho_2/\varrho_1 = 7$, the same result for equilibrium radiative shocks with relativistic effects ignored. The velocity ratio and all other quantities can readily be obtained by a numerical procedure (see for example ref. [12] or [31]).

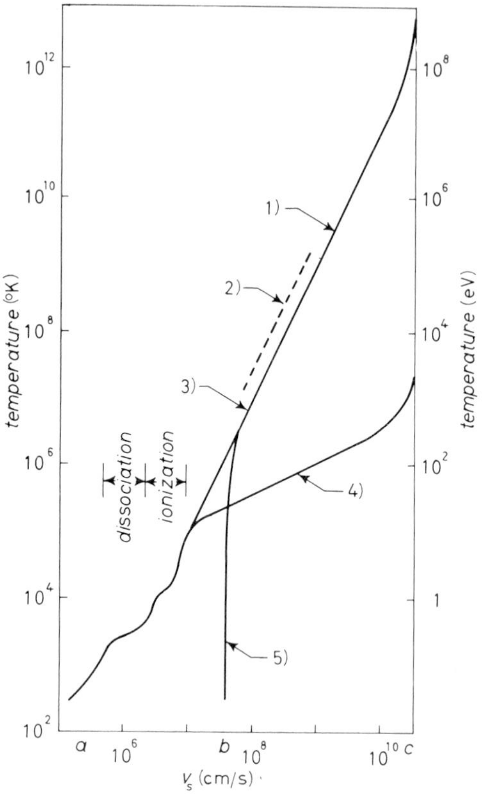

Fig. 8. – The temperature created by a strong-shock wave in hydrogen (from ref. [12]): 1) relativistic electrons; 2) reflected shock; 3) optically thin shock; 4) optically thick shock; 5) switch-on plasma shock. $p_1 = 0.10$ Torr, $T_1 = 300$ °K.

MASANI [31] has shown that, if pair production is included, the density ratio reaches a value of approximately 14 in the nnorelativistic case and in the relativistic case, the density ratio increases without limit. The gas does not become degenerate because of the accompanying increase in temperature.

When $\varrho_2 \sim 10^{12}$ g/cm³ account must be taken of transformation into a free neutron gas and for $\varrho_2 \sim 10^{14}$g/cm³ the phenomenology of nuclear matter becomes important.

The influence of pair production, in the equilibrium case, can be accounted for by noting that [32].

$$\text{(69)} \qquad \begin{cases} n^+ = n^- = 4\left(\dfrac{mkT}{2\pi\hbar}\right)^3 \exp\left[\dfrac{-2mc^2}{kT}\right], & \text{for } \alpha \gg 1, \\ n^+ = n^- = 0.183\left(\dfrac{kT}{\hbar c}\right)^3, & \text{for } \alpha \ll 1, \end{cases}$$

where n^+ is the positron number density. The energy density per unit volume of positrons is

$$\text{(70)} \qquad e^+ = e^- = \frac{7\pi^2(kT)^4}{120(\hbar c)^2}\,.$$

The energy density of e^+ is seven-eighths the energy density of black-body radiation.

The temperature and density ratio of strong relativistic shocks in hydrogen are shown in Fig. 8 and 9.

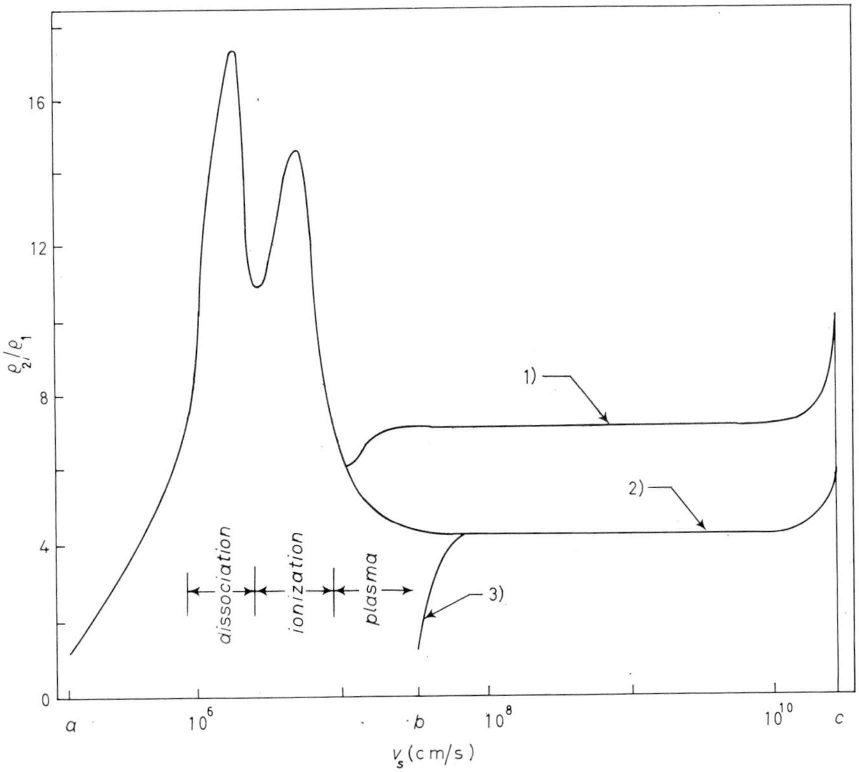

Fig. 9. – The density ratio for strong shock waves in hydrogen (from ref. [12]): 1) radiative equilibrium optically thick shock; 2) optically thin shock; 3) switch-on plasma shock; $p_1 = 0.10$ Torr, $T_1 = 300$ °K.

6. – Examples of recent research.

There is a wide spectrum of research in progress studying very strong shock waves in gases and the physical effects generated by them. A great deal of this research is motivated by programs seeking to obtain controlled thermonuclear fusion, interest in astrophysical phenomena, and a general fascination with high-temperature physics.

In this Section a summary is given of the theoretical structure of a thermonuclear detonation wave and then the present status of laboratory research with strong-shock waves is surveyed.

6·1. *Thermonuclear detonation wave structure.* – A strong-shock wave heats and compresses the gas through which it moves and if that gas is deuterium or tritium and the temperature is sufficiently high ($T_2 \sim 10^8$ °K), thermonuclear reactions occurs in the gas. The structure of such a wave, called a thermonuclear detonation wave, has been studied by FULLER and GROSS [35] and their results are briefly summarized here.

The order of magnitude of the temperature behind a thermonuclear detonation wave, if it propagates at its Chapman-Jouguet speed ($\sim 10^9$ cm/s), is about 10^{10} °K. The thickness of such a wave is very large since the reaction probability $\langle \sigma v \rangle$ is 10^{-16} cm^3/s at 10^{10} °K. Therefore, the mean time for a thermonuclear reaction is 10 s at a plasma particle density of 10^{15} cm^{-3} and this implies a wave thickness of the order of 10^{10} cm. For an initial density of the order of solid hydrogen ($n \sim 5 \cdot 10^{22}$ cm^{-3}), the wave thickness is of the order of ~ 20 cm for a substantial fraction of the particles to undergo fusion.

The ratio of the mean time for a two-particle Coulomb 90 degree collision to the mean thermonuclear reaction time is, for these post-shock conditions, much less than one. Hence, it is reasonable to consider the shock structure to occur first and to consider it as separate from the thermonuclear reaction zone. This concept of separate shock and deflagration (reaction) regions is called the von Neumann, Zel'dovich, Doring model of a detonation. It is applicable to thermonuclear detonations and was used in the analysis of FULLER and GROSS [35]. They also show that in the deflagration region the effects of viscosity, thermal conduction and diffusion are negligible. The two-particle relaxation time is, however, of the order of the mean fusion reaction time and hence the electrons will not be in equilibrium with the ions. The neutrons, on the other hand, are roughly in equilibrium with the tritium ions, their cross-section for scattering being about one barn at these energies. Essentially all the bremsstrahlung radiation is optically thin and leaves the reaction zone, thus cooling the plasma. Figure 10 shows a sketch of the structure of a fusion detonation wave.

If a strong shock in a laboratory-size apparatus propagates through a deuterium-tritium gas, at 10^8 cm/s, FULLER and GROSS predict a neutron flux of 10^{17} neutrons $cm^{-1}\,s^{-1}$ and a fusion power density of about 2 kW/cm³ in a

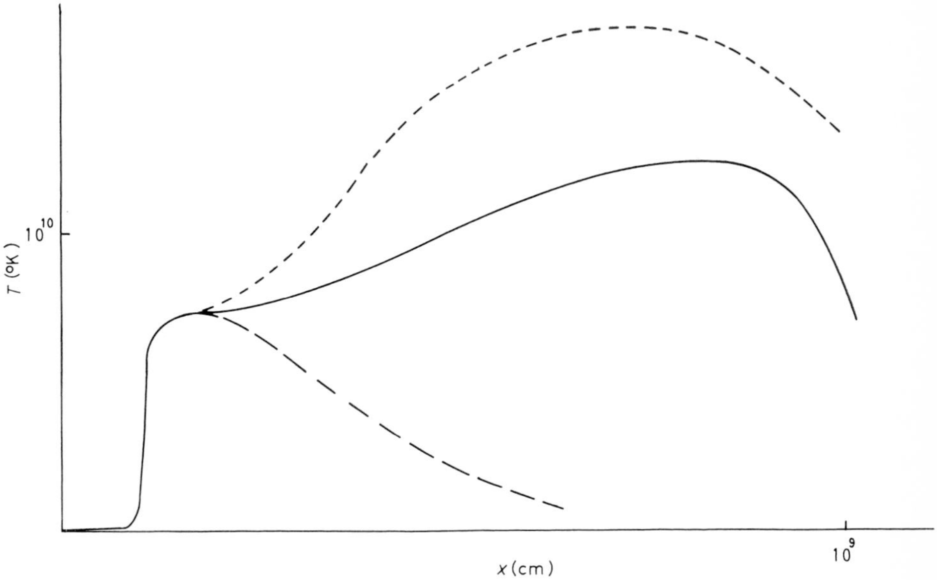

Fig. 10. – The temperature structure of a thermonuclear detonation wave in a plasma with density $n = 10^{15}$ cm⁻³. At $x = 10^9$ cm, the percentage of fusion reaction that has occured is about sixty percent. (Adapted from ref. [35].) - - - Thermonuclear reactions and no radiation losses; ——— detonation wave structure with fusion reactions and radiation losses; — — — optically thin radiation and no thermonuclear reactions.

plasma with $n = 10^{16}$ cm⁻³. At this shock speed the radiation power loss exceeds the released fusion power, the ratio being about 1.25. At a shock speed of $4 \cdot 10^8$ cm/s ($T_2 \sim 40$ keV), the reaction zone has its minimum length.

6'2. *Laboratory strong gas shock wave experiments.* – Nearly all laboratory experiments with high-speed shock waves employ electromagnetic devices. The technology of these devices and their performance has been discussed in detail by GROSS and MILLER [36]. Only some of the very recent experimental results concerning strong shock waves are discussed here.

Theta-pinch devices, with large (megajoule) energy-storage systems capable of discharging in several microseconds, have been used to produce strong gas shock waves. A typical theta-pinch (collision-dominated) shock speed is 10^7 cm/s into hydrogen whose number density is of the order of 10^{15} cm⁻³. Similar results have been achieved in Z-pinch devices and in co-axial electromagnetic shock tubes. For example, in the plasma research laboratory at

Columbia University, shock speeds up to 30 cm/μs have been obtained in co-axial electromagnetic shock tubes. Such a shock speed in room temperature hydrogen ($M \sim 240$) should produce an equilibrium temperature of 100 eV (10^6 °K). There is to date little solid evidence that such equilibrium temperatures have in fact been achieved

Laser-produced sparks have been used to generate a hot plasma, which upon expansion, generates a blast wave. These shocks move at speeds $\sim 10^7$ cm/s during their early development.

Of course, as previously shown, the wave speed is very dependent upon the power per unit area delivered to the shock front and the initial gas density. In attempts to achieve higher shock speeds and temperatures, lower initial pressures have been used and the realm of collisionless shock waves is encountered. Shock speeds of nearly 10^8 cm/s in deuterium at $n \sim 10^{12}$ cm^{-3}, were reported by KURTMULLAEV and colleagues at a fusion conference in 1966 [37]. At the seventh International Shock Tube Symposium in June of 1969, LEVINE and colleagues at the U.S. Naval Research Laboratory reported [38] on their experimental study of fast collisionless shock waves. They used a Blumlein line energy storage and transmission system (see Table II, Item 3). They observed collisionless shock waves propagating into a low-density argon plasma ($n_e \sim 10^{12}$ cm^{-3}, $T \sim 1$ eV) at speeds up to $(100 \div 150)$ cm/μs and they measured X-ray emission indicative of electron temperatures of 10 to 100 keV (*i.e.* up to about 10^9 °K). A typical time scale in these experiments was 10 nanoseconds. These are the fastest laboratory-produced shocks of which the author is aware. It is doubtful that an equilibrium condition can be achieved at these energies and time durations. Equilibrium temperatures for hydrogen shock speeds of $1 \cdot 10^8$ and $3 \cdot 10^8$ cm/s are 1 and 10 keV respectively, the latter being of definite interest in controlled-fusion research.

REFERENCES

[1] R. COURANT and K. O. FRIEDRICKS: *Supersonic Flow and Shock Waves* (New York, 1948).

[2] H. W. EMMONS (ed.): *Fundamentals of Gas Dynamics* (Princeton, 1958).

[3] H. W. LIEPMANN and A. ROSHKO: *Elements of Gas Dynamics* (New York, 1957).

[4] YA. B. ZEL'DOVICH and YU. P. RAIZER: *Physics of Shock Waves and High-Temperature Hydrodynamic Phenomena* (New York, 1966).

[5] A. G. GAYDON and I. R. HURLE: *The Shock Tube in High Temperature Chemical Physics* (New York, 1963).

[6] D. GILBARG and D. PAOLUCCI: *Journ. Ratl. Mech. Anal.*, **2**, 617 (1953).

[7] H. M. MOTT-SMITH: *Phys. Rev.*, **82**, 885 (1951).

[8] M. T. CHAHINE and R. NARASHIMA: *Rarefied Gas Dynamics, Fourth Symp.*, edited by J. H. DE LEEUW (New York, 1965), p. 140.

[9] G. I. TAYLOR: *Proc. Roy. Soc.*, A **201**, 159 (1950).
[10] L. I. SEDOV: *Prikl. Mat. Mekh.*, **10**, 241 (1946); also see the book: *Similarity and Dimensional Methods in Mechanics*, edited by M. HOLT, 4-th edition (New York, 1959).
[11] D. D. LAUMBACH and R. F. PROBSTEIN: *Journ. Fluid Mech.*, **35**, 1, 53 (1969).
[12] R. A. GROSS: *Rev. Mod. Phys.*, **37**, 724 (1965).
[13] C. K. CHU and R. A. GROSS: *Shock Waves in Plasma Physics*, in the book *Advances in Plasma Physics*, vol. **2**, edited by A. SIMON and W. B. THOMPSON (New York, 1969).
[14] D. A. TIDMAN: *Phys. Rev.*, **111**, 1439 (1958).
[15] M. Y. JAFFRIN and R. F. PROBSTEIN: *Phys. Fluids*, **7**, 1658 (1964).
[16] W. MARSHALL: *Proc. Roy. Soc.*, A **233**, 367 (1955).
[17] R. A. GROSS: *Phys. Fluids*, **10**, 1853 (1967).
[18] D. L. CHUBB: *Phys. Fluids*, **11**, 2363 (1968).
[19] R. A. GROSS: *Phys. Fluids*, **7**, 1078 (1964).
[20] Papers presented in Sect. A of the *Proc. of the Second IAEC Conf. on Plasma Physics and Controlled Thermonuclear Research* (Novosibirsk, 1968), to be published.
[21] S. CHANDRASEKHAR: *Radiative Transfer* (New York, 1960).
[22] W. G. VINCENTI and C. H. KRUGER jr.: *Introduction to Physical Gas Dynamics* (New York, 1965).
[23] L. SPITZER jr.: *Physics of Fully Ionized Gases*, 2nd ed. (New York, 1962).
[24] P. A. KOCH and R. A. GROSS: *Phys. Fluids*, **12**, 1182 (1969).
[25] P. A. KOCH: *Phys. Fluids*, **8**, 2140 (1965).
[26] D. H. SAMPSON: *Radiative Contributions to Energy and Momentum Transport in a Gas* (New York, 1965).
[27] C. ECKARD: *Phys. Rev.*, **58**, 919 (1940).
[28] A. H. TAUB: *Phys. Rev.*, **74**, 328 (1948).
[29] L. D. LANDAU and E. M. LIFSHITZ: *Fluid Mechanics*, (New York, 1959).
[30] J. L. SYNGE: *The Relativistic Gas* (Amsterdam, 1957).
[31] A. MASANI, V. BORLA, A. FERRARI and A. MARTIN: *Nuovo Cimento*, **48** B, 326 (1967).
[32] L. D. LANDAU and E. M. LIFSHITZ: *Statistical Physics*, vol. **10**, Chapt. XI (Reading, Mass., 1958).
[33] P. A. KOCH: *Phys. Rev.*, **140**, A 1161 (1965).
[34] L. H. THOMAS: *Quarterly Journ. Math. Oxford*, **1**, 239 (1930).
[35] A. L. FULLER and R. A. GROSS: *Phys. Fluids*, **11**, 534 (1968).
[36] R. A. GROSS and B. MILLER: *Plasma heating by strong shock waves*, in *Methods in Experimental Physics* (*Plasma Physics*), vol. **9**, Part A, edited by H. R. GRIEM and R. H. LOVBERG (New York).
[37] R. KH. KURTMULLAEV, YU. E. NESTERIKHIN, V. I. PIL'SKII and R. Z. SAGDEEV: Session V, *Proc. 1965 Culhan Conf. Plasma Physics and Controlled Nuclear Fusion Research*, International Atomic Energy Agency, *1965.*
[38] L. S. LEVINE, I. M. VITKOVITSKY and A. C. KOLB: *Experimental study of fast collisionless shock waves*, in *Proceedings of the Seventh International Shock Tube Symposium, Toronto, Canada, June 1969* (to be published).

High-Temperature and Plasma Phenomena Induced by Laser Radiation.

O. N. Krokhin

Lebedev Physical Institute - Moscow

The wide range of the laser flux density variation which can be obtained at the different laser power and focusing conditions makes it possible to change strongly the thermodynamical state of the various substances. In this case the laser light interaction with absorbing media results in the changing of the internal energy, pressure and density of the substance permitting to use this phenomenon for a study of the substance's behaviour at high-temperature and high-pressure conditions.

A problem of interaction of laser radiation with the absorbing condensed media may be somehow considered from two aspects, characterized by different physical processes: *a*) there is a low-temperature region, where the substance is practically not ionized, and *b*) a high-temperature region, where the substance passes into the plasma state. In the low-temperature region corresponding to comparatively small laser radiation flux densities ($< 10^9$ W/cm^2), the laser light interaction with matter results mainly for the substance in the process of phase transition from condensed to gas state. In connection with this there appears the possibility to study the mechanism of phase transition, evaporation kinetics, equilibrium phase curves, thermodynamical and, in particular, critical constants of any substances including those having a high boiling temperature. In the cases considered the phase transitions are characterized by a strong dynamical effect with a fast transition of a significant amount of substance from one state into another within a short period of time. This circumstance demands to develop a special methodics of investigation based upon the correct theoretical understanding of the whole process.

In a high-temperature region corresponding to high laser radiation flux densities ($> 10^9$ W/cm^2), a hot pulsed plasma may be produced. Opaque and transparent substances in this region of fluxes behave in the same way because the primary transparent dielectrics, (solid and liquid hydrogen for instance) pass into an absorbing state due to the tremendous « optical » breakdown and the consequent heating in a very short period of time. Pulsed high-temperature plasma is an interesting object for a separate physical investigation. Here we investigate mainly the dynamics of plasma formation, plasma interaction with a powerful light radiation, kinetics of ionization and recombination as well as other relaxation processes, and the radiation spectra of strongly excited multi-ionized atoms.

On the other side, investigations of laser plasma have mutual points of reference to works carried out in the field of thermonuclear plasma. Laser methods employed for production and heating of plasma may be somehow used for the filling of magnetic traps and for other thermonuclear devices.

It is essential to mention here also that a pulsed laser plasma is the source of high-energy concentration with pressures up to 10^9 atm. and can be used therefore to produce in both cases strong shock waves in the condensed and gaseous substances.

1. – An approximate mathematical description of the problem.

Laser radiation absorption on the surface of an opaque condensed substance leads to heat liberation in the thin layer under the surface. This causes the increase of surface temperature and the appearance of heat waves propagating inside the substance. As the velocity of the heat-wave propagation reduces with time, the process of heat-diffusivity cannot provide a sufficient heat flow-out of the absorbing layer. Thus, the temperature of the substance will be increasing up to the moment when the energy removal from the surface due to evaporation of the substance sets in. From this moment the surface temperature will be fully determined by the evaporation mechanism. Thermodiffusion stops to play a significant part and will only lead to the existence of a comparatively thin heated layer in the condensed substance which is attached to the surface of evaporation—« evaporation wave ».

This condition is fufilled at sufficiently large light fluxes. In the opposite case the substance will not be heated sufficiently for the evaporation process to begin to play the main part. The minimal value of laser flux q_s may be evaluated in the following way. The thickness of the heated zone at time t equals the order of magnitude $(at)^{\frac{1}{2}}$, where a, is the thermal diffusivity coefficient. The thermal energy estracted to this moment is $q_0 t$ where q_0, is the laser radiation flux density. Thus the specific energy of the heated zone is equal to $q_0 t/(\varrho_0^2 at)^{\frac{1}{2}}$, where ϱ_0 is the substance density. To entertain the process of evaporation, the specific thermal energy should be equal to the specific energy of the substance evaporation (sublimation) U and time t should not exceed the duration of the laser pulse τ. From this the condition follows:

(1) $$q_0 > q_e \approx \varrho_0 U a^{\frac{1}{2}}/\tau^{\frac{1}{2}} .$$

For substances like metals ($a \approx 0.1$ cm²/s, $U \approx 10^{11}$ erg/g, $\varrho_0 \approx 10$ g/cm³) the value q_s is approximately equal to 10^{13} erg/cm² s $= 10^6$ W/cm².

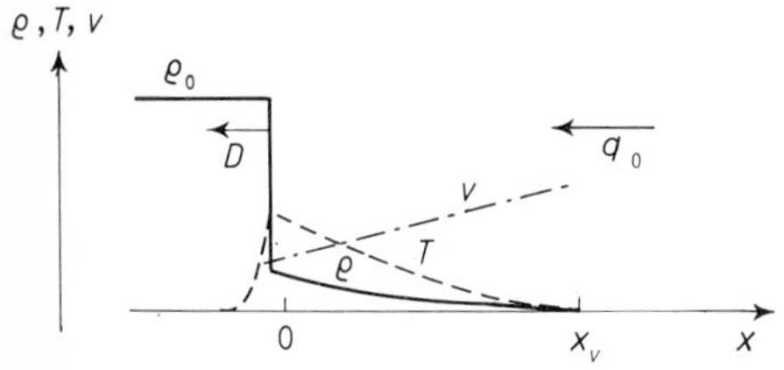

Fig. 1. – Profiles of distributions of temperature T, density ϱ and velocity v at the evaporation of the substance by laser radiation.

Thus, at $q_0 > q_e$ the evaporation of the condensed substance will play the main role. In this case the increased amount of laser flux energy causes the breaking of the bounds in the condensed media and the further heating of evaporated matter. The velocity of the evaporation wave propagation $|D|$ is of the order of magnitude of $q_0/(U + CT_0)\varrho_0$, where the coefficient C equals approximately the double of the vapour's specific thermal capacity, and T_0 is the temperature on the evaporation wave front (Fig. 1).

The thickness of the heated zone of condensed substance and temperature profile may be defined by the solution the quasi-stationary equation of thermodiffusivity (if the laser radiation flux q_0 does not vary very fast with time)

(2) $$|D| \frac{\partial T}{\partial x} = \frac{\partial}{\partial x} a \frac{\partial T}{\partial x} + \frac{\partial q}{\partial x} ,$$

where x is the co-ordinate in the direction of the evaporation wave propagation, $q = q_0 \exp[+kx]$ is the radiation flux in the condensed substance, K, the absorption coefficient. The solution of eq. (2) for independence of temperature is represented in works of READY [1] and ANISIMOV *et al.* [2]. However, as it is was mentioned above, the heat diffusion at $q_0 > q_e$ does not play a very important role. Therefore for an estimation of thickness of the heated layer one may be confined to the approximation $K \to +$, the solution of eq. (2) can be written in the form

(3) $$\left\{ \begin{aligned} & x = \int\limits_{T_0}^{T} \frac{a(T)\,\mathrm{d}T}{|D|\,T} , \\ & T(x) = T_0 \exp[ax/|D|] , \end{aligned} \right. \qquad x < 0 .$$

The effective thickness of the heated layer is of the order of $a/|D| \approx 10^{-3}$ cm at $q_0 \approx 10^7$ W/cm², while K has for metals a value about ($10^4 \div 10^5$) cm⁻¹.

Evaporation of the substance leads to mechanical acceleration of the different parts of the substance. In turn the acceleration has an influence on the rate of the evaporation process because the last one has a strong dependence on the pressure. From this it is clear that the whole problem should be investigated on the basis of the full system of dynamical equations.

The three gas-dynamical conservation equations, the equation of Euler and those of the energy, have the form

$$(4)\qquad \begin{cases} \dfrac{\partial \varrho}{\partial t} + \boldsymbol{\nabla}(\varrho \boldsymbol{v}) = 0\,, \\[2ex] \dfrac{\partial \boldsymbol{v}}{\partial t} + (\boldsymbol{v}\boldsymbol{\nabla})\boldsymbol{v} + \dfrac{1}{\varrho}\boldsymbol{\nabla} p = 0\,, \\[2ex] \dfrac{\partial}{\partial t}\left(\varrho\varepsilon + \dfrac{\boldsymbol{v}^2}{2}\right) + \boldsymbol{\nabla}\left[\varrho \boldsymbol{v}\left(\varepsilon + \dfrac{\boldsymbol{v}^2}{2} + \dfrac{p}{\varrho}\right)\right] + \boldsymbol{\nabla}\boldsymbol{q} = 0\,, \end{cases}$$

where $\boldsymbol{v}$, is the substance velocity, ϱ, the density, p, the pressure, ε, the specific energy, $\boldsymbol{\nabla}$, a vector with the corresponding components $\partial/\partial x$, $\partial/\partial y$, $\partial/\partial z$.

The fact that at $q_0 > q_e$ the quantity of evaporated mass appears to be larger in respect to that in the thin heated zone of condensed media where the process of vaporization takes place, significantly simplifies the investigation of eqs. (4). Physically it is clear that the dynamics of the whole process will be defined by the larger of the masses, *i.e.* by the mass of the evaporated substance. So one may neglect the motion in the thin heated layer, *i.e.* this layer may be considered as infinitely thin. This position is fully analogous to the case of the shock waves, and represents a standard method of introducing breaks in gas-dynamics when inside the break region, instead of the solutions of the gas-dynamical equation, are inserted on both sides of the break theoretic ratios of magnitudes as the boundary conditions of the equations of motion. In the case considered the gas-dynamical break region represents the front of the evaporation wave of the substance.

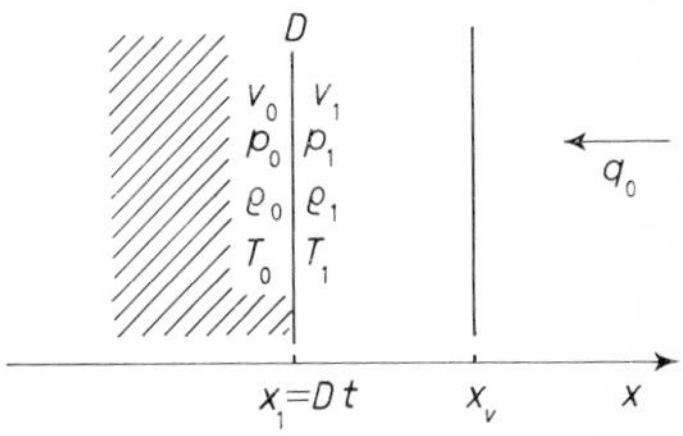

Fig. 2. – Meanings of gas-dynamical parameters on the front of the evaporation wave. The front's position is x_1. The evaporated substance is situated between x_1 and x_v. The condensed phase is on the left of x_1.

The conditions on the evaporation wave front express the laws of conservation of mass, momentum and energy fluxes. The x-axis is directed perpendicular to the substance's surface. Laser radiation falls from the side $x = \infty$ on the substance's surface which fills the region $x \leqslant 0$ at the initial moment of time $t = 0$ (Fig. 2). Evaporated matter flows from the surface with density ϱ_1 and velocity v_1 directed along the positive values of the x-axis. Thus the evaporated substance carries away momentum flux. Due to the law of conservation, this momentum should be compensated by a momentum flux directed inside the condensed substance. The latter value equals $p_0 + \varrho_0' v_0^2$ where p_0, v_0, ϱ_0' corresponds to pressure, velocity and density in the condensed substance (ϱ_0'-differing from the initial density of the condensed phase ϱ_0 due to movement) on the left of the evaporation wave front. If the pressure p_0 is small in comparison with a parameter $\varrho_0 c_0' \approx 10^6$ atm where c_0 is the sound velocity in the condensed substance, ϱ_0' should be close to ϱ_0 and v_0 should be small. In fact,

$$(5)\qquad \left\{\begin{array}{l} \dfrac{\varrho_0' - \varrho_0}{\varrho_0} \approx p_0 \Big/ \left(\dfrac{\partial p}{\partial \varrho}\right) \varrho_0 \approx p_0/\varrho_0 c_0^2 \ll 1 \,, \\ v_0/c_0 \approx (\varrho_0' - \varrho_0)/\varrho_0 \ll 1 \,. \end{array}\right.$$

This correlations express the fact of the small compressibility of the condensed media. Another simple statement is

$$(6)\qquad v_1 \ll v_0 \,,$$

which is the result of the obvious condition $\varrho_1 \ll \varrho_0$. From expressions (5), (6) it follows that one can neglect the term $\varrho_0' v_0^2$ in the momentum flux as well as in the energy flux in the condensed phase of media.

Noting this, finally the conditions on the evaporating wave front can be written in following form:

$$(7)\qquad \left\{\begin{array}{ll} -\varrho_0 D = \varrho_1 (v_1 - D) \,, & \\ p_0 & = p_1 - \varrho_0 D v_1 \,, \\ q_1 & = -\varrho_0 D(\varepsilon_1 + v_1^2/2) + p_1 v_1 \,, \end{array}\right.$$

where q_1 is the part of laser radiation flux absorbed in the evaporation wave front, D is the velocity of the evaporation wave ($D < 0$). The index « 0 » indicates the value before the front of the evaporation wave (in condensed phase), « 1 » behind the front (in evaporated substance). The conditions (7) are highly analogous to those describing the combustion process [3].

As was already mentioned, correlations (7) are the boundary conditions for the system of eq. (4) on the surface of the substance's phase division. $p = 0$ (vacuum) at large distance from the surface $x \to \infty$ is another boundary condition. One should also add the state equation of the substance $\varepsilon = \varepsilon(p, \varrho)$ and the transfer equation for the laser radiation $\boldsymbol{q}(\boldsymbol{x})$. The system of eqs. (4) describes now the behavior of the evaporated mass.

To simplify the further analysis we should also suppose the form of the laser pulse to be rectangular, so that

$$q_0 = \begin{cases} 0 & t<0\,, \quad t>\tau\,, \\ q_0 & 0<t<\tau\,. \end{cases} \tag{8}$$

2. – An evaporation process.

In the range of flux values $q_0 > q_c$ (but not very large ones), the temperature of the evaporated substance will be small in comparison with the temperature at which significant excitation and ionization of atoms begin. Thus in this range of fluxes, evaporated substance represents a gas transparent for a falling radiation. Correspondingly the term $\nabla \boldsymbol{q}$ in the system of eq. (4) will equal zero, and the gas-dynamical movement of evaporated substance will be adiabatic. An adiabatic movement is somehow an analogue to the inertial mechanical movement which is completely defined by initial conditions. In the considered case it means that gas dynamics will be determined by physical conditions on the evaporation surface, *i.e.* by the solution of ratios (7) consistently with the state equation relating to $p_1, v_1, \varrho_1, \varepsilon_1$. However, ratios (7) contain two more unknown magnitudes p_0 and D to find which one should add another two equations. One of these equations strictly follows from the movement properties of the evaporated substance, and expresses the fact that expansion proceeds in vacuum. For stationary movement (*i.e.* independent of time parameters in the evaporation wave front) this equation should indicate equality of propagation velocities of the evaporation wave and a perturbations in a gas (Chapman-Jouguet condition).

$$v_1 - c_1 = D\,, \tag{9}$$

where $c_1 = (\gamma p_1/\varrho_1)$ is the sound velocity in a gas near the surface, γ the ratio of specific-heat capacities.

The latter equation should take into account the character of the evaporation process on the surface of phase division. In fact, this equation should imply the equation of the kinetics of phase transition, as the velocity of evaporation is quite large in the considered case. The characteristic time for establishing equilibrium in condensed phase has a value $\omega_D^{-1}\exp[-u/kT]$ where $\omega_D \approx 10^{13}\,\text{s}^{-1}$, Debye frequency, u is a magnitude close to the bounding energy of atoms in condensed phase. A quasi-equilibrium condition of the evaporation process lies in the fact that the evaporation time is small in comparison with the lifetime of a surface atom which equals $b/|D|$ where b is the interactomic distance. As $b\omega_D \approx c_0$ this condition may be expressed in the form

$$|D| < c_0 \exp[-u/kT] \approx (10^4 \div 10^5)\ \text{cm/s}\,, \tag{10}$$

i.e. $q_0 < c_0 \varrho_0 U \exp[-u/k]T \approx 10^9\ \text{W/cm}^2$. In fluxes of more than $10^9\ \text{W/cm}^2$ where condition (10) is not fulfilled phase transition is absent as in this region the temperature of the substance exceeds the critical one. Thus, a phase transition may be considered, in first approximation to be an equilibrium in the whole range of fluxes where temperature does not reach the critical value.

To connect gas-dynamical parameters with the equation of the phase equilibrium curve one should investigate the nature of vapour formation in the heated substance layer. Two mechanisms of vapour formation are possible in the considered case: bulk vaporization (boiling and expansion of the heated layer) and surface vaporization. Bulk vapour formation demands overheating of the condensed substance, to provide the existence of the noticeable numbers of vaporization centers (bubbles) with additional energy equal to the energy of the surface tension $4\pi r^2 \alpha_0$, where r is the bubble's radius, α_0 the coefficient of the surface tension. At a certain degree of overheating with increasing radius the total energy for bubbles formation increases due to the increase of its surface, reaching a maximum at some $r = r_1$, and then reduces on account of the negative volume term. (FOLMER [4], FRENKEL [5]). The maximal value of the bubbles' energy is equal to $4\pi\alpha_0/3kT$, and, consequently, their concentration N_b due to the Boltzman distribution is described by formula

$$N_b \approx N_0 \exp[-4\pi\alpha_0 r_1^2/3kT]\,, \tag{10}$$

where N_0 is the atomic density in condensed phase. Volume vaporization will obviously take place if the total surface of bubbles in the heated substance layer equals the free surface of condensed phase, *i.e.*

$$4\pi r_1^0 N_0 \frac{a}{|D|} \exp[-4\pi\alpha_0 r/3kT] = 1\,, \tag{11}$$

where $a|D|$ is the thickness of the heated layer (3).

The maximum of the expression on the left in (11) is realized at $r_1^* = (3kT/4\pi\alpha_0)^{\frac{1}{2}}$, and inserting the dimensionless variable $y = (r_1/r_1^*)^2$ we obtain instead of (11)

$$N_0 \frac{a}{|D|} \frac{3kT}{\alpha_0} y \exp[-y] = 1\,. \tag{12}$$

Numerical estimates at $a/|D| = 10^{-3}$cm and $N_0 = 10^{23}\ \text{cm}^{-3}$ show that $y \approx 13$, and the value r_1, hence, comprises several interatomic distances. It leads to the fact that vapour pressure in bubbles increases due to the surface curvature on the larger magnitude $2\alpha_0/r_1$ in comparison with pressure in condensed media. This pressure should be compensated by means of increase of the medium's temperature (overheating) to make the pressure of saturated vapours exceed the value $p_0 + 2\alpha_0/r_1$. The values given for r_1 correspond to the degree of media overheating up to temperatures of the order of 0.3 u/k, *i.e.* when the temperature of the condensed phase is close or formally even exceeds the critical one. Thus, the case of volume vaporization in the considered conditions is not realized because surface evaporation provides a rather large rate of vaporization without significant overheating of media. (AFANASIEV, KROKHIN [6]).

Hence, the equation describing the evaporation process on the boundary of the phase division in the frame of the equilibrium theory expresses the equality of the condensed substance to the pressure of saturated vapours p_s, *i.e.*

$$p_s = p_1 + \varrho_0 |D| v_1 \, . \tag{13}$$

Let us write down solutions of the gas-dynamical equation (4) in the one-dimensional plane case. The system of equations (4) at $\nabla \boldsymbol{q} = 0$ with boundary conditions (7) permits solutions in the form of simple rarefaction waves (see LANDAU and LIFSHITZ [7]) which is a self-similar solution in relation to the variable x/t. We introduce for convenience a self-similar variable in the form

$$\lambda = U^{-\frac{1}{2}} x/t \, . \tag{14}$$

Correspondingly, the gas-dynamical functions will have the form:

$$v = U^{\frac{1}{2}} V(\lambda) \, , \qquad \varrho = q_1 U^{-\frac{3}{2}} R(\lambda) \, , \qquad p = q_1 U^{-\frac{1}{2}} P(\lambda), \tag{15}$$

where $V(\lambda)$, $R(\lambda)$, $P(\lambda)$ are functions of only one variable λ. After inserting (15) in (4) the system of equations will have the form

$$\begin{cases} \dfrac{\mathrm{d}}{\mathrm{d}\lambda}(RV) - \lambda \dfrac{\mathrm{d}R}{\mathrm{d}\lambda} = 0 \, , \qquad (V-\lambda)\dfrac{\mathrm{d}V}{\mathrm{d}\lambda} + \dfrac{1}{R}\dfrac{\mathrm{d}R}{\mathrm{d}\lambda} = 0 \, , \\ (V-\lambda)\dfrac{\mathrm{d}}{\mathrm{d}\lambda}\left(\dfrac{P}{R}\right) + (\gamma - 1)\dfrac{P}{R}\dfrac{\mathrm{d}V}{\mathrm{d}\lambda} = 0 \, , \end{cases} \tag{16}$$

where the specific energy ε is excluded by the state equation $\varepsilon = p/(\gamma-1)\varrho + U$. Solutions of these equations satisfying the condition of equality to zero of pressure and density on the boundary with the vacuum have the form:

$$\begin{cases} R(\lambda) = R(\lambda_1)[1 - (\lambda - \lambda_1)/(\lambda_2 - \lambda_1)]^{2/(\gamma-1)} \, , \\ V(\lambda) = [2\lambda + (\gamma - 1)\lambda_2]/(\gamma + 1) \, , \\ P(\lambda) = R(\lambda_1)\dfrac{(\gamma-1)^2(\lambda_2-\lambda_1)^2}{\gamma(\gamma+1)^2}[1 - (\lambda-\lambda_1)/(\lambda_2-\lambda_1)]^{2\gamma/(\gamma-1)} \, , \end{cases} \tag{17}$$

where λ_1 and λ_2 correspond to the boundaries of evaporated substance with the surface of condensed substance and vacuum. The velocity of the evaporation wave D and the velocity of the boundary of the evaporated substance with vacuum v_v are equal corresponding (Fig. 3) to:

$$D = \lambda_1 U^{\frac{1}{2}}, \qquad v_v = \lambda_2 U^{\frac{1}{2}}. \tag{18}$$

Solutions (17) also satisfy eq. (9) at $\lambda = \lambda_1$. The algebraic system of expressions (7) and the equation of the phase equilibrium curve (13) allow us to determine the values $R(\lambda_1)$, λ_1, λ_2 in (17). For this purpose it is necessary to apply the self-similar functions by formulae (15) in these expression. Then the first of equations (7) will have the form

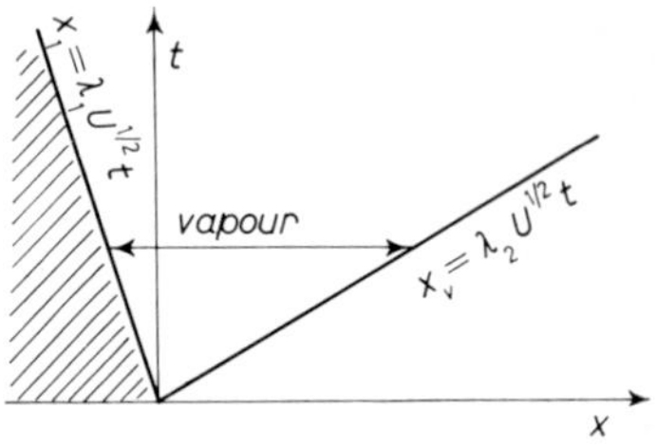

Fig. 3. – (x-t) diagram of the steady adiabatic flow of the evaporated substance. x_1, the position of the evaporation wave front, x_v, boundary of the evaporated substance with vacuum.

$$R(\lambda_1) = (k+1)\lambda_1/\eta(\gamma-1)(\lambda_2-\lambda_1), \tag{19}$$

where $\eta = q_1/\varrho_0 U^{\frac{3}{2}}$, $q_1 = q_0(1-\phi)$, ϕ is the reflection coefficient.

The equation of the phase equilibrium curve relating to the saturated density ϱ_s may be taken in the form

$$\varrho_s = A \exp[-U\varrho_s/p_s], \tag{20}$$

or

$$p_s = \frac{p_s}{\varrho_s} A \exp[-U\varrho_s/p_s]. \tag{20a}$$

At temperature equality of both phases of a substance, $p_s/\varrho_s = p_1/\varrho_1$ is valid. Taking this into account and combining the second of eqs. (7) with (13), (15), (17) and (19) we obtain

$$\frac{(\gamma+1)^2\lambda_1[\lambda_1+(\gamma-1)\lambda_2]}{(\gamma-1)^2(\lambda_1-\lambda_1)^2} = \alpha_1 \exp\left[\frac{\gamma(\gamma+1)^2}{(\gamma-1)^2(\lambda_2-\lambda_1)^2}\right], \tag{21}$$

where $\alpha_1 = A/\varrho_0$. If to include in the state equation the «potential» energy going for the break of bonds in the condensed substance U, *i.e.* to adopt $\varepsilon = p/(\gamma-1)\varrho + U$, taking into account (15) and (17), the last of expressions (7) gives

$$\lambda_1\left[1+\frac{\lambda_1^2}{\gamma(\gamma+1)}+\frac{(\gamma-1)^2\lambda_2^2}{2(\gamma-1)}+\frac{\lambda_1\lambda_2(\gamma-1)}{\gamma(\gamma+1)}\right] = \eta. \tag{22}$$

Numerical solution of the system of eqs. (21) and (22) relating to λ_1 and λ_2 fully solves the problem. However, using the obvious condition $|\lambda_1| \ll \lambda_2$ one may obtain analytical expressions for all gas-dynamical parameters. Thus, if we neglect the terms containing λ_1^2 and λ_1^3 in eqs. (21) and (22) the solution may be expressed in the form

$$\left\{\begin{aligned} \lambda_2 &= B/[B_1 - \ln(\eta/\alpha_1)]^{\frac{1}{2}}\,, \\ \lambda_1 &= -B_2\eta\,, \end{aligned}\right. \tag{23}$$

where

$$\left\{\begin{aligned} B &= \gamma(\gamma+1)^2/(\gamma-1)^2\,, \\ B_1 &= \ln\left[\frac{(\gamma-1)^2\lambda_2^3}{2(\gamma+1)^3} + \frac{(\gamma-1)\lambda_2}{(\gamma+1)^2}\right] \approx \text{const}\,, \\ B_2^{-1} &= [1 + (\gamma-1)^2\lambda_2^2/2(\gamma+1)] \approx \text{const}\,. \end{aligned}\right. \tag{24}$$

As the numerical analysis shows, such approximation turns out to be exact enough at the changing of value of q_0 by 2 or 3 orders. Finally, for values ϱ_1, v_1, p_1, D, B_1 we obtain the following expressions

$$\left\{\begin{aligned} \varrho_1 &= \frac{B_2(\gamma+1)q_1}{B^{\frac{1}{2}}(\gamma-1)U^{\frac{3}{2}}}[B_1 - \ln(\eta/\alpha_1)]^{\frac{1}{2}}\,, \\ v_1 &= \frac{(\gamma-1)B^{\frac{1}{2}}U^{\frac{1}{2}}}{(\gamma+1)[B_1 - \ln(\eta/\alpha_1)]^{\frac{1}{2}}}\,, \\ p_1 &= \frac{B_2B^{\frac{1}{2}}(\gamma-1)q_1}{\gamma(\gamma+1)U^{\frac{1}{2}}[B_1 - \ln(\eta/\alpha_1)]^{\frac{1}{2}}}\,, \\ D &= -B_2q_1/\varrho_0U\,, \\ T_1 &= A_0U/R_0[B - \ln(\eta/\alpha_1)]\,, \end{aligned}\right. \tag{25}$$

where A_0, is the gram-atom weight, R_0, is the molar gas constant. For typical condensed media ($A \approx 10^4$ g/cm³; $U = 10^{10} - 10^{11}$ erg/g) at $\gamma = \frac{5}{3}$ the values B, B_1, B_2, are respectively equal to 26.7; 1.9; 0.75. As follows from (25), the temperature on the front of the evaporation wave B_1 is proportional to the bounding energy of the atom in the condensed media u and increases logarythmically with increasing flux density q_0. The velocity of the evaporated mass stream from the surface of the condensed phase is of the order of magnitude of the sound velocity $c_1 = (\gamma p_1/\varrho_1) \sim T_1^{\frac{1}{2}}$ (see eq. (9)).

The density ϱ_1 is less the saturated one ϱ_s

$$\varrho_1 = \varrho_s\frac{p_1}{p_s} = \varrho_s\frac{p_1}{p_0} = \varrho_s\frac{p_1}{p_1 + \varrho_0|D|v_1} \approx \varrho_s/(\gamma+1)\,. \tag{26}$$

The application of the method considered here may be restricted by two factors: 1) Appearance of radiation absorption in the evaporated substance at some density ϱ_1^*. 2) Approaching of the substance's temperature to the critical value T_c, *i.e.* approaching of the density ϱ_s to the critical density $\varrho_c \approx \varrho_0/3$, ($\varrho_1^{**} \approx \varrho_0/3(k+1)$). Substituting ϱ_1, in the first of expressions (25) for the maximum of flux density q_c one may obtain the following approximate expression:

$$q_c = \frac{(\gamma-1)\,BU^{\frac{3}{2}} \times (\varrho_1^* \text{ or } \varrho_1^{**})}{(\gamma+1)\,B_2[B_1 - \ln(\eta/\alpha_1)]} \approx 0.57\,U^{\frac{3}{2}} \times (\varrho^* \text{ or } \varrho^{**}), \tag{27}$$

where the value $T_c \approx u/10k$ is taken for the critical temperature.

As it was noted above the formulae here obtained are based on the assumption that the evaporation process is in equilibrium. But in fact, the macroscopic flow of substance from one phase into another, does not take place at phase equilibrium, *i.e.* $-\varrho_0 D = 0$. Therefore, a certain overheating of the substance is necessary to have macroscopic evaporation, *i.e.* excess of saturated vapour pressure p_s over p_0.

The equilibrium position is justified by the fact that the rate of substance evaporation sharply depends on the difference $p_s - p_0$ as is proved in Volmer's [4] and Frenke's [5] approximate kinetical theory. In this theory the evaporation rate (*i.e.* the flow of substance mass $-\varrho_0 D$) is

$$\varrho_0 |D| = \left(\frac{m_a}{2\pi k T}\right)^{\frac{1}{2}} (p_s - p_0), \tag{28}$$

where m_a is the atom mass.

This formula allows one to determine the degree of divergence of the real temperature T_0' from the calculated one, T_0, according to the equilibrium theory. From formula (25) it follows, that

$$\varrho_0 |D| = \frac{\gamma^{\frac{1}{2}}}{(\gamma+1)} \left(\frac{m_a}{kT_0}\right)^{\frac{1}{2}} p_0 = \left(\frac{m_a}{4.2\,kT_0}\right)^{\frac{1}{2}} p_0 . \tag{29}$$

By comparison of (28) and (29) we obtain

$$\Delta p_0 \approx 1.23 (T_0'/T_0)^{\frac{1}{2}} p_0 \tag{30}$$

for a pressure difference $\Delta p_0 = p_0 - p_s$ and

$$\frac{\Delta T_0}{T_0} \approx 1.3 \left(\frac{kT_0}{u} + \frac{k\,\Delta T_0}{2u}\right) \approx 1.23\,\frac{kT_0}{u} \tag{31}$$

for a temperature difference $\Delta T_0 = T_0' - T_0$. Thus, as at $T_0 < T_c$ the value of $kT_0 < 0.1u$, we have $\Delta T_0/T_0 \leqslant 10\,\%$:

An experimental measurement of the condensed phase temperature with 10% accuracy meets certain difficulties. Estimations may be carried out for example, on the basis of the metallographic study of samples after the focused laser radiation action. Another method may be based on the direct measurement of the velocity of vapour flow from the surface, because the latter equals the local sound velocity. However, a more realistic way is the registration of the shock wave created by the evaporated substance under laser radiation action in the regime of regular spikes. As the velocity of vapour flow and the velocity of shock waves are analytically connected the unknown velocity v_1 may be calculated from experimental data.

Thus, the model of quasi-equilibrium evaporation allows to study the equations of phase equilibrium curves by means of simultaneous measurements of pressure p_0 and temperature T_0. However, there also exists an equivalent method based on the measurement of pressure only and usage of the law of energy conservation, which may be expressed from (25) in the form

$$q_1 = \frac{\gamma^{\frac{1}{2}}}{(\gamma+1)}\left(\frac{m_a}{kT_0}\right)p_0\left[\frac{\gamma(\gamma+1)}{\gamma-1}\,\frac{KT_0}{m_a} + U\right], \tag{32}$$

where q_1 is the absorbed part of radiation flux. This method demands, in fact, additional measurements of the reflection coefficient ϕ. One may use expression (32) to determine critical parameters of substances, basing on a sharp reduction of the reflection coefficient. It corresponds to the fact that usual Fresnel reflection in the regime of phase transition changes for a reflection on the diffusive distribution of density and, consequently of refraction index, when evaporation takes place above the critical point [8]. It is necessary to take into account here that in some cases, when the potential of atom ionization is small, the reduction of reflection coefficient may be also connected with partial radiation absorption by the evaporated substance vapours. Therefore the analysis of experimental data may be carried out taking into account its dependence on

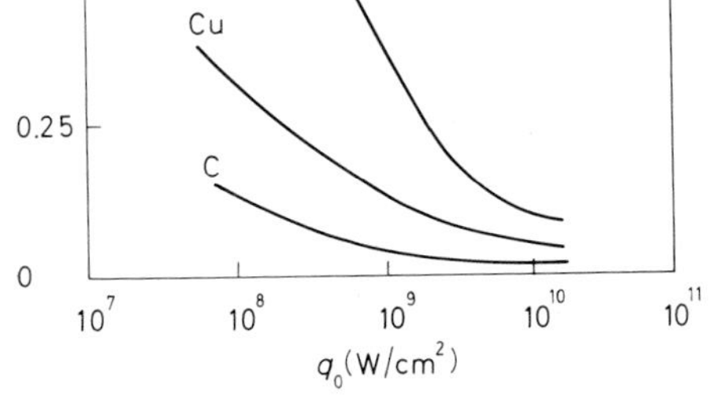

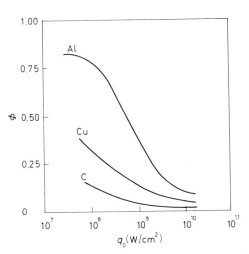

Fig. 4. – The dependence of reflectivity ϕ on laser radiation flux density.

time or on the pulse duration. In a common case 10^{-7} s pulses may be proved to be optimal when applying formula (32) to the determination of critical parameters. The dependence of reflection coefficient on the density of falling flux is illustrated in Fig. 4. Pulse duration was of $1.5\cdot 10^{-8}$ s [25].

3. – The process of plasma formation. Plane flow.

At high values of flux density, when the substance temperature is high enough to cause noticeable ionization and atom excitation the evaporated substance begins to absorb laser radiation. Absorption leads to plasma temperature increase and mainly determines the dynamics of the whole process.

As the relaxation processes in the considered case are determined by times significantly shorter than the duration of laser radiation, the coefficient of light absorption may be calculated on the basis of the equilibrium theory. According to ZEL'DOVICH and RAIZER [9] the absorbing coefficient k in plasma with average ion charge Z taking into account the reverse bremsstrahlung and photo-effects has the form:

$$K = \frac{4(2\pi)^{\frac{5}{2}}}{3\sqrt{3}} \frac{e^6 (Z+1)^2 Z N_0^2}{\hbar c \omega^3 m^{\frac{3}{2}} (kT)^{\frac{1}{2}}} \left(\exp\left[\frac{\hbar\omega}{kT}\right] - 1 \right), \tag{33}$$

where N_0 is the initial atom density, ω the radiation frequency.

At $\hbar\omega \ll kT$, the only reverse bremsstrahlung effect contributes to the absorption process, and in this case the absorption coefficient has the form:

$$K = 0.70 \frac{Z(Z+1)^2 N_0^2}{\omega^2 T^{\frac{3}{2}}} \text{cm}^{-1}. \tag{34}$$

At nanosecond laser pulses the condition for existence of a well-developed gas-dynamical plasma movement is fulfilled, because the characteristic time of plasma acceleration at its evaporation from the surface is $a/|D|c_1 \approx 10^{-10}$ s at a sound velocity $c_1 \approx 10^7$ cm/s.

Thus, the problem of hot plasma formation under the action of laser radiation on the surface of condensed substance is formally described by a system of gas-dynamical equations (4) with $\nabla \boldsymbol{q} \neq 0$ and boundary conditions (7). The equation for laser flux has the form

$$\nabla \boldsymbol{q} = -k q. \tag{35}$$

Considering the one-dimensional plane problem with a condensed substance in a semi-space $x < 0$ one may write

$$(36)\qquad \begin{cases} -\nabla \boldsymbol{q} = \partial q/\partial x = Kq\,, \\ q(x) = q_0 \exp\left[-\int_0^\infty k(x')\,dx'\right] \end{cases}$$

instead of (35).

Solution of eqs. (4) and (35) in a general case may be only fulfilled by numerical integration. However, together with a good agreement with the real conditions, by introducing some ad hoc assumption one may also obtain analytical solutions [10, 11].

A physical basis for this is a stable asymptotic behaviour of a plasma, produced by laser radiation near the surface of a dense substance, so-called « self-matched regime » (KROKHIN [12]). Here the optical plasma thickness tends to the definite value about 1 independent of time, which is the result of the dependence of the absorbing coefficient on the plasma density and temperature in expression (34). If, for example, the plasma density proves to be very small or its temperature proves to be very high, the absorbing coefficient will be small which causes reduction of the heating rate and increase of the evaporating rate. As a result, the absorbing coefficient will increase. The inverse assumption for density and temperature will lead to reduction of the absorbing coefficient. Thus, there should exist an optimal value of the optical plasma thickness under laser heating. This circumstance simplifies the obtaining of expressions for average gas-dynamical values in the functions of flux density and time.

Let us assume the following formula for the absorbing coefficient:

$$(37)\qquad k(\varrho, \varepsilon) = k_0 \varrho^\alpha \varepsilon^\beta\,,$$

where instead of temperature the specific internal energy is introduced in (34). Then, taking a value of the order of 1 for the optical thickness and assuming that geometrical thickness of the plasma layer equals the sound velocity $c = [k(\gamma - 1)\varepsilon]^{\frac{1}{2}} \approx \varepsilon^{\frac{1}{2}}$ multiplied by the time t, we obtain

$$(38)\qquad k_0 \varrho^\alpha \varepsilon^{\beta+\frac{1}{2}}\, t = 1\,.$$

The law of energy conservation (one may neglect here the bounding specific energy U as $T \approx u/k$) gives:

$$(39)\qquad M(\varepsilon + v^2/2) = q_0 t\,,$$

where M is the plasma mass, $M \approx \varrho\varepsilon^{\frac{1}{2}} t$. Combination of (38) and (39) leads to the following final expressions:

$$
(40) \qquad \begin{cases} \varepsilon \approx (k_0 t q_0^{\alpha})^{2/(3\alpha-2\beta-1)} , \\ \varrho \approx [(k_0)^3 q_0^{2\beta+1}]^{-1/(3\alpha-2\beta-1)} , \\ M \approx (k_0^{-2} q_0^{\alpha-2\beta-1} t^{3\alpha-2\beta-3})^{1/(3\alpha-2\beta-1)} . \end{cases}
$$

In the particular case of fully ionized gases

$$
\alpha = 2 , \qquad \beta = -\tfrac{3}{2} , \qquad k_0 = 2.18 \cdot 10^{29} Z^3 (Z+1)^{\frac{3}{2}} A_0^{-\frac{7}{2}} \text{ cm}^5 \text{ g}^{-\frac{7}{2}} \text{ erg}^{\frac{3}{2}} ,
$$

where A_0 is the gramm-atom weight, expressions (40) give

$$
(41) \qquad \begin{cases} \varepsilon \approx k_0^{\frac{1}{4}} q_0^{\frac{1}{2}} t^{\frac{1}{4}} , \\ \varrho \approx k_0^{-\frac{3}{8}} q_0^{\frac{1}{4}} t^{-\frac{3}{8}} , \\ M \approx k_0^{-\frac{1}{4}} q_0^{\frac{1}{2}} t^{\frac{3}{4}} . \end{cases}
$$

These expressions permit to estimate the mean plasma parameters averaged on the thickness of the plasma layer. However, the obtained results show the possibility of constructing the exact solutions of eq. (4) in a self-similar form, *i.e.* to find the profile functions of the gas-dynamical variables.

As it follows from expressions (41) the final results depend on the two-dimensional parameters k_0 and q_0 and, corresponding to the general theory of dimensionality (SEDOV [13]), from four variables k_0, q_0, x, t an only undimensional self-similar variable may be formed, reducing eq. (4) to the equations in the full derivatives.

To introduce the self-similar variable it is convenient to express the absorption coefficient through density and pressure, substituting p in (37) to ε according to the state equation $\varepsilon = p/(\gamma-1)\varrho$

$$
(42) \qquad k = k_0(\gamma-1)^{-\beta} \varrho^{\alpha-\beta} p^{\beta} \equiv k_1 \varrho^{\delta} p^{\beta} .
$$

The unidimensional variable λ may be represented then in the form

$$
(43) \qquad \begin{cases} \lambda = \xi x t^s , \\ \xi = (k_1 q_0^{\delta+\beta})^{1/(1-3\delta-\beta)} , \\ s = (3\delta+\beta)/(1-3\delta-\beta) . \end{cases}
$$

By substituting this variable in the sets of the gas-dynamical equations (4) and the transfer equations (36) we obtain the system of four equations in the full derivatives of λ. In contrast to the case of adiabatic motion, these equations determine the dynamics of the plasma behavior, *i.e.* the connection of the temperature growth (and acceleration) with the plasma mass. But these equations contain a single arbitrary parameter (plasma mass or its temperature) which determines the optical thickness of the plasma layer. To find this quantity, we need to add to the set of eqs. (4) and (36), the single boundary condition expressing the energy conservation law, *i.e.* the third of eqs. (7).

Omitting the details of the calculations, we represent here the approximate solutions of the system of self-similar gas-dynamical equations, which give profile functions describing the space-time dependence of the interested parameters. For the case of a fully ionized plasma, *i.e.* for $\alpha = 2$, $\delta = \frac{7}{2}$ $\beta = -\frac{3}{2}$, the profile functions for the velocity v, density ϱ and pressure p have the form

$$\text{(44)} \qquad \begin{cases} v \sim V(0)[1 + \sigma x/x_v]\,, \\ \varrho \sim R(0)[1 - x/x_v]^{\frac{5}{2}}\,, \\ p \sim P(0)[1 - x/x_v]^{\frac{7}{2}}\,, \end{cases}$$

where $V(0)$, $R(0)$, $P(0)$ and σ are the numerical factors, x_v, the co-ordinate of the plasma edge corresponding to some value of the self-similar variable λ_2

$$\text{(45)} \qquad x_v = \lambda_2 k_0^{\frac{1}{8}} q_1^{\frac{1}{4}} t^{\frac{9}{8}}\,.$$

The numerical values of $V(0)$, $R(0)$, $P(0)$, σ and λ_2 are equal respectively to 0.77, 0.46, 0.34, 5.25 and 3.6. The ratio of the absorbed part of laser radiation in a hot plasma to initial intensity equals

$$\text{(46)} \qquad \frac{q_0 - q_1}{q_0} = 0.25\,.$$

The maximal values of the plasma density and temperature correspond to the boundary of the phase division at $x = 0$. Finally, for the gas-dynamical plasma parameters at $x = 0$ we obtain the following expressions:

$$\text{(47)} \qquad \begin{cases} \varrho_1 = 0.46\, k^{-\frac{3}{8}} t^{-\frac{3}{8}} q_0^{\frac{1}{4}}\,, \\ v_1 = 0.77\, k_1^{\frac{1}{8}} t^{\frac{1}{8}} q_0^{\frac{1}{4}}\,, \\ p_1 = 0.34\, k_1^{-\frac{1}{8}} t^{-\frac{1}{8}} q_0^{\frac{3}{4}}\,, \\ T_1 = 0.74\, k_1^{\frac{1}{4}} t^{\frac{1}{4}} q_0^{\frac{1}{2}} A_0/R_0\,, \\ M = 0.47\, k_1^{-\frac{1}{4}} t^{\frac{3}{4}} q_0^{\frac{1}{2}}\,. \end{cases}$$

As is seen from (47) and (41) the obtained formulae differ only in the numerical multipliers.

This ratios allow to define the dependence of the main plasma properties on the parameters of the light pulse, *i.e.* on flux density q_0 and duration τ. Plasma temperature is proved to increase slowly with the increase of laser pulse duration, (as $\tau^{\frac{1}{4}}$). The plasma mass increases more rapidly (as $\tau^{\frac{3}{4}}$). This reflects the fact, that the optical plasma thickness is comparatively small, *i.e.* heating consumes a comparatively small amount of pulse energy. The degree index in expression $T \sim t^{\nu}$ exactly corresponds to the fraction of energy absorbed by the plasma $\nu = q_0 - q_1/q_0$, and in case of fully ionized plasma equals $\frac{1}{4}$. Thus, a rather good transparency of plasma layer leads to the effect of faster grow of its mass $M(q_0\tau)$. Temperature-dependence on radiation flux density q_0 is more strong and is represented by a ratio $T \sim q_0^{\frac{1}{2}}$.

Formulae (47) may be represented through the energy density E and the duration of the laser pulse, as $E = q_0\tau$. Then, for example, the expression for plasma temperature will have the form

$$T = 0.74 A_0 k_1^{\frac{1}{4}} E^{\frac{1}{2}} / \tau_0^{\frac{1}{4}} R_0 \,. \tag{48}$$

Note, that kinetical and thermal plasma energy are approximately equal fractions of its full energy $q_0\tau$. Therefore, the gas-dynamical motion itself does not lead to significant losses of thermal energy. Reduction of the optical thickness of the plasma due to its rarefaction is the main function of gas dynamics. It is this circumstance that preserves plasma from more effective heating, *i.e.* from temperature increase and reduction of plasma mass.

The analytical relations considered here are correct at two obvious conditions

$$\begin{cases} \varepsilon \gg U \,, \\ \varrho \ll \varrho_0 \,. \end{cases} \tag{49}$$

Substitution of expressions (47) in these formulae instead of ε and ϱ determines the limits of application of the theory. This gives us

$$\begin{cases} q_0 > 4U^2(k_1 t)^{-\frac{1}{2}} \,, \\ q_0 < 20\varrho_0^4(k_1 t)^{\frac{3}{2}} \,. \end{cases} \tag{50}$$

The generalization of the formulae obtained here to the case of dependence on time radiation fluxes, is of a certain interest, as the leading front of a real laser pulse has a final rise time. If we approximate a leading pulse front by a power function of the form

$$q_0 = \varphi t^{\psi} \tag{51}$$

the total system of equations allows a self-similar solution corresponding to the self-matched regime as in the case $q_0 = \text{const}$. The formal analysis of the equations is fully analogous to the case $q_0 = \text{const}$, and the final result for $\psi = 2$ can be represented in the form

$$\left\{\begin{aligned} \varrho_1 &= 0.63\, k_1^{-\frac{3}{8}}\, t^{-\frac{3}{8}}\, q_0(t)^{\frac{1}{4}}\,, \\ v_1 &= 0.73\, k_1^{\frac{1}{8}}\, t^{\frac{1}{8}}\, q_0(t)^{\frac{1}{4}}\,, \\ T_1 &= 0.55\, k_1^{\frac{1}{4}}\, t^{\frac{1}{4}}\, q_0(t)^{\frac{1}{2}}\, A_0/R_0\,, \\ M &= 0.26\, k_1^{-\frac{1}{4}}\, t^{\frac{3}{4}}\, q_0(t)^{\frac{1}{2}}\,. \end{aligned}\right. \tag{52}$$

As is seen from the obtained expressions, the consideration of the final rise time of the pulse increase leads to insignificant changes of the numerical multiplier. As it should be expected from simple physical considerations the density increases to a certain extent, while the temperature reduces in comparison with the case $q_0 = \text{const}$.

4. – Plasma heating by focused laser radiation.

At pulse durations longer than 10^{-9} s, strong focusing of laser the radiation influences the dynamics of movement and plasma heating. In this case interaction of laser radiation with plasma corresponds to the one-dimensional plane flow only in the time interval $t \ll r_0/c$, where r_0 is the radius of a focal spot and c the sound velocity equaling as to order of magnitude the velocity of the moving plasma. A real picture of the mouvement of a plasma when a laser is focused on a surface of condensed matter is two-dimensional and symmetric about the axis, which is perpendicular to the matter's surface and passes through the center of the focusing spot.

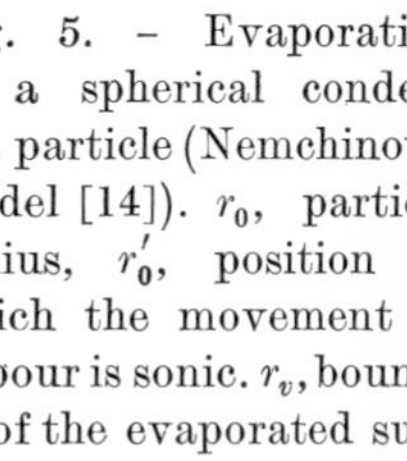

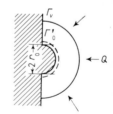

Fig. 5. – Evaporation of a spherical condensed particle (Nemchinov's model [14]). r_0, particle radius, r_0', position in which the movement of vapour is sonic. r_v, boundary of the evaporated substance against vacuum.

Investigation of the two-dimensional problem presents some mathematical difficulties. In connection with this a one-dimensional spherical model has been proposed and investigated by NEMCHINOV [14]. Physically this model corresponds quite well to the nature of movement at the focusing of radiation on the surface. It is assumed in this model that laser radiation of power Q with uniform spherical symmetrical distribution falls on the surface of the condensed substance of spherical form with radius r_0 (Fig. 5).

It is obvious that at radiation focusing, the side plasma expansion leads to quasi-spherical movement, and the plasma density is reduced faster than in the one-dimensional plane case. Therefore, an external plasma region will be transparent for impinging radiation, and noticeable absorption will occur only in the layer adjoining the surface where the movement has still a plane character. The thickness of this layer is determined by geometrical parameters only, and it should have dimensions of the order of the radius of a sphere of condensed matter r_0, where the movement still conserves its planarity.

As in the case of the plane one-dimensional problem here also should exist a self-matched regime of evaporation and heating of matter, *i.e.* constancy of absorption (optical thickness) with relation now to the layer of fixed thickness r_0. This indicates the fact that gas-dynamical parameters must have steady-state values which are the physical maximum of their values in the plane one-dimensional problem.

According to NEMCHINOV [14] the analysis of the system of gas-dynamical equation (4) in the steady-state case leads to a condition to be imposed on the value of the absorbtion coefficient on the evaporation front (to be more exact on the external boundary of the gas-dynamical break, where plasma velocity and sound velocity are equal (see (9)). This condition has the form

$$r_0' k(\varrho, \varepsilon) = 4/(\gamma + 1 + 2U/\varepsilon)\,, \tag{53}$$

where r_0' is the co-ordinate of the external boundary of the evaporation (in the considered case r_0' slightly exceeds r_0, $r_0' = 1.2 r_0$, [14]).

In our case $\varepsilon \gg U$, we have

$$r_0 k(\varrho, \varepsilon) \approx 1\,. \tag{54}$$

This expression is analogous to (38) for a plane one-dimensional movement, however, remember that in this case $r_0 = \text{const}$.

Let us consider that expression (54) may be applied in the case of radiation focusing with total power Q in the spot of radius r_0 on the substance's surface. The energy flux through the cross-section on the distance r_0 from the surface (*i.e.*, there where absorption approximately disappears) equals

$$\frac{\mathrm{d}M}{\mathrm{d}t}\left(\varepsilon + \frac{v^2}{2} + \frac{p}{\varrho}\right) = Q\,, \tag{55}$$

where $Q = \pi r_0^2 q_0$—full radiation flux, $\mathrm{d}M/\mathrm{d}t$—mass stream (rate of condensed substance evaporation):

$$\frac{\mathrm{d}M}{\mathrm{d}t} = \pi r_0^2 \varrho v\,. \tag{56}$$

Assuming the plasma velocity in this cross-section to be close to the sound velocity $v \approx c = [\gamma(\gamma-1)\varepsilon]^{\frac{1}{2}} \approx \varepsilon^{\frac{1}{2}}$ we obtain from (55) and (56)

$$\varrho \approx Q/2\pi r_0^2 \varepsilon^{\frac{3}{2}} . \tag{57}$$

Taking expression (54) and assuming, as before, $k = k_0 \varrho^\alpha \varepsilon^\beta$ for gas-dynamical parameters of interest we obtain

$$\left\{ \begin{aligned} \varepsilon &\approx [(2\pi)^\alpha k_0^{-1} r_0^{2\alpha-1} Q^{-\alpha}]^{2/(2\beta-3\alpha)} , \\ \varrho &\approx \frac{Q}{2\pi r_0^2} [(2\pi)^\alpha k_0^{-1} r_0^{2\alpha-1} Q^{-\alpha}]^{-3/(2\beta-3\alpha)} , \\ M &\approx \tfrac{1}{2} t Q [(2\pi)^\alpha k_0^{-1} r_0^{2\alpha-1} Q^{-\alpha}]^{-2/(2\beta-3\alpha)} . \end{aligned} \right. \tag{58}$$

For a fully ionized gas, *i.e.* when $\alpha = 2$, $\beta = -\frac{3}{2}$, expressions (58) transform into the following ones:

$$\left\{ \begin{aligned} \varepsilon &\approx 0.45 k_0^{\frac{2}{9}} r_0^{-\frac{2}{3}} Q^{\frac{4}{9}} , \\ \varrho &\approx 0.53 k_0^{-\frac{1}{3}} r_0^{-1} Q^{\frac{1}{3}} , \\ M &\approx 1.1 \ k_0^{-\frac{2}{9}} r_0^{\frac{2}{3}} Q^{\frac{5}{9}} . \end{aligned} \right. \tag{59}$$

As is well seen from (59), internal plasma energy (*i.e.* temperature) for the fixed value r_0 increases proportionally to $Q^{\frac{4}{9}} \sim q_0^{\frac{4}{9}}$, *i.e.* in a fashion very similar to the case of a plane flow, where $\varepsilon \sim q_0^{\frac{1}{2}}$. However, the value ε also significantly depends on the condition of radiation focusing and it increases with a spot radius decreasing as $r_0^{-\frac{2}{3}}$. Expressions (59) are, as was noted above, the limiting values of gas-dynamical parameters, which are independent of time at $t \ll r_0/c \approx r_0/\varepsilon^{\frac{1}{2}}$. The characteristic time interval t' at which the plane gas-dynamical flow transforms into a two-dimensional steady-state flow, may be defined by equaling the values from (41) and (59)

$$t' \approx k_0^{-\frac{1}{9}} r_0^{\frac{8}{9}} q_0^{-\frac{2}{3}} . \tag{60}$$

Numerical estimations for the case of a fully ionized gas at

$$Q = 10^9 \text{ w} , \quad r_0 = 2\cdot 10^{-2} \text{ cm} , \quad q_0 = 10^{12} \text{ w/cm}^2$$

give $t \approx 1\cdot 10^{-9}$ s.

Let us rewrite formulae (59) taking 10^9 W as a unit of power and 10^{-2} cm as a unit of length. For hydrogen ($Z = A_0 = 1$) we have

$$
(61)\qquad
\left|\begin{aligned}
\varepsilon_{\mathrm{H}} &\approx 5.3\cdot 10^{14}\, r_f^{-\frac{2}{3}} Q_f^{\frac{4}{9}}\ \mathrm{erg/g}\,,\\
T_{\mathrm{H}} &\approx 2.1\cdot 10^{3}\ r_f^{-\frac{2}{3}} Q_f^{\frac{4}{9}}\ {}^{\circ}\mathrm{K}\,,\\
\varrho_{\mathrm{H}} &\approx 1.3\cdot 10^{-3} r_f^{-1} Q_f^{\frac{1}{3}}\ \mathrm{g/cm^3}\,,\\
N_{\mathrm{H}} &\approx \ \ 8\cdot 10^{20}\ r_f^{-1} Q_f^{\frac{1}{3}}\ \mathrm{cm^{-3}}\,,\\
M_{\mathrm{H}} &\approx \ \ 9\cdot t\, r_f^{\frac{2}{3}} Q_f^{\frac{5}{9}}\ g,
\end{aligned}\right.
$$

where $r_f = r_0\,[\mathrm{cm}]\cdot 10^2$, $Q_f = Q\,[\mathrm{W}]\cdot 10^{-9}$, N_{H}, concentration of hydrogen ions in the plasma.

Neglecting the energy for ionization (which is right for light atoms and large radiation fluxes) the dependence of gas-dynamical parameters on the mass number of an element A_0 and its charge Z may be represented in the form

$$
(62)\qquad
\left|\begin{aligned}
\varepsilon &= \varepsilon_{\mathrm{H}} Z^{\frac{2}{3}} \left(\frac{Z+1}{2}\right)^{\frac{1}{3}} A_0^{-\frac{7}{9}}\,,\\
T &= 2^{-\frac{1}{3}} T_{\mathrm{H}} Z^{\frac{2}{3}}\, (Z+1)^{-\frac{2}{3}}\, A_0^{\frac{2}{9}}\,,\\
\varrho &= \varrho_{\mathrm{H}} Z^{-1} \left(\frac{2}{Z+1}\right)^{\frac{1}{2}} A_0^{\frac{7}{6}}\,,\\
N &= N_{\mathrm{H}} Z^{-1} \left(\frac{2}{Z+1}\right)^{\frac{1}{2}} A_0^{\frac{1}{6}}\,,\\
M &= M_{\mathrm{H}} Z^{-\frac{2}{3}} \left(\frac{2}{Z+1}\right)^{\frac{1}{3}} A_0^{\frac{7}{9}}\,.
\end{aligned}\right.
$$

From these formulae follows that temperature and density of ions enhance with increase of the mass number A_0, which is connected with the larger inertia of the expanding plasma (reduction of flow velocity). The full quantity of ions is reduced by this. Increasing the ion charge, the ion temperature initially increases, which is the consequence of the numerical multiplier increasing in the absorption coefficient, and then at $Z \gg 1$ it reaches the constant value independent of Z. By this, ion density reduces as $Z^{-\frac{3}{2}}$, while the absorption coefficient (34) maintains the constant value, independent of Z.

Thus, for deuterium (D), tritium (T), helium (He), and lithium (^{6}Li) we will correspondingly have

$$(63)\qquad \begin{cases} T_{\rm d} \;\;= 1.17\,T_{\rm H}\,, & N_{\rm d} \;\;= 1.12\,N_{\rm H}\,, \\ T_{\rm t} \;\;= 1.28\,T_{\rm H}\,, & N_{\rm t} \;\;= 1.2\;\;N_{\rm H}\,, \\ T_{\rm He} = 1.68\,T_{\rm H}\,, & N_{\rm He} = 0.51\,N_{\rm H}\,, \\ T_{\rm Li} \;= 1.9\;\;T_{\rm H}\,, & N_{\rm Li} \;= 0.32\,N_{\rm H}\,. \end{cases}$$

Note in conclusion, that the analysis here considered is correct at the obvious condition $\varrho < \varrho_0$ (for internal energy there is no condition because in the stady-state equation (53), evaporated energy U may be added to ε). Using the second expression of (59) it gives

$$(64)\qquad \begin{cases} Q < 8\varrho_0^3 k_0 r_0^3\,, \\ q_0 < 3\varrho_0^3 k_0 r_0\,. \end{cases}$$

The theory on plasma heating by focused laser radiation presented here may be also applied to the problem of heating a small particle. If the particle's diameter is large comparing with the radiation free path (*i.e.* a particle is untrasparent in the initial state), the values of gas-dynamical plasma parameters may be evaluated following eqs. (59), where the value of the full radiation flux Q should be substituted for a flux on the particle surface $Q' = Qd^2/r_0^2$, where d is the particle radius. The time of particle evaporation may be evaluated from the third of equations (59) where M is the particle mass. At small particle-dimension eqs. (59) give the upper value for plasma temperature. Heating particles the symmetrization of plasma parameters will occur after the substance is completely evaporated. This circumstance allows to develop an approximate theory of spherically symmetrical plasma, the asymptotic parameters of which will depend on the total absorbed energy.

Spherically symmetrical plasma gas dynamics under laser heating with application of parameters averaged in volume has been considered by BASOV and KROKHIN [15] and DAWSON [16]. A more detailed analysis has been made in works of BASOV *et al.* [17] and HAUGHT and POLK [18]. Investigation was carried out there on the basis of self-similar solutions of gas-dynamical equations for a spherically symmetrical case.

5. – Plasma heating by ultrashort laser pulses.

In the previous Section we considered dynamics of plasma heating by laser radiation in the case when pulse duration exceeds the caracteristic time of gas-dynamical evolution. Therefore plasma properties and their dependence on pulse parameters were mainly determined only by gas-dynamics of the whole process. At sufficiently high temperature a factor of electron thermoconductivity in plasma appears which should be taken into account in the energy equation of the system (4). But electron thermoconductivity leads to redistribution of temperature, density and plasma velocity without changing their average values, which, as is seen from comparison of (41) and (47), do not significantly differ from these values on the boundary with condensed substance.

This is explained by the fact that the velocity of heat propagation v_t reduces with time according to the law $t^{-\frac{1}{2}}$, and therefore, gas-dynamical perturbations propagating with sound velocity c, will reach at some moment of time, the heating wave. This case corresponds to the problems considered above (see expression (1)).

However, at strong shortening of pulse duration to the range of picosecond pulses delivered by mode-locked lasers, the thermal period of substance heating will prevail even at large radiation power.

The coefficient of electron thermoconductivity a_e equals (SPITZER [19])

$$a_e = \frac{40\sqrt{2}\,k(kT)^{\frac{5}{2}}\mu}{\pi^{\frac{3}{2}}\,m^{\frac{1}{2}}\,e^4 ZL}\,, \tag{65}$$

where L is the Coulomb logarithm (here $L \approx 10$). Factor μ considers corrections connected with plasma quasi-neutrality and electron-electron collisions, and equals ≈ 0.2 for small Z. Thus, for the value a_e we may assume

$$\begin{cases} a_e = \chi T^{\frac{5}{2}}\,, \\ \chi = 1.8\cdot 10^{-6}\ \text{erg cm}^{-1}\,\text{s}^{-1}\,{}^\circ\text{K}^{-\frac{7}{2}}\,. \end{cases} \tag{66}$$

The penetration depth of the heating wave in the substance is of the order of $l_t \approx (\chi T^{\frac{5}{2}} t/\varrho_0 C)^{\frac{1}{2}}$ where C is the specific thermocapacity. Equaling this expression to the distance of gas-dynamical perturbation propagation during the time t, $l_g \approx (CT)^{\frac{1}{2}} t$, we obtain the following expression for the time of the substitution of the heating regime by a gas-dynamical one:

$$t_1 = \frac{\chi T^{\frac{3}{2}}}{\varrho_0 C^2}\,. \tag{67}$$

Using the law of energy conservation $l_t \varrho_0 CT \approx q_0 t$ and excluding temperature T we obtain

$$\begin{cases} t_1 \approx (q_0 t)^{\frac{3}{2}} \chi^{\frac{1}{2}}/\varrho_0 C^{\frac{7}{4}} & \text{at } t_1 > \tau\,, \\ t_1 \approx q_0 \chi/\varrho_0^2 C^{\frac{7}{2}} & \text{at } t_1 < \tau\,. \end{cases} \tag{68}$$

At $t < t_1$, the heating regime will prevail, at $t > t_1$, the gas-dynamical one.

As it was mentioned by CARUSO and GRATTON [20] (see (67)) time t_1, in order of magnitude, equals the energy exchange between electrons and ions equipartition time). In fact, the coefficient of electron thermoconductivity a_e in order of magnitude equals $a_e \approx l_e v_e \varrho_0 C$, where l_e and v_e correspond to the mean free path and velocity of the electrons. Inserting the duration of electron-ion collisions (momentum relaxation time) $\tau_e = l_e/v_e$ and substituting v_e for the ion velocity v_i due to the obvious relation $v_e = v_i(M_i/m)^{\frac{1}{2}}$ after simple calculations we have

$$\frac{M_i}{m}\tau_e \approx \chi T^{\frac{3}{2}}/\varrho_0 C^2 \,. \tag{69}$$

The value on the left of this relation indicates the time of energy transition from electrons to ions with the accuracy of the numerical multiplier. Thus, the thermal stage of plasma heating by laser radiation takes place during the time of thermalization of electrons and ions. Therefore, within the time $t < t_1$, equilibrium state of electrons and ions may be strongly violated, and the ion temperature may occur to be significantly lower than the electron temperature. Therefore in further descriptions of heat propagation, the temperature T will represent the electron temperature.

The qualitative analysis of the problem in question based on the process of hot-electron diffusion from the region of laser radiation absorption is carried out in works by BASOV *et al.* [21], CARUSO and GRATTON [20] and FABRE [22].

We assume that dense substance fills a semi-space, laser radiation falls from the positive values of axis x being absorbed in the negligibly thin layer. The equation of thermodiffusivity in the one-dimensional plane case has the form

$$\varrho_0 C\frac{\partial T}{\partial t} = \frac{\partial}{\partial x}\chi T^{\frac{5}{2}}\frac{\partial T}{\partial x} + \frac{\partial q}{\partial x}\,. \tag{70}$$

As the solution of eq. (70) let us take an expression of the form (see ZEL'DOVIĆ, RAIZER [9])

$$T(x, t) = T_0(t)[1 - x/x_1(t)]^{\frac{2}{5}}\,, \tag{71}$$

which describes wave propagation inside the substance with velocity $v_t = \mathrm{d}x_1/\mathrm{d}t$. Strictly speaking, expression (71) rather exactly describes the temperature distribution in the region of the wave front. It also well reflects the temperature behaviour of heat propagation in a medium with a nonlinear thermodiffusivity coefficient which sharply increases with temperature increase. In this case the more heated regions have a higher coefficient of thermoconductivity, and consequently, heat flows out of these regions faster, smoothing out the temperature profile.

Using the solution of (71) one may obtain equations for two functions $T_0(t)$, and $x_1(t)$. The first equation describes the energy conservation law and is the result of integrating the thermodiffusion equation by the co-ordinate x

$$\frac{5}{7}\varrho_0 C \frac{\mathrm{d}}{\mathrm{d}t}(T_0 x_1) = q_0 . \tag{72}$$

The second equation may be obtained by substituting (71) in (70) and eliminating $x \to x_1$. After simple calculations, the equation will have the form

$$x_1 \frac{\mathrm{d}x_1}{\mathrm{d}t} = \frac{2}{5}\frac{\chi}{\varrho_0 C} T^{\frac{5}{2}} . \tag{73}$$

Solution of (72) and (73) is simple at any value of $q_0(t)$. Integrating the first equation with regard to the initial conditions $x_1(0) = T_0(0) = 0$, and then substituting the value of T in eq. (77), we obtain an equation for definition of $x_1(t)$

$$x_1^{\frac{7}{2}} \frac{\mathrm{d}x_1}{\mathrm{d}t} = \frac{2}{5}\left(\frac{7}{5}\right)^{\frac{5}{2}} \frac{\chi}{(\varrho_0 C)^{\frac{7}{2}}} \left\{\int_0^t q_0(t')\,\mathrm{d}t'\right\} . \tag{74}$$

Integration of this equation leads to the final result

$$x_1 = \left(\frac{9}{5}\right)^{\frac{7}{9}} \left(\frac{7}{5}\right)^{\frac{5}{9}} \frac{\chi^{\frac{2}{9}}}{(\varrho_0 C)^{\frac{7}{9}}} \left\{\int_0^t \mathrm{d}t' \left[\int_0^{t'} q_0(t'')\,\mathrm{d}t''\right]^{\frac{5}{2}}\right\}^{\frac{2}{9}} . \tag{75}$$

Let us consider, as an example, an infinitely short pulse ($\tau \to 0$) then

$$x_1 = \left(\frac{9}{5}\right)^{\frac{2}{9}} \left(\frac{7}{5}\right)^{\frac{5}{9}} \frac{\chi^{\frac{2}{9}}}{(\varrho_0 C)^{\frac{7}{9}}} t^{\frac{2}{9}} E^{\frac{5}{9}} , \tag{76}$$

where

$$E = \int_0^\tau q_0(t)\,\mathrm{d}t .$$

The temperature is determined directly by means of expression (72)

$$T_0 = \left(\frac{7}{5}\right)^{\frac{4}{9}} \left(\frac{5}{9}\right)^{\frac{2}{9}} (\varrho_0 C \chi)^{-\frac{2}{9}} t^{-\frac{2}{9}} E^{\frac{4}{9}} . \tag{77}$$

The time of development of the gas-dynamical movement is determined by the first of expressions (68). As is seen from expressions (76) and (77), the propagation of the heating wave front is slowed down with time $v_t = \mathrm{d}x_1/\mathrm{d}t \sim t^{-\frac{7}{9}}$. Temperature drops comparatively slow with time $T_0 \sim t^{\frac{2}{9}}$ and enhances with increase of laser pulse energy as $E^{\frac{4}{9}}$. This latter dependence is rather close to the results obtained for nanosecond pulses with regard to gas dynamics. The comparatively slow fall of temperature up to the moment t_1, when the movement envelops the whole region of hot plasma, was a basis for CARUSO and GRATTON [20] to assume that t_1 expresses the lifetime of hot plasma. As, however, this time approximately equals the time of energy exchange between electrons and ions, then natural difficulties arise in connection with obtaining high values of ion temperature.

Let us further consider a radiation pulse of rectangular form: $q_0 = 0$ at $t < 0,\ t > \tau$ and $q_0 = \text{const}$ at $0 < t < \tau$. Having performed the integrations in (75) we obtain

$$x_1 = \left(\frac{9}{5}\right)^{\frac{4}{9}} \left(\frac{7}{5}\right)^{\frac{2}{9}} \frac{\chi^{\frac{2}{9}}}{(\varrho_0 C)^{\frac{7}{9}}} \begin{cases} \left(\dfrac{2}{7}\right)^{\frac{2}{9}} q_0^{\frac{5}{9}} t^{\frac{7}{9}}\,, & t < \tau\,, \\ q_0^{\frac{5}{9}} \tau^{\frac{5}{9}} \left(t - \dfrac{5}{7}\tau\right)^{-\frac{2}{9}}\,, & t > \tau\,. \end{cases} \tag{78}$$

Analogously, for temperature T_0 we shall have

$$T_0 = \left(\frac{7}{5}\right)^{\frac{4}{9}} \left(\frac{5}{9}\right)^{\frac{2}{9}} (\chi q \varrho_0 C)^{-\frac{2}{9}} \begin{cases} \left(\dfrac{7}{2}\right)^{\frac{2}{9}} q_0^{\frac{4}{9}} t^{\frac{2}{9}}\,, & t < \tau\,, \\ q_0^{\frac{4}{9}} \tau^{\frac{4}{9}} \left(t - \dfrac{5}{7}\tau\right)^{-\frac{2}{9}}\,, & t > \tau\,. \end{cases} \tag{79}$$

From these expressions it directly follows that at constant flux q_0, plasma temperature still increases in time though rather slowly. It is interesting to note that in gas-dynamical regime at constant flux radiation, dependence of temperature and value of the heated mass of substance (in our case $M \approx \varrho_0 x_1(t)$) on flux density and time practically fully coincides with results given in formulae (78) and (79). This circumstance allows one to hope that the dependence pointed out remains the same at a more complicated transitional stage as well, when one should simultaneously consider thermoconductivity and gas dynamics.

As was noted above, formulae (77) and (79) may be applied for $t < t_1$. However, in these expressions one may approximately consider ϱ_0 as a parameter, assuming that at $t \approx t_1$, the value of ϱ_0 is a function of time. As it follows from the obtained relations, at reduction of substance density ϱ_0 owing to movement, the thickness of the heated layer increases almost as ϱ_0^{-1}, and temperature is practically invariable. The reason for this is that the coefficient of thermodiffusivity does not depend on plasma density, and the dependence of the considered values on density is connected only with reduction of the thermal heat capacity (per unit of volume) of the substance. In other words, a substance with the lesser value of thermal heat capacity is more «lightly» heated by the heating wave. Thus, if the radiation pulse has duration $\tau > t_1$, then no drop of temperature at $t > t_1$, may be expected, *i.e.* there is not any reason to consider the value t_1 as the lifetime of hot plasma.

As it was pointed above, t_1 equal as to order of magnitude the time of thermalization of electrons and ions, therefore condition $\tau > t_1$ will provide an effective heating of ions. This condition itself is an obvious criterion for the existence of quasi-equilibrium laser plasma, and the heating regime here considered does not bring any new points in explanation of this condition.

Note in conclusion, that the problem considered here of laser light interaction with matter is related only to the phenomena of energy concentration, when the light power is transformed into internal substance energy. Thus, the considered subject does not envelope the whole problem of laser light-matter interaction which includes also, for example such a broad field as electrodynamics of matter in strong electrical and magnetic fields. Besides, in view of future developments of the laser technique, some possibilities for the observation of fundamental physical phenomena like interactions due to vacuum polarization, might be also discussed. In this sense the observation of photon-photon scattering is unlikely to be of any possibility due to very small cross-section of the process in the optical region. The attempt to reach the process of pair production in a pure vacuum by a concentrated laser beam having a special mode structure in which $E^2 - H^2 \neq 0$, $\boldsymbol{EH} \neq 0$ is more probable. This process represents the electron tunneling through the forbidden gap $2mc^2$ and needs electrical field strength approximately one order of magnitude less than $m^2c^3/\hbar e$ to be quite observable.

A set of another possible electrodynamical processes, including the third body in the reaction (a particle or hard quantum) was considered in papers of RITUS [23] and JAKOVLEV [24].

This brief remarks show that the problem of laser light energy concentration can be of interest in the different branches of physical research. Nevertheless, they are connected with and depend on the future development in laser physics and laser technical progress.

REFERENCES

[1] J. F. READY: *Journ. Appl. Phys.*, **36**, 462 (1965).

[2] S. I. ANISIMOV, A. M. BONCH-BRUEVICH, M. A. ELJASHEVICH, YA. A. IMAS, G. S. ROMANOV and N. A. PAVLENKO: *Sov. Phys., Tech. Phys.* (Engl. transl.), **11**, 945 (1967).

[3] B. LEWIS and G. VON ELBE: *Combustion, Flames and Explosions of Gases* (New York, London, 1961).
[4] O. VOLMER: *Zeits. f. Elektrochem.*, **35**, 555 (1929).
[5] YA. I. FRENKEL: *Statisticheskaya Fizika*, ch. VII, Sect. **5** (Moscow, 1948), p. 183; ch. IX, Sect. **8** (Moscow, 1948), p. 257.
[6] YU. V. AFANASIEV and O. N. KROKHIN: *Sov. Phys. JETP* (Engl. transl.), **25**, 639 (1967).
[7] L. D. LANDAU and E. M. LIFSHITZ: *Fluid Mechanics*, Chapt. X, Sect. **92** (London, 1959).
[8] L. P. PRESNJAKOV and I. I. SOBELMAN: *Proceedings of the High School, Radiophysics*, **8**, 57 (1965).
[9] YA. B. ZEL'DOVICH and YU. P. RAIZER: *Physics of Shock Waves and High-Temperature Hydrodynamic Phenomena*, Chapt. V, Sect. **2-8** (New York, 1966).
[10] YU. V. AFANASIEV, V. M. KROL, O. N. KROKHIN and I. V. NEMCHINOV: *Appl. Math. and Mech.* (in Russian), **30**, 1022 (1966).
[11] A. CARUSO and B. BERTOTTI: *Nuovo Cimento*, **45** B, 176 (1966).
[12] O. N. KROKHIN: *Sov. Phys., Tech. Phys.* (Engl. transl.), **9**, 1024 (1965).
[13] L. I. SEDOV: *Method of the Similarity and Dimensioness in Mechanics* (Moscow, 1967).
[14] I. V. NEMCHINOV: *Appl. Math. and Mech.* (in Russian), **31**, 300 (1967).
[15] N. G. BASOV and O. N. KROKHIN: *The condition of plasma heating by the optical generator*, in *Quantum Electronics, Proceedings of the Third International Congress*, edited by P. GRIVET and N. BLOEMBERGEN, vol. **2** (Paris, New York, 1964), p. 1373.
[16] J. M. DAWSON: *Phys. Fluids*, **7**, 981 (1964).
[17] N. G. BASOV, V. A. BOIKO, YU. P. VOINOV, E. YA. KONONOV, S. L. MANDELSHTAM and G. V. SKLIZKOV: *JETP Lett.*, **6**, 291 (1967).
[18] A. F. HAUGHT and O. N. POLK: *Phys. Fluids*, **9**, 2047 (1966).
[19] L. SPITZER jr.: *Physics of Fully Ionized Gases*, ch. 5 (New York, London, 1956).
[20] A. CARUSO and R. GRATTON: *Interaction of short laser pulses with solid materials*, Report of Laboratori Gas Ionizzati LG-I 6911, Frascati, Roma (1969).
[21] N. G. BASOV, P. G. KRIUKOV, S. D. ZAKHAROV, YU. V. SENATSKY and S. V. TCHEKALIN: *IEEE Journ. Quant. Electron.*, Qe-**4**, 864 (1968).
[22] E. FABRE: *Dynamique de la formation d'un plasma par irradiation laser de cibles solides*, Report of Laboratoire de Physique des Milieux Ionisés, Ecole Polytechnique, Paris, PMI 391 R (1969).
[23] A. M. NIKISHOV and V. I. RITUS: *Sov. Phys. JETP* (Engl. transl.), **19**, 1191, 559 (1964); **20**, 622, 757 (1965).
[24] V. P. JAKOVLEV: *Sov. Phys. JETP* (Engl. transl.), **22**, 223 (1966); **24**, 411 (1967); **26**, 592 (1968).
[25] N. G. BASOV, V. A. BOIKO, O. N. KROKHIN, O. G. SEMENOV and G. V. SKLIZKOV: *Sov. Phys., Tech. Phys.* (Engl. transl.), **13**, 1581 (1969).

Interaction of Intense Photon and Electron Beams with Plasmas (*).

R. E. KIDDER

Lawrence Radiation Laboratory - Livermore, Cal.

1. – Introduction.

The giant pulse or Q-spoiled laser, first demonstrated by MCCLUNG and HELLWARTH in 1962 [1], has since been developed to the point where extremely high light intensity can be achieved by focusing its output beam with a lens. It is this property that accounts for the relevance of focused laser beams to the *physics of high energy-density*. Indeed, sufficient intensity can be achieved not only to permit the heating of plasmas to temperatures of many kilovolts, but also to bring about a variety of nonlinear intensity-dependent effects.

The plan of these lectures is first to describe some of the properties of focused laser beams such as the achievable electromagnetic-energy density, electric- and magnetic-field intensity, and effective beam temperature, to demonstrate the relevance of laser beams to the physics of high-energy density in a hopefully convincing manner. Then the *linear* theory of the interaction of light beams with plasmas together with some simple hydrodynamic considerations is employed to derive scaling relations for the production of high-temperature and high-pressure plasma, to determine the pressure due to light-induced plasma blow-off, and to consider light-wave-driven detonation waves. These more general topics are then followed by a specific consideration of the production and properties of tenth-microgram multikilovolt deuterium and mercury laser-produced plasmas, including the results of detailed computer calculations.

Next to be considered are some of the *nonlinear* effects that can occur in the interaction of very intense light beams with plasmas. They are divided into two classes: *single-particle* nonlinear effects such as the intensity-depend-

(*) The work has been performed under the auspices of the U.S. Atomic Energy Commission.

ence of the refractive index, optical-absorption coefficient, and ion-electron equipartition time of the plasma; and *collective* nonlinear effects such as the parametric excitation of plasma oscillations by a light beam, parametric amplification of light waves, and second harmonic light reflection at a plasma surface.

Finally, we consider briefly the application of high current pulsed relativistic electron beams to the achievement of high-energy density in matter. Machines have recently been built which can provide extremely powerful bursts (0.1 MA) of electrons having energies of a few MeV [2-4]. These pulses are not capable of being focused with the facility with which laser pulses can be focused, but nevertheless interestingly high beam intensities may be achievable. In any case, the penetrating power of energetic electrons through matter contrasts so markedly with that of light beams that we may expect electron pulses and photon pulses to be complementary rather than competing ways of achieving high-energy density.

2. – Properties of laser beams, plasmas, and their interaction.

2·1. *Properties of focused laser beams.* – If the output beam of a laser is focused with a lens of speed F and solid angle Ω subtended at its focus by the clear aperture of the lens, quantities which are related according to the expression

$$\Omega = \pi/(2F)^2 , \tag{2.1}$$

the intensity at the focus is given by

$$I = \Omega B , \tag{2.2}$$

where B is the brightness of the laser beam. At the focus of a fast ($F=0.9$) lens that subtends an angle Ω of one steradian, it follows that the intensity is numerically equal to the brightness B.

Another property of the beam that is related to its brightness is its *effective temperature*, which is defined to be the temperature of a black-body radiator with the same brightness as the laser beam, *i.e.*

$$\pi B = \sigma T^4_{\text{eff}} , \tag{2.3}$$

where σ $(=1.03\cdot 10^5\ \text{W/cm}^2(\text{eV})^4)$ is the Stefan-Boltzmann radiation constant.

The brightest laser beam so far reported [5] has a brightness B of $2\cdot 10^{17}\ \text{W/cm}^2\ \text{sr}$, a remarkable achievement. This brightness implies an effective beam temperature of 1.6 keV, and an achievable ($F = 0.9$) focused intensity of $2 \cdot 10^{17}$ W/cm². The electric and magnetic field intensity at the focus are given by,

$$\langle E^2 \rangle = 4\pi I/c \,, \tag{2.4}$$

$$E_{\text{r.m.s.}} = \sqrt{\langle E^2 \rangle} = 8.7 \text{ GV/cm} \,, \tag{2.5}$$

$$B_{\text{r.m.s.}} = 29 \text{ MG} \,. \tag{2.6}$$

For comparison, the electric field at a distance of one Bohr radius a_0 from the nucleus of a hydrogen atom,

$$E_{\text{atomic}} = e/a_0^2 = 5.2 \text{ GV/cm} \,, \tag{2.7}$$

is only 60 % of that at the beam focus; and the magnetic-field intensity is higher than that achieved by any other means, including high explosive field compression.

The electromagnetic energy density is given by

$$U = I/c = 67 \text{ Mb} \,. \tag{2.8}$$

Other properties of interest are the mean kinetic energy of oscillation of an electron in the electric field of the focused light beam,

$$\langle KE \rangle_{\text{osc}} = (e^2/2m\omega^2)\langle E^2 \rangle = U/2N_{\text{ec}} = 21 \text{ keV} \,, \tag{2.9}$$

where N_{ec} is the critical electron density for light of angular frequency ω, *i.e.* the value for which the electron plasma frequency ω_p is equal to ω,

$$N_{\text{ec}} = m\omega^2/4\pi e^2 \,; \tag{2.10}$$

and the drift velocity v_d, due to the Lorentz force, of an electron in the focused beam,

$$v_d/c = \langle KE \rangle_{\text{osc}}/mc^2 = 0.04 \,. \tag{2.11}$$

From these latter results we conclude that even at these extremely high focused intensities, relativistic effects will be observable but not dominant.

These properties of focused laser beams are summarized in Table I below and, we think, serve to establish the relevance of pulsed laser beams to the physics of high energy density.

TABLE I. – *Properties of focused laser beans* (F:0.9 Lens).

Brightness (B)	$2\cdot10^8$ GW/cm$^2\cdot$sr
Effective temperature (T_{eff})	1.6 keV
Electric field ($E_{\text{r.m.s.}}$)	8.7 GV/cm
Magnetic induction ($B_{\text{r.m.s.}}$)	29 MG
EM energy density (U)	67 Mb
Electron kinetic energy ($\langle KE\rangle_{\text{osc}}$)	21 keV
Electron drift velocity (v_{o})	(0.04)c cm/s

2·2. *Linear absorption and reflection of light by a plasma.* – The dispersion relation for the propagation of a light wave ($\boldsymbol{k}$, ω) in a plasma with plasma frequency ω_p is

$$(ck/\omega)^2 = 1 - (\omega_p/\omega)^2[1 + i(\nu_c/\omega)]^{-1}\,, \tag{2.12}$$

where ν_c is the electron-ion collision frequency for momentum transfer defined in terms of the mean « viscous drag » force $\langle \boldsymbol{F}_c\rangle$ acting on an electron moving through the plasma with velocity $\boldsymbol{u}$,

$$\langle \boldsymbol{F}_c\rangle = - m\nu_c\boldsymbol{u}\,. \tag{2.13}$$

The absorption coefficient $\varkappa$, refractive index n, and dielectric constant ε are obtained from the real and imaginary parts of the complex wave-number k according to,

$$n = \sqrt{\varepsilon} = (c/\omega)\,\text{Re}\,(k)\,, \qquad \varkappa = 2\,\text{Im}\,(k)\,. \tag{2.14}$$

If the collision frequency is much less than the light frequency ($\nu_c \ll \omega$) and the plasma is underdense ($\omega_p < \omega$), we obtain for the dielectric constant and absorption coefficient

$$\varepsilon = n^2 = 1 - (\omega_p/\omega)^2\,, \tag{2.15}$$

$$\varkappa = \varkappa_0/\sqrt{\varepsilon}\,, \tag{2.16}$$

$$\varkappa_0 = (\omega_p/\omega)^2(\nu_c/c)\,. \tag{2.17}$$

The classical value of the optical (high frequency, $\omega \gg \nu_c$) absorption coefficient $\varkappa$ is readily obtained from the Boltzmann equation by treating the electric field of the light wave as a small perturbation of an otherwise equilibrium plasma with temperature T. We shall merely quote the result here, postponing further discussion until later when we consider nonlinear effects at high intensity, *i.e.*

$$\varkappa_0 = \frac{(4\pi)^3 (Ze^3)^2 N_e N_i}{3\omega^2 c(2\pi mkT)^{\frac{3}{2}}} \ln \Lambda , \tag{2.18}$$

where $\ln \Lambda$ denotes the usual Coulomb logarithm [6].

The quantum theory Born approximation value of the absorption coefficient $\varkappa_0$ is valid at high temperature ($kT \gg Z^2$ Ryd) where Z is the ionic charge and Ryd is the Rydberg (13.6 eV), and is obtained from the classical value given above by the substitution

$$\ln \Lambda \rightarrow (1/\alpha) \sinh (\alpha) K_0(\alpha) , \tag{2.19}$$

where

$$\alpha = \hbar\omega/2kT , \tag{2.20}$$

and $K_0(\alpha)$ denotes the modified Bessel function of zero order and second kind. For optical frequencies $\hbar\omega \ll \text{Ryd} \ll kT$, so that $\alpha \ll 1$ and relation (19) becomes simply,

$$\ln \Lambda \rightarrow \ln (2/\alpha) - 0.577 , \tag{2.21}$$

which illustrates the striking but not unexpected similarity between the classical and quantum results.

It will be convenient to define the absorption coefficient $\varkappa$ in terms of φ,

$$\varphi = N_e/N_{ec} = (\omega_p/\omega)^2 , \tag{2.22}$$

the ratio of the free-electron density to the critical electron density for light propagation. In terms of φ we have

$$\varkappa = \left(\varphi^2/\sqrt{1-\varphi}\right) \varkappa_{0c}(\omega, T) , \tag{2.23}$$

where $\varkappa_{0c}$ is the value of $\varkappa_0$ at the critical electron density,

$$\varkappa_{0c} = \varkappa_0(N_e = N_{ec}) = (1/c)\nu_c(N_e = N_{ec}) = \varkappa(\varphi = 0.7245) \propto \omega^2 . \tag{2.24}$$

For the important case of light produced by neodymium-doped glass lasers ($\lambda_L = 1.06$ μm, $\hbar\omega_L = 1.17$ eV) we have:

$$N_{ec} = 10^{21} \text{ electrons/cm}^3 , \qquad \varkappa_{0c} \simeq 30\, Z/T_e^{\frac{3}{2}} \text{ cm}^{-1} , \tag{2.25}$$

where the slowly varying factor $\ln \Lambda$ has been set equal to ten (at $T = 10$ keV, $\ln \Lambda = 9.86$ according to eq. (21)), and T_e is the electron temperature in keV.

The electron heat energy w_{ec} contained in a volume one cm² in area and one absorption mean free path λ_{ac} $(=1/\varkappa_{0c})$ thick, at the critical electron density, is given by

(2.26) $$w_{ec} = \tfrac{3}{2} N_{ec} kT_e \lambda_{ac} = 8T_e^{\frac{5}{2}}/Z \text{ kJ/cm}^2 .$$

For deuterium plasma at 10 keV, $\lambda_{ac} = 1$ cm, $w_{ec} = 2.5$ MJ/cm² (a lot of energy!).

The Thomson scattering mean free path at the critical electron density is *15 meters* so that scattering is entirely negligible in comparison with absorption.

2·2.1. Reflection and absorption at a sharp plasma boundary. – We consider the reflection of light at normal incidence from a hypothetical plane surface of discontinuity between a uniform plasma and vacuum. The general expression for the reflection coefficient r and penetration depth δ are somewhat cumbersome, but in the high-frequency limit $(\omega \gg \nu_c)$ reduce to the following simple forms:

(2.27)

r	δ	
$(1+x)^{-1}$	$(2\omega_p/c)^{-1}$	$\omega_p \gg \omega$
$(1+2\sqrt{x})^{-1}$	$[\sqrt{x}(\omega_p/c)]^{-1}$	$\omega_p = \omega$
$(\omega_p/2\omega)^4$	$\varkappa_0^{-1}$	$\omega_p \ll \omega$

where

(2.28) $$x = 2\nu_c/\omega_p = 2(\varkappa_{0c}/k_L)\sqrt{\varphi} \, .$$

The ratio φ has been defined previously in eq. (22); k_L is the wave number (ω/c) of the incident light in vacuum.

For the case of the reflection of 1.06 μm light $(k_L = 6\cdot 10^4\text{ cm}^{-1})$ by a strongly overdense plasma $(\varphi \gg 1)$, we have

(2.29) $$r = (1+x)^{-1}, \qquad \delta = (2k_L\sqrt{\varphi})^{-1},$$

(2.30) $$x = Z\sqrt{\varphi/(100\,T_e)^3} \, .$$

Evaluating r and δ for liquid deuterium $(Z=1,\ \varphi=50)$ at a temperature of 1 keV we find that

$$r = 99.3\,\% \text{ (an excellent mirror)},$$

$$\delta = 120\text{ Å} \quad (\sim 1/100 \text{ wavelength of light}).$$

The reflection coefficient (actually $1-r$) is shown as a function of temperature in Fig. 1 for fully ionized D, LiD, C and Al at their normal solid density. All of these materials are rather good mirrors at temperatures above 1 keV.

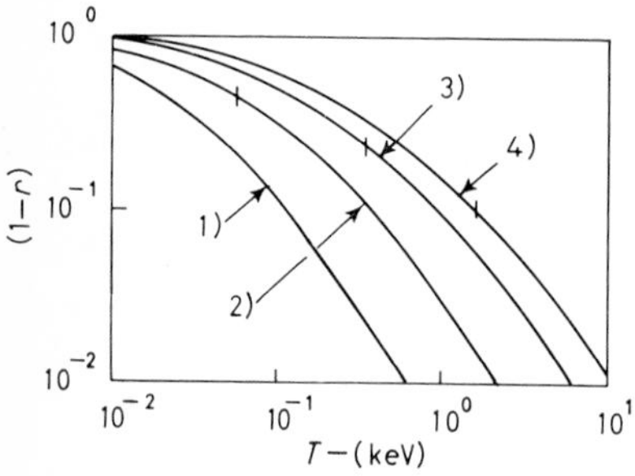

Fig. 1. – Reflectivity r of surfaces of 1) deuterium, 2) lithium deuteride, 3) carbon and 4) aluminum at their normal solid density in vacuum. (The materials are assumed to be fully ionized, which will hold for a given material only if the temperature exceeds that indicated by the vertical mark on its reflectivity curve).

2·2.2. Reflection and absorption at a diffuse plasma boundary. – We shall assume that the electron density is an exponentially decreasing function of distance z, with scale-height H, passing through the critical value at $z=z_c$, *i.e.*

$$\varphi = N_e/N_{ec} = \exp[(z_c - z)/H]\,. \tag{2.31}$$

Assuming further that the rate of change of density is slow when measured in wavelengths of light ($Hk_L \gg 1$), one can calculate the fraction $1-r$ of a light beam, incident from $z=\infty$, that is absorbed by the plasma by simple geometrical optics, *i.e.*

$$1-r = 1-\exp[-2S]\,, \tag{2.32}$$

where S is the optical thickness of the underdense ($\varphi<1$) region of the plasma. Expressing the absorption coefficient $\varkappa$ in terms of φ in accordance with eq. (23), and assuming that the electron temperature is uniform throughout the underdense region of the plasma, we obtain for the optical thickness

$$S = \int_{z_c}^{\infty} \varkappa\, dz = \tfrac{4}{3} H \varkappa_{0c}\,. \tag{2.33}$$

This result slightly understimates S because we have assumed perfect reflection at the critical surface ($\omega_p = \omega$). The absorption at the critical surface can be calculated in the WKB approximation, with the result that we should replace S as given by eq. (33) by the slightly larger value S',

$$S' = S\left(1+\sqrt{\varkappa_{0c}/2k_L}\right)\,. \tag{2.34}$$

We may also note by comparing eqs. (27), (28), (32), and (33) that the reflection from a *diffuse* plasma boundary with scale height H is roughly equivalent to that from a *sharp* plasma boundary at the same temperature and with an overdensity ratio φ given by

(2.35) $$\sqrt{\varphi} = \tfrac{4}{3} Hk\,(\gg 1)\,.$$

2˙2.3. Time required to form a diffuse absorbing layer. – If a semi-infinite overdense plasma with an initially sharp plane surface expands isothermally into a vacuum, the density profile is exponential with a scale height H that increases linearly with time according to

(2.36) $$H = c_s\, t\,,$$

where

(2.37) $$c_s = \sqrt{ZkT_e/M} = 3\cdot 10^7 \sqrt{(Z/A)\,T_e}\ \text{cm/s}\,,$$

is the isothermal sound speed in the plasma (we are assuming the ion pressure to be small compared with the electron pressure). The optical thickness S of the undense layer increases with time as

(2.38) $$S = 1.2\sqrt{Z^3/A}\; T_e^{-1} t\,,$$

where t is in nanoseconds. Defining the time required to form an absorbing layer τ_{abs} as the time required for the reflection coefficient to be reduced to $1/e$, we have

(2.39) $$\tau_{abs} = 0.4\sqrt{A/Z^3}\; T_e\ \ \text{ns}\,.$$

For deuterium ($A = 2$, $Z = 1$) at 10 keV, for example, $\tau_{abs} = 6$ ns.

2˙3. *Heating of plasma ions by electrons: The shortest possible heating time.* – The equilibration of the ion and electron temperatures in a plasma is described by the relaxation equation [6]

(2.40) $$\mathrm{d}T_i/\mathrm{d}t = (T_e - T_i)/\tau_{ei}\,,$$

where τ_{ei} is the ion-electron relaxation time. At the critical electron density of 10^{21} electrons/cm^3 τ_{ei} is given by

(2.41) $$\tau_{ei} = 10\, A T_e^{\frac{3}{2}}/Z^2 \ln \Lambda\ \ \text{ns}\,.$$

Neglecting the slow variation of $\ln \Lambda$ with temperature, it can be shown from eqs. (40) and (41) that the maximum instantaneous rate of ion heating occurs if $T_e = 3T_i$. The minimum time τ_i required to heat ions from zero temperature to temperature T_i by means of binary Coulomb encounters with electrons can therefore be found by assuming that at every instant the electrons are three times as hot as the ions. One then obtains the result

$$\tau_i = \sqrt{3}\, \tau_{ei}(T_e = T_i)\,. \tag{2.42}$$

The minimum time required to heat deuterons to 10 keV by electrons at the critical density is 110 ns (a long time!), and the electrons must achieve a temperature of 30 keV in order that this minimum time be realized.

2·4. *Scaling relations for the production of high-temperature and high-pressure plasmas.* – We consider an isolated sphere of homogeneous plasma with an initial radius r_0. This plasma is to be heated by a light pulse of power P, duration τ, energy W $(=P\tau)$, and optical frequency ω. The heated plasma is characterized by a temperature T, pressure p, free electron density N_e, nuclear charge number Z, fractional ionization f $(=N_e/ZN_i)$, sound speed c_s, and absorption coefficient $\varkappa$ (for light of frequency ω). If we assume that the cooling of the plasma is due to hydrodynamic expansion rather than radiation, then we require the dimensionless quantity $(r_0/c_s\tau)$ to be an invariant in order that the predominant cooling process be properly scaled. If in addition we require the dimensionless quantity $(\varkappa r_0)$ to be a scale invariant so that the fraction of the incident laser light that is absorbed by the plasma remains fixed, then the plasma temperature will scale according to the simple relation

$$T \propto P\tau/N_e r_0^3\,. \tag{2.43}$$

Using the relation

$$\varkappa \propto fZN_e^2/\omega^2 T^{\frac{3}{2}} \tag{2.44}$$

for the optical absorption coefficient (valid for $N_e < (1/2)\, N_{ec}$) together with the invariance of the quantities $(r_0/c_s\tau)$ and $(\varkappa r_0)$, we obtain from eq. (43) the following scaling relation for the plasma temperature:

$$T \propto (Z^2/f^3)^{\frac{1}{12}}(W/\omega^{\frac{1}{3}})/\tau^{\frac{5}{6}}\,. \tag{2.45}$$

It is clear from this result that the duration τ of the light pulse should be made as small as possible to achieve the highest plasma temperature. At the same time, however, the free electron density scales according to the relation

$$N_e \propto (W\omega^5)^{\frac{1}{6}}/f^{\frac{7}{8}} Z^{\frac{5}{12}} \tau^{\frac{11}{12}}, \tag{2.46}$$

so that as τ becomes smaller N_e increases and will eventually exceed the critical free electron density N_{ec}, at or above which the incident light wave can no longer propagate into the plasma. Instead, if the plasma is hot ($T > 1$ keV), the light will be mainly reflected at the plasma surface.

Since we desire the free electron density to be as large as possible, consistent with reasonably efficient absorption of the incident light pulse, we shall subject our scaling relations to the constraint that

$$N_e \sim N_{ec} \; (\propto \omega^2) . \tag{2.47}$$

We then obtain the constrained scaling relations

$$T \propto (fZ)^{\frac{6}{11}} (W\omega^4)^{\frac{2}{11}}, \tag{2.48}$$

$$p \propto \omega^2 T . \tag{2.49}$$

From these results we conclude that to achieve the highest possible temperature and pressure with a given pulse-energy W, one should employ *high-Z target materials*, and *laser light with as short a wavelength as possible*. Having chosen ω as large as possible, the target density is selected such that $N_e \sim N_{ec}$. The pulse duration τ is then made small enough to allow the pulse energy to be absorbed before the plasma can cool appreciably by expansion.

The scaling of the quantity $\psi = N_e \tau T^n$ is also of interest since it is this quantity, with an appropriate value of the exponent n, that determines the x-ray or neutron emission by the plasma. (For the case of neutron production by a DT plasma, for example, $n \geqslant 8/3$ if $T < 10$ keV.) It can be shown from the scaling relations we have derived that

$$(\partial\psi/\partial\tau)_W \leqslant 0 , \qquad \text{if } n \geqslant \tfrac{1}{10}, \tag{2.50}$$

$$(\partial\psi/\partial\omega)_W \geqslant 0 , \qquad \text{if } n \geqslant -1, \; (N_e = N_{ec}) . \tag{2.51}$$

Hence the criteria listed for maximizing the temperature and pressure (ω large, $N_e \sim N_{ec}$, τ small) will also maximize the yield ψ of processes for which $n \geqslant \frac{1}{10}$.

3. – Hydrodynamics of laser-produced plasmas: some simple analytical results.

3·1. *Pressure resulting from light-induced blow-off.* – If a sufficiently intense beam of light is directed against the surface of a solid in vacuo, the surface will melt, vaporize and ionize. The ionized plasma will blow off into the vacuum and its reaction will drive a shock into the solid. A simple model characterizing the processes that take place is that of a deflagration preceded by a shock wave and followed by a centered simple rarefaction wave [7]. The energy supplied to the deflagration is due to the absorption of light.

We will assume that the light beam is a collimated beam of parallel light of effectively infinite lateral extent, propagating in a direction perpendicular to the plane surface of the solid. The flow will then be a one-dimensional, plane-parallel flow. We will also assume that the rarefaction is adiabatic, and that the plasma is an ideal ($\gamma = \frac{5}{3}$) gas with an absorption coefficient $\varkappa_0$ proportional to the square of the plasma density ϱ and inversely proportional to the cube of the adiabatic sound speed c_a ($c_a = \sqrt{\gamma} c_s$), *i.e.*

$$\varkappa_0 = \beta \varrho^2 / c_a^3 , \tag{3.1}$$

a high-temperature ($\alpha = \hbar\omega/2kT \ll 1$) approximation to the free-free absorption coefficient given by eqs. (2.18) and (2.21) which neglects the slow variation with temperature of the log term appearing in eq. (2.21).

Using the expression for $\varkappa_0$ given by eq. (3.1), the optical thickness S of the rarefaction to the incident light is readily found to be

$$S = \varkappa_0 c_a t , \tag{3.2}$$

where $\varkappa_0$ and c_a are evaluated at the head of the rarefaction. (The fact that the dielectric constant ε may differ from unity at the head of the rarefaction has been disregarded in evaluating S, with the result that eq. (3.2) will underestimate its true value.)

Assuming that all of the incident light energy is utilized to support the adiabatic expansion, it follows that

$$I = \varrho u [h + \tfrac{1}{2} u^2] = \mu \varrho c_a^3 , \tag{3.3}$$

where $\mu = M(M^2 + 3)/2$, $M (= u/c_a)$ is the Mach number of the flow, u is the flow speed, h $(= e + pv)$ is the specific enthalpy, again all evaluated at the head of the rarefaction, and I is the intensity of the incident light. Solving eqs. (3.1), (3.2) and (3.3) for ϱ and c_a, we obtain

$$\varrho = (\beta t/S)^{-\frac{3}{8}} (I/\mu)^{\frac{1}{4}} , \tag{3.4}$$

$$c_a = (\beta t/S)^{\frac{1}{8}} (I/\mu)^{\frac{1}{4}} . \tag{3.5}$$

The pressure p_r at the head of the rarefaction is then

$$p_r = \varrho c_a^2/\gamma = (3/5)(\beta t/S)^{-\frac{1}{8}}(I/\mu)^{\frac{3}{4}} . \tag{3.6}$$

The shock pressure p exceeds the pressure p_r at the head of the rarefaction by the pressure drop $\gamma M^2 p_r$ across the deflagration, or

$$p = (1 + \gamma M^2) p_r . \tag{3.7}$$

We observe that the shock pressure p resulting from the light intensity I is determined if we specify the Mach number M, and either the density ϱ at the head of the rarefaction or the optical thickness S. We now assert that:

1) The deflagration is a *Chapman-Jouguet* deflagration; that is, the Mach number M of the downstream flow is unity.

2) The rarefaction is *self-regulating* such that:

   i) The density ϱ is maintained at the critical value ϱ_c until the time t_1 when the optical thickness S of the rarefaction becomes equal to unity.

   ii) The optical thickness S is maintained equal to unity thereafter. (It should be noted that the shock pressure p is rather insensitive to the value assigned to S since it is proportional only to the eighth root of S. If S were to vary by as much as a factor of 200, p would vary by less than a factor of 2.)

It follows that the shock pressure p is given by the relations

$$p = \begin{cases} p_c , & t \leqslant t_1 , \\ (t_1/t)^{\frac{1}{8}} p_c , & t > t_1 , \end{cases} \tag{3.8}$$

where the « obscuration time » t_1 is given by

$$t_1 = (I/2)^{\frac{2}{3}}/\beta \varrho_c^{\frac{8}{3}} , \tag{3.9}$$

and the pressure p_c is given by

$$p_c = \tfrac{8}{5}\, \varrho_c^{\frac{1}{3}} (I/2)^{\frac{2}{3}} . \tag{3.10}$$

For the important case of 1.06 μm light, and assuming for simplicity that $A \sim 2Z$, we have $\varrho_c = 3.3 \cdot 10^{-3}$ gm/cm³,

$$t_1 = 2 \cdot 10^{-3} I^{\frac{2}{3}}/Z \ \text{ns} , \tag{3.11}$$

and

$$p_c = 7\,I^{\frac{2}{3}}\ \text{kb}\,, \tag{3.12}$$

where I is expressed in units of GW/cm². Values of Zt_1 and p_c obtained from eqs. (3.11) and (3.12) are given in Table II below for selected values of the ligh intensity I.

TABLE II. – *Shock pressure* (p_c), *and obscuration time* (t_1) *times* Z, *vs. light intensity* (I).

I (GW/cm²)	Zt (ns)	p_c (Mb)
10^4	1	3
10^6	20	70
10^8	400	1500

The assumptions that the blow-off is adiabatic, that all of the incident light is absorbed and none is reflected, and that a negligible amount of energy is required to support the shock wave, all lead to an *overestimate* of the shock pressure p. If we had assumed the blow-off to be isothermal instead of adiabatic, the shock pressure would be reduced by 26%.

An upper limit to the shock pressure can readily be obtained by assuming that all of the incident-light energy is utilized to support the shock wave. In this case we have,

$$I = p_{\max} u = p_{\max}\sqrt{2p_{\max}/(1+\gamma_s)\varrho_0}\,, \tag{3.13}$$

or

$$p_{\max} = [(\gamma_s + 1)\,\varrho_0/2]^{\frac{1}{3}} I^{\frac{2}{3}}\,, \tag{3.14}$$

where ϱ_0 is the density of the material ahead of the shock (assumed to be large compared with ϱ_c), and γ_s is its effective γ. Comparing p_c given by eq. (3.10) with $p_{\max}$, we obtain the result,

$$p_c/p_{\max} \simeq (\varrho_c/\varrho_0)^{\frac{1}{3}} \ll 1\,. \tag{3.15}$$

For the case of 1.06 μm light ($\varrho_c = 3.3\cdot 10^{-3}$) directed against the surface of liquid deuterium ($\varrho_0 = 0.17$) it follows from eq. (3.15) that the shock pressure p_c amounts to $\sim 30\%$ of the maximum pressure possible.

The pressure and temperature profiles obtained from detailed computer calculations [7] for the case in which light of intensity 10^4 GW/cm² is directed against deuterium gas with an initial specific volume of 100 cm³/gm ($\varrho_0 = 3\varrho_c$) are shown in Fig. 2 at a time of 16 ns after the light beam is turned on ($t = 16t_1$). The shock pressure is seen to be 1.7 Mb. For comparison, the shock pressure p obtained from eq. (3.8) together with Table II is $3/16^{\frac{1}{3}} \simeq 2$ Mb, a remarkably good result considering the crudeness of our model.

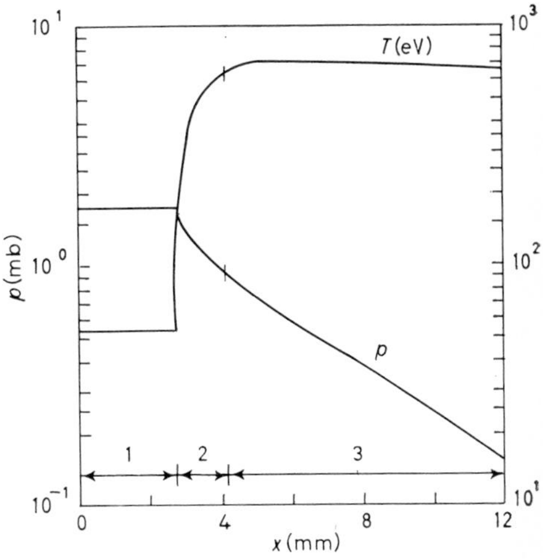

Fig. 2. – Light-induced shock in deuterium. Pressure p and temperature T *vs.* distance x. Indicated regions are the precompression shock (1), Chapman-Jouguet deflagration (2), and isothermal expansion (3). ($I = 10^4$ GW/cm², $v_0 = 100$ cm³/g, $t = 16$ ns).

3·2. *Homogeneous expansion of an isothermal sphere into a vacuum.* – A useful analytic model of a hot spherically symmetric plasma expanding in vacuum is provided by a homogeneously expanding isothermal sphere. Such a model is useful in describing the later stages of expansion of a small speck of matter suspended in vacuum and heated to a high temperature by an intense focused laser pulse [8].

A homogeneous expansion is one in which the element of volume d^3x, in the Lagrangian description in which $\boldsymbol{x}(t)$ denotes the instantaneous position vector of a mass point, increases everywhere at the same rate. That is,

$$d^3x = h^3(t)\, d^3x_0\,, \tag{3.16}$$

where $\boldsymbol{x}_0 = \boldsymbol{x}\,(t = 0)$. Writing eq. (3.16) in spherical co-ordinates we obtain immediately,

$$r(r_0, t) = r_0\, h(t)\,, \tag{3.17}$$

$$u(r_0, t) = \dot{r} = r_0\, \dot{h}(t)\,. \tag{3.18}$$

From the law of mass conservation we may write

$$dm = \varrho r^2\, dr = dm_0 = \varrho_0 r_0^2\, dr_0\,, \tag{3.19}$$

which together with eq. (3.17) implies that

$$\varrho(r_0, t) = \varrho_0(r_0)/h^3(t)\,. \tag{3.20}$$

The equation of motion is,

(3.21) $$\dot{u} = -r^2(\mathrm{d}p/\mathrm{d}m) = -(r^2/\varrho_0 r_0^2)(\mathrm{d}p/\mathrm{d}r_0)\,.$$

We wish to write $\mathrm{d}p/\mathrm{d}r_0$ in terms of $\mathrm{d}\varrho_0/\mathrm{d}r_0$. For that purpose we need the (ideal gas) equation of state,

(3.22) $$e = pv/(\gamma - 1)\,, \qquad (v = 1/\varrho),$$

and the isothermality condition,

(3.23) $$e(r_0, t) = e_0\, a(t)\,, \qquad \mathrm{d}e_0/\mathrm{d}r_0 = 0\,.$$

Making use of these relations the equation of motion can be written

(3.24) $$h\ddot{h}/(\gamma - 1)\, e = -(1/r_0\, \varrho_0)\,(\mathrm{d}\varrho_0/\mathrm{d}r_0) = 1/R^2 = \text{constant}\,,$$

where the r.h.s. of eq. (3.24) is a function of r_0 alone and the l.h.s. is a function of t alone. Since r_0 and t are independent Lagrangian variables, each side of the equation must be constant. Integrating the equation governing ϱ_0 we obtain

(3.25) $$\varrho_0 = \varrho(0)\exp\left[-r_0^2/2R^2\right].$$

That is, the kinematic assumption of a homogeneous isothermal expansion is compatible with the dynamical equation of motion if and only if *the density (and pressure) profile is Gaussian.* The total mass M of the sphere determines the constant R by the relation

(3.26) $$2\pi R^2 = [M/\varrho(0)]^{\frac{2}{3}}\,.$$

To find the *rate* of expansion we need to employ the equation of energy conservation,

(3.27) $$\dot{e} + p\dot{v} = \dot{q} = v^{1-\gamma}\,(\mathrm{d}(ev^{\gamma-1})/\mathrm{d}t)\,.$$

The assumption of homogeneous isothermal expansion implies that the l.h.s. of eq. (3.27), and hence the source term $\dot{q}$, must be independent of r. If isothermality is the result of *high thermal conductivity*, however, the heat sources contributing to $\dot{q}$ need not be independent of r since thermal conduction will appropriately distribute the heat sources anyway. In the absence of heat sources ($q = 0$) eq. (3.27) implies that

(3.28) $$ah^{3(\gamma-1)} = 1\,.$$

Using this result we may write eq. (3.24) as:

(3.29) $$\ddot{h} = (c_0/R)^2/h^{3\gamma-2}\,,$$

where

$$c_0^2 = (\gamma - 1)\, e_0 \,. \tag{3.30}$$

Assuming that $\gamma = \frac{5}{3}$ and that $\dot{h}(0) = 0$, the solution to eq. (3.29) is:

$$h^2 = 1 + \tau^2 \,, \qquad (\tau = c_0 t/R), \tag{3.31}$$

so that

$$r = r_0 \sqrt{1+\tau^2} \,, \tag{3.32}$$

$$e = e_0/(1+\tau^2) \,, \tag{3.33}$$

$$\varrho = \varrho(0)(1+\tau^2)^{-3/2} \exp[-r_0^2/2R^2] = \varrho(0)(1+\tau^2)^{-3/2} \exp[-r^2/2(R^2 + c_0^2 t^2)] \,, \tag{3.34}$$

that is, for $\tau \gg 1$ the sphere expands at a constant speed equal to its initial isothermal sound speed c_0.

Although this simple result provides a good description of the later stages of plasma expansion after isothermality has had time to take place, the situation is much more complex at earlier times in the history of a laser-heated plasma. An example of this behavior at early times in the heating of a small sphere of deuterium gas in vacuum is shown in Fig. 3 and 4.

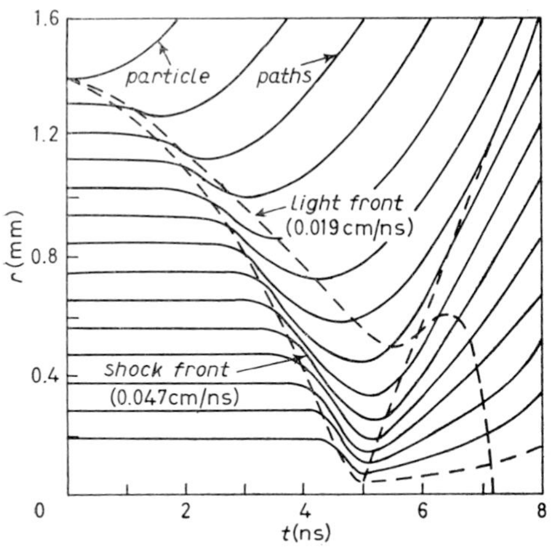

Fig. 3. – Subsonic heating of deuterium gas sphere. Radius r vs. time t plot of particle paths, shock front, and light front. (10 kJ in 5 ns. D_2: $v_0 = 185\ cm^3/g$).

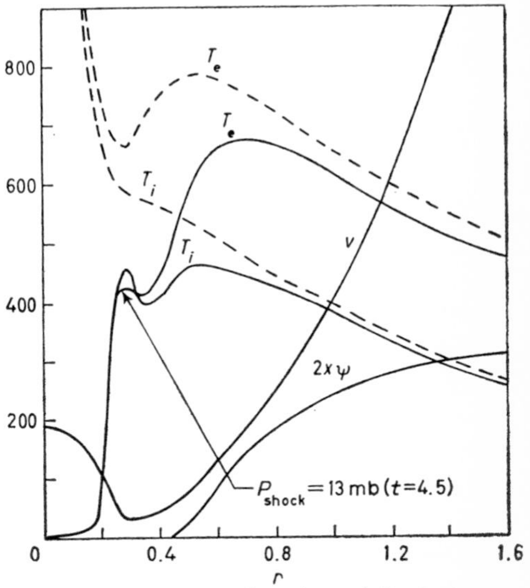

Fig. 4. – Subsonic heating of deuterium gas sphere. Ion and electron temperatures T_i and T_e, specific volume v (cm^3/g), and light power ψ (GW/st) vs. radius r. (10 kJ in 5 ns. D_2: $v_0 = 185\ cm^3/g$). ——— $t = 4.5$ ns; — — — $t = 5.0$ ns.

3'3. *Light-supported detonations.* – The detonation velocity U resulting from the shock-initiated release of q units of energy per unit mass in an ideal gas with adiabatic exponent γ is given by the simple relation

$$U^2 = 2(\gamma^2 - 1)\, q \,. \tag{3.35}$$

In deriving this relation it is assumed that the detonation satisfies the Chapman-Jouguet condition: that the detonation propagates at sonic velocity c_a relative to the matter behind it,

$$U = u + c_a \,. \tag{3.36}$$

If a light beam of intensity I is directed at a shock wave under conditions such that the light beam is completely absorbed at the shock front, then the shock wave becomes a « light-supported detonation » with an effective energy release q given by

$$q = (v_0/U)\, I \,, \tag{3.37}$$

where v_0 is the specific volume of the undisturbed fluid ahead of the detonation. The detonation speed of the light-supported detonation, obtained from eqs. (3.35) and (3.37), is given by

$$U = [2(\gamma^2 - 1)\, v_0 I]^{\frac{1}{3}} \,, \tag{3.38}$$

and the detonation pressure p is given by

$$p = \varrho_0\, U^2/(\gamma + 1) \,. \tag{3.39}$$

The preceding result applies to the case of a plane detonation wave supported by an antiparallel plane light wave. Another case of interest is that of a spherically diverging detonation supported by a spherically convergent light beam. In this case the light intensity decreases with the square of the radius r according to

$$I = P/4\pi r^2 \,, \tag{3.40}$$

where P is the power of the light beam. The radius of the divergent detonation wave then satisfies the equation,

$$\dot{r} = U = [(\gamma^2 - 1)\, v_0 P/2\pi r^2]^{\frac{1}{3}} \,. \tag{3.41}$$

Integrating this equation with respect to time, and assuming that $r(t = 0) = 0$, we obtain

$$r = t^{\frac{2}{5}}[(\tfrac{5}{3})^3(\gamma^2 - 1)\,v_0\,W/2\pi]^{\frac{1}{5}}\,, \tag{3.42}$$

where the energy W ($=Pt$) contained in the detonation wave is assumed to increase linearly with time.

For comparison, the corresponding relation for a Taylor blast wave is

$$r = t^{\frac{2}{5}}[(\tfrac{5}{2})^2 v_0\,W/B]^{\frac{1}{5}}\,, \tag{3.43}$$

where W is the energy that is assumed to have been released instantaneously at the origin when $t=0$, and B is a numerical constant depending on γ ($B = 5.33$, 3.08, when $\gamma = \frac{7}{5}$, $\frac{5}{3}$, respectively). Comparing eqs. (3.42) and (3.43), we find that the radius of the light-supported detonation is very closely equal to that of a Taylor blast wave with two-thirds as much energy, other things (t, v_0, γ) equal.

These results have been employed to describe the behavior of a small spark or ionized region within a homogeneous volume of gas that is being heated by the focused output of a giant-pulse laser [9]. That is, while the light pulse is present the spark expands according to eq. (3.42) as a light-supported detonation, and thereafter as a Taylor blast wave. This description is a good one provided that the light intensity is not so large as to cause sparking or « optical breakdown » of the undisturbed gas ahead of the expanding shock wave, and that the spark is not so small and hot that it expands as a supersonic thermal diffusion wave.

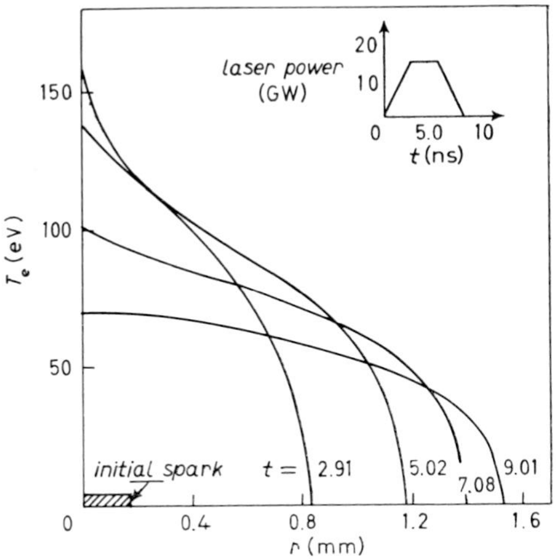

Fig. 5. – Laser-heated spark. Electron temperature T_e *vs.* radius r at several times t. (100 J in 5 ns. D_2: $v_0 = 5600\ cm^3/g$.)

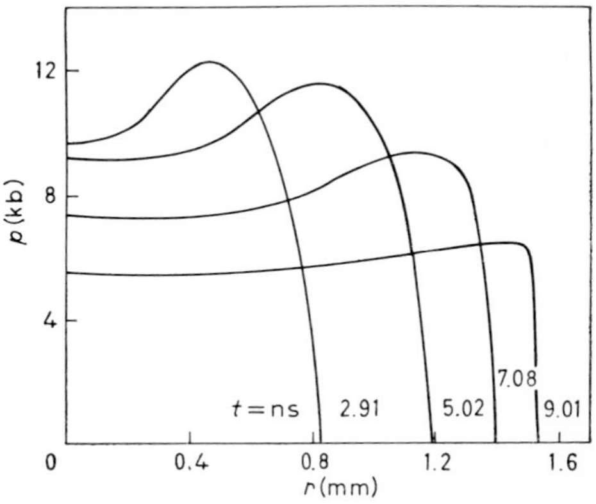

Fig. 6. – Laser-heated spark. Pressure p *vs.* radius r at several times t. (100 J in 5 ns. D_2: $v_0 = 5600\ cm^3/g$.)

The process by which light is absorbed by the spark is that of free-free absorption by the free electrons of the spark plasma. If the plasma is not opaque to the incident light, it is not obvious that a description in terms of a detonation wave is legitimate. Nevertheless, most of the absorption that does occur takes place near the shock front where the plasma is cooler and more dense than in the interior, so that the treatment as a detonation is still a reasonable approximation. Of course, the energy W contained in the detonation is then only that fraction of the incident light that is actually absorbed by the plasma.

The results of computer calculations [7] of the temperature, pressure, and specific volume profiles, and spark radius *vs.* time of a laser-heated spark in STP deuterium gas are shown in Fig. 5, 6, 7, and 8. These calculations apply to the heating of a small ionized region, arbitrarily given an initial temperature of 2 eV (see Fig. 5), by a 100 J spherically convergent light pulse with a full width at half maximum of 5 ns and peak power of 20 GW.

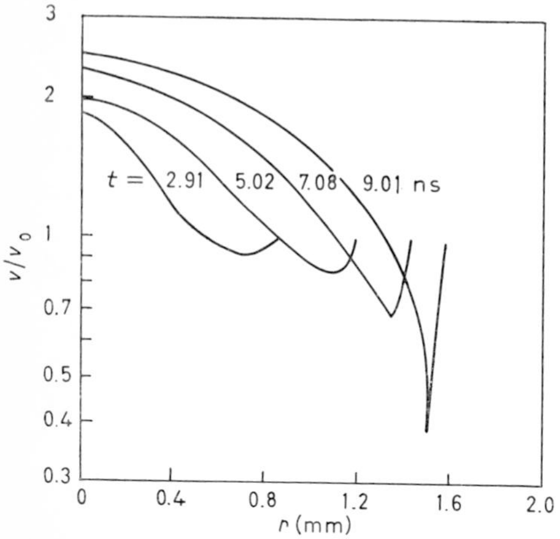

Fig. 7. – Laser-heated spark. Expansion ratio v/v_0 *vs.* radius r at several times t. (100 J in 5 ns. D_2: $v_0 = 5600\ cm^3/g$.)

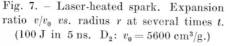

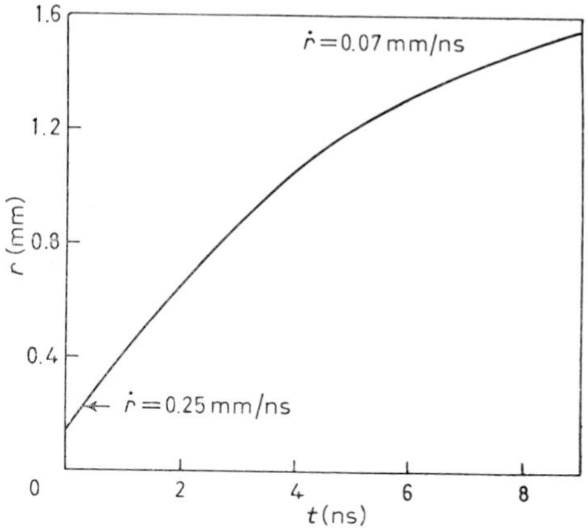

Fig. 8. – Laser-heated spark. Radius r of fireball *vs.* time. (100 J in 5 ns. D_2: $v_0 = 5600\ cm^3/g$.)

4. – Production of multikilovolt tenth-microgram deuterium or mercury plasmas.

The scaling relations discussed earlier have shown that to produce the highest-temperature plasma with a light pulse of limited energy W, one should use short-wavelength light to heat high-Z plasma. The pulse duration τ should be made short enough that the plasma does not have time to cool appreciably by expansion during the heating period. The pulse length required depends on the pulse energy available and the desired plasma temperature. For example, if temperatures of a few keV are desired to be produced with 100 J pulses, we will see that the duration of these pulses must be on the order of 1/10th nanosecond.

Light pulses of such short duration have not been available until recently, a few nanoseconds being the limit previously achievable with giant-pulse lasers. However, the study of the problem of locking together in phase many modes of oscillation of a laser oscillator [10] has led to the development of lasers capable of producing light pulses with pulse widths as short as a few picoseconds [11] in duration. Subnanosecond pulses have also been produced by reverse-pumped Raman lasers [12], so we may assume that sufficiently short light pulses are now a practical reality.

4·1. *Laser requirements to produce multikilovolt temperatures in tenth-microgram plasmas.* – The requirements that a laser must meet if it is to be capable of heating a tenth-microgram deuterium or mercury target potentially (*i.e.* in the absence of reflection or other losses) to a temperature of 10 keV are determined by the target properties listed in Table III below.

TABLE III. – *Properties of tenth-microgram deuterium and Mercury.* Targets at 10 keV.

	D_2	Hg	Units
Charge number Z	1	80	—
Specific volume v_c	300	240	cm^3/gm
Radius r_0	190	180	μm
Internal energy W	150	80	J
Cooling time τ_E	0.2	0.3	ns

The target is assumed to be a homogeneous spherical plasma whose number density of free electrons N_e is equal to the critical value N_{ec} ($=10^{21}\ cm^{-3}$) for light produced by a neodymium-glass laser (wavelength $= 1.06\ \mu m$). The cooling time τ_E is defined as the time required for an isothermal rarefaction to propagate from the surface of the plasma to its center,

$$\tau_E = r_0/c_s\,. \tag{4.1}$$

It follows that the laser used to heat the plasma must be capable of delivering ~ 100 J in 0.2 ns and of being *focused to a spot* 100 μm *in diameter.* (This implies a brightness of $\sim 10^{16}\ W/cm^2\ sr$, which has already been exceeded in practice [5].) These requirements are roughly the same whether the target plasma is deuterium ($Z = 1$) or mercury ($Z = 80$).

A single spatial mode laser beam of wavelength λ_L can be focused to a spot of radius $r_s \sim F\lambda_L$ by a lens of speed F (focal length/diameter). Use of an F:5 lens, for example, will allow a 1.06 μm beam to be focused on a spot 50 μm in radius if the angular spread of the beam does not exceed ten times that of a single-mode beam, a readily achievable beam quality.

4·2. *Properties of deuterium and mercury plasmas at* 10 keV. – We shall assume that the free electron density of the spherical plasma decreases exponentially at its surface with scale-height H, attaining the critcal density for light propagation at radius $r = r_c$, *i.e.*

$$\varphi = N_e/N_{ec} = \exp[(r_c - r)/H], \qquad r \geqslant r_c. \tag{4.2}$$

If the mass M of the plasma is small the scale height H will also be small, with the result that it may be difficult to achieve efficient absorption of the incident light pulse. For a given mass of plasma, the maximum scale height is obtained when the critical density is just achieved at the center of the plasma so that $r_c = 0$. Specifically, we find that

$$H \leqslant r_0/6^{\frac{1}{3}} \simeq r_0/2, \qquad r_c \geqslant 0, \tag{4.3}$$

where r_0 is the radius of a homogeneous sphere of plasma of the same mass at the critical specific volume v_c,

$$r_0 = (3Mv_c/4\pi)^{\frac{1}{3}}. \tag{4.4}$$

TABLE IV. – *Properties of tenth-microgram deuterium and mercury plasmas at* 10 keV.

	D_2	Hg	Units
Scale height H	100	100	μm
Electron pressure p_e	16	16	Mb
Light absorption α	2.4	84	%
Electron MFP λ_e	9	0.12	r_0
Ion heating time τ_i	110	0.7	ns
Brems cooling time τ_b	1400	18	ns

In Table IV below we list some properties of small, hot deuterium and mercury plasmas.

In this Table the scale height H is obtained from eqs. (4.3) and (4.4); the light absorption α $(= 1 - r)$ from eqs. (2.25), (2.32) and (2.33); and the ion heating time τ_i from eqs. (2.41) and (2.42).

The electron mean free path λ_{be} for scattering by at least 90° in a single collision with a plasma ion is given by the expression

$$\lambda_{be} = 0.14\, T_e^2/Z \ \text{cm}. \tag{4.5}$$

When the effects of many distant collisions are taken into account this result is diminished by the factor [6] $8 \ln \Lambda$ to give the mean free path λ_e for deflection through 90°;

$$\lambda_e = \lambda_{be}/8 \ln \Lambda \,, \tag{4.6}$$

the Coulombian logarithm $\ln \Lambda$ having typically a value of ~ 10.

The bremsstrahlung cooling time τ_b is defined as the time required for the plasma electrons to radiate an amount of energy $(3/2)kT_e$ per free electron by free-free emission at constant electron temperature and at the critical electron density of 10^{21} electrons/cm³;

$$\tau_b = 450\, T_e^{\frac{1}{2}}/Z \text{ ns} \,. \tag{4.7}$$

We may draw the following *conclusions concerning tenth-microgram amounts of deuterium or mercury plasma at* 10 keV from the properties listed in Table IV:

i) The mercury plasma can effectively absorb the incident light (84 % absorption); the deuterium plasma cannot (2 % absorption).

ii) The electron temperature will be strictly uniform throughout the deuterium plasma ($r_0 \ll \lambda_e$), and nearly so in the mercury plasma.

iii) The ions will remain « cold » ($\ll 1$ keV) in the deuterium plasma ($\tau_i \gg \tau_E$). They will get hot in the mercury plasma, but remain considerably cooler than the electrons.

iv) The mercury plasma will radiate $\sim 1\%$ of the incident light pulse as bremsstrahlung ($\tau_b \sim 100\,\tau_E$); the deuterium plasma only $\sim 0.01\%$.

4·3. *Results of computer calculations.* – The results of computer calculations of the heating of a 0.08 microgram spherical mercury plasma, having an initial density profile as shown in Fig. 9, are presented in Table V below. Results for the heating of an equal mass of deuterium plasma with the same initial density profile are included for comparison. The calculations are based on a one-dimensional (spherical symmetry), single-fluid, two-temperature (ion, electron) model of the plasma, which accounts for hy-

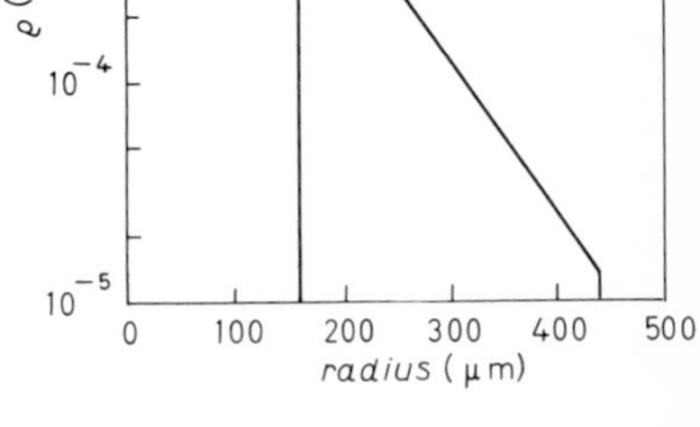

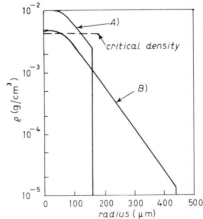

Fig. 9. – Density profiles of a spherical mercury plasma ($M = 0.08$ μg). *A*) Initial profile ($t = 0$). *B*) Profile at $t = 0.34$ ns. (Pulse width $\tau = 0.25$ ns).

drodynamic motion, ionic and electronic heat conduction, Coulomb coupling between ions and electrons, the emission of bremsstrahlung, and the absorption and reflection of the (spherically convergent) incident laser light [13].

TABLE V. – *Computer results for the heating of deuterium and mercury plasmas by subnanosecond light pulses.*

Plasma	D_2	Hg		
Laser pulse (J/ns)	46/0.25	60/0.25	60/0.1	240/0.1
Max electron temp (keV)	0.72 (*)	4	6	12
Bremsstrahlung (%)	0.01	0.8	0.4	0.1
Internal energy (%)	9.3	40	**63**	29
Kinetic energy (%)	6·0	**50**	20	23
Reflected energy (%)	**84.7**	10	17	**48**

(*) The maximum ion temperature is 0.26 keV.

The mercury plasma is assumed to be a fully ionized ideal gas at all times, no account being taken of the ionization process other than the energy required to remove the electrons.

The incident laser pulse was assumed to have a trapezoidal shape (see insert in Fig. 5) with a full-width τ at half-maximum intensity. Calculations were done of the heating of mercury plasma with 60 J light pulses 0.25 and 0.1 ns in width (τ), and with a 240 J pulse 0.1 ns in width; and of the heating of deuterium plasma with a 46 J 0.25 ns pulse. In the case of fully ionized mercury, approximately 14 J of the total of 60 supplied represents the binding energy of its 80 electrons, leaving 46 J available as electronic and ionic kinetic energy. For this reason a 46 J light pulse was used to heat the deuterium for comparison, the binding energy of its electron being negligible at the temperatures considered.

The percentages of the incident light energy that appear as bremsstrahlung emission and as internal, kinetic, and reflected energy are listed in Table V at the time that the intensity of the incident light returns to zero.

For the mercury plasma it can be seen that appreciable cooling by expansion occurs for a pulse duration as long as 0.25 ns (50 % of the pulse energy appears as kinetic energy of the expanding plasma). Reducing the pulse duration to 0.1 ns lead to a reasonably efficient (63 %) conversion of light energy into heat. If four times as much light energy is supplied in the same time (0.1 ns) the peak plasma temperature doubles, but the efficiency drops to 29 % due to a large increase in reflection losses.

For the deuterium plasma the reflection losses are predominant, only 15% of the incident light being absorbed. The peak ion temperature is only one-third that of the electrons demonstrating the weakness of the ion-electron coupling.

The penetration of the light pulse into the deuterium and mercury plasmas is shown in Fig. 10 and 11, respectively.

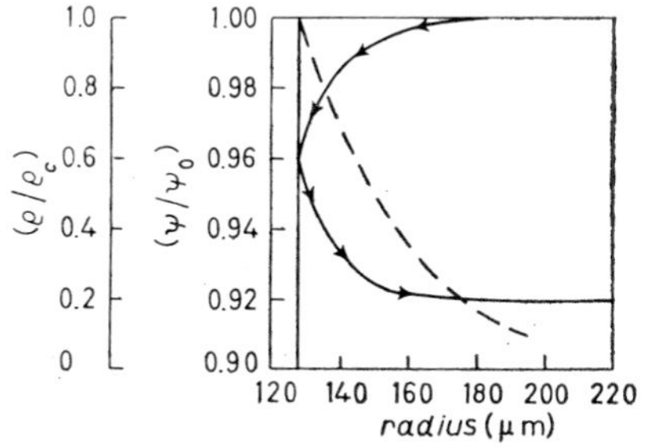

Fig. 10. – Absorption and reflection of light at the surface of deuterium plasma. Relative beam power ψ/ψ_0 and relative plasma density ϱ/ϱ_c *vs.* radial distance r (micron). $t = 0.24$ ns; ——— ψ/ψ_0; – – – ϱ/ϱ_c.

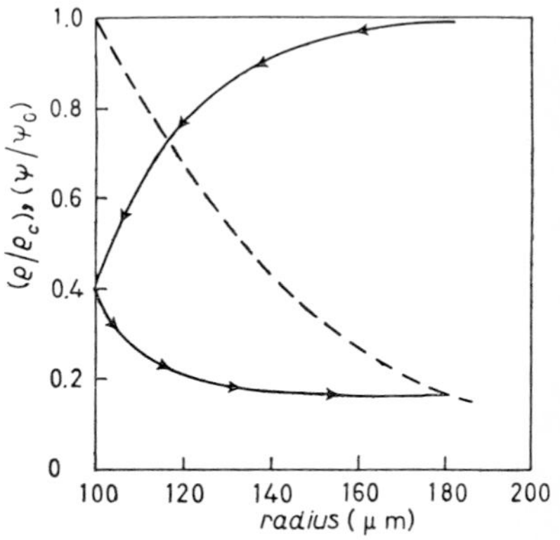

Fig. 11. – Absorption and reflection of light at the surface of mercury plasma. Relative beam powe1 ψ/ψ_0 and relative plasma density ϱ/ϱ_c *vs.* radial distance r (micron). $t = 0.24$ ns; ——— ψ/ψ_0; – – – ϱ/ϱ_c.

5. – Nonlinear optical effects in plasmas.

No consideration has been given thus far to the possibility that the interaction of a light pulse with a plasma may be significantly modified by nonlinear optical effects arising from the high intensity of the incident light. Nonlinear effects will be expected to occur when the mean kinetic energy $\langle KE\rangle_{\text{osc}}$ of a free electron undergoing forced oscillation in the electric field of a light wave of intensity I (see eqs. (2.8) and (2.9)) becomes comparable with the mean kinetic energy due to thermal motion, *i.e.* when

$$I \sim I_{NL} = 2N_{\text{ec}}\,ckT_{\text{e}}\,, \tag{5.1}$$

or equivalently when

$$p_{\text{rad}} \sim 2(\omega/\omega_p)^2 p_{\text{e}}\,, \tag{5.2}$$

where p_{rad} $(= U = I/c)$ denotes the radiation pressure of the light wave. That is, the nonlinearity is a high-intensity low-temperature effect.

Nonlinear effects associated with light intensities of magnitude I_{NL}, for example, are the intensity dependence of the optical absorption coefficient $\varkappa$, the ion-electron relaxation time τ_{ei}; and other temperature-dependent plasma properties; and the unstable excitation of coupled electron-optical and ion-acoustic plasma modes.

At still higher intensities such that $\langle KE \rangle_{osc} \sim mc^2$, relativistic nonlinearities that originate with the Lorentz force acting on the plasma electrons, or with their relativistic mass increase, will become important. Such effects, are the parametric amplification of light waves in a plasma, the production of second-harmonic light at a plasma surface, and the intensity dependence of the refractive index of a plasma. The characteristic intensity I_{rel} for these relativistic effects is

$$I_{rel} = 2mc^3 N_{ec} \,. \tag{5.3}$$

Using the value $N_{ec} = 10^{21}\ \text{cm}^{-3}$ appropriate to 1.06 μm light, and an electron temperature of 1 keV, we find that

$$I_{NL} = 10^{16}\ \text{W/cm}^2 \qquad (T_e = 1\ \text{keV}),$$

$$I_{rel} = 5\cdot 10^{18}\ \text{W/cm}^2 \,.$$

A focused intensity 20 times as large as this value of I_{NL}, and therefore well into the region of strong nonlinear optical effects, can be achieved with existing high brightness lasers [5]. Although this intensity is only 1/25-th as large as I_{rel}, relativistic nonlinearities should be easily observable.

We shall discuss some of these nonlinear optical effects in plasmas, considering first single-electron effects and then collective effects involving plasma oscillations.

5·1. *Single electron nonlinear optical effects.*

5·1.1. Intensity-dependence of the free-free absorption coefficient. The intensity-dependence of the optical absorption coefficient of a plasma has been discussed by Rand [14] and also by Hughes and Nicholson-Florence [15]. The treatment given below is due to Scofield [16], and yields results similar to those of Rand.

We begin with the expression for the time-averaged rate $\langle \dot{W} \rangle$ at which an external force $\boldsymbol{F}$ $(= m\dot{\boldsymbol{u}})$ does work on the plasma electrons (per unit volume):

$$\langle \dot{W} \rangle = \int \langle f\boldsymbol{F} \rangle \cdot \boldsymbol{v}\, \mathrm{d}^3 v \,, \tag{5.4}$$

where $f(\boldsymbol{v}, t)$ is the time-dependent distribution of electron velocities $\boldsymbol{v}$ satisfying the Boltzmann equation

$$\partial f/\partial t + (\dot{\boldsymbol{u}}\cdot\boldsymbol{\nabla})f = \dot{f}_{\text{coll}}\,, \tag{5.5}$$

and the normalization condition

$$\int f(\boldsymbol{v}, t)\,\mathrm{d}^3 v = N_{\mathrm{e}}\,. \tag{5.6}$$

In the case of light absorption by the plasma, the external force $\boldsymbol{F}$ is due to the electric field $\boldsymbol{E}$ of the light wave,

$$\boldsymbol{F} = m\dot{\boldsymbol{u}} = e\boldsymbol{E}(t)\,, \qquad \boldsymbol{E}(t) = \boldsymbol{E}\exp[-i\omega t]. \tag{5.7}$$

The absorption coefficient $\varkappa$ is obtained from the relation

$$\varkappa = \langle \dot{W} \rangle / I\,, \tag{5.8}$$

where I is the intensity of the light

$$I = (c/4\pi)\langle E^2 \rangle\,. \tag{5.9}$$

Equation (5.4) is not directly useful in the form given because the effects of collisions are implicitly contained in the unknown distribution function $f(\boldsymbol{v}, t)$. However it can be transformed into the following well-known form:

$$\langle \dot{W} \rangle = m\int \nu_m \langle f\boldsymbol{u} \rangle \cdot \boldsymbol{v}\,\mathrm{d}^3 v\,, \tag{5.10}$$

in which the effect of collisions appears explicitly in the coefficient $\nu_m(v)$, the electron-ion collision frequency for momentum transfer. That is, the external force $m\dot{\boldsymbol{u}}$ appearing in eq. (5.4) can be replaced by an equivalent « viscous drag » force $m\nu_m\boldsymbol{u}$ to obtain eq. (5.10).

The collision frequency ν_m is defined in terms of the differential scattering cross-section $\sigma(v, \theta)$ by the relations

$$\nu_m(v) = N_i v \sigma_m(v)\,, \tag{5.11}$$

where

$$\sigma_m(v) = \int_{4\pi} \sigma(v, \theta)(1 - \cos\theta)\,\mathrm{d}\Omega\,. \tag{5.12}$$

For the case of Coulomb scattering of the plasma electrons by ions the appropriate cross-section is the Rutherford cross-section,

$$\sigma(v, \theta) = (\sigma_B/4\pi) \sin^{-4}(\theta/2) , \tag{5.13}$$

where σ_B is the backward scattering ($> 90°$) cross-section,

$$\sigma_B = \pi(Ze^2/mv^2)^2 . \tag{5.14}$$

Evaluating the integral in eq. (5.12), we obtain

$$\sigma_m = - 4\sigma_B \ln \sin (\theta_{min}/2) , \tag{5.15}$$

where θ_{min} is the cut-off at small scattering angles or large impact parameters that is needed to prevent divergence. The scattering angle θ is related to the impact parameter b according to,

$$\mathrm{tg}^2 (\theta/2) = \sigma_B/\pi b^2 , \tag{5.16}$$

so that we obtain,

$$\sigma_m = 2\sigma_B \ln [1 + (\pi b^2_{max}/\sigma_B)] = 2\sigma_B \ln [1 + \Lambda^2] \simeq 4\sigma_B \ln \Lambda , \tag{5.17}$$

where b_{max} is taken equal to the Debye length λ_D, and $\ln \Lambda$ denotes the usual Coulomb logarithm [6]. From eqs. (5.11), (5.14), and (5.17) we obtain from ν_m the result

$$\nu_m = A/v^3 , \tag{5.18}$$

$$A = 4\pi N_i (Ze^2/m)^2 \ln \Lambda . \tag{5.19}$$

All that remains to be done is to determine the distribution function $f(\boldsymbol{v}, t)$. If the light intensity is very high and the electron temperature is low we expect that the electrons will interact more strongly with each other than with the ions, relative to which they are in very rapid oscillatory motion. It follows that the electrons will be in approximate thermal equilibrium among themselves in the oscillating frame of reference, so that we may write

$$f(\boldsymbol{v}, t) = f[\boldsymbol{v} - \boldsymbol{u}(t)] \tag{5.20}$$

$$= (m/2\pi kT)^{\frac{3}{2}} N_e \exp [- m(\boldsymbol{v} - \boldsymbol{u})^2/2kT] , \tag{5.21}$$

where we have disregarded the (small) collisional damping of the oscillatory motion of the electrons in identifying their oscillatory component of velocity with $\boldsymbol{u}$, the oscillatory velocity of a free electron in the electric field of the light wave.

From eqs. (5.10), (5.18), and (5.20) we obtain after some manipulations,

$$\langle \dot{W} \rangle = 4\pi m A \left\langle \int_0^1 (uy)^2 f(uy)\, \mathrm{d}y \right\rangle . \tag{5.22}$$

A curious property of this result is that *only those electrons whose thermal velocity is less than their oscillatory velocity contribute to the absorption of light.* If for simplicity we restrict our attention to the case of circularly polarized light, then

$$u^2 = \text{const} = (4\pi e^2/\omega^2 mc)\, I . \tag{5.23}$$

Making use of the specific form of f given by eq. (5.21) and evaluating the integral in eq. (5.22) we obtain finally the result

$$\varkappa_0(I) = \varkappa_0 F(\alpha) , \tag{5.24}$$

where $\varkappa_0$ is the absorption coefficient given earlier in eq. (2.18), and

$$F(\alpha) = \frac{3}{4\alpha}\left\{\sqrt{\frac{\pi}{\alpha}}\,\mathrm{erf}\,(\sqrt{\alpha}) - 2\exp[-\alpha]\right\}, \tag{5.25}$$

$$\alpha = I/I_{NL} , \qquad I_{NL} = 2N_{ec}\, ckT . \tag{5.26}$$

The function $F(\alpha)$ is a decreasing function of α with the properties

$$F(\alpha) = \begin{cases} 1 - \frac{3}{5}\alpha , & \alpha \ll 1 , \\ 0.568 , & \alpha = 1 , \\ 3\sqrt{\pi}/4\alpha^{\frac{3}{2}} , & \alpha \gg 1 . \end{cases} \tag{5.27}$$

I_{NL} is therefore a value of the light intensity such that nonlinear intensity-dependent effects have reduced the absorption coefficient to approximately half (0.568) its low-intensity value $\varkappa_0$.

The power absorbed from the light wave per unit volume of plasma is given by

$$\langle \dot{W} \rangle = \varkappa_0 I_{NL}\, \alpha F(\alpha) , \tag{5.28}$$

attaining a maximum value when the light intensity I equals approximately

$(2.3) I_{NL}$. A graph of the function $\alpha F(\alpha)$ showing this maximum is given in Fig. 12.

The case of plane-polarized light has also been considered, the absorption coefficient being at most 20% smaller than that given above for circularly polarized light for $\alpha < 100$, as have corrections to the theory which account for the departure of the electron distribution function from the Maxwellian form assumed in eq. (5.21) [16]. These latter corrections lead to a decrease in the absorption coefficient given by eq. (5.24) amounting to at most 24% the largest decrease occurring when $\alpha \sim 1$.

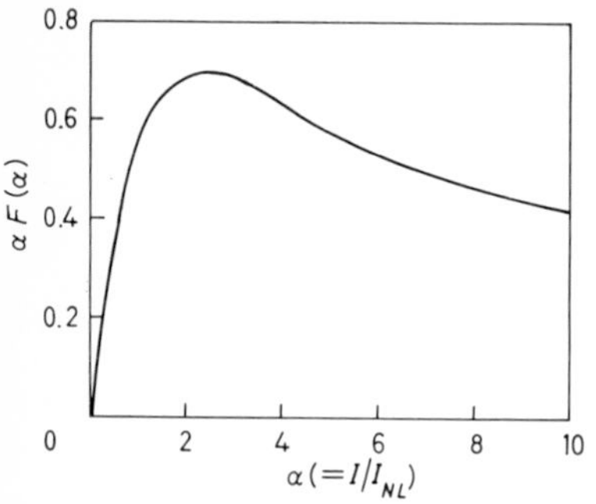

Fig. 12. – Dimensionless plot of the power absorbed from a light wave per unit volume of plasma as a function of the light intensity.

It has been assumed in the foregoing treatment that the intensity-dependence of the absorption coefficient arises only through the intensity-dependence of the electron velocity distribution function, there being no change in the ion-electron scattering cross-section itself in the presence of the intense light wave. This is justifiable if the electric field of the light wave changes only slightly during the time $\Delta\tau_c$ during which an electron is scattered by an ion. Only in this case can one speak of an ion-electron encounter in which the electron has a well-defined velocity $\boldsymbol{v}$ of encounter with the ion. Consequently we require that

$$\omega\,\Delta\tau_c \ll 1\,. \tag{5.29}$$

The duration of a collision $\Delta\tau_c$ is the range r_B of the interaction divided by the speed of the electron. Defining the range r_B in terms of the backward scattering cross-section σ_B according to

$$\pi r_B^2 = \sigma_B\,, \tag{5.30}$$

we have

$$\Delta\tau_c = r_B/v\,, \qquad r_B = Ze^2/mv^2\,, \tag{5.31}$$

or

$$\Delta\tau_c = (2\cdot 10^{-20})\,Z/T_e^{\frac{3}{2}}\ \text{s}\,, \tag{5.32}$$

where T_e is the electron temperature in keV. For 1.06 μm light ($\omega = 1.78\cdot$ $\cdot 10^{15}$ rad/s) we have

$$\omega\,\Delta\tau_c = (4\cdot 10^{-5})\,Z/T_e^{\frac{3}{2}}\ \text{rad}\,, \tag{5.33}$$

so that except for rather low temperatures the requirement that $\omega\Delta\tau_c$ be small compared with unity is amply satisfied.

Even though the scattering cross-section itself may not be significantly intensity-dependent, the cut-off at large impact parameters will be, since it depends on intensity-dependent screening effects. The correct value of this cut-off remains to be determined, but the results are not expected to be very sensitive to its value bacause it enters logarithmically.

5'1.2. Intensity-dependence of the refractive index. – If the collision frequency is much less than the light frequency ($\nu_c \ll \omega$) and the plasma is underdense ($\omega_p < \omega$), the refractive index n and dielectric constant ε can be written

$$\varepsilon = n^2 = 1 - (\omega_p^*/\omega)^2 , \tag{5.34}$$

$$(\omega_p^*)^2 = 4\pi e^2 N_e/m^* , \tag{5.35}$$

where m^* is the effective mass of the plasma electrons. The intensity-dependence of m^* arises through the relativistic increase of mass with energy, and through the excitation of longitudinal electron oscillations at frequency 2ω.

The relativistic equation of motion of an electron in an electromagnetic field is,

$$\mathrm{d}(\gamma m\boldsymbol{v})/\mathrm{d}t = e\{\boldsymbol{E} + (\boldsymbol{v}\times\boldsymbol{B})/c\} , \tag{5.36}$$

$$\gamma = (1-\beta^2)^{-\frac{1}{2}} , \qquad \beta = v/c, \tag{5.37}$$

and may be written in the dimensionless form,

$$\boldsymbol{\beta} = \sqrt{1-\beta^2}\int(\tilde{\boldsymbol{E}} + \boldsymbol{\beta}\times\tilde{\boldsymbol{B}})\,\mathrm{d}\tau , \tag{5.38}$$

where

$$\boldsymbol{\beta} = (1/c)\,\boldsymbol{v} , \qquad \tau = \omega t, \tag{5.39}$$

$$\tilde{\boldsymbol{E}}, \tilde{\boldsymbol{B}} = (e/mc\omega)(\boldsymbol{E}, \boldsymbol{B}) . \tag{5.40}$$

We now choose $\tilde{\boldsymbol{E}}$ and $\tilde{\boldsymbol{B}}$ so as to represent a plane light wave propagating in the positive z-direction, *i.e.*

$$\tilde{\boldsymbol{E}}, \tilde{\boldsymbol{B}} = (\tilde{\boldsymbol{E}}_0, \tilde{\boldsymbol{B}}_0)\cos(\tau - kz) , \tag{5.41}$$

$$\tilde{\boldsymbol{E}}_0\cdot\tilde{\boldsymbol{B}}_0 = 0 , \qquad |\tilde{\boldsymbol{B}}_0| = \sqrt{\varepsilon}\,|\tilde{\boldsymbol{E}}_0| , \tag{5.42}$$

$$kc = \sqrt{\varepsilon}\,\omega , \tag{5.43}$$

and solve eq. (5.38) for $\boldsymbol{\beta}$ as an expansion in powers of the dimensionless field amplitudes $\tilde{\boldsymbol{E}}$ and $\tilde{\boldsymbol{B}}$. We obtain

$$\boldsymbol{\beta} = \sum_i \boldsymbol{\beta}_i\,, \qquad \boldsymbol{\beta}_0 = 0\,, \tag{5.44}$$

$$\boldsymbol{\beta}_1 = \int_0^\tau \tilde{\boldsymbol{E}}\,\mathrm{d}\tau = \tilde{\boldsymbol{E}}_0 \sin\tau\,, \tag{5.45}$$

$$\boldsymbol{\beta}_2 = \int_0^\tau (\boldsymbol{\beta}_1 \times \tilde{\boldsymbol{B}})\,\mathrm{d}\tau = \tfrac{1}{4}(\tilde{\boldsymbol{E}}_0 \times \tilde{\boldsymbol{B}}_0)[1 - (1/\varepsilon_2)\cos 2\tau]\,, \tag{5.46}$$

$$\boldsymbol{\beta}_3 = -(\beta_1^2/2)\boldsymbol{\beta}_1 + \int_0^\tau (\boldsymbol{\beta}_2 \times \tilde{\boldsymbol{B}})\,\mathrm{d}\tau + \int_0^\tau \{\tilde{\boldsymbol{E}}[\tau - \varphi(\tau)] - \tilde{\boldsymbol{E}}(\tau)\}\,\mathrm{d}\tau\,. \tag{5.47}$$

$\varphi(\tau)$ is the change in phase of the light wave in the rest frame of an electron resulting from its longitudinal motion;

$$\varphi(\tau) = kz(\tau) = \sqrt{\varepsilon_1}\int_0^\tau \beta_2\,\mathrm{d}\tau = \alpha[\tau - (1/2\varepsilon_2)\sin 2\tau]\,, \tag{5.48}$$

$$\varepsilon_1 = \varepsilon_0(\omega) = 1 - (\omega_p/\omega)^2\,, \qquad \omega_p = \omega_p^*\,(m^* = m)\,, \tag{5.49}$$

$$\varepsilon_2 = \varepsilon_0(2\omega)\,, \qquad \alpha = \varepsilon_1\,\tilde{E}_0^2/4\,. \tag{5.50}$$

In assuming $\boldsymbol{\beta}_0$ to be zero we are disregarding any electron motion other than that due to the presence of the light wave. The factor $(1/\varepsilon_2)$ appearing in eq. (5.46) is included to properly account for the dynamic response of the longitudinal electron oscillations to the Lorentz force driving them at frequency 2ω. (We assume that $k \ll k_{\mathrm{D}}$ = reciprocal Debye length.)

Evaluating the first two terms on the right of eq. (5.47) and discarding terms with frequency 3ω, we obtain:

$$\boldsymbol{\beta}_3^{(1)} = -[1 + (3/2\varepsilon_1) - (1/2\varepsilon_2)]\alpha\tilde{\boldsymbol{E}}_0 \sin\tau\,. \tag{5.51}$$

Evaluating the field $\tilde{\boldsymbol{E}}$ in the rest frame of the electrons, we have,

$$\begin{aligned}\tilde{\boldsymbol{E}}[\tau + \varphi(\tau)] &= \tilde{\boldsymbol{E}}_0 \cos\{(1-\alpha)\tau + (\alpha/2\varepsilon_2)\sin 2(1-\alpha)\tau\} = \\ &= \tilde{\boldsymbol{E}}_0 \left\{\left(1 - \frac{\alpha}{4\varepsilon_2}\right)\cos(1-\alpha)\tau + \left(\frac{\alpha}{4\varepsilon_2}\right)\cos 3(1-\alpha)\tau\right\}.\end{aligned} \tag{5.52}$$

The factor $(1-\alpha)$ represents a red-shift of the frequency of the light wave, as seen by the electrons, resulting from their steady drift in the direction of propagation of the light wave. Discarding the component with frequency 3ω (in the laboratory frame), we obtain for the third term on the right side of eq. (5.47) the result,

$$\boldsymbol{\beta}_3^{(2)} = [1 - (1/4\varepsilon_2)]\alpha\tilde{\boldsymbol{E}}_0 \sin\tau \,, \tag{5.53}$$

and adding this to $\boldsymbol{\beta}_3^{(1)}$ we obtain for the transverse part of $\boldsymbol{\beta}$ at frequency ω,

$$\boldsymbol{\beta} = \{1 - [3 - (\varepsilon_1/2\varepsilon_2)](\tilde{E}_0^2/8)\}\,\tilde{\boldsymbol{E}}_0 \sin\tau \,, \tag{5.54}$$

where

$$\varepsilon_0(\omega) = \varepsilon_1 < \varepsilon_2 = \varepsilon_0(2\omega) \,.$$

We note that the predominant nonlinear effect is the relativistic mass increase. It should also be noted that the effect of the steady electron drift has cancelled out upon adding $\boldsymbol{\beta}_3^{(1)}$ and $\boldsymbol{\beta}_3^{(2)}$, so that our results will be unchanged if, for example, the electron drift should be reduced to zero by an opposing electrostatic field.

It follows from eq. (5.54) that the effective mass m^* is given by,

$$m^*/m = 1 + \pi[3 - (\varepsilon_1/2\varepsilon_2)](e/mc\omega)^2 U \,, \tag{5.55}$$

where U is the energy density of the light wave in the plasma,

$$U = \langle E^2\rangle/4\pi = E_0^2/8\pi \,. \tag{5.56}$$

Substituting eq. (5.55) into eq. (5.34) yields the result:

$$\varepsilon = 1 - \left\{\left(\frac{\omega}{\omega_p}\right)^2 + \tfrac{3}{4}[1 - (\varepsilon_1/6\varepsilon_2)]\frac{U}{N_e mc^2}\right\}^{-1} . \tag{5.57}$$

Since $\mathrm{d}\varepsilon/\mathrm{d}U > 0$ it follows that focusing rather than spreading of intense collimated light beams will occur in a plasma. If the frequency ω is large compared with the plasma frequency ω_p, eq. (5.57) can be written,

$$\varepsilon = \varepsilon_1 + \delta\varepsilon \,, \tag{5.58}$$

$$\delta\varepsilon = \tfrac{5}{4}(\omega_p/\omega)^4(I/I_{\text{rel}}) \,, \tag{5.59}$$

where I $(= cU)$ is the intensity of the light wave and I_{el} is the characteristic intensity for the occurrence of significant relativistic effects described earlier in eq. (5.3).

The effect of collisions has been disregarded from the outset in the preceding treatment. If the effect of collisions on the dielectric constant is taken into account, still assuming that $\omega_p < \omega$ and that $\nu_c \ll \omega$ the expression for ε becomes,

$$\varepsilon = 1 - [(\omega/\omega_p^*)^2 + (\nu_c/\omega_p)^2]^{-1} . \tag{5.60}$$

According to eq. (17), the collision frequency ν_c is proportional to the absorption coefficient $\varkappa_0$ which we have seen is intensity-dependent (eq. (5.24)), so that the collision term in the expression for ε above will contribute to the intensity-dependence of the dielectric constant. Making use of the expansion of $F(\alpha)$ for small α given in eq. (5.27) we can combine the results of eq. (5.57) and eq. (5.60) to obtain

$$\varepsilon = 1 - \left\{\left(\frac{\omega}{\omega_p}\right)_0^2 + \left(\frac{\nu_c}{\omega_p}\right)_0^2 + \left[\frac{3}{2}\left(\frac{kT_e}{mc^2}\right) - \frac{6}{5}\left(\frac{\nu_c}{\omega_p}\right)_0^2\right]\left(\frac{I}{I_{NL}}\right)\right\}^{-1}, \tag{5.61}$$

where the subscript (0) denotes evaluation at zero light intensity, and where the small term involving the ratio $\varepsilon_1/\varepsilon_2$ has been discarded.

The coefficient of I in the expression for ε will be positive, and hence self-focusing rather than defocusing will occur, at sufficiently high temperature or low density that

$$kT_e/mc^2 > \tfrac{4}{5}(\nu_c/\omega_p)_0^2 , \tag{5.62}$$

or when

$$T_e > (0.1)\, Z^{\frac{1}{2}}\varphi^{\frac{1}{4}}\ \text{keV} , \tag{5.63}$$

where

$$\varphi = N_e/N_{ec} , \qquad N_{ec} = 10^{21}\ \text{cm}^{-3} .$$

5·2. *Collective nonlinear optical effects.* – A number of interesting nonlinear optical effects can be discussed within the framework of the simple two-fluid (electrons and ions) hydrodynamic model of a plasma. The two fundamental modes of oscillation of such a plasma are the high-frequency electron oscillations (optical branch) in which the ions do not participate significantly due to their large mass, and the low-frequency ion oscillations (acoustic branch) which exhibit approximate electrical neutrality because the electrons can readily respond to the relatively slow ion oscillations and neutralize any imbalance of charge. The dispersion relations (neglecting damping) for these two modes of oscillation are illustrated in Fig. 13 and are given by [6]

$$\omega^2 \simeq \begin{cases} \omega_e^2 + V_e^2 k^2 , & \text{electron (optical branch)} , \\ \dfrac{\omega_i^2\, V_e^2 k^2}{\omega_e^2 + V_e^2 k^2} & \text{electron-ion (acoustic branch)} , \end{cases} \tag{5.64}$$

where ω_e is the electron plasma frequency,

$$\omega_e^2 = 4\pi e^2 N_{e0}/m \, , \tag{5.65}$$

ω_i is the ion plasma frequency,

$$\omega_i^2 = (Zm/M)\,\omega_e^2 \, , \tag{5.66}$$

and V_e is proportional to the r.m.s. thermal velocity of the electrons, *i.e.*

$$V_e^2 = \gamma_e K T_e / m \, . \tag{5.67}$$

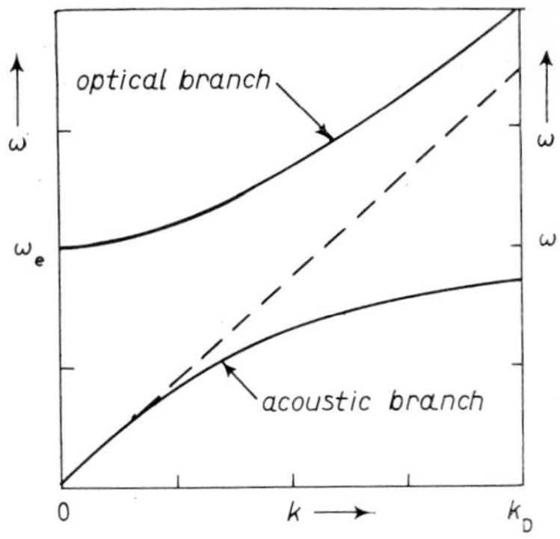

Fig. 13. – Dispersion curves of frequency ω *vs.* wavenumber k for the optical and acoustic modes of plasma oscillation. The scale at the left applies to the optical branch; that at the right to the acoustic branch. These scales are very different because $\omega_i = \sqrt{Zm/M}\,\omega_e \ll \omega_e$. ($k_D$ is the Debye wavenumber.)

In these relations N_{e0} denotes the electron density of the undisturbed plasma, m and M are the electron and ion masses, respectively, and γ_e is the (frequency-dependent) adiabatic exponent for the compression of the plasma electrons. For the high-frequency electron mode γ_e has the value (3) corresponding to an effectively one-dimensional ideal electron gas; for the low-frequency acoustic mode γ_e equals $\frac{5}{3}$ corresponding to a 3-dimensional gas.

If the wave number k of the oscillation is small compared with the Debye wave number k_D, *i.e.*, $k^2 \ll k_D^2/\gamma_e$ $(= \omega_e^2/V_e^2)$, then the dispersion relations become simply

$$\begin{cases} \omega_e \ \omega \ (\text{optical}) \, , \\ Vk \ (\text{acoustic}) \, , \end{cases} \tag{5.68}$$

where the acoustic wave velocity V is given by the self-evident relation

$$V^2 = (Z\gamma_e K T_e + \gamma_i k T_i)/(M + Zm) \, . \tag{5.69}$$

An intense light wave $(\boldsymbol{k}_L, \omega_L)$ propagating in a plasma can interact with either the electronic $(\boldsymbol{k}_v, \omega_v)$ or acoustic $(\boldsymbol{k}_a, \omega_a)$ modes, or both, giving rise to some interesting 3-wave nonlinear effects. If the intense light wave is scattered by the electron oscillations of the plasma one has the plasma equivalent of the *stimulated Raman effect*. If it is scattered by the ion-acoustic oscillations one obtains the *stimulated Brillouin effect*. In both cases the third wave is the scattered (Stokes) light wave. In these cases, as well as in the case in which it interacts with both an electron wave and an ion-acoustic wave, the intense light wave parametrically couples the other two waves together in such a fashion that, if the intensity is high enough, the amplitude of the parametrically-driven waves grows exponentially with time, energy being supplied to the waves faster than it can be dissipated.

We shall illustrate this nonlinear parametric excitation process by considering in some detail the coupling of the optical and acoustic plasma modes, and also the stimulated Raman effect in a plasma. Another nonlinear optical effect mentioned earlier is the excitation of electron plasma oscillations at twice the frequency of the incident light wave, the coupling between the transverse light wave and the longitudinal electron oscillation being provided by the Lorentz force. Although these longitudinal waves cannot radiate electromagnetic energy in the interior of a plasma, they *can* radiate at the *surface* of a plasma giving rise to the production of second harmonic light [17].

5.2.1. Parametric coupling and excitation of electron optical and electron-ion acoustic modes by an intense light wave. – This nonlinear parametric interaction appears to have first been studied by DuBois and Goldman [18] using the Feynman-diagram approach of quantum statistical mechanics. The treatment we shall give closely follows the simple two-fluid hydrodynamic approach subsequently given by Lee and Su [19].

The two-fluid equations governing the fluctuation in the electron and ion densities n and N, respectively, are the following:

$$\ddot{n} + \nu_e \dot{n} - V_e^2 \nabla^2 n = -\omega_e^2 (n - ZN) + (\nabla \cdot \boldsymbol{F}), \tag{5.70}$$

$$\ddot{N} + \nu_i \dot{N} - V_i^2 \nabla^2 N = (\omega_i^2/Z)(n - ZN), \tag{5.71}$$

where ν_e and ν_i are phenomenological damping coefficients,

$$V_i^2 = \gamma_i K T_i / M, \tag{5.72}$$

$$\boldsymbol{F} = (e/m) n \boldsymbol{E}_L + N_e (\boldsymbol{v} \cdot \nabla) \boldsymbol{v} - \dot{n} \boldsymbol{v} + N_e (e/mc)(\boldsymbol{v} \times \boldsymbol{B}_L), \tag{5.73}$$

$$N_e = N_{e0} + n, \qquad N_i = N_{i0} + N, \tag{5.74}$$

$\boldsymbol{v}$ is the velocity of the electrons, and $\boldsymbol{E}_L$ and $\boldsymbol{B}_L$ denote the electric and magnetic fields of the light wave:

$$\boldsymbol{E}_L(\boldsymbol{x}, t) = \boldsymbol{E}_L \exp [i(\boldsymbol{k}_L \cdot \boldsymbol{x} - \omega_L t)], \qquad (\boldsymbol{k}_L \cdot \boldsymbol{E}_L) = 0. \tag{5.75}$$

Equations (5.70) and (5.71) are derived from the equations of motion and charged-particle conservation for electrons and ions together with Poisson's equation, assuming scalar electron and ion pressures. The equation for the ion motion has been linearized and any direct effect of the light wave on the ions has been disregarded in obtaining eq. (5.71). The nonlinear coupling effects all originate with the source term $(\nabla\cdot\boldsymbol{F})$ which drives the oscillations.

Using eq. (5.70) to eliminate the electrostatic coupling term on the r.h.s. of eq. (5.71) we obtain:

$$(5.76)\qquad (\ddot{n}+\nu_e\dot{n}-V_e^2\nabla^2 n)+(M/m)(\ddot{N}+\nu_i\dot{N}-V_i^2\nabla^2 N)=(\nabla\cdot\boldsymbol{F}).$$

We now consider the excitation of the electron mode $(\boldsymbol{k}_v, \omega_v)$ and the acoustic mode $(\boldsymbol{k}_a, \omega_a)$, where:

$$(5.77)\qquad n(\boldsymbol{x},t)=n_a\exp[i(\boldsymbol{k}_a\cdot\boldsymbol{x}-\omega_a t)]+n_v\exp[i(\boldsymbol{k}_v\cdot\boldsymbol{x}-\omega_v t)],$$

$$(5.78)\qquad N(\boldsymbol{x},t)=N_a\exp[i(\boldsymbol{k}_a\cdot\boldsymbol{x}-\omega_a t)],$$

$$(5.79)\qquad \boldsymbol{k}_L=\boldsymbol{k}_v+\boldsymbol{k}_a,\qquad \omega_L=\omega_v+\omega_a^*.$$

Note that we disregard the high-frequency component N_v of the ion fluctuation.

a) Acoustic mode: $(n_a\simeq ZN_a)$.

For the acoustic mode we obtain directly from eq. (5.76) the result,

$$(5.80)\qquad (\omega_a^2-V_a^2k_a^2+i\nu_a\omega_a)N_a=-i(m/M)\boldsymbol{k}_a\cdot\boldsymbol{F}(\omega_a),$$

where

$$(5.81)\qquad \nu_a=\nu_i+(mZ/M)\nu_e,$$

$$(5.82)\qquad V_a^2=V_i^2+(mZ/M)V_e^2=(\gamma_i KT_i+Z\gamma_e KT_e)/M,$$

$$(5.83)\qquad \boldsymbol{F}(\omega_a)=(e/m)n_v^*\boldsymbol{E}_L-i\omega_v^*n_v^*\boldsymbol{v}_L+iN_{e0}[(\boldsymbol{v}_v^*\cdot\boldsymbol{k}_L)\boldsymbol{v}_L-(\boldsymbol{v}_L\cdot\boldsymbol{k}_v)\boldsymbol{v}_v^*],$$

disregarding the Lorentz force (last term on r.h.s. of eq. (5.73)). However,

$$(5.84)\qquad \boldsymbol{v}_L=-i(e/m\omega_L)\boldsymbol{E}_L;$$

and from the (linearized) equation of electron conservation,

$$(5.85)\qquad \dot{n}+N_{e0}(\nabla\cdot\boldsymbol{v})=0,$$

it follows that

$$(5.86)\qquad N_{e0}\boldsymbol{v}_v^*=(\omega_v^*/k_v^2)\boldsymbol{k}_v n_v^*.$$

Substituting eqs. (5.84) and (5.86) into eq. (5.83) we obtain,

$$(5.87)\qquad -(m/M)\boldsymbol{k}_a\cdot\boldsymbol{F}(\omega_a)=(e/M)(\omega_v^*/\omega_L)(\boldsymbol{k}_a\cdot\boldsymbol{E}_L)\{1-(\omega_a/\omega_v^*)-2(\boldsymbol{k}_L\cdot\boldsymbol{k}_v)/k_v^2\}n_v^*.$$

Taking $\omega_v/\omega_L \simeq 1$, $\omega_a/\omega_v \ll 1$, and $k_L/k_v \ll 1$, we can combine eqs. (5.80) and (5.87) to obtain the *acoustic mode equation*:

$$D_a(\omega_a)\,N_a = i(e/m)(\boldsymbol{k}_a\cdot\boldsymbol{E}_L)\,n_v^*\,, \tag{5.88}$$

where

$$D_a(\omega_a) = \omega_a^2 - V_a^2 k_a^2 + i\nu_a\omega_a\,. \tag{5.89}$$

b) Optical mode:

For the high-frequency electron optical mode we obtain directly from eq. (5.70),

$$(\omega_v^2 - \omega_e^2 - V_e^2 k_v^2 + i\nu_e\omega_v)\,n_v = -i\boldsymbol{k}_v\cdot\boldsymbol{F}(\omega_v)\,. \tag{5.90}$$

In the same manner that the result of eq. (5.87) was obtained we find that

$$\boldsymbol{k}_v\cdot\boldsymbol{F}(\omega_v) = (Ze/m)(\omega_v/\omega_L)(\boldsymbol{k}_v\cdot\boldsymbol{E}_L)\{1-(\omega_a^*/\omega_v)[1-2(\boldsymbol{k}_L\cdot\boldsymbol{k}_a)/k_a^2]\}\,N_a^*\,. \tag{5.91}$$

Our assumption that $k_L/k_v \ll 1$ implies, by eq. (5.79), that $\boldsymbol{k}_a \simeq -\boldsymbol{k}_v$ so that k_L/k_a is also $\ll 1$. Again employing the relations $\omega_v/\omega_L \simeq 1$, $\omega_a/\omega_v \ll 1$, and combining eq. (5.90) with (5.91) we obtain the *optical-mode equation*,

$$D_e(\omega_v)\,n_v = -i(Ze/m)(\boldsymbol{k}_v\cdot\boldsymbol{E}_L)\,N_a^*\,, \tag{5.92}$$

where

$$D_e(\omega_v) = \omega_v^2 - \omega_e^2 - V_e^2 k_v^2 + i\nu_e\omega_v\,. \tag{5.93}$$

We now determine the *threshold light intensity for instability*. Eliminating N_a between eqs. (5.88) and (5.92) we obtain the following equation for the propagation of n_v:

$$[D_e(\omega_v) - (\Omega^2/|D_a(\omega_a^*)|^2)\,D_a(\omega_a)]\,n_v = 0\,, \tag{5.94}$$

where

$$\Omega^2 = (Ze^2/mM)|\boldsymbol{k}_v\cdot\boldsymbol{E}_L|^2 = (Ze^2/mM)\,k_v^2|\boldsymbol{E}_L|^2\cos^2\theta\,, \tag{5.95}$$

and we have used the property that $(\boldsymbol{k}_v\cdot\boldsymbol{E}_L) = -(\boldsymbol{k}_a\cdot\boldsymbol{E}_L)$.

The threshold condition (zero net damping of n_v and N_a) is obtained by equating to zero the imaginary part of the coefficient of n_v in eq. (5.94), *i.e.*

$$\nu_e\omega_v = (\Omega^2/|D_a(\omega_a)|^2)\,\nu_a\omega_a\,, \tag{5.96}$$

ω_v and ω_a now being real. Writing

$$Q_e = \omega_v/\nu_e\,, \qquad Q_a = \omega_a/\nu_a\,, \tag{5.97}$$

$$I_L = \sqrt{\varepsilon_L}\,c\,U_L\,, \qquad U_L = |E_L|^2/2\pi\,, \tag{5.98}$$

$$\varepsilon_L = 1 - (\omega_e/\omega_L)^2\,(\ll 1)\,, \tag{5.99}$$

$$\text{Min}\,|D_a(\omega_a)|^2 = |D_a(\omega_a)|^2_{\omega_a = V_a k_a} = (\nu_a\,\omega_a)^2\,, \tag{5.100}$$

where I_L is the intensity and U_L is the energy density of the light wave *in the plasma*, and again using the relations $\omega_v/\omega_L \sim 1$, $k_a/k_v \sim 1$, we find the *threshold light intensity for instability* to be:

$$I_{\text{th}} = \sqrt{\varepsilon_L}(\gamma/Q_a Q_e)(1 + T_i/ZT_e)\,I_{NL}\,. \tag{5.101}$$

In obtaining this result we have let $\cos\theta = 1$, and $\gamma_e = \gamma_i = \gamma$.

Since $\sqrt{\varepsilon_L}$ is less than unity and the product $Q_a Q_e$ can be large compared with unity (especially if the electrons are very hot and $T_i \ll T_e$), the threshold for this instability can be quite small compared with the characteristic intensity for nonlinear optical effects I_{NL} defined earlier in eq. (5.1). Hence there are conditions in which this instability will occur at lower light intensities than those required ($I \sim I_{NL}$) to produce significant intensity dependence (reduction) in the optical absorption coefficient of a plasma. Unless $\omega_L \sim \omega_e$, however, the instability cannot be excited.

5·2.2. Parametric amplification of light waves in a plasma. The stimulated Raman effect in a plasma [17] is a parametric amplification process in which a light wave ($\boldsymbol{k}_s$, ω_s) (Stokes wave) is parametrically coupled to an electron plasma oscillation (k_v, ω_v) by a second (intense) light wave (k_L, ω_L) in such a way as to be amplified as it propagates through the plasma. In the terminology of parametric amplifiers the intense light wave is the « pump », the Stokes wave is the « signal » to be amplified, and the electron oscillation is the « idler ». These three waves satisfy the relations,

$$(\boldsymbol{k}_v, \omega_v) + (\boldsymbol{k}_s, \omega_s) = (\boldsymbol{k}_L, \omega_L)\,. \tag{5.102}$$

To investigate this parametric process we make use of eq. (5.70) to describe the electron plasma oscillations, and disregard any fluctuation in ion density, *i.e.* we set N equal to zero. Also we need only consider the Lorentz force term in the source $\boldsymbol{F}$ given by eq. (5.73), in which the coefficient N_e can be replaced by its unperturbed value N_{e0}. Equations (5.70) and (5.73) then become,

$$\ddot{n} + \nu_e \dot{n} + \omega_e^2 n - V_e^2 \nabla^2 n = (\boldsymbol{\nabla}\cdot\boldsymbol{F}), \tag{5.103}$$

$$\boldsymbol{F} = N_{e0}(e/mc)(\boldsymbol{v}\times\boldsymbol{B}). \tag{5.104}$$

We now let

$$\boldsymbol{v}(\boldsymbol{x}, t) = \boldsymbol{v}_v \exp[i(\boldsymbol{k}_v\cdot\boldsymbol{x} - \omega_v t)] + \\ + \boldsymbol{v}_L \exp[i(\boldsymbol{k}_L\cdot\boldsymbol{x} - \omega_L t)] + \boldsymbol{v}_s \exp[i(\boldsymbol{k}_s\cdot\boldsymbol{x} - \omega_s t)], \tag{5.105}$$

$$\boldsymbol{B}(\boldsymbol{x}, t) = \boldsymbol{B}_L \exp[i(\boldsymbol{k}_L\cdot\boldsymbol{x} - \omega_L t)] + \boldsymbol{B}_s \exp[i(\boldsymbol{k}_s\cdot\boldsymbol{x} - \omega_s t)], \tag{5.106}$$

$$n(\boldsymbol{x}, t) = n_v \exp[i(\boldsymbol{k}_v\cdot\boldsymbol{x} - \omega_v t)], \tag{5.107}$$

where

$$\boldsymbol{k}_v \times \boldsymbol{v}_v = (\boldsymbol{k}_L\cdot\boldsymbol{v}_L) = (\boldsymbol{k}_s\cdot\boldsymbol{v}_s) = 0. \tag{5.108}$$

We now calculate the source term $(\boldsymbol{\nabla}\cdot\boldsymbol{F})$ that drives the plasma oscillations at the frequency ω_v. We have,

$$(\boldsymbol{v}\times\boldsymbol{B})_{\omega_v} = (\boldsymbol{v}_s^*\times\boldsymbol{B}_L + \boldsymbol{v}_L\times\boldsymbol{B}_s^*)\exp[i(\boldsymbol{k}_v\cdot\boldsymbol{x} - \omega_v t)], \tag{5.109}$$

so that

$$\boldsymbol{\nabla}\cdot(\boldsymbol{v}\times\boldsymbol{B}) = i\boldsymbol{k}_v\cdot(\boldsymbol{v}_s^*\times\boldsymbol{B}_L + \boldsymbol{v}_L\times\boldsymbol{B}_s^*). \tag{5.110}$$

In addition, we have the transverse motion of the plasma electrons (neglecting damping)

$$i\omega_{L,s} m\boldsymbol{v}_{L,s} = e\boldsymbol{E}_{L,s}. \tag{5.111}$$

From eqs. (5.110) and (5.111), and using Faraday's law

$$\omega\boldsymbol{B} = c(\boldsymbol{k}\times\boldsymbol{E}), \tag{5.112}$$

we obtain the result:

$$D_e(\omega_v)\, n_v = (\boldsymbol{k}_v\cdot\boldsymbol{G})/4\pi m, \tag{5.113}$$

$$\boldsymbol{G} = (\omega_e^2/\omega_L\omega_s)[\boldsymbol{E}_s^*\times(\boldsymbol{k}_L\times\boldsymbol{E}_L) - \boldsymbol{E}_L\times(\boldsymbol{k}_s\times\boldsymbol{E}_s^*)], \tag{5.114}$$

where $D_e(\omega_v)$ is given by eq. (5.93). For simplicity we now assume that the pump wave and the Stokes wave have the same direction of polarization so that eq. (5.113) describing the excitation of the plasma oscillations becomes finally,

$$D_e(\omega_v)\, n_v = (\omega_e^2 k_v^2/4\pi m\omega_L\omega_s)\, E_L E_s^*. \tag{5.115}$$

Next, we consider the propagation of the Stokes wave through the plasma. From Maxwell's equations we have,

$$\nabla \times \nabla \times \boldsymbol{E}_s = (\omega_s^2/c^2)(\boldsymbol{E}_s + 4\pi P_s)\,, \qquad (\nabla \cdot \boldsymbol{E}_s = 0): \tag{5.116}$$

Separating out the nonlinear contribution $\boldsymbol{P}^{NL}(\omega_s)$ to the polarization of the plasma $\boldsymbol{P}_s$ at the Stokes frequency, eq. (5.115) may be written:

$$(\nabla^2 + k_s^2)\,\boldsymbol{E}_s = -\,4\pi(\omega_s/c)^2\boldsymbol{P}^{NL}(\omega_s)\,, \tag{5.117}$$

$$k_s^2 = \varepsilon_s \omega_s^2/c^2\,, \qquad \varepsilon_s = 1 - (\omega_e/\omega_s)^2\,. \tag{5.118}$$

The nonlinear polarization $\boldsymbol{P}^{NL}(\omega_s)$ arises from the fact that the dielectric constant ε of the plasma is (linearly) dependent on the electron density $N_e = N_{e0} + n_v$, which is fluctuating at the frequency (ω_v) with which the electrons are oscillating. This modulation of the dielectric constant causes sidebands of the pump wave ω_L to appear at frequencies $\omega_L \pm \omega_v$, the lower of which is equal to the Stokes frequency ω_s. (In this treatment we are systematically ignoring the presence of the anti-Stokes light wave at frequency $\omega_L + \omega_v$. This is justifiable except for scattering near the forward direction where the anti-Stokes coupling is significant. The effect of the anti-Stokes wave and also the depletion of the pump wave are considered in detail by SHEN and BLOEMBERGEN [20]). Taking into account the dependence of ε on n_v, the nonlinear polarization source of Stokes radiation is easily found to be,

$$\boldsymbol{P}^{NL}(\omega_s) = -\,(e^2/m\omega_L\omega_s)\,n_v^*\boldsymbol{E}_L\,, \tag{5.119}$$

so that the Stokes wave satisfies the inhomogeneous wave equation

$$(\nabla^2 + k_s^2)\,E_s = (\omega_s/\omega_L)(4\pi e^2/mc^2)\,E_L n_v^*\,. \tag{5.120}$$

Solving eq. (5.115) for n_v^* and substituting the result in eq. (5.120) we can write the propagation equation for the Stokes wave in the form

$$\nabla^2 E_s + (\omega_s/c)^2[\varepsilon_s + 4\pi\chi_{NL}|E_L|^2] = 0\,, \tag{5.121}$$

where the nonlinear susceptibility χ_{NL} of the plasma is given by:

$$4\pi\chi_{NL} = -\,(e\omega_e k_v/m\omega_L\omega_s)^2/D^*(\omega_v)\,. \tag{5.122}$$

The absorption coefficient α_s for the Stokes wave is (for intensity):

$$\alpha_s = (4\pi/\sqrt{\varepsilon_s})(\omega_s/c)\,\chi''_{NL}|E|^2\,, \tag{5.123}$$

where

$$\chi''_{NL} = \mathrm{Im}\,(\chi_{NL}) < 0\,. \tag{5.124}$$

The fact that the absorption coefficient is negative implies that the plasma exhibits *gain* for the Stokes wave. Maximum gain occurs at resonance: $\mathrm{Re}\,[D_e(\omega_v)] = 0$, *i.e.* when

$$\omega_v^2 = \omega_e^2 + V_e^2 k_v^2\,, \tag{5.125}$$

and we then obtain for the absorption coefficient the result

$$\alpha_s = -g_s I_L\,, \qquad I_L = (c/2\pi)\sqrt{\varepsilon_L}\,|E_L|^2\,, \tag{5.126}$$

where the coefficient g_s is given by

$$g_s = \frac{2\pi e^2}{m^2 c^4}\left(\frac{k_v^2}{k_L k_s}\right)\left(\frac{\omega_e^2}{\omega_L \omega_v \nu_e}\right). \tag{5.127}$$

Equation (5.127) together with the relation

$$k_v^2 = k_L^2 + k_s^2 - 2k_L k_s \cos\theta_{Ls}\,, \tag{5.128}$$

shows that maximum gain is achieved in the *backward* direction ($\theta_{Ls} = \pi$). We then have $k_v = k_L + k_s \simeq 2k_L$ (assuming $\omega_L \gg \omega_e$) and the absorption coefficient given by eqs. (5.126) and (5.127) becomes for the case of backward scattering

$$\alpha_{sb} = -4Q_e k_L (I_L/I_{rel})\,, \tag{5.129}$$

where I_{rel} is the characteristic intensity for relativistic effects defined by eq. (5.3). The ratio of forward to backward gain is

$$g_{sf}/g_{sb} = (k_L - k_s)^2/(k_L + k_s)^2 \simeq (\omega_e/2\omega_L)^2\,. \tag{5.130}$$

If a diffraction-limited light beam is brought to a focus, one can show that the product of the intensity at the focus I_L times the length L of the focal region (region of high intensity) is given by the relation

$$LI_L \sim k_L P_L/2\pi_L\,, \tag{5.131}$$

where P_L is the power of the light beam. Note that this relation is not dependent on the focal length of the lens or mirror used to focus the beam.

Using the relation together with eq. (5.129) we find that the gain G_{sb} experienced by a Stokes wave traveling backward through the focal region is given by

$$G_{sb} = \exp\left[-\alpha_{sb} L\right] = \exp\left[4Q_e(P_L/P_{rel})\right], \tag{5.132}$$

where

$$P_{rel} = m^2c^5/e^2 = 8.71\ \text{GW}. \tag{5.133}$$

A laser capable of producing one GW of diffraction-limited power, when focused in a hot (homogeneous) plasma in which the electron oscillations are only weakly damped ($Q_e \sim 100$), will therefore produce a gain in the focal region of e^{46}; enough to create from spontaneously scattered light an appreciable amount of backward-traveling Stokes radiation.

The power P_{rel} is not only the natural unit of power associated with stimulated Raman scattering in a plasma, as shown by eq. (5.132), but it is also the natural unit of power associated with the self-focusing of light in a plasma. The critical power P_c for self-focusing is related to the variation $\delta\varepsilon$ in the dielectric constant due to the presence of a light wave with intensity I according to the expression [21]

$$P_c = 2\cdot 5.76(c/\omega)^2(I/\delta\varepsilon). \tag{5.134}$$

Substituting the value of $\delta\varepsilon$ obtained earlier in eq. (5.59) into this expression, we obtain for the critical power P_c the result

$$P_c = 1.47(\omega/\omega_e)^4 P_{rel}. \tag{5.135}$$

6. – Plasma heating by means of an intense burst of relativistic electrons.

We have dwelt at some length on the application of powerful light pulses produced by lasers to the production of high-energy density in matter, and on some of the accompanying nonlinear optical interactions. This application of lasers is by now well known and is being vigorously pursued in many laboratories. A more recent and less widely known approach to the achievement of high-energy density in matter is the utilization of powerful bursts of relativistic electrons for this purpose. This latter approach has been made possible by the development within the last few years of machines capable of producing extremely energetic electron pulses. In 1967 machines were built that could provide pulses of MeV electrons with a pulse energy of a few kilojoule [2]-[4]. Development of these machines has since proceeded to the point where it is now possible to produce megajoule pulses of relativistic electrons [22]. Let us see what this implies for the production of high-energy density in matter.

If a burst of electrons is focused on an area A of a target material, the specific internal energy of the matter in which the electrons are stopped will increase by an amount

$$q = (1/w)(\mathrm{d}w/\varrho\,\mathrm{d}x)(W/A)\,, \tag{6.1}$$

where w is the mean kinetic energy of an electron in the burst

$$w = (\gamma - 1)\,mc^2\,, \qquad \gamma = (1-\beta^2)^{-1/2}\,, \tag{6.2}$$

and W is the total energy of the burst. The collisional component of the energy loss rate $\mathrm{d}w/\varrho\,\mathrm{d}x$ of electrons in matter is approximately independent of electron energy in the range of 0.4 to 10 MeV, and is a slowly decreasing function of the atomic number Z ranging from a value of 2.2 MeV/(g/cm^2) for deuterium to a value of 1.2 MeV/(g/cm^2) for uranium, having an intermediate value of 1.5 MeV/(g/cm^2) for iron [23]. If we now assume, for example, that a megajoule of 10 MeV electrons could be focused on an area as small as 0.1 cm^2 of an iron target, we obtain from eq. (6.1) the result

$$q = (1/10)(1.5)(1/0.1) = 1.5\ \mathrm{MJ/g}\,.$$

Employing the Thomas-Fermi-Dirac equation of state results of Cowan and Ashkin [24], we find that a specific internal energy of 1.5 MJ/g in iron at normal density (7.85 g/cm^3) implies a temperature of 50 eV and a pressure of 50 Mb. This is a very respectable energy-density indeed, being some two orders of magnitude greater than that produced in high explosives.

A pulse of 10 MeV electrons with an energy of 1 megajoule is expected to have a time duration of 100 ns [22], implying an electron current of 10^6 A. With currents this large, we may anticipate that the electromagnetic energy associated with the beam may be comparable or larger than its kinetic energy, and that beam self-forces will then strongly influence its manner of propagation. A detailed theoretical understanding of the propagation characteristics of such high-current beams does not yet exist, but considerable insight can nonetheless be achieved by the consideration of a simple idealized model of the beam such as that employed by Lawson [25].

We consider an idealized electron beam of radius a, current J, and uniform electron density N_e, propagating in a homogeneous background medium in which no currents are flowing. The return current will be assumed to flow on a cylindrical conductor of radius b $(> a)$ coaxial with the beam, and the electron beam will be assumed to be fully charge-neutralized by the medium through which it passes so that the electric field vanishes in the laboratory frame. The electromagnetic energy W_{em} per unit length of beam is then just the magnetic energy per unit length

$$W_{em} = \tfrac{1}{2}\psi J^2 , \tag{6.3}$$

where ψ is a dimensionless geometrical factor

$$\psi = \tfrac{1}{2} + 2\ln(b/a) \tag{6.4}$$

representing the inductance per unit length of beam. The kinetic energy W_k per unit length of beam can be written

$$W_k = (\gamma - 1)(mc^2/e\beta)J , \tag{6.5}$$

and hence the ratio of electromagnetic to kinetic energy is

$$W_{em}/W_k = [\beta\psi/2(\gamma - 1)]J/J^* , \tag{6.6}$$

where

$$J^* = mc^2/e = 17\ \text{kA} . \tag{6.7}$$

For relativistic electrons $\beta \sim 1$, and eq. (6.6) can be written

$$W_{em}/W_k \sim J/(\gamma - 1)J^* , \tag{6.8}$$

where we have taken $\psi \sim 2$.

We conclude that the characteristic beam current J_{em} associated with significant beam self-forces is given by

$$J_{em} = (\gamma - 1)mc^2/e . \tag{6.9}$$

The corresponding beam power is given by

$$P_{em} = (\gamma - 1)P_{rel} , \tag{6.10}$$

where P_{rel}, defined by eq. (5.133), is the same unit of power, 8.71 GW used earlier in the discussion of the self-focusing and stimulated Raman scattering of light in a plasma. In fact, *there is a striking similarity between the critical power* P_c *for self-focusing of light in a plasma*, as given by eq. (5.135), *and the power* P_{em} *required for the occurrance of significant self-forces in an electron beam.* The principal difference is that the former is proportional to the *fourth power* of the energy of the *photons*, whereas the latter is proportional to the *first power* of the energy of the *electrons*.

Evaluating J_{em} for 10 MeV electrons we obtain a value of 0.32 MA, a factor of three smaller than that of the megajoule pulse assumed earlier in our pressure calculations. Therefore, electromagnetic self-forces will be expected to have an important effect on the propagation of such a pulse, unless the conductivity of the medium in which the pulse is propagating is sufficiently high that the transient counter-currents induced in the medium by the beam, which we have so far ignored, can effectively neutralize the beam current during the passage of the pulse.

Finally, let us explore briefly the individual electron trajectories implied by our idealized model of the beam [25]. Relaxing the assumption of charge neutralization, the circumferential induction B and the radial electric field E are given by

$$B = (2J/a)\times\begin{cases}(r/a)\,, & 0<r<a\,,\\(a/r)\,, & a<r<b\,,\end{cases} \tag{6.11}$$

$$E = (1-f)\,B/\beta\,, \tag{6.12}$$

where f is the ratio of the charge density of the beam itself to the positive charge density of the background medium within the region of the beam. The radial force F acting on a beam electron is therefore,

$$F = (2eJ/\beta ca^2)(1-f-\beta^2)\,r\,, \tag{6.13}$$

and the equation for the x-component of transverse motion of a beam electron, disregarding collisions, is that of a simple harmonic oscillator,

$$\ddot{x} + \omega^2 x = 0\,, \tag{6.14}$$

where the frequency of oscillation is given by

$$\omega^2 = \omega_\beta^2[1-(1-f)/\beta^2]\,, \tag{6.15}$$

ω_β being the «betatron» frequency,

$$\omega_\beta^2 = 2\pi N_e e^2\beta^2/\gamma m = \tag{6.16}$$

$$= (2\beta/\gamma)(c/a)^2(J/J^*)\,. \tag{6.17}$$

In writing eq. (6.14) we have assumed that the transverse velocity $v_\perp$ of the beam electrons is small compared with their longitudinal velocity v along the direction of the beam so that γ may be considered constant during the oscillation.

For the case of helical orbits with radius r we have

$$v_\perp = \omega r \,. \tag{6.18}$$

Averaging over the assumed uniform density of beam electrons we obtain

$$\langle v_\perp^2 \rangle = \omega^2 \langle r^2 \rangle = \tfrac{1}{2}\,\omega^2 a^2 \,, \tag{6.19}$$

or, employing eqs. (6.15) and (6.17), we have

$$\langle v_\perp^2 \rangle / v^2 = (J/\beta\gamma J^*)[1-(1-f)/\beta^2] \,. \tag{6.20}$$

Again assuming a fully neutralized beam ($f=1$) of relativistic electrons ($\beta \sim 1$), eq. (6.20) becomes simply

$$\langle v_\perp^2 \rangle / c^2 = J/\gamma J^* \,. \tag{6.21}$$

Defining the transverse beam « temperature » $T_\perp$ according to

$$kT_\perp^2 = \tfrac{1}{2}\gamma m \langle v_\perp^2 \rangle \,, \tag{6.22}$$

and using eq. (6.2), we can write eq. (6.21) in the form

$$kT_\perp / w = J/2J_{\text{em}} \,. \tag{6.23}$$

We see that unless $J \ll J_{\text{em}}$ the transverse component of the beam energy becomes significant, and the assumption that the electrons are only slightly deflected by beam self-forces from the direction of propagation of the beam is no longer true.

The power P_{em} therefore marks a transition between low-power pulses, in which the propagation of the pulse is dominated by inertial forces and the motion of a given electron is only slightly perturbed by the other electrons in the pulse; and high-power pulses, in which the pulse propagation is dominated by electromagnetic forces. In the former case one may speak of an electron pulse carrying along some electromagnetic field; in the latter of an electromagnetic pulse carrying along some electrons. A consequence of the high transverse temperature accompanying high-power pulses is that they can propagate through a plasma without exciting beam-plasma instabilities [26], the motion of the electrons of the pulse being too chaotic to couple effectively with the collective oscillations of the plasma electrons.

REFERENCES

[1] P. J. McClung and R. W. Hellwarth: *Journ. Appl. Phys.*, **33**, 828 (1962).

[2] W. T. Link: *IEEE Trans.*, *NS*-**14**, 777 (1967).

[3] S. E. Graybill and S. V. Nablo: *IEEE Trans.*, *NS*-**14**, 782 (1967).

[4] F. M. Charbonnier, J. P. Barbour, J. L. Brewster, W. P. Dyke and F. J. Grundhauser: *IEEE Trans.*, *NS*-**14**, No. 3, 789 (1967).

[5] M. Cox *et al.*: *Laser Focus*, **3**, 21 (1967).

[6] L. Spitzer jr.: *Physics of Fully Ionized Gases*, 2nd Ed. (New York, 1961).

[7] R. E. Kidder: *Nucl. Fusion*, **8**, 3 (1968).

[8] W. J. Fader: *Phys. of Fluids*, **11**, 2200 (1968).

[9] S. A. Ramsden and P. Savic: *Nature*, **203**, 1217 (1964).

[10] L. E. Hargrove, R. L. Fork and M. A. Pollack: *Appl. Phys. Lett.*, **5**, 4 (1964).

[11] J. A. Armstrong: *Appl. Phys. Lett.*, **10**, 16 (1967).

[12] W. H Culver, J T. A. Vanderslice and V. W. T. Townsend: *Appl. Phys. Lett.*, **12**, 189 (1968).

[13] R. E. Kidder and W. S. Barnes: *WAZER: a one-dimensional, two-temperature hydrodynamic code*, UCRL-50583, Jan. 31, 1969.

[14] S. Rand: *Phys. Rev.*, **136**, B 231 (1964).

[15] T. P. Hughes and M. B. Nicholson-Florence: *Journ. Phys.* A, *Proc. Phys. Soc.*, **1**, 588 (1968).

[16] J. Scofield: to be published.

[17] N. Bloembergen and Y. R. Shen: *Phys. Rev.*, **141**, 298 (1966).

[18] D. F. Du Bois and M. V. Goldman: *Phys. Rev. Lett.*, **14**, 544 (1965).

[19] Y. C. Lee and C. H. Su: *Phys. Rev.*, **152**, 129 (1966).

[20] Y. R. Shen and N. Bloembergen: *Phys. Rev.*, **137**, A 1787 (1965).

[21] N. Bloembergen: *Am. Journ. Phys.*, **35**, 989 (1967).

[22] *Design development of the Aurora facilities*, PISR-127-1, Oct. 15, 1968, Physics International Company, San Leandro, Calif.

[23] *Studies in penetration of charged particles in matter*, Publication 1133, Nat. Acad. of Sc.-Nat. Res. Council, Wash. D.C. (1964).

[24] R. D. Cowan and J. Ashkin: *Phys. Rev.*, **105**, 144 (1957).

[25] J. D. Lawson: *Journ. Electron. and Control*, **5**, 146 (1958).

[26] S. A. Bludman, K. M. Watson and M. N. Rosenbluth: *Phys. of Fluids*, **3**, 747 (1960).

Interaction of Intense Light Pulses with Solid Materials.

A. CARUSO

Laboratori Gas Ionizzati (Associazione EURATOM-CNEN) - Frascati (Roma)

1. – Introduction.

By using lasers the creation of light pulses of several joule and very different durations is possible; recently the duration of such pulses has been decreased to the picosecond range [1]. In this lecture we shall consider the interaction of such short pulses with a solid material. Generally speaking to study the interaction of intense laser pulses with solids it is convenient to consider two limiting situations:

1) the light pulse is so short that during the absorption of the light the heated matter has no time to move; the energy diffuses in the target by heat conduction and this process is stopped by the expansion of the heated matter at a time $t_0 > \tau$;

2) the expansion process occurs during the absorption of the light and heat conduction is not important.

To evaluate when we are in the first or in the second limiting situation let us assume that the surface of a semi-infinite solid body absorbs energy at a rate φ ($\varphi = \text{erg cm}^{-2}\,\text{s}^{-1}$); at the instant t the heated layer has a thickness $l_H \approx \sqrt{\chi t}$, χ being the thermal conduction; generally χ depends on the temperature T. Using the law $\chi = bT^n$ (with $n > 0$) and the energy balance $\varphi t \sim Tl_H$ we get

$$l_H \sim t^\alpha \,, \qquad \alpha = \frac{n+1}{n+2} < 1 \,. \tag{1}$$

On the other hand from the surface starts an expansion wave; at the instant t its boundary is located at a distance $l_E \approx c_S t$, being the velocity of the sound. We have

$$l_E \approx C_S t \sim \sqrt{T} \sim t^\beta \,, \qquad \beta = \frac{2n+5}{2n+4} > 1 \,. \tag{2}$$

The behaviour of l_H and l_E is sketched in Fig. 1: the expansion wave reaches the boundary of the heat wave in a finite time t_0 at a distance l_0 from the initial boundary surface. If the light pulse lasts for a time $\tau < t_0$ the dynamics is unimportant during the absorption process; on the contrary if $\tau \gg t_0$ the expansion process is important and the light interacts with a continuously changing part of the target.

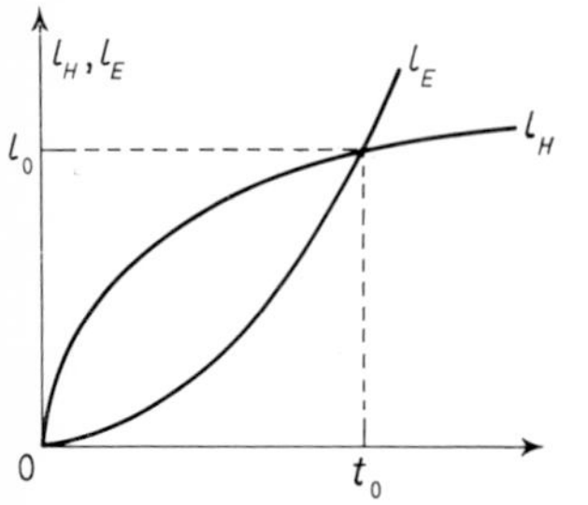

Fig. 1. – Time behavior of l_H and l_E.

To evaluate t_0 and l_0 we represent the initial target as a cold plasma with the density of a solid; this is reasonable if the typical temperatures involved in the interaction process are such that $KT/M \gg \varepsilon$, M being the ion mass and ε the binding energy per unit mass of the solid. It is clear that when the previous inequality is verified all the ionization and dissociation processes are only marginally important; so it is appropriate to use the transport coefficients calculated for a plasma. For heat diffusion we have (for $z = 1$):

$$\chi = bT_e^{\frac{5}{2}}, \qquad b = 2\cdot 10^{-6} \text{ c.g.s. °k}.$$

From the equations

$$\varphi t \approx \tfrac{3}{2} nKT_e l_H, \tag{3}$$

$$\frac{\frac{3}{2} nKT_e}{t} \approx \frac{bT_e^{\frac{5}{2}} T_e}{l_H^2} \tag{4}$$

it follows that

$$l_H \approx b^{\frac{2}{9}} (\tfrac{3}{2} nK)^{-\frac{7}{9}} \varphi^{\frac{5}{9}} t^{\frac{7}{9}}, \tag{5}$$

$$T_e \approx b^{-\frac{2}{9}} (\tfrac{3}{2} nK)^{-\frac{2}{9}} \varphi^{\frac{4}{9}} t^{\frac{2}{9}}, \tag{6}$$

in (3)-(6) n is the electronic density ($\approx 5\cdot 10^{22}$ cm^{-3}) and is Boltzmann's constant. From (6) it is easily seen that

$$l_E \approx \sqrt{\frac{2KT_e}{M}}\, t \sim \varphi^{\frac{2}{9}} t^{\frac{10}{9}},$$

The value of t_0 and l_0 follows from $l_0 = l_H(t_0) = l_E(t_0)$:

$$t_0 = \left(\frac{M}{2K}\right)^{\frac{3}{2}} b (\tfrac{3}{2} Kn)^{-2} \varphi = 2.2\cdot 10^{13} A^{\frac{3}{2}} n^{-2} \varphi, \tag{7}$$

$$l_0 = \left(\frac{M}{2K}\right)^{\frac{7}{6}} b (\tfrac{3}{2} Kn)^{-\frac{7}{3}} \varphi^{\frac{4}{3}} = 2.4\cdot 10^{21} A^{\frac{7}{6}} n^{-\frac{7}{3}} \varphi^{\frac{4}{3}}; \tag{8}$$

in these formulae A is the mass number and c.g.s. units are used. From (7) and (8) we see that both t_0 and l_0 increase with φ. Let us consider a few numerical examples:

φ (erg cm^{-2} s^{-1})	10^{19}	10^{21}	10^{23} ,
t_0 s	10^{-13}	10^{-11}	10^{-9} ,
l_0 cm	$6\cdot10^{-7}$	$3\cdot10^{-4}$	0.1 .

We assumed $A=1$. It is important to remark also that, for $\varphi < 10^{19}$ erg cm^{-2} s^{-1}, l_0 is smaller than the depth of penetration of the light in a plasma with $n = 5\cdot10^{22}$ cm^{-3} and electron-ion collision frequency $\nu[T_e(t_0)]$. The case $\tau \gg t_0$ is fully covered by Krokhin's lectures. In what follows we shall consider the opposite case, that is $\tau \leqslant t_0$ [2].

2. – Irradiation of a massive body by ultrashort laser pulses.

In the previous Section we tacitly assumed that the quantity φ, which is the flux of power absorbed at the surface, is equal to the flux of power of the light; this is not necessarily the case because a part of the incident light could be reflected. The problem of the light absorption is particularly crucial in the regime we are considering in this Section. To see this, let us consider the parameters of a typical light beam. To a specific power $\varphi \geqslant 10^{23}$ erg cm^{-2} s^{-1} corresponds an electric field $E = (8\pi\varphi/c)^{\frac{1}{2}} \approx 3\cdot10^9$ V cm^{-1}; such pulses have been produced by means of neodimium glass lasers so the angular frequency ω is $1.7\cdot10^{15}$ s^{-1}. The time duration is $\tau < 10^{-11}$ s and the energy deposited per unit surface ε is of the order of 10^{12} erg cm^{-2}. Due to the large electric fields of the electromagnetic wave, when such a pulse reaches the surface of the solid, the ionization process begins without time lag: the ionization time for a field of $3\cdot10^9$ V cm^{-1} is less than 10^{-16} s. So the solid is covered by a layer of ionized matter with a density about equal to that of a solid. The plasma frequency $\omega_p = (4\pi n l^2/m)^{\frac{1}{2}}$ corresponding to the solid density is about 10ω. The energy of an electron oscillating in the electric field of the beam is $\frac{1}{2}m(lE/m\omega)^2 \approx$ ≈ 2.5 keV; the corresponding collision frequency is

$$\nu \approx 3\cdot10^{13}\ \text{s}^{-1} = 1.6\cdot10^{-2}\omega\ .$$

Since ν is smaller than ω_p and ω, most of the light should be reflected in a distance $\delta \geqslant \delta_0 = c/\omega_p \approx 2\cdot10^{-6}$ cm. However a large light absorption has been observed in (1) and (3). An explanation for this can be based on the following considerations. If we assume that a reflecting layer is formed in the first stages of the light-solid interaction, an important requirement, to have reflection in all subsequent instants, is the stability of the motion of the electrons reflecting the light. If because of instabilities, the motion becomes turbulent, we can have anomalous absorption even in the absence of collisions: the electrons collide with the turbulent electric fields. In what follows we shall indicate some of the possible sources of turbulence.

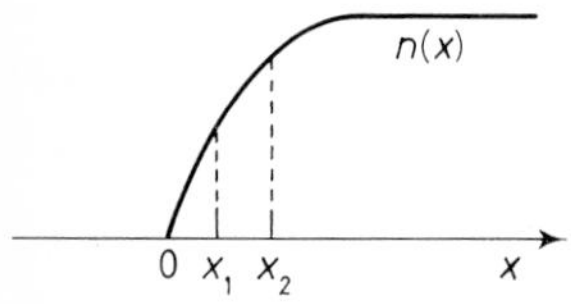

Fig. 2. – A qualitative density profile.

Let us consider a density profile $n(x)$ like that of Fig. 2: the density is a function of x only and reaches its asymptotic value $n(\infty)$ in a distance d of the order of or smaller than a few wavelengths of the light, $K_0^{-1} = c/\omega$. We consider such scale length for the following reason: if the solid is a crystal, initially the boundary is sharp; however it seems reasonable to take d of the order or less than $v_i\tau$ where v_i is the ion velocity and τ the pulse duration. This distance d results comparable with the wavelength of the light $c/\omega \approx \approx 10^{-4}$ cm for $v_i \geqslant 10^7$ cm s^{-1} $\tau \leqslant 10^{-11}$ s. If we assume initially a complete reflection, a standing wave is formed in the boundary in a time of about ω^{-1}. Let us consider a linearly polarized wave with the electric field H_z along the y-axis and the magnetic field H_z along z; if we take $E_y = E_0(x)\cos\omega t$, it results, from Maxwell's equations,

$$H_2 = -\frac{c}{\omega}\frac{\mathrm{d}E_0}{\mathrm{d}x}\sin\omega t\,.$$

The electronic velocity, v_y is given by

$$v_y = V(x)\sin\omega t\,. \tag{9}$$

The function $V(x) = -eE_0(x)/m\omega$ satisfies the equation

$$\frac{\mathrm{d}^2V}{\mathrm{d}x^2} + K_0^2\left(1 - \frac{\omega_P^2(x)}{\omega^2}\right)V = 0\,; \tag{10}$$

in order to have reflection from a monotonic profile the existence of a point x_1, in which $\omega_P(x_1) = \omega$, is required. On the other hand in this point the conditions for a growth of electrostatic oscillations due to the parametric resonance are satisfied. The study of this phenomenon for the case of a uniform plasma is due to SILIN [4]: the range of frequency to be considered is that in which ω is appreciably greater than the ion plasma frequency but comparable with that of the electrons. A relevant quantity in the problem is the component of the wave number of the perturbation along the direction in which the electrons oscillate (the y-axis in our case). When $K_y^{-1} \approx r$ amplitude of the electron oscillation, the maximum growth rate for instability occurs. Precisely if $\omega_P = N\omega$, N being an integer, the growth rate is $\gamma = (m/M)^{\frac{1}{3}}\omega_P$; if $\omega_P \neq N\omega$, γ is about the ion plasma frequency. For wavelengths $K_y^{-1} > r$ the growth rates are reduced by a factor $(K_y r)^{\frac{2}{3}}$. In the case we are considering, since $\omega_P(\infty) \approx 10\omega$ we can have more than one point in which the condition $\omega_P = N\omega$ is verified. We must remember, however, that this theory holds for resonances in a uniform plasma, so its application to our case should be taken with caution; at any rate it seems very plausible that similar phenomena are relevant in our problem.

Another source for nonlaminar motion may be the nonlinear coupling between transverse and longitudinal oscillations [5]. The physical mechanism is as follows: on the electrons acts a force

$$-\frac{e}{c}\, v_y H_2 = -\frac{e^2}{4m\omega^2}\,\frac{\mathrm{d}E_0^2}{\mathrm{d}x}\,(1-\cos 2\omega t)$$

directed along the x-axis; if E_0 is appreciable in the point x_2 where $\omega_P(x_2) = 2\omega$, then the oscillating part of the force,

$$f = \frac{e^2}{4m\omega^2}\,\frac{\mathrm{d}E_0^2}{\mathrm{d}x}\cos 2\omega t \equiv f_0(x)\cos 2\omega t$$

resonates with the electrons near x_2. If the oscillations start at the instant $t = 0$, the region in which there is resonance is given by $\Delta\omega_P \approx 1/t$ or $(\Delta x)^{-1} = (\mathrm{d}\omega_P/\mathrm{d}x)_{x_2} t$; the mean power absorbed per unit surface by the resonating electrons is $\varphi_a \approx (f_0^2(x_2)/m)\, t n(x_2)\Delta x$ or

(11) $$\varphi_a \approx \frac{f_0^2(x_2)}{m}\,\frac{n(x_2)}{|\mathrm{d}\omega_P/\mathrm{d}x|_{x_2}}\,.$$

If we approximate the derivatives by $\mathrm{d}E_0/\mathrm{d}x \approx E_0/\delta$ and $\mathrm{d}\omega_P/\mathrm{d}x \approx \omega_P/d$ it results that

(12) $$\frac{\varphi_a}{\varphi} \approx \frac{d\lambda}{\delta^2}\,\frac{\omega_P}{\omega}\,\frac{v_{osc}^2}{c^2}\,,$$

where λ is the wavelength of the light, $v_{osc} = eE_0/m\omega$ and $\varphi \approx cE_0^2$ is the flux of power of the incident light. We can take as growth rate for this process the quantity γ given by $\varphi_a/\gamma = E_0^2\delta$; we have

$$\frac{1}{\gamma} = \frac{\delta^3}{d\lambda^2}\left(\frac{c}{v_{\text{osc}}}\right)^2 \frac{1}{\omega_p}\,. \tag{13}$$

In the regime we are considering $\omega_p \approx 10^{16}\,\text{s}^{-1}$ and $v_{\text{osc}} \geqslant 3\cdot 10^9\,\text{cm}\,\text{s}^{-1}$, so γ can be quite large.

Among the possible sources of turbulence it is perhaps relevant to consider the well-known Kelvin-Helmholtz instability. This instability can arise when in a stationary parallel flow, the velocity changes in a direction perpendicular to that of the flow itself. This problem has been studied in hydrodynamics [6], and also when such kind of flow occurs in a cold collisionless electron gas, the ions being infinitely massive [7]. In this case the equations of the motion are coupled with Poisson's equation, because charge separation, is allowed. However, if we consider perturbations with wavelengths longer than V/ω_p, V being a typical velocity, the oscillations are quasi-neutral (the electrostatic potential plays the role of the pressure) and we can use the results found in hydrodynamics.

When the motion is unstable, the typical growth rates are about V/δ, δ being the distance in which the velocity changes; these growth rates occur for wavelengths comparable with δ. In our case the velocity changes over a distance which is found by solving (10); since the minimum δ is $\delta_0 = c/\omega_p \gg V/\omega_p$ it is consistent to search for quasi-neutral perturbations. The velocity given by (9) is oscillating with frequency ω; so, we can expect instability only if $\delta\omega < V$. This implies $\delta < \lambda$ a condition verified, at least, in the limiting case $\delta = \delta_0 = c/\omega_p$.

We extended the hydrodynamical calculations to the case in which the unperturbed situation is described by

$$n = n(x)\,, \qquad v_x = v_z = 0\,, \qquad v_y = V(x)\sin\omega t\,. \tag{14}$$

We found the following necessary conditions for instability:

$$\begin{cases} 1)\ \text{excursion}\ \dfrac{V}{\omega} > \delta\,, \\[2ex] 2)\ \dfrac{\mathrm{d}}{\mathrm{d}x}\left(\dfrac{1}{n}\dfrac{\mathrm{d}V}{\mathrm{d}x}\right) = 0\ \text{in some point}\,. \end{cases} \tag{15}$$

The condition (15) is the generalization of the well-known one $\mathrm{d}^2V/\mathrm{d}x^2 = 0$, valid when $\mathrm{d}n/\mathrm{d}x = \omega = 0$.

The preceding list of instabilities, very likely not complete, renders improbable the existence of a laminar electronic motion and can be important in the explanation of the observed light absorption.

Let us now try to evaluate the parameters of the plasma produced when a light pulse is absorbed from the surface of a solid. Since we are considering the case $t_0 > \tau$, it is clear that a part of the diffusion process must be studied by assuming not a constant flux at the boundary but constant energy per unit surface $\varepsilon = \varphi\tau$. Since, after the end of the laser pulse, the turbulence is damped by collisions in a time much shorter than t_0, we can use the thermal conductivity of a Maxwellian non-turbulent plasma.

The solution of the resulting diffusion equation is given in [2]; the thickness of the heated region turns out to be

$$l_H = 1.485\, b^{\frac{2}{9}} (\tfrac{3}{2} nK)^{-\frac{7}{9}} \varepsilon^{\frac{5}{9}} t^{\frac{2}{9}}\,, \tag{16}$$

whereas the space-averaged temperature is

$$\langle T_e \rangle = 0.674\, b^{\frac{2}{9}} (\tfrac{3}{2} nK)^{-\frac{2}{9}} \varepsilon^{\frac{4}{9}} t^{-\frac{2}{9}}\,. \tag{17}$$

In obtaining (16) and (17) we assumed $T_e \gg T_1$; this point will be discussed later. The time t_0 at which the dynamics becomes important is evaluated from the equation

$$l_H(t_0) = \gamma \left(\frac{3k\langle T_e(t_0)\rangle}{M}\right)^{\frac{1}{2}} t_0\,, \tag{18}$$

where γ is a number of the order of unity. Solving for t_0 we get

$$t_0 = 8\cdot 10^6 \gamma^{-\frac{3}{2}} A^{\frac{3}{4}} n^{-1} \varepsilon^{\frac{1}{2}}\,, \tag{19}$$

$$l_0 = 4.5\cdot 10^{12} \gamma^{-\frac{1}{3}} A^{\frac{1}{6}} n^{-1} \varepsilon^{\frac{2}{3}}\,, \tag{20}$$

$$\langle T_e(t_0)\rangle = 10^3 \gamma^{\frac{1}{3}} A^{-\frac{1}{6}} \varepsilon^{\frac{1}{3}}\,. \tag{21}$$

We see from the previous equations that only t_0 is rather dependent on γ. It is interesting to observe that the time t_0 is, in principle, of the same order of magnitude of the relaxation time of the energy from the electrons to the ions. This can be seen in the following way: at the instant t the thickness of the heated region is $l_H \approx v_{th}\sqrt{t/\nu}$, where v_{th} is an average electronic thermal velocity and ν is an average collision frequency; on the other hand, due to the electronic pressure, the plasma expands with a velocity about equal to $\sqrt{(mv_{th}^2/M)}\,t_0$, m and M being the electronic and the ionic masses. The time t_0 is given by the equation

$$l_H(t_0) \approx v_{th}\sqrt{\frac{t_0}{\nu}} = \sqrt{\frac{mv_{th}^2}{M}}\, t_0\,;$$

solving for t_0 we get $t_0 = (M/m)(1/\nu)$. So t_0 is of the same order of magnitude of the equipartition time t_{equ}; since $t_{equ} \sim T_e^{\frac{3}{2}}$, using (19) and (21) we find that $t_{equ}/t_0 \sim \gamma^2$. These considerations show that the ratio t_{equ}/t_0 cannot be determined, by these order-of-magnitude estimates, with the precision required in calculating, for instance, the number of neutrons produced in a shot, when the target is made by fusionable materials. Also, the model itself is rather sensitive to the value of t_{equ}/t_0 because, if $t_{equ}/t_0 < 1$ ion heating should be taken into account; in this case the study of the diffusion process is more complicated. In any case the previous formulae should give the right order of magnitude for the quantities of interest, as can be seen by considering the limiting case in which $T_e = T_i$ during all the diffusion process. Of course the treatment of the case $T_e \approx T_i$ is analogous to that in which $T_e \gg T_i$.

In conclusion, if U joule are released in a spot of radius R on the surface of a massive body we expect the production of a plasma formed by

$$\mathcal{N} = \pi R^2 2 n l_H(t_0) = 6 \cdot 10^{17} A^{\frac{1}{6}} R^{\frac{2}{3}} U^{\frac{2}{3}}, \tag{22}$$

particles, the mean ion asymptotic expansion energy, K, being

$$K = \tfrac{3}{2}\langle T_e \rangle = 2 \cdot 10^5 A^{-\frac{1}{6}} R^{-\frac{2}{3}} U^{\frac{1}{3}} \;(^\circ\mathrm{K}) . \tag{23}$$

We assumed $\gamma^{\frac{1}{3}} \approx 1$ and $l_0 < R$; the effects of a three-dimensional diffusion are considered in Gratton's lecture.

The scaling laws (22) and (23) seem to be in good agreement with the experiments [4].

3. – Irradiation of solid pellets.

In this Section we shall discuss the conditions to be satisfied in order to heat efficiently a solid speck by means of a short laser pulse. By «efficient heating» we mean that all the energy of the pulse is distributed uniformly throughout the entire pellet before any significant expansion.

This requirement can be satisfied only if the radius R of the pellet verifies the inequality

$$l_0 > 2R . \tag{24}$$

If inequality (24) is not verified only a part of the target is burned. On the other hand, if the pellet is too small, the energy per particle is large and the target explodes during the time τ in which the energy is supplied; to avoid this, we must have

$$2R > \left(\frac{W\tau}{\frac{4}{3}\pi R^3 n A m_P}\right)^{\frac{1}{2}} \tau \,, \tag{25}$$

W being the mean power during the time τ. From (25) we obtain

$$R > \left(\frac{3}{16\pi}\right)^{\frac{1}{5}} \left(\frac{W\tau}{nAm_P}\right)^{\frac{1}{5}} \tau^{\frac{2}{5}} \,. \tag{26}$$

Assuming $\varepsilon \approx W\tau/R^2$ in evaluating l_0, it results from (24), (26):

$$2\cdot 10^5 A^{-\frac{1}{14}} n^{-\frac{3}{7}} (W\tau)^{\frac{2}{7}} > R > 8.2\cdot 10^4 A^{-\frac{1}{5}} n^{-\frac{1}{5}} \tau^{\frac{2}{5}} (W\tau)^{\frac{1}{5}} \,. \tag{27}$$

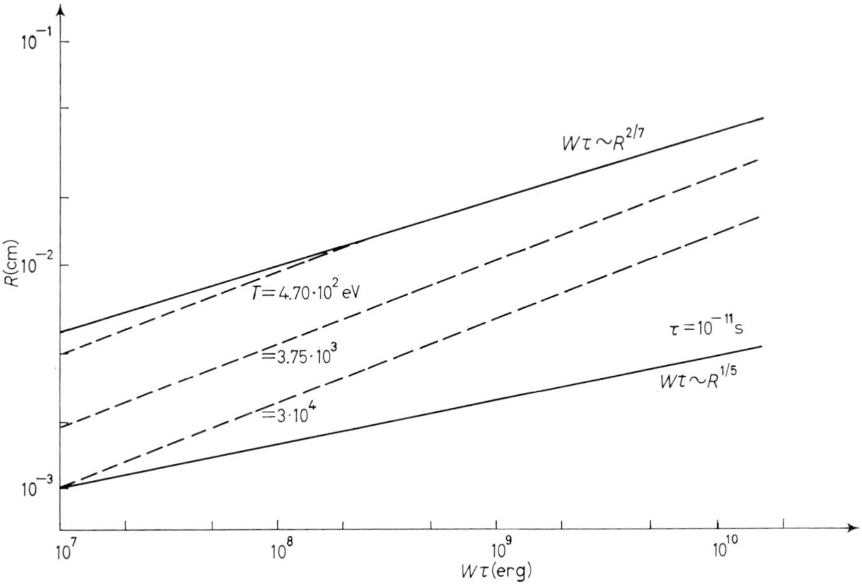

Fig. 3. – The useful range for efficient heating of small solid pellets.

The inequality (27) is represented in Fig. 3 for the case

$$\tau \approx 10^{-11}\,\text{s}\,, \qquad n = 5\cdot 10^{22}\,\text{cm}^{-3}\,, \qquad A = 1\,;$$

the values of R satisfying (27) are those between the continuous lines. In this regime the temperature can be estimated simply by taking the ratio between the energy $W\tau$ and the number of the particles contained in the pellet.

4. – Concluding remarks.

We have studied the plasma production from a solid by means of a short light pulse ($\tau < 10^{-11}$ s) with energies of several joule. The physics of the process is very different from that arising in the case of the nanosecond pulses.

The main reason for this is the absence of gas-dynamical expansion during the interaction light-solid.

It has been found that, in the irradiation of a massive target, the lifetime of the hot spot at densities of the order of that of a solid is, in principle, comparable with the equipartition time for the energy between ions and electrons. In the framework of this theory it is not possible to determine the exact ratio between these two times. However it is possible to give a good estimate for the average energy and the total number of the heated particles; it results that the energy increases as $\varepsilon^{\frac{1}{3}}$ and the total number as $\varepsilon^{\frac{2}{3}}$, where ε is the energy absorbed per unit surface. If we are interested in obtaining very high temperatures, the irradiation of small solid specks seems promising. If the radius of the speck R has a value in the range determined by the straight lines in Fig. 3, the gas-dynamical expansion does not occur during the heating process and all parts of the speck are heated. The useful range depends on the energy $W\tau$ released on the target. In this regime the value of the energy per particle is obtained simply by dividing $W\tau$ for the number of particles contained in the speck. It is easy to see that temperatures of several keV can be obtained.

In these regimes, the problem of the light absorption by the target is very important. In Sect. **2** we discussed some possible reasons for anomalous absorption. From a theoretical point of view, it seems very difficult to obtain definitive answers to this problem, mainly because we ignore the density profile in the region where the light interacts with the solid. Experiments are very likely required to have a reliable answer to this question.

REFERENCES

[1] N. G. Basov, S. D. Zaharov, P. C. Kriukov, U. V. Senatski and S. V. Tsekalin: *Int. Quantum Electronic Conf.* (Miami, 1968).

[2] A. Caruso and R. Gratton: Laboratori Gas Ionizzati, Frascati, Internal Report LGI 69/1 (1969), to be published in *Plasma Phys.*

[3] A. Caruso, A. De Angelis, G. Gatti, R. Gratton and S. Martellucci: *Phys. Lett.*, **29** A, 316 (1969).

[4] V. P. Silin: *Sov. Phys. JETP*, **21**, 1127 (1965).

[5] A. Caruso: *Plasma Phys.*, **10**, 963 (1968).

[6] C. C. Lin: *The Theory of Hydrodynamic Stability* (Cambridge, 1965).

[7] A. Caruso and F. Gratton: *Nuovo Cimento*, **37**, 62 (1965).

Plasma Produced by Subnanosecond Laser Pulses.

R. Gratton

Laboratori Gas Ionizzati (Associazione EURATOM-CNEN) - Frascati (Roma)

This seminar will be divided in three Sections:

1) we shall consider some problems arising in testing the model [1] by an experiment.

2) we shall report the results of an experiment recently performed at Laboratorio Gas Ionizzati of Frascati.

3) we shall give an estimate of the parameters of the plasmas produced by subnanosecond pulses, based on the scaling laws of [1].

1. – On the experimental test of model [1].

Model [1] contains two physical hypotheses which need experimental verification:

a) It was assumed that most of the incoming energy U is absorbed by the target.

b) To carry on simple calculations it was necessary to assume either $T_i = 0$ or $T_i = T_e$ during the lifetime t_0 of the hot region. This second point is related to the value of the constant γ introduced in [1]. It is clear that because t_0 goes as $\gamma^{-\frac{2}{3}}$ and the equipartition time is of the same order of t_0 the ions may be heated or not depending on the actual value of γ.

It is interesting to see how the above hypotheses influence the parameters of the produced plasma, for instance the ion's asymptotic expansion energy K and the total number N of ions contained in the hot region.

Regarding K, with the hypothesis $T_i = 0$, we have for a deuterium target:

$$K = 15.5\gamma^{\frac{1}{3}} R_L^{-\frac{2}{3}})(\alpha U)^{\frac{1}{3}}\ (\mathrm{eV})\,, \tag{1}$$

where R_L is the focal spot radius, U is the laser output energy in joule and α is the fraction of light absorbed at the target surface. If, instead, we assume $T_i = T_e$, we must multiply the previous value by a factor of $2^{\frac{1}{6}}$. This is due to the fact that if the energy goes both to electrons and ions, less matter is heated and hence t_0 is considerably reduced.

On the other hand this effect is likely to be compensated or overcompensated, because, in fact, longer lifetimes are required for the case $T_i = T_e$. In other words we cannot reasonably assume the same value of γ for both cases, and therefore the $2^{\frac{1}{6}}$ factor has no practical meaning.

We must conclude that the determination of K gives no information about the above hypothesis; the only result that we can expect from the measurement of the ion's asymptotic expansion energy is the test of the one-third dependence on U.

The number N of energetic ions produced in the case $T_i = 0$ is given by

$$N = 3.4 \cdot 10^{17} \gamma^{-\frac{1}{3}} R_L^{\frac{2}{3}} (\alpha U)^{\frac{2}{3}} \tag{2}$$

while for the case $T_i = T_e$ we have to multiply the previous value by $2^{-\frac{7}{6}}$. However, the above argument is valid also for N.

Therefore the measurement of N gives little information on the ion heating and on the value of α and only the dependence of N on U may be verified. Strictly speaking, even the two-thirds dependence of N on U is a simple consequence of the one-third law for K, because of the relation $KN \sim U$.

We see, therefore, that from an experimental point of view it is rather easy to test the scaling laws $K \sim U^{\frac{1}{3}}$ and $N \sim U^{\frac{2}{3}}$ but it is another question to determine α with accuracy and to establish if the ions are heated or not in the hot region. Information about these points should be obtained by a direct measurement of the light reflected by the target and by the determination of the number $\mathcal{N}$ of neutrons produced through nuclear reactions.

To illustrate the problems arising in interpreting the measurement of $\mathcal{N}$, we use the following semi-empirical formula:

$$\mathcal{N} = \pi R_L^2 l(t_0)_0 \frac{n^2}{4} \langle \sigma v \rangle \simeq \pi R_L^2 (t_0)_0 n^2 \frac{7.5 \cdot 10^{-10}}{T_i^{\frac{2}{3}}} \exp\left[-\frac{4.25 \cdot 10^3}{T_i^{\frac{1}{3}}}\right] (T_i \text{ in } {}^\circ\text{K}),$$

where a Maxwellian distribution have been assumed for the ions (a reasonable assumption in the present problem) because the ion-ion collision time is of the order of $(m_e/m_i)^{\frac{1}{2}} t_0$. Introducing $T_i = \eta T_e$, with $n_0 = 5 \cdot 10^{22}$ and $l(t_0)$, t_0 and T_e given by model [1] with $\gamma = 1$, we get:

$$\mathcal{N} = 3.0 \cdot 10^{14} \frac{(\alpha U)^{\frac{17}{18}}}{\eta^{\frac{2}{3}} R^{\frac{7}{9}}} \exp\left[-83 \frac{R_L^{\frac{2}{9}}}{\eta^{\frac{1}{3}} (\alpha U)^{\frac{1}{9}}}\right].$$

In Fig. 1 $\mathcal{N}$ is reported as a function of (αU) with R_L as a parameter for $\eta = 1$ (for lithium deuteride targets the values of $\mathcal{N}$ are lesser by an order of magnitude). It should be noted that $\mathcal{N}$ depends critically not only on η, but also on R_L and α, so that a small error in the evaluation of α and R_L would lead to a strong difference on $\mathcal{N}$. Therefore, in order to interpret correctly the results of a measurement of $\mathcal{N}$ it is necessary to obtain with great accuracy, using independent methods, the values of α and R_L.

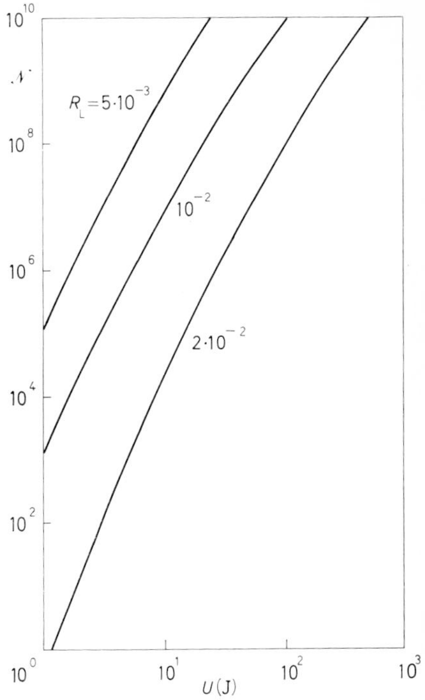

Fig. 1. – Number $\mathcal{N}$ of neutrons produced per shot, with R_L as a parameter, for $\eta = 1$.

2. – An experiment of plasma production by subnanosecond laser pulses.

We shall describe here an experiment performed at Laboratori Gas Ionizzati (CNEN-Euratom) Frascati (see also [2]).

A neodymium-doped glass laser was used, formed by a mode-locked oscillator and three amplifier stages.

The oscillator produces a train of about 15 light pulses, separated by 7 ns. Energy and duration of the higher pulses were, respectively, of the order of 10^{-2} J and 10^{-11} s. The oscillator was followed by a « pulse selector », analogous to that described by BASOV *et al.* [3].

It was formed by two crossed polarysers with a Kerr cell between them. This device acts as a very fast optical shutter when a voltage pulse is applied to the cell electrodes, so that only one pulse of the train can pass. The cell is triggered by the laser itself, in such a way that one of the higher pulses is selected . The selected pulse is then amplified by a factor of the order of 10^2 through the amplification states. As a matter of fact the laser output energy strongly fluctuates from shot to shot and during the experiment the value of U ranged from 0.1 to 1.3 joule.

The laser beam was focussed by a 3.8 cm diameter lens, having a focal length of 12 cm. The focal spot radius has been indirectly estimated of the order of 10^{-2} cm by measuring the beam's divergence.

The pure deuterium targets employed in this experiment were produced in a cryostat cooled by circulating liquid helium [4]. They were small cylinders with a diameter of $2\cdot10^{-2}$ cm and a height of 1 mm. The atomic density in the target has been measured to be very near $5\cdot10^{22}$ cm^{-3}, which is the value assumed for deuterium targets throughout all this work. The laser beam was directed orthogonally to the target's axis.

We measured K and N using a set of electrostatic probes. K was obtained comparing the signals given by two probes placed along the same direction but at different distances from the target, whereas N was estimated from the amplitude of the time-integrated probe's signals. The results for K and N are shown in Fig. 2; the agreement with the theoretical dependences is rather

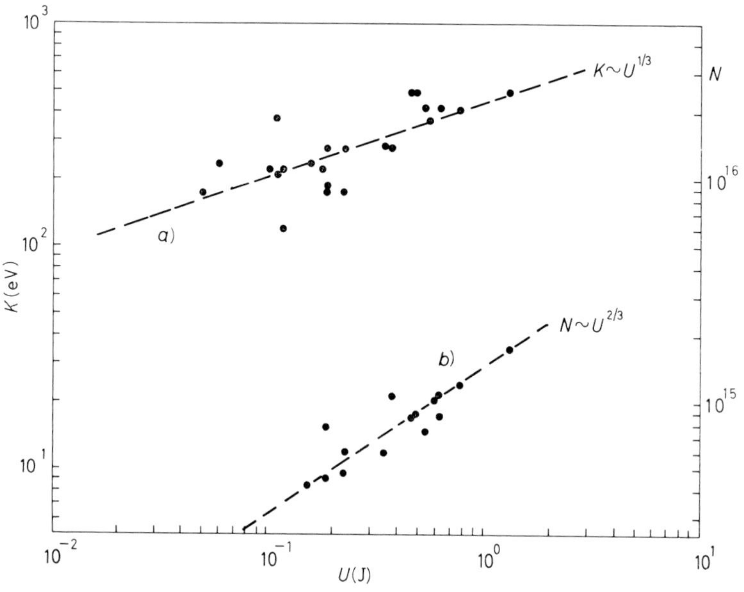

Fig. 2. – *a*) Ion energy K as a function of U. *b*) Total number of particles N as a function of U.

good. Furthermore, the numerical values of K are very near the predicted ones, whereas the values of N are smaller than those given by model [1] (for the case $\alpha=\gamma=1$ and $T_i=T_e$) by a factor of 4 or 5.

We also measured the energy back-scattered by the target into a 0.08 sterad solid angle centered with respect to the laser beam axis. This measurement gives results strongly fluctuating from shot to shot, but the fraction of back-scattering energy was always less than 0.01. If we assume that the light is isotropically scattered in a semisphere, the above result would lead to a value of α of the order of 0.5, whereas, if a geometrical reflection from the target surface is assumed, the resulting value of α would be of the order of 0.1.

From the results of this experiment, we may conclude that, within the covered range of U:

a) The scaling laws of model [1] are correct,

b) More than fifty per cent of the incoming light is absorbed.

On the other hand nothing can be said about the ion heating in the hot region.

It should be noted that even for the relatively low energy pulses employed in this experiment, the little amount of energy reflected cannot be explained simply by assuming absorption through Coulombian collisions between electrons and ions. In fact, an effective collision frequency of the order of the plasma frequency seems to be required.

3. – Parameters of the plasmas which may be produced by subnanosecond laser pulses focused on large deuterium targets.

Even if the scaling laws and the main hypotheses of model [1] have been only partially tested, it is interesting to use them in order to estimate the parameters of the plasmas which may be produced by short laser pulses. For simplicity we shall report the results only for pure deuterium targets, assuming $\alpha = 1$, $\gamma = 1$ and $T_i = T_e$.

The formulae given by [1] for this situation are:

$$t_0 = 1.4 \cdot 10^{-13}\, U^{\frac{1}{2}} R_L^{-1}\ (\text{s})\,, \tag{3}$$

$$l(t_0) = 0.97 \cdot 10^{-6}\, U^{\frac{2}{3}} R_L^{-\frac{4}{3}}\ (\text{cm})\,, \tag{4}$$

$$K(t_0) = 17.2\, U^{\frac{1}{3}} R_L^{-\frac{2}{3}}\ (\text{eV})\,, \tag{5}$$

$$N(t_0) = n_0 \pi R_L^2\, l(t_0) = 1.5 \cdot 10^{17}\, U^{\frac{2}{3}} R_L^{\frac{2}{3}}\,. \tag{6}$$

These formulae have been derived assuming a one-dimensional geometry, therefore they may be used only if $l(t_0) \leqslant R_L$; if this is not the case, the above relations cannot be used and must be modified in some way.

In the case $l(t_0) > R_L$, the plane approximation is valid only to a time $t^*(U, R_L)$ given by

$$l(t^*) \approx R_L\,,$$

after t^* the expansion of the hot region will be three-dimensional, that is the number of heated particles will go as l^3 and the temperature as Ul^{-3}. The focal spot radius obviously disappears as a relevant parameter, its place being taken by l.

In other words the plasma parameters must be calculated assuming the energy U as deposited on a point instead of on a finite surface.

By a simple dimensional analysis we obtain for this situation the following scaling laws:

$$l(t_0) \sim U^{\frac{2}{7}}, \tag{7}$$

$$K(t_0) \sim U^{\frac{1}{7}}, \tag{8}$$

$$N(t_0) \sim U^{\frac{6}{7}}. \tag{9}$$

This point has an important practical consequence.

Let us assume to perform an experiment with a fixed value for R_L and with U increasing from shot to shot. Relations (4), (5) and (6) are the correct ones as long as U is less than a critical value U_{crit}, obtained by solving the equation $t^*(R_L, U) = t_0$ which gives:

$$U_{crit} \approx 10^9 R_L^{\frac{7}{2}}. \tag{10}$$

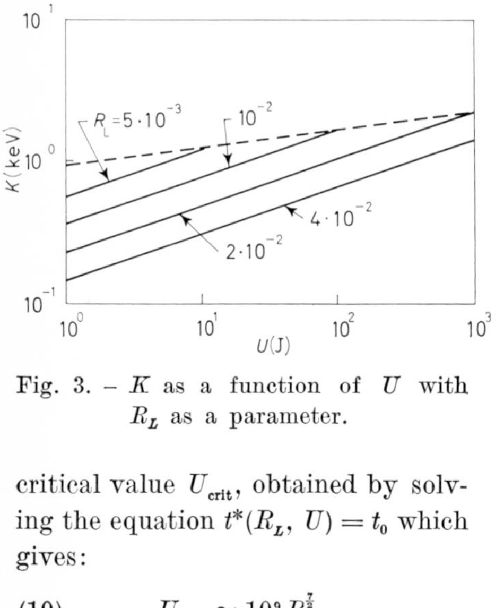

Fig. 3. – K as a function of U with R_L as a parameter.

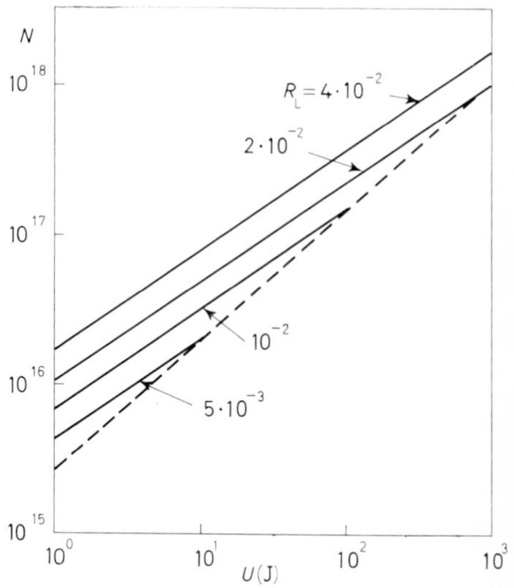

Fig. 4. – Number of ions contained in the hot region as a function of U, with R_L as a parameter.

On the other hand, for $U > U_{crit}$, K increases only as $U^{\frac{1}{7}}$, so a very large energy increment is required to vary considerably K.

If we repeat the experiment with a smaller R_L, we get larger values for K in the regime $U \leqslant U_{crit}$ but now U_{crit} is considerably less than in the previous case, so that there is an improvement in K only if small values of U are used, while for large values of U we should get the same results. This situation is illustrated in Fig 3, where the full lines give K as function of U with R_L as a parameter for the regime $U < U_{\text{crit}}$; the lines are stopped for $U = U_{\text{crit}}$ and the dashed line joining the upper extremes gives K for $U > U_{\text{crit}}$.

From the above argument it is clear that if we want to produce very energetic ions by irradiating large targets we need a large amount of energy even if a small focal spot radius could be obtained. The physical reason for this lies in the fact that the energy diffuses in the target in an uncontrolled form, so that if we increase the energy we get an almost proportional increase of the number of particles heated and only a very slow increase of the particle energy.

In Fig. 4 N is reported as a function of U, with R_L as a parameter. The full lines and the dotted line must be interpretated as in Fig. 3.

REFERENCES

[1] A. Caruso: this volume, p. 353.

[2] A. Caruso, A. De Angelis, G. Gatti, R. Gratton and S. Martellucci: *Phys. Lett.*, **29** A, 316 (1969).

[3] N. G. Basov, S. D. Zaharov, P. C. Kronkov, U. V. Senatsky and S. V. Tsekalin: *Sov. Phys. JETP Lett.*, **8**, 114 (1968).

[4] A. Cecchini, A. De Angelis, R. Gratton and F. Parlange: *Journ. Sci. Instr.* (*Journ. Phys. E.*) **2**, 1 (1968).

Production of Dense Thermonuclear Plasmas by Intense Relativistic Electron Beams (*).

F. WINTERBERG

University of Nevada System - Las Vegas, Nev.

Introduction.

Over the past two decades considerable effort has been expended towards the goal of achieving the controlled release of thermonuclear energy. The greatest part of this effort has been directed at the problem of stably confining a thermonuclear plasma in a magnetic field. It is well known that a multitude of plasma instabilities, some of them unexpected, has so far prevented success in these attempts. In fact, it is not even known at the present time whether a practical solution to the problem of stable plasma confinement under thermonuclear conditions does exist at all. In 1963 it was proposed [1, 2] that the desired goal might possibly be achieved in a totally different way by igniting a thermonuclear micro-explosion through the action of a small projectile (also called a macro-particle or macron) accelerated up to velocities of 10^8 cm/s impinging on a dense thermonuclear target. This approach necessitated that the projectile be larger than 1 mm. In this case the projectile would create at impact on the thermonuclear material a shock wave strong enough to trigger a small thermonuclear explosion.

The major difficulty encountered in this approach is that of accelerating projectiles having the dimensions of a milimeter to the required high velocities. The most obvious idea is to accelerate the macrons electrostatically. However, it turns out that in order to accelerate a macron of millimeter size an accelerator longer than 100 km would be needed. Since smaller macrons require shorter accelerator lengths, it was therefore proposed that a beam of microparticles

---

(*) Work supported in part by the National Aeronautics and Space Administration, USA, under grant NGR 29-001-016.

be accelerated. When focussed onto a thermonuclear target, such a beam would act at impact like one large, solid particle.

An alternative to the electrostatic acceleration is the magnetic acceleration of a superconducting projectile by a travelling magnetic wave [3-5]. Although this method promises the acceleration of one large projectile, the accelerator capable of imparting a velocity of 10^8 cm/s to the macron would still have to be longer than 10 km.

Also in 1963 a different proposal for igniting a thermonuclear micro-explosion was put forth [6-8], one in which the thermonuclear target would be irradiated by the focused photon beam from a giant pulsed laser. The difficulty involved with this proposal is that of the minimum energy of $\simeq 10^{14}$ erg required to trigger a thermonuclear micro-explosion. Apart from the poor efficiency of lasers requiring input energies much larger than the value of 10^{14} erg, this would inevitably lead to a laser of enormous and at present unfeasible dimensions.

Although quite different in nature, both of the preceding proposals, the first based on the acceleration of macro-particles and the second based on the irradiation of a thermonuclear target by an intense laser beam, have the important feature in common that the ignition of a small thermonuclear explosion would be effected by the bombardment of a target by an intense particle beam whose kinetic energy would be converted into heat at impact. From this point of view it is clear that the macro-particle accelerator and the laser represent the two extremes of the available particle spectrum, leaving open a wide area, especially electrons and ions. It was shown in a recent theoretical study [9] that with the aid of intense electron and ion beams the ignition of a thermonuclear micro-explosion (micro-bomb) might indeed become possible, and that the acceleration of beams of electrons and ions with the required intensity and energy seems to be feasible. This approach circumvents the great difficulties which otherwise seem to exist for the two extremes (macrons and photons) of the available particle spectrum.

The situation is summarized in Table I, where the various modes of acceleration for the different particles in the available spectrum are shown, ranging from photons (laser) to the superconducting macron.

For the electrostatic acceleration of electrons and ions, an energy storage system which can discharge an energy in excess of 10^{14} erg in about 10 ns is required. One such energy storage system consists of a large capacitor bank used in conjunction with the Marx-circuit Blumlein-line high-voltage technique. Alternatively, a novel system [9] based on the electrostatic charging of a levitated, superconducting, highly magnetized ring to gigavolt potentials, if technically feasible, would give a very powerful and compact energy storage system avoiding the large and expensive capacitor banks otherwise required.

The fact that with such a novel electrostatic energy storage system intense relativistic electron beams of many million ampere and up to GeV energies

TABLE I. – *Ignition of a small thermonuclear explosion by beam kinetic energy conversion.*

Type of accelerator	Type of particle	Rest mass (g)	Energy per particle	Number of particles	Target heating	Type of acceleration	Length of accelerator	Type of energy storage	Volume of energy storage (m^3)	Storage efficiency	Advantage	Disadvantage
Laser	photon	0	1 eV	10^{26}	electromagnetic shock wave	virtual	0	optical pumping	10^2	1%	good focusing	10^{16} erg energy for optical pumping
Relativistic electron beam	electron	10^{-27}	10 MeV	10^{19}	collisionless electrostatic and hydromagnetic shock wave	electrostatic	10 cm	Marx circuit capacitor bank	$10^2 \simeq 10^3$	high	relativistic focusing; magnetic reduction of fusion product range	enhanced energy loss by bremsstrahlung and turbulent electronic heat conduction
			GeV	10^{17}				levitated super-conducting ring	2			
Ion beam	ion	10^{-24}	10 MeV	10^{19}	collisionless and hydrodynamic shock wave	electrostatic	10 cm	Marx circuit capacitor bank	$10^2 \simeq 10^3$	high	shorter range in target	difficulty in suppressing electron field emission
			1 GeV	10^{17}				levitated super-conducting ring	2			
Microparticles	dust	10^{-12}	10^2 erg	10^{12}	hydrodynamic shock wave	electrostatic	10 cm	levitated super-conducting ring	2	high	short range in target	difficulty in suppressing field emission
Macron	super conducting solenoid	1	10^{16} erg	1	hydrodynamic shock wave	magnetic traveling wave	10 km	capacitor banks a transmission line	$10^4 \simeq 10^5$ long	high	short range in target	accelerator too long

will be attainable deserves special attention quite apart from its potential application to controlled fusion.

1. – Conditions for the ignition of a small thermonuclear explosion.

In order to obtain an idea about the required beam intensities and energies, we shall make some estimates about the minimum dimensions and energy input data of the thermonuclear target.

In order to neglect expansion losses of the hot plasma ball, it will be assumed that the target is surrounded by heavy, high-atomic-number material. It seems advantageous to use ordinary uranium for this purpose, since the fast neutrons set free in the thermonuclear reaction will cause fast fission processes in the uranium, thereby substantially increasing the total power output [10]. Ordinary uranium has the further advantage of being inexpensive. We assume a liquid or solid spherical T-D target with an atomic number density of $N = 5\cdot 10^{22}\ \text{cm}^{-3}$. In the following we shall always express the temperature in keV units.

In order to ignite a small thermonuclear explosion a number of conditions have to be fulfilled. Firstly, the energy has to be supplied to the target in a time shorter than the shortest energy loss time. The energy losses result from radiation (bremsstrahlung), electronic heat conduction and expansion.

The loss time for radiation is given by

$$\tau_R = 1.8\cdot 10^{-8}\sqrt{T}\ \text{s}, \tag{1.1}$$

and for electronic heat conduction by

$$\tau_c = 5.75\cdot 10^{-5} r_0^2 T^{-\frac{5}{2}}\ \text{s}, \tag{1.2}$$

where r_0 is the radius of the T-D sphere. If the T-D sphere is surrounded by a high-A-number material, for example uranium, the expansion losses are determined by the velocity of sound at several keV in those materials. One can safely assume a value less than 10^7 cm/s for this velocity. The loss time due to expansion is then given by

$$\tau_{ex} \simeq 10^{-7} r_0\ \text{s}. \tag{1.3}$$

For a target radius of the order of a few millimeters, it is obvious that the most important losses result from radiation and heat conduction. However, it will be seen that a large portion of the fusion products will go into a thin shell of the surrounding high-A-number material. The hot layer thus formed acts as a natural insulator against the heat-conduction losses, thereby reducing them substantially. Furthermore, the high magnetic field associated with the intense particle beam can result in a substantial reduction of the heat-conduction losses. For these various reasons we shall neglect the losses by expansion and electronic heat conduction in order to obtain a simple estimate for the critical dimensions of the thermonuclear target. Secondly, in order to obtain a positive energy balance, the Lawson criterion for the required minimum confinement time

$$\tau_L > 2 \cdot 10^{-9}\,\mathrm{s} \tag{1.4}$$

has to be fullfilled.

Thirdly, for a small thermonuclear explosion the charged fusion products have to be stopped within the reacting volume. This is the condition for detonation. If the range of the fusion products is λ, the probability P for the fusion products to be stopped inside a spherical volume of radius r_0 is given by

$$P = \frac{3\lambda}{4\pi r_0^3} \int\limits_{\underline{r}_1} \int\limits_{\underline{r}_2} \frac{\exp[-|\underline{r}_1 - \underline{r}_2|/\lambda]}{4\pi|\underline{r}_1 - \underline{r}_2|^2}\, \mathrm{d}\underline{r}_1\, \mathrm{d}\underline{r}_2\,. \tag{1.5}$$

This double integral has to be taken over the entire sphere. With $x = r_0/\lambda$, this integration yields

$$P(x) = 1 - \frac{3}{4x^3}\left[x^2 - \tfrac{1}{2} + \left(\tfrac{1}{2} + x\right) \exp[-2x]\right]. \tag{1.6}$$

The condition for detonation must be combined with the energy losses to obtain the critical size for a thermonuclear micro-explosion and the minimum (*i.e.* critical) energy. It should be emphasized that the minimum energy is not attained at the smallest possible ignition radius, since the smallest possible radius may be connected with a higher ignition temperature that takes into account the higher losses of charged fusion products from the reaction region.

The energy production of the T-D, reaction can be expressed in the form

$$E_f = 5.4 \cdot 10^{-18} N^2 T^{-\frac{2}{3}} \exp[-19.9 \cdot T^{-\frac{1}{3}}]\ \mathrm{erg/cm^3\,s}\,. \tag{1.7}$$

In order to determine the fusion energy output reabsorbed by the target, one must multiply the right-hand side of eq. (1.7) by the probability $P(x)$ that the fusion reaction products will be stopped within the target. The energy input from the fusion products has then to be balanced by the bremsstrahlung losses, which is the condition for ignition.

For a fully ionized hydrogen plasma the bremsstrahlung losses are given by

$$E_r = 5.35 \cdot 10^{-24} N^2 T^{\frac{1}{2}}\ \mathrm{erg/cm^3\,s}\,. \tag{1.8}$$

We thus obtain as the condition for the ignition and detonation of a T-D thermonuclear reaction the equation

$$1.01\cdot 10^{6}\,T^{-\frac{7}{6}}\exp\,[-19.9\,T^{-\frac{1}{3}}]\,P(x)=1\,, \tag{1.9}$$

where, as before, $x=r_0/\lambda$. The range λ of the ^{4}He fusion product from the T-D reaction is given by

$$\lambda=3.2\cdot 10^{-2}\,T^{\frac{3}{2}}\ \text{cm}\,. \tag{1.10}$$

In a situation where detonation can occur, eq. (1.9) has two roots. The smaller root, which gives the ignition temperature, approaches $T=4.3$ keV for $r_0=\infty$ and $P(x)=1$. The solution of eq. (1.9) gives the function $T(r_0)$ connecting the temperature and radius at which detonation occurs.

The energy input which must be supplied from outside to reach the ignition temperature is given by

$$E_0=3NkT\cdot\frac{4\pi}{3}\,r_0^3=10^{15}\,r_0^3\,T(r_0)\ \text{erg}\,. \tag{1.11}$$

The functions $T(r_0)$ and $E(r_0)$ are shown in Fig. 1. The input energy for the T-D reaction (Fig. 1*b*)) has a sharp minimum at $r_0=2.1$ mm with an energy

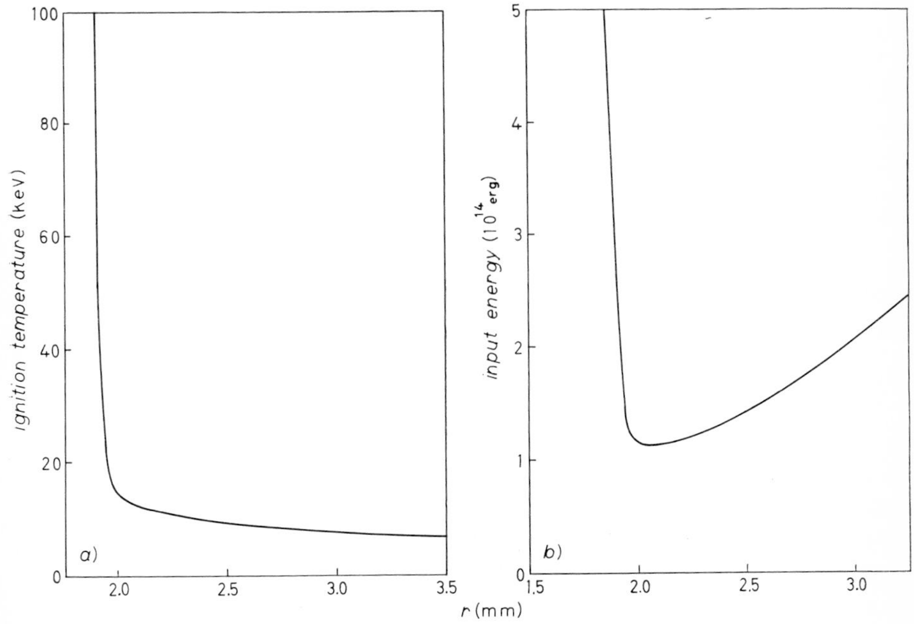

Fig. 1.

of $1.12\cdot10^{14}$ erg. The ignition temperature at this point is 12.1 keV and $P(x)=0.108$. About 90% of the fusion product energy is dissipated into the volume surrounding the reacting region. If the target is surrounded by some high-A-number material, the stopping power for the fusion products escaping from the reacting region will be very large. Accordingly, the fusion products dissipate their energy into a thin shell surrounding the thermonuclear target. According to eq. (1.1) the input energy of $1.12\cdot10^{14}$ erg has to be supplied to the thermonuclear target in a time less than 20 ns. The expansion loss time, according to eq. (1.3), is $2\cdot10^{-8}$ s, putting us safely above the Lawson minimum value of $2\cdot10^{-9}$ s given by eq. (1.4). To supply an energy of $\sim10^{14}$ erg in 10 ns would, for example, require a beam of $2\cdot10^{7}$ A at an electron energy of 50 MeV.

We would like to add two remarks. Firstly, since the ignition of the thermonuclear target is effected by an intense relativistic electron beam of many MA, very large magnetic fields may occur. These fields can substantially quench the range of the charged fusion products, thereby facilitating the ignition of the thermonuclear micro-explosion [10]. Although a perfectly conducting plasma can magnetically neutralize an intense relativistic electron beam [11] by backstreaming electrons, it is to be expected that at least part of the beam magnetic field will be trapped inside the target, since the assumption of perfect electrical conductivity is not valid during the initial cold phase of the heating process. The significance of the magnetic quenching effect can be estimated as follows. An electric current with a beam radius smaller than the target radius will produce at the target surface a magnetic field given by (I in ampere)

$$H=0.2I/r_0\,. \tag{1.12}$$

The Larmor radius of the charged fusion products is given by

$$r_{\mathrm{L}}=\frac{A}{Z}\frac{Mvc}{eH}\,, \tag{1.13}$$

where v is the velocity of the fusion product. Eliminating H from eqs. (1.12) and (1.13) and inserting the values for the ^{4}He T-D fusion products yields

$$\frac{r_{\mathrm{L}}}{r_0}\simeq0.7\cdot10^{6}/I\,. \tag{1.14}$$

From eq. (1.14) it follows that the Larmor radius becomes smaller than the target radius for beam currents in excess of 10^6 A. Since beam currents up to 10^8 A are required for ignition, this implies that a magnetic field trapping of only 1% would make magnetic quenching an important effect.

A second remark to be made is with regard to the effect which the strong magnetic field has on the electronic heat conduction.

The reduction in the coefficient of the electronic heat conduction due to a strong magnetic field is given by [12]

$$\frac{\chi_\perp}{\chi} = 6.76 \cdot 10^{-32} (N/H)^2 T^{-3} . \tag{1.15}$$

If, for example, $I = 10^7$ A and $r_0 = 0.2$ cm, it follows that $H = 10^7$ G. With $N = 5 \cdot 10^{22}$ cm^{-3} and $T = 10$ keV one thus obtains from eq. (1.15) $\chi_\perp/\chi \simeq \simeq 1.7 \cdot 10^{-3}$. The inverse of the characteristic time for the energy losses by electronic heat conduction, given by eq. (1.2), is then reduced by the same factor. For $r_0 = 0.2$ cm and $T = 10$ keV one obtains $\tau_c \simeq 0.7 \cdot 10^{-8}$ s, which would be changed to $\tau_{c_\perp} \simeq 4 \cdot 10^{-6}$ s. This supports the approximation in which the heat condition losses are neglected compared to the bremsstrahlung losses.

Both the fusion product range quenching and the reduction in electronic heat conducting will reduce the ignition energy for the micro-explosion.

2. – The interaction of the intense relativistic electron beam with the target.

A single electron with an energy of several MeV, striking a hydrogen target of $5 \cdot 10^{22}$ particles/cm^3 at thermonuclear temperatures, has a range of the order of 10^4 cm. It is therefore apparent that single-particle interactions of electrons are insufficient for heating a T-D target a few millimeters across up to thermonuclear temperatures. However, in plasma physics it is experimentally well known, and to a considerable degree also theoretically understood, that through the collective action of electric and magnetic fields the effective interaction range of a particle beam can be substantially reduced by collisionless dissipation effects. The collective plasma instabilities are of special significance in this respect. This situation is in contrast to the conventional magnetic-plasma-confinement approach towards controlled thermonuclear fusion, where one tries to avoid as many plasma instabilities as possible. In the different approach outlined here, the opposite is the case; it is important to have fast growing instabilities which can rapidly deposit the beam energy into the target.

An electron beam interacts strongly with a plasma target through the phenomena known as the counterstream instability. This is one of a class of electrostatic instabilities which, in general, grow more rapidly than other instabilities. The maximum growth rates of these streaming instabilities have been discussed by BUNEMAN [13] for cold, collisionless plasmas for nonrelativistic, monoenergetic electron energies and by BLUDMAN, WATSON, and ROSENBLUTH [14] for relativistic electron streams with finite beam temperature and with collisions.

The target electron and target ion plasma frequencies are given respectively by

(2.1) $$\omega_e = (4\pi n_1 e^2/m)^{\frac{1}{2}}$$

and

(2.2) $$\omega_i = (4\pi n_1 e^2/M)^{\frac{1}{2}} = (m/M)^{\frac{1}{2}}\omega_e ,$$

where m is the electron mass and M is the ion mass. The beam electron plasma frequency ω_b' is given by

(2.3) $$\omega_b' = (\omega_{b\perp}^2 \sin^2\theta + \omega_{b\parallel}^2 \cos^2\theta)^{\frac{1}{2}} ,$$

where θ is the angle between the wave number vector and the beam direction, and where

(2.4) $$\omega_{b\perp} = \omega_b \gamma^{-\frac{1}{2}} ,$$

(2.5) $$\omega_{b\parallel} = \omega_b \gamma^{-\frac{3}{2}} ,$$

(2.6) $$\omega_b = (4\pi n_2 e^2/m)^{\frac{1}{2}} ,$$

$$\gamma \equiv (1 - v^2/c^2)^{-\frac{1}{2}} .$$

In this, $\omega_{b\perp}$ and $\omega_{b\parallel}$ are the beam plasma frequencies for wave propagation respectively perpendicular and parallel to the beam direction. Note that in the perpendicular beam plasma frequency the transverse electron mass enters, whereas in the parallel plasma frequency it is the longitudinal mass that is involved. Clearly one has $\omega_{b\perp} = \gamma\omega_{b\parallel}$. It has been shown [14] that the fastest developing instabilities occur in waves propagating obliquely to the beam direction at the angle $\theta \simeq \pi/4$. For $\gamma \gg 1$, which is always true for the ranges of interest here, one has

(2.7) $$\omega_b' \simeq 2^{-\frac{1}{2}} \omega_b \gamma^{-\frac{1}{2}} \simeq 0.71 \omega_b \gamma^{-\frac{1}{2}} .$$

Since we are interested in the fastest growing instability, we shall herein always use this value for ω_b'.

Introducing the electron collision frequency [12] ν for a T-D target with an atomic number density of $N = 5 \cdot 10^{22}$ cm^{-3},

(2.8) $$\nu = 4.27 \cdot 10^{13} T^{-\frac{3}{2}} ,$$

we can distinguish the three frequency regions,

(2.9) I) $$\nu \ll (\omega_b'/\omega_e)^{\frac{2}{3}} \omega_e ,$$

(2.10) II) $$(\omega_b'/\omega_e)^{\frac{2}{3}} \omega_e \ll \nu \ll \omega_e ,$$

(2.11) III) $$\nu \gg \omega_e .$$

In region I), collisions can be neglected, and the growth rate is given by

$$\sigma \simeq 0.7(\omega_b'/\omega_e)^{\frac{2}{3}}\omega_e \simeq 0.55(n_2/n_1)^{\frac{1}{3}}\omega_e\gamma^{-\frac{1}{3}}\,. \tag{2.12}$$

In region II) the growth rate is given by

$$\sigma \simeq (\omega_b'/\omega_e)(\omega_e/2\nu)^{\frac{1}{2}}\omega_e \simeq 0.5(n_2/n_1)^{\frac{1}{2}}(\omega_e/\nu)^{\frac{1}{2}}\omega_e\gamma^{-\frac{1}{2}}\,. \tag{2.13}$$

In this region the effect of the collisions is to slow down the growth rate. In region III) the electron plasma oscillations are inhibited. However, there is a fast growing instability due to ion plasma oscillations for which the growth rate is

$$\sigma \simeq 0.7(\omega_b'/\omega_i)^{\frac{1}{3}}\omega_i \simeq 0.62(m/M)^{\frac{1}{3}}(n_2/n_1)^{\frac{1}{6}}\omega_e\gamma^{-\frac{1}{6}}\,. \tag{2.14}$$

The transition between regions II) and III) takes place at $\omega_e = \nu$. Using the expression for ν given by eq. (2.8), one computes for this frequency the temperature $T \simeq 0.15$ keV $\simeq 1.7\cdot10^6$ °K.

After a value for the growth rate σ has been determined, the beam range λ_D for beam velocities $v \simeq c$ can be computed from the equation

$$\lambda_D \simeq c/\sigma\,. \tag{2.15}$$

In the case of a T-D target one has $n_1 = 5\cdot10^{22}$ cm^{-3}, $\omega_e = 1.26\cdot10^{16}$ s^{-1}, and $M = 2.5\,M_H$, where M_H is the mass of the hydrogen atom; hence $m/M = 2.18\cdot10^{-4}$.

The number density n_2 for the electrons in the beam can be expressed in terms of the electric current density as

$$j \simeq n_2 ec \simeq 14.4 n_2\,, \tag{2.16}$$

where j is measured in electrostatic units. If j is given in A/cm^2, one has

$$n_2 = 2.1\cdot10^8 j\,. \tag{2.17}$$

Assuming that the beam has a diameter of 0.1 cm, which seems technically feasible, one has $j = 10^2 I$, where I is the total current. Thus with I given in MA, we have

$$n_2 = 2.1\cdot10^{16} I\,. \tag{2.18}$$

For $I = 20$ MA, one has, for example, $n_2 \simeq 4\cdot10^{17}$ cm^{-3}. Inserting the values for n_1, n_2, ω_e, and M/m into eqs. (2.12), (2.13), and (2.14) and solving for the range λ_D yields

$$\text{(2.19)} \quad \text{I)} \qquad \lambda_D = 5.8\cdot 10^{-4}(\gamma/I)^{\frac{1}{3}},$$

$$\text{(2.20)} \quad \text{II)} \qquad \lambda_D < 7.4\cdot 10^{-3}(\gamma/I)^{\frac{1}{2}},$$

$$\text{(2.21)} \quad \text{III)} \qquad \lambda_D = 7.4\cdot 10^{-4}(\gamma/I)^{\frac{1}{6}}.$$

We have assumed for region II) the smallest possible growth rate, and hence the largest possible range, by putting $\omega_e = \nu$. The actual range will be smaller because the transition into region III) takes place at $\omega_e = \nu$. The transition from region II) into region III) implies that the growth rate formula for region II) at the transition point is no longer accurate. Since the range in region III) is smaller than the value given by eq. (2.20), this implies that the range at $\omega_e = \nu$ is actually smaller, the actual range in region II) will most likely be between the values for λ_D in regions I) and III).

For the 50 MeV electrons attainable with a Marx generator, one has $\gamma \simeq 100$ and $I = 20$ MA. For $\gamma \simeq 10^3$, $I = 2$ MA. The value $\gamma = 10^3$ can be attained with the levitated superconducting ring energy storage system which will be discussed in detail below. It is obvious that in the latter case the ranges become larger than for 50 MeV electrons. But even with $\gamma/I \simeq 10^3$ the largest value for λ_D is reached at II with $\lambda_D \simeq 0.23$ cm. For $\gamma/I \simeq 5$, which is the case for 50 MeV electrons, region II) would give $\lambda_D \simeq 1.6\cdot 10^{-2}$ cm. The general property expressed by eqs. (2.19) to (2.21) is that the range decreases with increasing beam current. Since a high current is needed anyway, this behavior is a very fortunate one.

The actual beam ranges will be larger by virtue of the fact that the ranges computed from eq. (2.15) would correspond to a change in the beam velocity by a factor e. Since the beam must be slowed down to thermal energies, the actual range has to be multiplied by the factor $\frac{1}{2}\ln(E_0/E_{th})$, where E_0 is the initial beam energy and E_{th} is the thermal energy. If $E_{th} \simeq 10$ keV and $E_0 \simeq 50$ MeV, then $\frac{1}{2}\ln(E_0/E_{th}) \simeq 4.2$. For $E_0 \simeq 1$ GeV, with an unchanged value for E_{th}, this factor is 6.8. This, therefore, cannot pose a serious problem in the stopping of the beam.

On the basis of these simplified calculations, one has the impression that the computed range of the beam seems to be sufficiently short to ensure complete dissipation of the beam energy for targets a few millimeters in diameter.

There is, however, another important effect which might affect the growth rate of the electrostatic instabilities and hence the beam stopping range. This effect results from a possible longitudinal and transverse temperature, respectively velocity, spread of the beam. Such a temperature spread can lead to strong Landau damping opposing the growth of the counterstream instability. At the present time it is not known whether Landau damping will pose a serious problem. One can, however, say that it will become less important with increasing beam energy since the overlapping of the target and beam velocity distributions, which is responsible for the Landau damping, decreases as the beam energy increases.

In case the collisionless energy dissipation for energetic electron beams should lead to serious problems, then one can still always employ an intense ion discharge [9], which will be discussed later.

It will also always be necessary to preheat the target up to 10 eV $\simeq 10^5$ °K in order to obtain a completely ionized plasma. There are a number of possible mechanisms by which preheating can take place. Firstly, through ordinary stopping power the beam will lose part of its energy to the target. To preheat the target to 10 eV, assuming a target radius of 0.2 cm, one needs an energy of about 10^{11} erg. The range of 10 MeV electrons in liquid hydrogen is about 10^2 cm, so a beam of 10^{14} erg would deposit $5\cdot 10^{11}$ erg, five times the amount required to preheat the target to 10 eV. Therefore, 20% of the beam energy would be used to preheat the target. This is a rather large fraction of the total beam energy, in view of the low temperature required for preheating. A more efficient mechanism for preheating, however, results from the high-A-number material surrounding the target. As the beam hits the high Z number material ($Z \simeq 100$), it is stopped over a range which can be as small as 10^{-2} cm. Accordingly, the surface of the material acquires a high temperature which can reach 10^6 °K as in the case of exploding wires. The beam thus generates a hot spot from which a large amount of ultraviolet light is emitted, thereby effectively preheating the target.

Finally, the energy of the target electrons must be transferred to the target ions via Coulomb collisions. Under our chosen target conditions the equipartition time for this process is given by [12]

$$t_{eq} = 5.5\cdot 10^{-11} T^{\frac{3}{2}} \text{ s.} \tag{2.22}$$

For $T = 4.3$ keV $= 5\cdot 10^7$ °K, one has $t_{eq} \simeq 5\cdot 10^{-10}$ s. This time is therefore sufficiently short to ensure that the ions are heated up to fusion temperatures within the required period.

3. – The method of bombarding a target by an intense relativistic electron beam from a pulsed field emission discharge.

The extremely short discharge time required for the ignition of a thermonuclear micro-explosion dictates the use of high voltages. Only with high voltages and short discharge times is high energy release possible. The disadvantage of high voltages is the low stopping power for energetic electrons, but fortunately this may be by-passed by employing the collisionless dissipation mechanisms discussed above. As already stated above, it is a fortunate coincidence that these collisionless dissipation mechanisms work at the very high current densities which are required for our purpose.

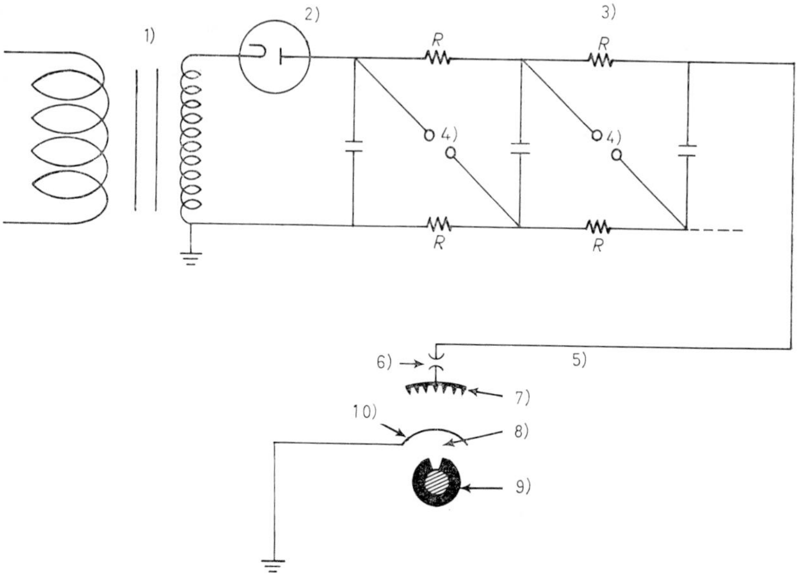

Fig. 2. – The Marx-circuit Blumlein-line method of bombarding the target. 1) High-voltage transformer; 2) rectifier; 3) Marx circuit; 4) spark gaps; 5) Blumlein-line; 6) triggered spark gap switch; 7) field emission cathode; 8) tenuos plasma for space-charge neutralization; 9) thermonuclear target; 10) thin anode window.

The principle of the idea of bombarding a target by an intense relativistic electron beam is shown in Fig. 2 and 3. In Fig. 2, in what is known as the Marx generator, a bank of parallel capacitors is charged up to a high voltage. The charging is performed by a high-voltage transformer and a rectifier. When the voltage across the spark gaps reaches a critical value, one spark gap closes, followed rapidly by all the others because of the sudden surge in the voltage across the other gaps. As a result the capacitors are suddenly switched into series. If the voltage before the spark gaps close is V and if there are n capacitors of equal capacitance, then the voltage after the spark gaps close rises to the value nV. Furthermore, if the discharge time of the bank with open spark gaps is τ, then the discharge time after the spark gaps close falls to the value τ/n. It is clear that, depending on the number n of the capacitors in parallel, very high voltages can be attained. For example, if $V = 500$ kV and $n = 100$, the voltage after the spark gaps close rises to 50 MV, and if the discharge time of the bank in parallel is $\tau = 10^{-5}$ s, it falls to 10^{-7} s for the series alignment.

The high-voltage terminal of the bank is connected with a low-inductance transmission line such as the Blumlein line explained below, which discharges the energy over a triggered spark gap switch in a time which can be as short as 10^{-8} s.

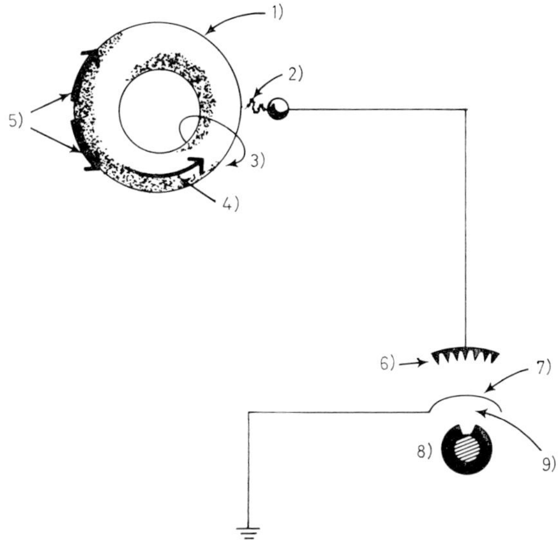

Fig. 3. – The levitated, superconducting ring method of bombarding the target. 1) Floating highly charged superconducting ring; 2) spark gap; 3) magnetic field; 4) high current; 5) electron current during discharge; 6) field emission cathode; 7) thin anode window; 8) thermonuclear target; 9) tenuous plasma for space-charge neutralization.

The high voltage at the end of the Blumlein line is applied to a brush of field emission cathodes arranged in a concave pattern around the thermonuclear target. The target itself is surrounded by a high-A-number material serving as a means of inertially confining the target as long as possible after ignition. Between the field emission brush and the thermonuclear target is a thin anode window. The electrons pass through this window with very little energy loss if their energy is high and the window sufficiently thin. In front of the target, behind the anode window, is a tenuous plasma which neutralizes the space charge of the beam. In this way the focusing of the beam can be effected by the self-magnetic forces with high efficiency because of the relativistic electron energies.

To make the discharge time as short as possible, the line connecting the bank with the load, which in our case is the field emission brush and the target, must have a low inductance. This can be done by the Blumlein line shown in Fig. 4. As we would use it, the Blumlein line would consist of a conical, coaxial, tapered wave guide. The capacitor bank is discharged onto the intermediate conical conductor, which discharges itself at the base of the wave guide over a circular spark gap onto the inner conical conductor. The grounded outer conical conductor serves as the outer wall for the whole wave guide, which is under high vacuum. The tapered end of the inner conical conductor

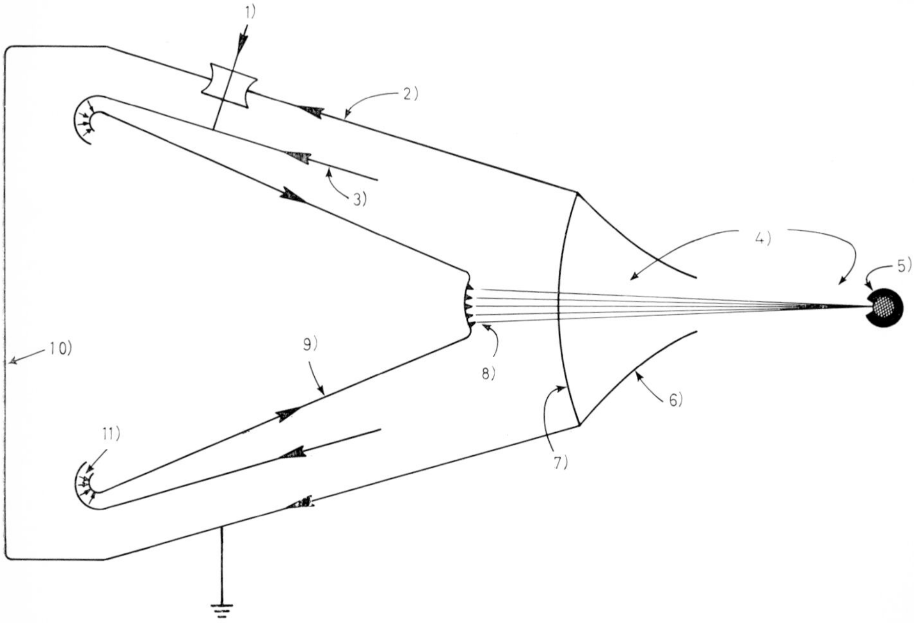

Fig. 4. – The low-inductance Blumlein line. 1) To capacitor bank; 2) outer cone serving as wall for vacuum chamber and as conductor for return current; 3) intermediate conical conductor; 4) tenuous-plasma; 5) thermonuclear target; 6) tapered conducting pipe for beam focussing; 7) anode window; 8) field emission cathode; 9) inner conical conductor; 10) vacuum tank; 11) circular spark gap.

is rounded off smoothly and is terminated by a concave depression containing the field emission brush. As the wave propagates down the transmission line, it suddenly hits the field emission brush. The field-emitted electrons are then accelerated towards the thin anode window at the right end of the outer conical conductor. After penetrating the anode window the electrons enter the region containing the tenuous space charge-neutralizing plasma located between the anode window and the thermonuclear target. In this region the beam pinches down to a small diamter before hitting the thermonuclear target. Beyond the window to the right side is a tapered, conducting tube which facilitates the focusing of the beam.

The dimension of the Blumlein line can be estimated as follows. For a given energy output E_{out} one has

$$E_{out} = CV^2/2 , \tag{3.1}$$

where C is the capacitance of the Blumlein line and V the voltage to which it is charged up by the high voltage terminal of the Marx generator.

The discharge time of the Blumlein line is one quarter of the Thomson time (in electrostatic c.g.s. units),

$$\tau = \frac{\pi}{2c}\sqrt{LC} = \frac{\pi}{2} ZC , \qquad Z = \sqrt{L/C}/c , \tag{3.2}$$

where L and Z are the inductance and impedance of the Blumlein line. From eqs. (3.1) and (3.2) we have

$$E_{out} = \tau V^2/\pi Z , \tag{3.3}$$

which demonstrates the significance of using high voltages, since for a given value of E_{out} and Z the discharge time is inversely proportional to V^2 and short discharge times are required for our purpose. The value of τ for given values of E_{out} and V can still be varied within certain practical limits by changing geometry-dependent impedance Z.

In order to calculate the capacitance and impedance of the Blumlein line we approximate the intermediate and outer conical conductors by two coaxial cylinders with inner and outer radii r_0 and r_1, forming a cylindrical capacitor.

The length of this cylindrical capacitor is approximately equal to the length of the intermediate conductor with the length D. Under these simplifying assumptions the capacitance and impedance are given by

$$C = D/2 \ln (r_1/r_0) , \tag{3.4}$$

$$Z = 2 \ln (r_1/r_0)/c . \tag{3.5}$$

The discharge time is then given by

$$\tau = \pi D/2c , \tag{3.6}$$

which is of the same order of magnitude as the time for an electromagnetic wave to propagate along a cylinder of length D. For $\tau \simeq 10^{-8}$ s one has $D \simeq 2$ m. From eqs. (3.3) and (3.5) one obtains

$$\ln (r_1/r_0) = \tau c V^2/2\pi E_{out} . \tag{3.7}$$

If $r_1 = r_0 + s$, $s \ll r_0$, one then has

$$r_0 \simeq 2\pi s E_{out}/c\tau V^2 . \tag{3.8}$$

Let us assume that the separation distance between the intermediate and outer conical conductors is $s \simeq 5$ cm. For $\tau = 10^{-8}$ s and $E_{out} = 10^{14}$ erg, one therefore has

$$r_0 = 0.94 \cdot 10^{18}/V^2 , \tag{3.9}$$

where V is expressed in volt. Let us take for example $V = 5 \cdot 10^7$ volt, which is attainable with Marx generators. One then has $r_0 = 3.8$ m. With decreasing voltage the dimension of the system increases rapidly, which again explains the importance of high voltages. Higher voltages and hence even smaller Blumlein lines may be attainable through use of the principle of high magnetic field insulation explained below.

It was shown above that to ignite a small thermonuclear explosion in T-D, some 10^{14} erg of energy must be dissipated into a sphere of liquid or solid T-D with the dimensions of cm in a time less than 20 ns. Beam energies up to $3 \cdot 10^{11}$ erg with a discharge time of 10 ns have already been achieved [15]. The electron energy was 10 MeV and the total beam current was $3 \cdot 10^5$ A. This is still short by almost three orders of magnitude from what is required to ignite a small fusion explosion, but Marx generators with energy outputs ranging from 10^{13} to 10^{14} erg directed into the critical dimensions required for our purpose are presently under construction.

The density of the field emission current is given by

$$j = 1.55 \cdot 10^{-6} (E^2/W) \exp[-6.9 \cdot 10^7 W^{\frac{3}{2}}/E] \qquad (\text{A/cm}^2) , \tag{3.10}$$

where E is the applied electric field in V/cm and W the work function in eV. For tungsten, for example, $W = 4.4$ eV. The electric field strength at each emitter tip is given by

$$E = V/r , \tag{3.11}$$

where V is the applied voltage and r the tip radius. We take as an example a field emission brush having 300 tungsten cathodes, each with a tip radius of 5 mm. The voltage applied to the brush is assumed to be 50 MV. One then calculates for each emitter a field emission current of $3.5 \cdot 10^5$ A. The total power output of each emitter will be $3.5 \cdot 10^{12}$ W and the energy delivered in 10 ns approximately 10^{14} erg. If the current in each cathode is too high, resulting in the melting of the cathode, then one can easily avoid this problem by increasing the number of cathodes. It seems technically feasible to increase the number of cathodes by one to two orders of magnitude. A higher increase can be accomplished indirectly by substituting for the field emission brush a concave tungsten electrode with sufficient surface irregularities for it to act like a field emission brush.

Because a capacitor bank capable of delivering energies of the order of 10^{14} erg might be considered uncomfortably large, a different scheme of energy storage based on the electric charging of a levitated, magnetized, superconducting ring to extremely high voltages has been proposed [9]. If strong electric currents flow inside the ring, as indicated in Fig. 3, they generate a strong azimuthal magnetic field. Figure 5 shows a cross-section through the superconducting torus. A steel jacket holds the superconducting cables together and pipes are provided for the liquid helium to cool the system down to the superconducting state. The breakdown in vacuum-insulated capacitors, caused by field electron emission, limits the attainable electric fields to approximately 10^5 V/cm. In the magnetized, levitated ring capacitor, losses resulting from field emission are prevented by the strong azimuthal magnetic field. If an electron is emitted from some surface irregularity, the combined action of the magnetic and electric fields forces it into a drift motion parallel to the direction of the circular axis of the toroidal superconductor, thus preventing its escape from the system.

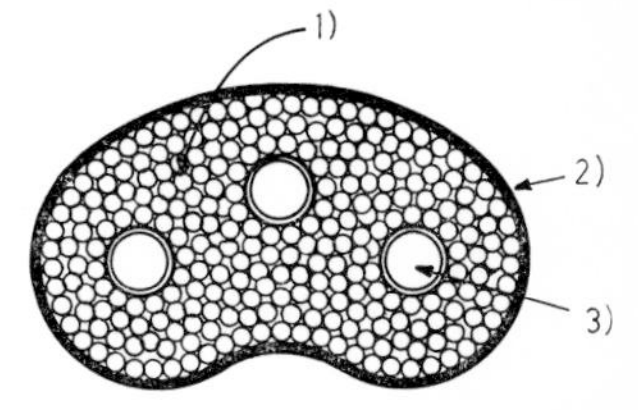

Fig. 5. – Cross-section through the superconducting torus, showing the superconducting cables and the pipes for the liquid helium. 1) Superconducting cables; 2) steel jacket; 3) cooling channel for liquid He.

In order for this system to work, a number of conditions must be satisfied. The first condition is that the electric pressure caused by the accumulated electric charges on the surface of the torus must not exceed the magnetic surface pressure resulting from the toroidal ring currents, requiring that

$$E < H , \tag{3.12}$$

where both E and H are measured in electrostatic units. With a maximum attainable magnetic field of $H = 3 \cdot 10^5$ G, this inequality means that $E < 10^8$ V/cm. The maximum value of $H = 3 \cdot 10^5$ G is consistent with the upper field limit of high-field superconductors and with the mechanical limitations set by the tensile strength of construction materials. The maximum electric field of 10^8 V/cm is also consistent with the upper limit of the tensile strength, since above 10^8 V/cm field ion emission becomes important.

The second condition to be satisfied is that the electron Larmor radius of the drift motion R_L be smaller than the principal torus radius R. This condition is necessary in order to prevent the so-called diocotron instability which otherwise would build up in the electron cloud above the surface of the magnetized and levitated ring. This electron cloud results from electron field emission from surface irregularities on the toroidal ring. The condition for preventing the diocotron instability is given in electrostatic c.g.s. units by

$$R \gg R_L = (mc^2 E_r)/(eH_\theta^2) , \tag{3.13}$$

where E_r is the radial electric surface field and H_θ is the azimuthal magnetic field.

Finally, the third condition requires that the electron Larmor radius of a field-emitted electron be smaller than the minimum distance to the external. levitating support, which is of the order of the inner torus radius r. This condition is given by

$$r \gg r_L = (mvc)/(eH_\theta), \tag{3.14}$$

where v is the electron velocity upon emission from the surface irregularity. This electron velocity will be rather small compared to the velocity of light. The estimates are valid for electrons not attaining high relativistic energies. For the electrons to gain relativistic energies would require that they be emitted into the space between the ring and the outside, zero-potential support. But even under these extreme conditions the inequalities eqs. (3.13) and (3.14) can be satisfied [9]. However, we believe that the electrons can never attain these high energies because the high magnetic field confines them so strongly to the toroidal ring that they most likely cannot reach large distances away from the torus. From eqs. (3.13) and (3.14) we obtain

$$r_L/R_L = (v/c)(H_\theta/E_r) . \tag{3.15}$$

The capacitance and inductance of the torus are given approximately by (electrostatic c.g.s. units)

$$C = \pi R/\ln(8R/r) , \tag{3.16}$$

$$L = 4\pi R[\ln(8/R) - \tfrac{7}{4}] . \tag{3.17}$$

The total electrostatic energy which can be stored on the torus is

$$E_s = \tfrac{1}{2} C V^2 , \tag{3.18}$$

where V is the total voltage of the ring expressed through the electric field strength by

$$V = 4rE_r \ln(8R/r) . \tag{3.19}$$

If the ring has dimensions of a few meters and is charged up to a surface field strength of 10^7 V/cm, which is one order of magnitude below the upper limit of 10^8 V/cm, the electric potential is of the order of many gigavolt. However, at the present time it is more likely that only a lower magnetic field is technically feasible, so a value of $3\cdot10^6$ V/cm is more realistic. Assuming principal and minor torus radii of $R=300$ cm and $r=50$ cm respectively, one obtains for the total voltage $V=2.3\cdot10^9$ V. For the capacitance and inductance one has $C=2.4\cdot10^2$ cm and $L=8.1\cdot10^3$ cm. The stored electric charge will be $Q=0.61$ coulomb, and the total stored electrostatic energy will be $E_s\simeq7\cdot10^{15}$ erg. Furthermore, assuming that $H_\theta=10^5$ G and that eq. (3.12) is satisfied, one has from eq. (3.13) $R_L=1.7\cdot10^{-3}$ cm, so $R\gg R_L$; because of eq. (3.15), then, the condition eq. (3.14) is also satisfied.

The projected ring voltage of $2.3\cdot10^9$ V seems very high, but even if it should prove possible to reach a total voltage of only 100 MV with the described system, this would represent a significant step towards an advanced electrostatic energy storage system greatly superior to conventional capacitor banks, since even then the total energy stored would still be $E_s=1.3\cdot10^{13}$ erg. The discharge time, which is $\frac{1}{4}$ the Thomson time given by

$$\tau_1=(\pi/2c)(LC)^{\frac{1}{2}}\,. \tag{3.20}$$

Inserting the values for L and C from our numerical example, we obtain $\tau_1\simeq7.5\cdot10^{-8}$ s $=75$ ns. The actual discharge time is shorter than this value for the following reason. The way in which the torus is discharged, as shown in Fig. 3, indicates that both halves of the torus to the left and right side of the spark gap are discharged simultaneously through the spark gap, which corresponds to the simultaneous discharge of two capacitors in parallel. This implies that one has to use half the values of L and C as given by eqs. (3.16) and (3.17) for this mode of discharge. This, in summary, will lead to a discharge time which is $\frac{1}{8}$ the Thomson time, *i.e.*

$$\tau_2=(\pi/4c)(LC)^{\frac{1}{2}}=38\text{ ns}\,. \tag{3.21}$$

This time is to be compared to the time it would take for an electromagnetic wave to travel the distance of one semicircle of the principal radius of the toroidal ring,

$$\tau_3=\pi R/c=31\text{ ns}\,, \tag{3.22}$$

which, as expected, is of the same order as the time computed from eq. (3.21).

It has been suggested by FORD [16] that the discharge time for the toroidal ring would be substantially reduced by using a ringlike discharge rather than just one spark gap. In this mode of discharging the ring, a second ring of the same principal radius would be brought close to the toroidal ring until breakdown occurs. The electromagnetic wave associated with the discharge of the ring would travel parallel to the minor torus circumference. This is in contrast to the other mode of discharge with one spark gap, where the electromagnetic wave travels parallel to the principal torus circumference. In the ringlike discharge we may think of the torus as being composed of many disks having radii which are equal to the minor torus radius. Each disk is discharged by one segment of the discharging ring, and the time for the electromagnetic wave to reach the discharging ring is equal to the time needed to travel one semicircle of the minor torus radius r. This time is

$$\tau_4 = \pi r / c = 5 \text{ ns} . \tag{3.23}$$

We would like to go one step further and suggest that the discharge be made not only ringlike but that the discharging ring form one end of a tapered, conical wave guide, as indicated in Fig. 6, with the field emission brush at the tapered terminal. In the tapered section of the wave guide the LC product per unit length decreases and thus steepens up the wave before it hits the field emission brush. The described form of the wave guide is very similar to the Blumlein

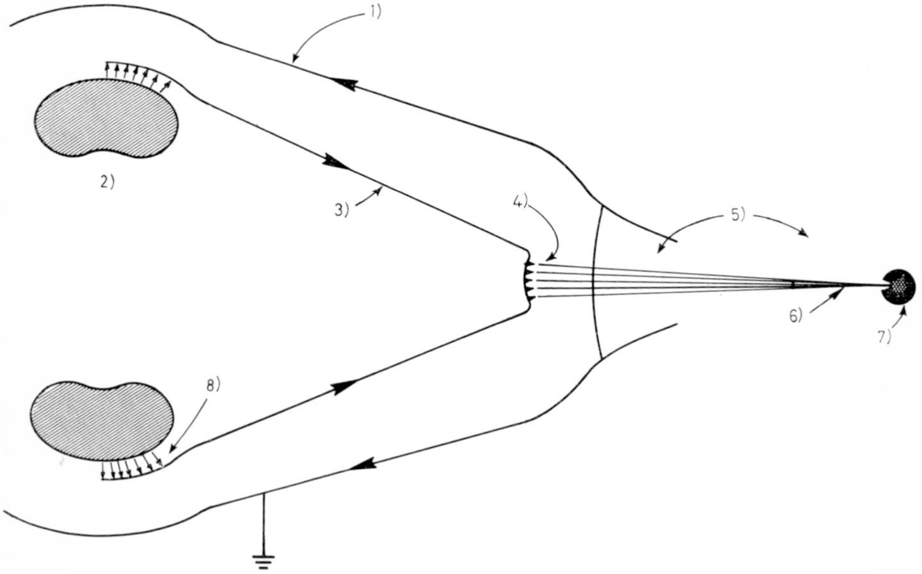

Fig. 6. – The ringlike discharge of the levitated torus. 1) Wall of vacuum chamber and outer return current conductor; 2) toroidal ring (cross-section); 3) tapered inner cone; 4) field emission brush; 5) tenuous plasma; 6) neutralized beam; 7) target; 8) triggered circular spark gap.

line described in Fig. 4, with the important difference that the intermediate cylinder can be omitted here.

If an energy of $E_s = 7\cdot10^{15}$ erg is discharged in 5 ns, this represents a power of $1.4\cdot10^{11}$ MW. The electric current would be $I = Q/\tau_4 = 1.2\cdot10^8$ A. The extremely short discharge time of 5 ns makes the ringlike discharge especially attractive for our purpose.

The charging of the ring can be performed in two different ways, as shown in Fig. 7 and 8. In the first scheme, Fig. 7, electrons coming from a variable-

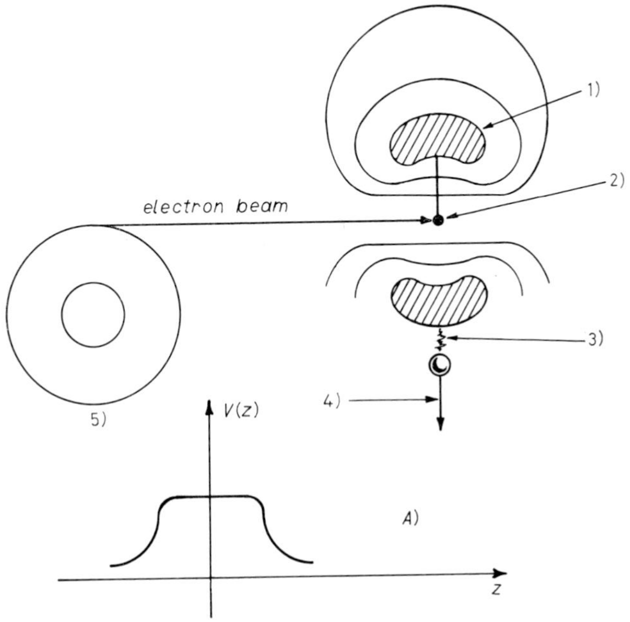

Fig. 7. – Charging the levitated torus by electron injection into a Faraday cage potential. 1) Levitated ring (cross-section); 2) collector electrode; 3) spark gap; 4) to load; 5) variable-energy betatron.

energy betatron are directed onto a collector electrode connected to the ring by a conducting bar. The electrons are shot along the z-axis of a cylindrical co-ordinate system centered at the ring. Since the magnetic field along the z-axis is everywhere parallel to the z-axis, the trajectories of the incoming electrons are unaffected by the field. If the energy of the electrons in the beam is kept just slightly above the ring voltage, they reach the collector electrode with zero velocity. Although the potential difference between the collector electrode and the ground may become very large, the ring can always be shaped in such a way as to form a Faraday cage potential along the z-axis, the only direction along which the electrons could escape. The Faraday-cage potential implies a higher-order flat point for the electric potential. Accordingly, the electric field along the z-axis will be small, preventing breakdown by field electron emission. The Faraday-cage potential can be improved by giving the ring a kidney-form cross-section or shaping it into somewhat of a cylindrical shell. If the voltage across the spark gap reaches a critical value, electric breakdown occurs and the discharge can be transmitted to the load. The time τ to charge the ring can be computed from the relation $\tau \simeq CV/I$. With $I = 10^{-3}$ A $= 3 \cdot 10^6$ e.s.u., $C = 2 \cdot 10^2$ cm, $V = 10^9$ V $= 3 \cdot 10^6$ e.s.u., one obtains $\tau \simeq 200$ s.

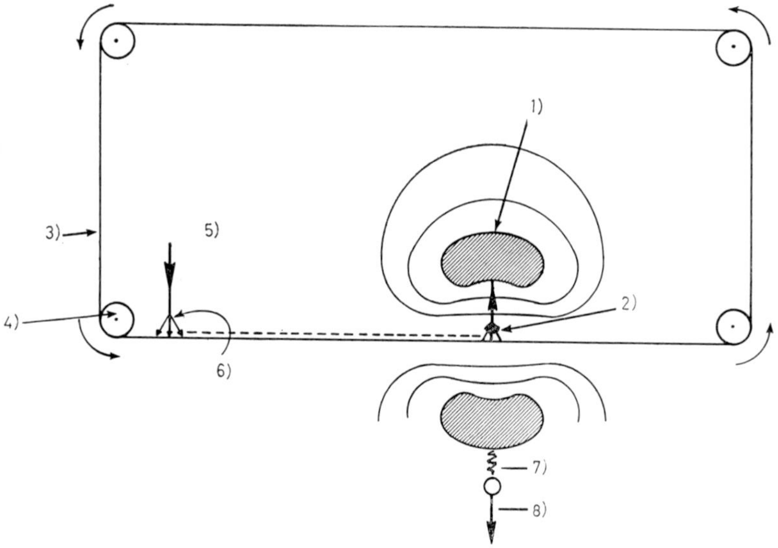

Fig. 8. – The Van de Graaff mode of charging the levitated torus. 1) Levitated ring (cross-section); 2) collector brush; 3) belt; 4) motor driving belt; 5) from high-voltage generator; 6) electrode spraying negative charges onto belt; 7) spark gap; 8) to load.

In the second mode of charging the ring, an insulating belt driven by a motor transports electric charges onto the collector electrode inside the Faraday-cage potential, as indicated in Fig. 8. This mode of charging up the ring works on the same principle as the Van de Graaff generator and will therefore be called the « Van de Graaff Mode » of charging the ring. An advantage of this mode is that it does not require the expensive and heavy electron accelerator. A disadvantage will be a lower maximum voltage attainable for the ring. As the belt passes through the region of the high electric field, the electrons will tend to be stripped off their positions, thus preventing them from reaching the inside of the Faraday cage, if the electric field along the belt becomes too large, to what limits is not known.

As the ring potential rises, the voltage across the spark gap reaches a critical value at which breakdown occurs. The stored electric energy is then discharged through the field emission process onto the thermonuclear target, as shown in Fig. 3.

It is also possible to charge the ring up to a high positive electric potential. In this case the field-emitted electrons from the outer support would be repelled from the ring by the high magnetic field. This mode of charging may be advantageous for the generation of intense ion beams discussed below. It is of course also possible to obtain a high positive electric potential with a Marx generator which can be utilized for the same purpose.

We would like to add that the principle of magnetic insulation can also be applied to the Blumlein line. In order to get short discharge times small Blumlein lines are required with resulting high voltages. The strong electric current of many mega-ampere flowing through the Blumlein line may impose partial magnetic insulation. It is, however, also possible to apply an external axial magnetic field to the Blumlein line for the same purpose.

4. – Converting the energy into useful power.

We shall briefly discuss some possible methods of converting the energy released in the thermonuclear reaction into useful power. For this purpose consider a chain of micro-explosions taking place inside a large spherical container, such as that illustrated in Fig. 9.

The interior of the container is permeated by a magnetic field produced by the surrounding coils. When a micro-explosion takes place at the center of the container the magnetic lines of force are pushed aside and a magnetic cavity is formed. Since the temperature is high, the electrical conductivity is high and the magnetic field does not penetrate into the magnetic cavity filled with hot plasma. Currents are induced in the external coils through the action of the expanding plasma ball on the magnetic field, resulting in a direct transformation of some of the thermonuclear energy into useful electrical energy. If the plasma ball creates a magnetic cavity of volume $V=(4\pi/3)R^3$, where R is the container radius, then the part of the energy transformed directly into electrical energy is given by

$$E_H = \frac{H^2}{8\pi} V = \frac{1}{6} H^2 R^3 . \tag{4.1}$$

As the hot plasma ball reaches the container wall a liquid film absorbs the shock wave of the spherical explosion and evaporates. The hot gas generated as a result of this evaporation can be channelled into a gas turbine generator of an electrical power plant. Finally, the heat absorbed by the container can be removed by a coolant circulating through cooling channels connected with a conventional electrical steam power plant.

Let us assume that the container wall can withstand a pressure of 10^3 atm $= 10^9$ dyn/cm², and that it is always filled with some residual gas left over from

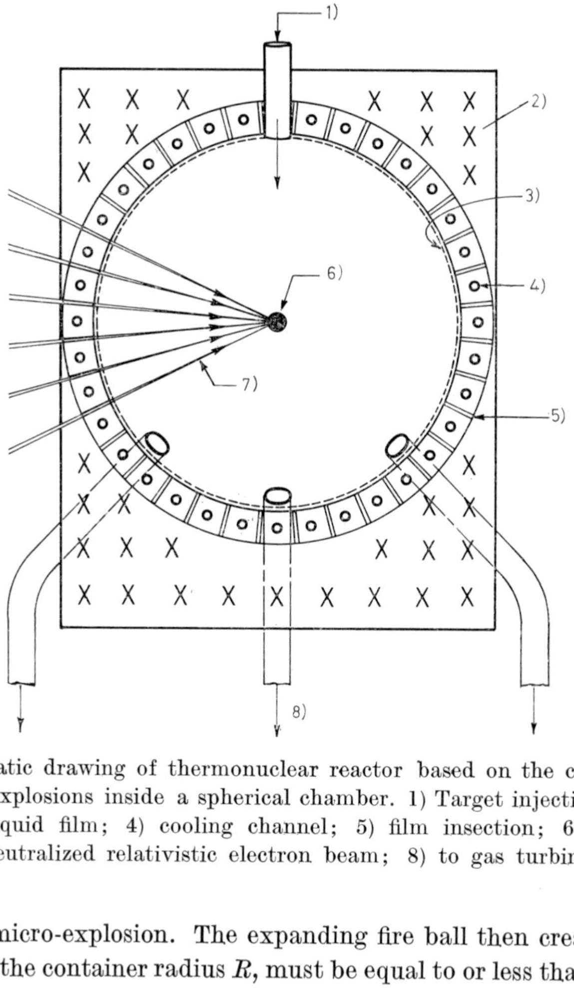

Fig. 9. – Schematic drawing of thermonuclear reactor based on the confinement of a chain of micro-explosions inside a spherical chamber. 1) Target injection; 2) magnetic field coils; 3) liquid film; 4) cooling channel; 5) film insection; 6) thermonuclear target; 7) neutralized relativistic electron beam; 8) to gas turbine generator.

the preceding micro-explosion. The expanding fire ball then creates a pressure wave which, at the container radius R, must be equal to or less than 10^9 dyn/cm². The pressure in the expanding wave decreases according to

$$p(R) = p_0(r_0)(r_0/R)^2 , \tag{4.2}$$

where $p_0 = 2NkT$ is the pressure in the original fire ball of radius $r_0 = 0.5$ cm. For $T = 5\cdot 10^7$ °K, one has $p_0 = 6.9\cdot 10^{14}$ dyn/cm². In order to have $p(R) = 10^9$ dyn/cm², one must have $R = 4.1$ m. If the externally applied magnetic field is $H = 10^4$ G, a typical value attainable with iron core magnets, one computes from eq. (4.1) that $E_H = 1.1 \cdot 10^{15}$ erg, which should represent a substantial fraction of the whole explosion energy. As a consequence, the pressure wave is reduced in intensity before reaching the wall. It should be kept in mind, however, that the estimate of the magnetohydrodynamic energy conversion may be too optimistic, since the magnetic cavity may form flutes and since part of the energy is set free in the form of radiation. The second is especially true for the case in which the thermonuclear target is surrounded by a fissionable blanket. The presence of a heavy metal blanket will lead to the emission of a large amount of black-body radiation by the fire ball. Whatever the energy balance may be, the amount of energy not converted directly into electrical power by magnetohydrodynamic-energy conversion can be absorbed by the liquid film covering the interior of the container and by the coolant in the container wall. The fact that a large fraction of the energy can be converted directly into electrical power, as indicated by the preceding estimates, indicates that good economical operation should be possible. If 10^{16} erg are released per micro-explosion, and if one assumes one micro-explosion per second, the total power output would be 10^3 MW.

5. – Rocket propulsion.

One of the most interesting aspects of controlled thermonuclear power is its potential use in rocket propulsion. The physics of the fission chain reaction indicates that the specific impulse of a nuclear rocket reactor system is not substantially greater than that of a chemical propulsion system. A gain by a factor of two is about the best that can be expected, and this is at least partially compensated for by the required heavy shielding and hardware of the fission engine. The situation might become quite different if the process of controlled thermonuclear energy release were to prove feasible. The specific impulse of a rocket engine is determined by the temperature of the gas being ejected. For thermonuclear machines these temperatures are in the range of 10^8 °K. This implies exhaust velocities for a thermonuclear rocket in the range of 10^3 km/s. As a consequence, one can expect maximum specific impulses larger by almost three orders of magnitude than those attainable with chemical rockets and two orders of magnitude above those attainable with solid core nuclear rockets. For an exhaust velocity $v \simeq 10^8$ cm/s (specific impulse $\sim 10^5$ s) and a mass flow $\mathrm{d}m/\mathrm{d}t \simeq 1$ g/s, the thrust would be $P = v\,\mathrm{d}m/\mathrm{d}t \simeq 10^8 \text{ dyn} \simeq$ $\simeq 0.1$ ton.

By some simple estimates we shall show that a very efficient rocket propulsion system might be constructed using the micro-explosions described previously. Consider a chain of micro-explosions occurring at the center of a reflector open on one side, as indicated in Fig. 10. For high specific impulse operation it is necessary to reflect the exhaust jet as much as possible and to avoid physical contact between the fire ball and the reflector. Physical contact of the fire ball with the reflector would lead to cooling problems which would drastically reduce the attainable thrust, since all the problems of a radiation cooling system known from ion propulsion would become significant. Furthermore, for the same reason, the idea of surrounding the target with fissionable material would raise similar problems, since the black-body radiation coming from the blanket would be absorbed by the reflector. However, it might prove advantageous to construct a propulsion system in which high thrust is achieved at the expense of specific impulse by evaporating a liquid film on the interior surface of the reflector. This film could be formed by injecting the liquid through many small nozzles, such as shown in Fig. 10, and would have to be re-established after each micro-explosion.

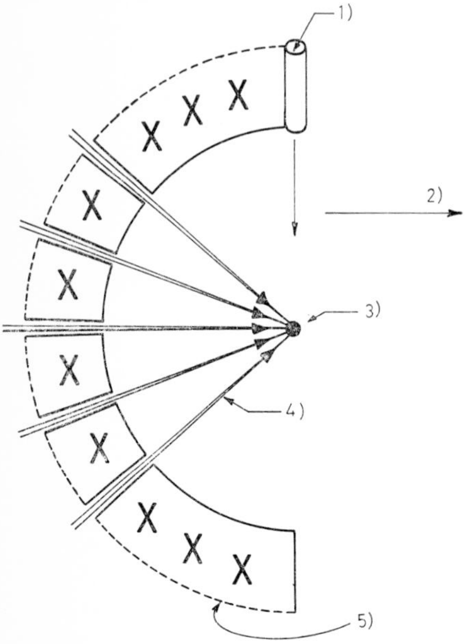

Fig. 10. – Schematic drawing of rocket propulsion by thermonuclear micro-explosions at the center of a reflector open on one side. 1) Target injection; 2) exhaust jet; 3) thermonuclear target; 4) neutralized relativistic electron beam; 5) superconducting magnetic-field coils.

If maximum specific impulse shall be achieved the reflector can be protected from the fire ball by a high magnetic field generated externally with superconducting coils.

Let us assume that the fire ball expands with the velocity of sound $v_0 \simeq 10^8$ cm/s at thermonuclear temperatures and that in this case the expansion takes place in a vacuum.

If the initial radius and density of the fire ball are r_0 and ϱ_0, then the stagnation pressure at the radius r is given by

$$p(r) = (\tfrac{1}{2})\varrho(r)\,v_0^2 = \tfrac{1}{2}\,\varrho_0(r_0/r)^3\,v_0^2\,. \tag{5.1}$$

If the stagnation pressure at the distance R, the radius of the reflector, is to be balanced by the magnetic pressure of the magnetic shield, then the condition

$$p(R) = \tfrac{1}{2}\,\varrho_0(r_0/R)^2\,v_0^2 = H^2/8\pi \tag{5.2}$$

must be satisfied. Thus

$$R = r_0 (v_0/v_A)^{\frac{2}{3}}, \tag{5.3}$$

where

$$v_A = H/\sqrt{4\pi\varrho_0}. \tag{5.4}$$

The mass density of T-D with a number density of $5 \cdot 10^{22}$ cm^{-3} is $\varrho_0 = 0.22$ g/cm^3. Let us assume that $H = 10^5$ G, a value attainable with present day technology. While $r_0 = 0.5$ cm we then have $R = 1.5$ m.

We would like to conclude this Section with a remark about the power supply needed for igniting the cycle of a chain of micro-explosions in the reflector. Initially an energy of 10^{14} erg $= 1$ kWh would be needed. This small amount of energy could be generated, for example, by a small nuclear reactor capable of delivering 1 kW of electric power. This power would be used to charge the capacitor bank or, preferably, the superconducting storage ring only once initially. After the first micro-explosion, sufficient electric power to recharge the storage system for the ignition of the next micro-explosion would be produced by magneto-hydrodynamic energy conversion through the external magnetic field coils. The levitated superconducting ring energy storage system might prove to be of special significance in rocket propulsion because of the necessity of keeping the total mass of the system as low as possible.

6. – The generation of intense ion beams.

It was pointed out above that if the stopping of an intense relativistic electron beam in a small target would pose a serious problem, the option of using an intense ion beam for heating and igniting a thermonuclear target still remains. The principal advantage of using an intense ion beam is the large stopping power that energetic ions have in dense matter. In fact, the range desired can be chosen, within a large margin, by properly choosing the nuclear charge Z of the ions.

An important difficulty encountered in using ions instead of electrons is the tendency of the electrons emitted from the cathode to discharge the stored electric energy before the ions have time enough to build up to an intense current. This is due to the fact that the electrons emerging from the cathode have speeds approximately ten times greater than the ions emitted from the anode. A cathode formed without sharp edges enhances the development of a ion current. However, at the high voltages involved, an intense electron current will always be produced. There is, though, a simple means of effectively suppressing the electron current in favor of the ion current. If a strong magnetic field is applied in front of the cathode in a direction perpendicular to the line connecting the anode with the cathode, such as shown in Fig. 11, then the electron Larmor radius can be made small compared to the distance separating the anode and cathode. This is in contrast to the ion Larmor radius, which can be made large or comparable to this distance. As a result of the applied magnetic field, the ions can still reach the cathode but the electrons are prevented from reaching

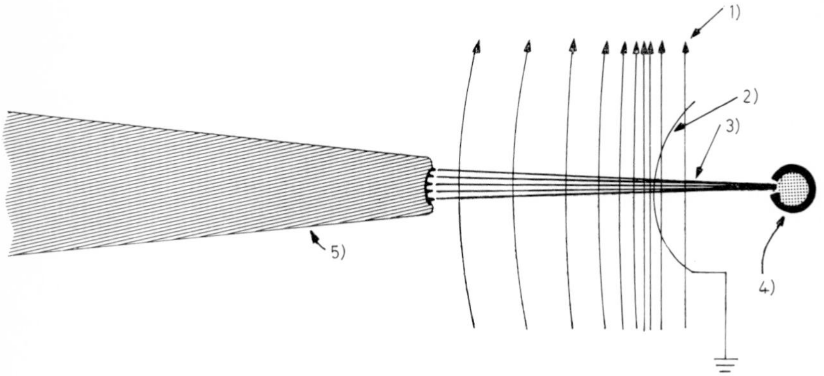

Fig. 11. – Showing the generation of intense ion beams and magnetic cathode insulation. 1) Magnetic lines of force; 2) cathode window; 3) tenuous plasma for beam neutralization; 4) thermonuclear target; 5) anode.

the anode. If there is a strong electric field acting between anode and cathode, the electrons coming from the cathode will drift in a direction perpendicular to the applied electric and magnetic fields. Near the anode $E > H$, so the expression for the drift motion is no longer valid and the ions can gain high energies before entering the strong magnetic field near the cathode. In this way the drift motion of the ions will be much smaller than that of the electrons. Alternately, we can say that the ratio of magnetic to electric pressure has to be large near the cathode and small near the anode, thereby confining the electrons to the cathode region. The electrons are forced into a drift motion right after leaving the cathode if in the expression for the drift velocity, $v = cE/H$, v is small compared to c near the cathode, implying that near the cathode $H > E$.

For 10 MeV, respectively 1 GeV, ions one would need a total of 10^{19}, respectively 10^{17}, ions to deposit an energy of 10^{14} erg into the thermonuclear target. This is a relatively small number of ions compared to the number of atoms in an anode a few centimeter in cross-section. It is, therefore, possible to produce the ions from the material of the anode itself. It has been suggested by DAWSON [17] that this could be accomplished by a pulsed laser irradiating the tip of the anode prior to discharge. If, for example, an energy of 10 eV is required to produce a ion, then 10^8 erg would be required for a 10 MeV discharge and 10^6 erg for a 1 GeV discharge, both of which are entirely feasible with pulsed lasers. By using a larger laser, one might even be able to multiply ionize the atoms to be accelerated.

The Larmor radii of the electrons and ions are given by

$$\text{Electrons:} \quad r_L^- = 1.1 \cdot 10^2 W_e^{\frac{1}{2}}/H \text{ (cm)}, \tag{6.1}$$

$$\text{Ions:} \quad r_L^+ = 4.6 \cdot 10^3 W_i^{\frac{1}{2}} A^{\frac{1}{2}}/ZH \text{ (cm)}, \tag{6.2}$$

where W_e and W_i are the particle energies in keV of the electrons and ions after they enter the strong magnetic field in front of the cathode, and Z is the net charge of a ion. If the electrode separation is D, then one demands that $r_L^- \ll D$ but $r_L^+ \gg D$, hence $r_L^-/r_L^+ \ll 1$. For the ratio we obtain

$$r_L^-/r_L^+ = 0.24(Z/A^{\frac{1}{2}})(W_e/W_i)^{\frac{1}{2}}. \tag{6.3}$$

This ratio can be made sufficiently small for a small W_e/W_i ratio, a large value for A and a small degree of ionization Z.

If the ions are accelerated to energies between 10 MeV and 1 GeV, their electron shells will be completely stripped off at impact on the target. In a thermonuclear target, the range of ions accelerated by a voltage V is easily calculated to be

$$\lambda_I = 1.7 \cdot 10^{18} (AV)^{\frac{1}{2}} T^{\frac{3}{2}}/NZ^2. \tag{6.4}$$

With $N = 5 \cdot 10^{22}$ cm^{-3}, one obtains

$$\lambda_I = 3.4 \cdot 10^{-5} (AV)^{\frac{1}{2}} T^{\frac{3}{2}}/Z^2. \tag{6.5}$$

The thermonuclear target has a diameter of the order of a few millimeters. Beam dissipation becomes effective if the range is of the order of 0.1 cm. Assuming a target temperature of $T = 10$ keV, one therefore obtains from the range formula

$$AV/Z^4 \simeq 10^4. \tag{6.6}$$

For intermediate-mass atoms one has $A \simeq 2Z$. Thus

$$Z \simeq 6 \cdot 10^{-2} V^{\frac{1}{3}}. \tag{6.7}$$

For $V \simeq 10^7$ V one obtains $Z \simeq 10$, for $V \simeq 10^8$ V $Z \simeq 35$, and for $V \simeq 10^9$ V $Z \simeq 60$. Hence the ions available are in the range from the light elements to iron. One could, of course, also use heavier ions, for which the range would be shorter and the target would be heated from the surface by shock waves.

For $V \simeq 10^9$ V and uranium ($Z = 92$, $A = 238$), the range would be 0.6 mm. For $W_i = 10$ MeV, the ion Larmor radius ($Z = 1$, $A = 20$) is $r_L^+ \simeq 20$ cm. As long as $E < H$ near the cathode and no energetic electrons are emitted from the cathode, the electron Larmor radius will be much smaller. The ratio E/H can be kept smaller than one for electric fields up to $3 \cdot 10^7$ V/cm if $H \geqslant 10^5$ G. If energetic ions hit the cathode, energetic electrons can be knocked out. For electron energies below the MeV range, assuming that $H \simeq 10^5$ G, the electron Larmor radius would still be below 0.1 cm. The situation is unchanged for GeV ions, for which the ion Larmor radius for an energy of 1 GeV ($Z = 1$, $A = 10$) is $r_L^+ \simeq 50$ cm. Here, however, the electric field will be larger, increasing the value of E/H near the cathode. This ratio must be kept below one. For a potential drop of 10^9 V over a distance of 50 cm, the electric field is $E = 6.6 \cdot 10^4$ e.s.u., so that for $H \simeq 10^5$ G the electron energy will still be small. Even if the GeV ions knock out of the cathode material electrons with energies in excess of 1 MeV, it is unlikely that such energies are produced as would allow the electrons to cross over the magnetic field barrier. Thus, if the separation distance between the anode and the cathode is of the order of 10 to 50 cm, the electrons can be effectively prevented from sparking to the anode.

The prospect of producing intense ion beams of many megaampere may have, of course, other important applications unrelated to controlled fusion.

* * *

The author would like to express his sincere thanks to Dr. L. SPIGHT, University of Nevada, for helping with the numerical calculations and for his many suggestions and critical reading of the manuscript. He would also like to express his special thanks to Dr. J. DAWSON, Princeton University, for his valuable comments and critical remarks.

REFERENCES

[1] F. WINTERBERG: *Zeits. Naturforsch.*, **19** a, 231 (1964).
[2] E. R. HARRISON: *Phys. Rev. Lett.*, **11**, 535 (1963).
[3] C. MAISONNIER: *Nuovo Cimento*, **42** B, 332 (1966).
[4] F. WINTERBERG: *Nucl. Fusion*, **6**, 152 (1966).
[5] F. WINTERBERG: *Plasma Phys.*, **8**, 541 (1966).
[6] N. G. BASOV and O. N. KROKHIN: in *Proceedings of the Third International Conference on Quantum Electronics, Paris, 1963*, edited by P. GRIVET and N. BLOUNBERGER (New York, 1964).
[7] J. M. DAWSON: *Phys. Fluids*, **7**, 981 (1964).

[8] A. G. Engelhardt: Westinghouse Research Laboratories Report No. 63-128-113-R2 (1963) (unpublished).
[9] F. Winterberg: *Phys. Rev.*, **174**, 212 (1968).
[10] F. Winterberg: University of Nevada Desert Res. Inst., preprint, No. 64 March 1969.
[11] T. G. Roberts and W. H. Bennett: *Plasma Phys.*, **10** 381 (1968).
[12] L. Spitzer: *Physics of Fully Ionized Gases* (New York, 1962).
[13] O. Bunemann: *Phys. Rev.*, **115**, 503 (1959).
[14] S. A. Bludman, K. M. Watson and M. N. Rosenbluth: *Phys. Fluids*, **3**, 747 (1960).
[15] F. C. Ford, D. Martin, D. Sloan and W. Link: *Bull. Am. Phys. Soc.*, **12**, 961 (1967).
[16] F. C. Ford: private communication.
[17] J. M. Dawson: private communication.

Relativistic Hydrodynamics in Supernovae.

E. TELLER

University of California, Lawrence Radiation Laboratory - Livermore, Cal.

In the summer of 1054 A.D., during the reign of Emperor Jen Tsung of the Sung Dynasty, during the Chih-Ho period, the Chief Calendrical Computer reported the appearance of a «guest star» near the configuration of Aldebaran. The star was so bright as to be visible in the daytime. The report seems to have been submitted with considerable trepidation because the predecessor of the imperial computer of stellar events lost his head for failing to predict a solar eclipse. The records of the guest star were carefully kept, so carefully indeed that the recorder (who seemed to have survived the event) has established the first series of quantitative data on a supernova. Today, a little less than a millenium later, the remnants of this stellar explosion are seen in the Crab nebula. Near the center of this nebula a recently discovered pulsar apparently rotating at the rate of 30 revolutions per second indicates the presence of a neutron star whose formation probably released the energy manifested as the guest star.

A supernova explosion cannot be discussed without the study of relativistic hydrodynamics. The discussion of such hydrodynamical equations is the main purpose of this lecture. At the same time, a few words should be said about supernovae themselves whose theory is at the present time by no means clarified.

Around 1930 the Swiss astronomer ZWICKY, working in Pasadena, recognized the existence of these extremely violent explosions. One reason for the suggestion was the bright nova observed in the Andromeda nebula. When it was recognized that Andromeda is a million light years from us it became evident that this star must have surpassed the emission of visible energy of the sun by more than a factor 10^{10}. ZWICKY then suggested that the Chinese guest star, as well as other novae observed by TYCHO BRAHE and by KEPLER, may be other examples of such supernovae. By surveying neighboring galaxies, ZWICKY found more examples. Supernova explosions are now reported at an average rate of about one per month, although they seem to occur in one galaxy only once in several hundred years.

The chief initiative toward the explanation of supernovae came a few years before Zwicky's suggestion. CHANDRASEKHAR investigated the stability of stars and found that this stability depends on the exponent gamma in the adiabatic pressure-density relation

$$P \approx \varrho^{\gamma} .$$

Stability is to be expected for $\gamma > \frac{4}{3}$, while instability occurs for $\gamma < \frac{4}{3}$. This is indeed easy to understand. The gravitational energy density varies as ϱ/r, where ϱ is the density and r the distance of the volume element in question from the center of the star. Since r in turn is inversely proportional to $\varrho^{-\frac{1}{3}}$ the gravitational energy density will change in an adiabatic process as $\varrho^{\frac{4}{3}}$. If γ is greater than $\frac{4}{3}$ then the pressure (which is that part of the thermal energy density that tends to re-establish the original position after an adiabatic compression) will increase more rapidly than the gravitational energy density, and the original condition will tend to be re-established. If, on the other hand, $\gamma < \frac{4}{3}$ then the gravitational energy density predominates and a disturbance displacing a volume element toward the center will be forced to continue this motion, eventually resulting in a phenomenon resembling free fall. In normal hot ionized gases γ has approximately the value $\frac{5}{3}$ and thus stars are stable. But at very high energy densities in which particles begin to move at relativistic speeds $\gamma = \frac{4}{3}$ is approached. At $\gamma = \frac{4}{3}$ the question of stability or instability is decided by the mass of the star and CHANDRASEKHAR showed that for stars somewhat more massive than the sun, collapse becomes unavoidable after the star has burned up most of its fuel. We believe now that a collapse of the type suggested by CHANDRASEKHAR does introduce the phenomenon of a supernova explosion. There are two questions that remain. What is the actual mechanism for γ becoming less than $\frac{4}{3}$ and in what manner does an original collapse turn into an explosion in which the outer regions of the collapsing star appear to be involved. Neither of these two questions have as yet a definite answer.

It appears that even before the thermal motion of the electrons becomes strongly relativistic there are two mechanisms which can cause a rapid collapse of the star. One, closely related to Chandrasekhar's original argument, was given by BURBIDGE, BURBIDGE, FOWLER and HOYLE in the 1950's.

These authors have given plausible arguments concerning the way in which nuclei are built up from hydrogen, up to the neighborhood of iron. As nuclear fuel of a smaller nuclear charge Z is used up, further reaction will depend on the establishment of a higher temperature which will permit the thermonuclear reaction to proceed at a higher Z-value. Finally, a temperature is reached which is high enough so that alpha-particles, which are exceptionally stable, can be split off from nuclei near the iron group. This process is similar to the process of molecular dissociation with which we are familiar in physical chemistry. In a gas in which dissociative equilibrium prevails the value of gamma in the relation $P \approx \varrho^{\gamma}$ tends toward one. The reason is that thermal energy produced by the compression of the gas does not appear exclusively as kinetic energy, in which case it would give rise to an adiabatic law with a normal gamma value. Instead, a considerable fraction of this energy is used up as energy of dissociation. It is to be noted that the pressure depends essentially only on the kinetic energy in the unit volume and not on the detailed questions of among how many particles of what type this kinetic energy is shared.

Thus the dissociative emission of alpha-particles is likely to lead to values of $\gamma < \frac{4}{3}$ and under these conditions the stellar collapse as described by CHANDRASEKHAR can proceed. In particular, it should be noted that this collapse need not extend to the whole star, but could occur only in the deep interior while the outer regions are temporarily not affected in a marked manner.

The second explanation for the collapse has been proposed almost two decades earlier by GAMOW. He called the process the URCA process (*). If the temperature of an interior region in the star has risen to a high value, maybe into the neighborhood of 100 000 eV it becomes energetically possible for a high-energy electron to be captured by a proton in the nucleus, converting this proton into a neutron. We assume that this conversion is energetically excluded unless the electron has additional kinetic energy. The neutrino which is emitted in the process escapes from the star and carries away some energy. Subsequently it is possible for the newly formed neutron to transform back into a proton while emitting an electron and an antineutrino. The antineutrino likewise escapes. This process can be repeated many times. While the decay processes we are discussing are essentially of a slow nature this fact can be compensated by the circumstance that the neutrinos move with light velocity and escape without hindrance. Thus energy may disappear from the star. The star contracts, the temperature rises and the URCA process can speed up, since the rate accelerates greatly as the temperature increases.

We have actually two processes which are satisfactory in a qualitative fashion to explain stellar collapse. Actually both of these processes contribute, although it is not clear in what order they initiate the collapse and what their relative quantitative contributions happen to be.

(*) URCA is the name of a night club in Rio de Janeiro. The historic origin of the name is connected with the fact that GAMOW made the observation in this night club that money was disappearing from his pocket in an unaccountable fashion. This inspired him to propose the theory which we are about to give in the text.

While the ideas concerning stellar collapse are generally accepted, the question how an explosion should follow the collapse remains open. BURBIDGE, BURBIDGE, FOWLER and HOYLE assumed that when densities exceeding those of nuclear densities are reached in the stellar interior nuclear repulsive forces will occur and cause a shock wave to be sent into the outer regions of the star. They then postulate that the heat of this shock wave will trigger thermonuclear reactions which account for the observed energy release of the supernova.

Actually, COLGATE initiated calculations in the Lawrence Radiation Laboratory at Livermore around 1962 which showed that this explanation is insufficient. Neutrinos carry away so much energy during the collapse that the process becomes highly inelastic and the shock wave that is sent out into the star is too feeble to account for the astronomical observations. He proposed, instead, an ingenious mechanism relying on the neutrinos themselves.

One has to assume that in the collapsing supernova nuclear densities, that is, densities of at least 10^{14} g per cubic centimeter will be reached. At these densities even neutrinos will have a limited mean free path, for which a value of about 10^4 cm can be estimated. This value is somewhat smaller than the radius of the stellar core which is formed and therefore the neutrinos, instead of escaping in a straight line, will carry energy away in an exceedingly rapid diffusion process. In the last collision of this diffusion process the neutrino is likely to deposite some energy in a somewhat less dense region of the star, where the temperature and the gravitational potential have not yet reached the extreme values that occur in the core. This less dense region which we shall call the mantle will therefore be heated, the origin of heat being the gravitational energy released near the center. The mantle, thereupon explodes and sends a shock wave into the outer regions of the star. This shock, according to COLGATE, is the source of the energy released by the supernova and of the high luminosity which is observed.

There is good reason to believe that the actual situation is even more complex. Recent unpublished calculations by WILSON, in the Lawrence Radiation Laboratory in Livermore, indicate that the collapse occurs too fast and that, therefore, not enough time is left for a sufficient number of neutrinos to escape and to deposit enough energy in the mantle. Nor can this difficulty be avoided by considering heavier stars. Although more energy is thereby involved, the collapse also occurs faster and there is at present no straightforward calculation which does explain the behavior of the supernova explosion. According to present calculations a « black hole » is produced, that is a general relativistic singularity from which no particles or energy can escape. Thus one gets neither a sufficiently big flare-up nor an observable remnant in the form of a neutron star.

I believe that the actual explanation of supernovae may be connected with the rotation of the star. During a collapse, angular momentum is conserved and this leads to a rapid increase of kinetic energy as the radius is decreasing. This can slow down the process of collapse to a sufficient extent so that in the central region the mechanism proposed by COLGATE can become effective. Unfortunately, the calculations which are required to verify this hypothesis must not only take into account the various effects of general relativity but must handle at the same time two-dimensional time-dependent hydrodynamics. The complexities are so great that no straightforward answer can be expected in the near future.

In the meantime, I would like to discuss a general model of supernovae which may be appropriate in describing the behavior in the outer regions of the star which, prior to the arrival of the shock wave, have not been influenced by the collapse of the interior. The main assumptions of this model are that the energy of the supernova has been suddenly released at the center and that the regions which we describe are not essentially influenced by rotation. The former of these two assumptions is highly plausible. The second may be justifiable because in the outer regions the energy of the rotation is negligible compared to the kinetic energy carried by the shock. At the same time, one should realize that the justification of this model is somewhat questionable since effects of rotation can be transmitted from the center to the outer regions, for instance, by coupling with magnetic fields if sufficient time is allowed. In spite of this doubt the considerations to be given seem to me instructive even though they are not absolutely convincing.

In particular, I would like to discuss another suggestion by COLGATE, according to which the shock will produce, in the outermost parts of the star, particles of exceedingly high energy which actually are thrown out into space and are observable as cosmic rays. With the help of a relativistic hydrodynamic theory worked out by MONTGOMERY JOHNSON the spectrum can be found if indeed cosmic rays are originated in the manner described. I will give a simplified version of this hydrodynamic theory.

We shall first write down the formulae for a strong shock which are valid both for the unrelativistic and for the relativistic case. We assume that the region to be shocked will be characterized by the subscript 2, while the shocked region is characterized by the subscript 1. The energy per cubic centimeter in region 1 can be written as

$$E_1 = n_1 mc^2 + \eta_1 , \tag{1}$$

where n_1 is the number of heavy particles per cubic centimeter (excluding light quanta, electrons and positrons), m is the average mass of these heavy particles, $n_1 mc^2$ is the rest mass of the shocked region and η_1 is the energy density in region 1 in excess of this rest mass as measured in the system 1.

In a strongly relativistic case one has the relation

$$\eta_1 = 3P_1 , \tag{2}$$

where P_1 is the pressure in the shocked region.

The assumption of a strong shock means that, apart from the rest energy in region 2, the additional internal energy density in this region, and also the pressure in this region are small compared to E_1. This condition for a strong shock is amply justified for the case that the undisturbed outer regions of a supernova are hit by the shock coming from the center.

In discussing shock equations it is the practice to use a co-ordinate system which moves with the shock front. In discussing strong shocks it is simpler to use a co-ordinate system which is at rest in the shocked material which carries the index 1.

The equation describing the conservation of heavy particles can be written as

$$n_1\beta_s = \frac{n_2(\beta_s + \beta_2)}{(1-\beta_2^2)^{\frac{1}{2}}} , \tag{3}$$

where β_s is the velocity of the shock front relative to the shocked material and divided by the light velocity c. Thus $n_1 c\beta_s$ is the number of heavy particles which are added per second per square centimeter to the shocked region. The expression on the right-hand side of eq. (3) is the number of heavy particles which leave region 2. This expression is more complicated for two reasons. One is that, as described by an observer in region 1, region 2 has a Lorentz contraction which increases the apparent density of heavy particles from n_2 to

$$\frac{n_2}{(1-\beta_2^2)^{\frac{1}{2}}} .$$

The symbol β_2 stands for the velocity with which the second region is approaching the first region, divided by c.

The second reason is that, as seen from region 1, particles from region 2 disappear into the shock with a velocity (in terms of c) which is equal to $\beta_s + \beta_2$. The shock velocity β_s is moving away from the observer while in region 2 the material velocity β_2 is moving toward him. In order to describe the relative velocities of the shock front and of region 2 one has simply to add the two velocities as described by the same observer in region 1.

The equation of conservation of energy can be written as

$$E_1\beta_s = \frac{mc^2 n_2(\beta_s + \beta_2)}{1-\beta_2^2} . \tag{4}$$

The equation is self-explanatory if one remembers that in region 2 only the rest energy of particles in region 2 must be taken into account by an observer in region 1, and that the energy of such a particle as viewed from region one will be

$$\frac{mc^2}{(1-\beta_2^2)^{\frac{1}{2}}}.$$

If we divide (4) by (3) and also consider (1) we obtain

$$\frac{E_1}{n_1} = mc^2 + \frac{\eta_1}{n_1} = \frac{mc^2}{(1-\beta_2^2)^{\frac{1}{2}}}. \tag{5}$$

This equation can also be written as

$$\frac{\eta_1}{n_1} = mc^2\left[\frac{1}{(1-\beta_2^2)^{\frac{1}{2}}} - 1\right]. \tag{6}$$

It means that in a strong shock the excess energy per particle in the shocked region is equal to the kinetic energy of the particle in the shocked region. Indeed, β_2 can be considered not only as the velocity of the unshocked region with respect to the shocked region but equally as the velocity of the shocked region with respect to the unshocked region. The equality of internal and kinetic energy for strong shocks is well known in the hydrodynamics of un-relativistic shocks. It is seen here that it is valid not only for all equations of state but also for all velocities including relativistic velocities.

The last shock equation, that of conservation of momentum, can be written as

$$P_1 = \frac{mc^2 \beta_2 n_2 (\beta_s + \beta_2)}{1-\beta_2^2}, \tag{7}$$

where P_1 is the pressure in the shocked region while the pressure in the unshocked region is neglected. The pressure P_1 just suffices to stop the momentum deposited per square centimeter and unit time in the shock. If one remembers that the momentum per particle is

$$\frac{mc\beta_2}{(1-\beta_2^2)^{\frac{1}{2}}}$$

while the velocities of the shock and of region 2 are $c\beta_s$ and $c\beta_2$, eq. (7) can be justified by similar considerations as eqs. (3) and (4).

Let us now assume an exceedingly strong shock so that $\eta_1 \gg n_1 mc^2$. Then it follows from eq. (6) that one can write for the relative velocity of unshocked and shocked regions

$$\beta_2 = 1 - \varepsilon\,, \tag{8}$$

where $\varepsilon \ll 1$. By dividing (7) by (4) one finds

$$\frac{P_1}{E_1} = \beta_s \beta_2\,, \tag{9}$$

which incidentally is valid for any strong shock whether or not the shock is relativistic (*). Neglecting ε and also neglecting the rest energy as compared to η_1 in eq. (1) one finds for the strong relativistic case using (2)

$$\beta_s = \tfrac{1}{3}\,. \tag{10}$$

This means that in the strong relativistic case the velocity of shock as viewed from the shocked region is $\frac{1}{3}$ of the light velocity.

In the following arguments it will be sufficient to retain only the leading terms. In any actual shock β_s will be less than $\frac{1}{3}$ and will converge to the value $\frac{1}{3}$ in the extremely relativistic case.

Continuing to consider the extreme relativistic case we can write for the energy density in the shocked region

$$E_1 = \frac{4n_2 mc^2}{1 - \beta_2^2} = \frac{2n_2 mc^2}{\varepsilon}\,, \tag{11}$$

where eqs. (4), (10) and (8) have been utilized.

This result establishes a connection between the energy density in the co-moving co-ordinate system behind the shock, the particle density in front of the shock and the quantity ε which characterizes the difference between the velocity of the shocked material and the velocity of light.

In order to obtain the needed consequences of the hydrodynamic treatment we shall require the Riemann invariant in its relativistic form. The invariant has been derived in the same context by M. JOHNSON several years ago. The simplified treatment given below is based on the idea that one may use in practically every part of the formalism a co-moving co-ordinate system.

The equations of motion are derived from the fact that the divergence of the relativistic energy-momentum-tension tensor is equal to 0. In the

(*) It is easy to show that (9) is valid for all shocks (rather than for strong shocks only) if P_1 is replaced by the pressure difference between the two sides of the shock.

co-moving system this tensor is

$$(12)\qquad \begin{bmatrix} P & 0 & 0 & 0 \\ 0 & P & 0 & 0 \\ 0 & 0 & P & 0 \\ 0 & 0 & 0 & nmc^2 + \eta \end{bmatrix}.$$

If we consider this tensor from a moving frame with a velocity (in terms of c) equal to β we obtain by the Lorentz transformation as applied to a tensor

$$(13)\qquad \begin{bmatrix} \dfrac{P + \beta^2(nmc^2 + \eta)}{1 - \beta^2} & 0 & 0 & \dfrac{\beta(P + nmc^2 + \eta)}{1 - \beta^2} \\ 0 & P & 0 & 0 \\ 0 & 0 & P & 0 \\ \dfrac{\beta(P + nmc^2 + \eta)}{1 - \beta^2} & 0 & 0 & \dfrac{nmc^2 + \eta + \beta^2 P}{1 - \beta^2} \end{bmatrix}.$$

Setting the space-time divergence of this tensor equal to zero one obtains

$$(14)\qquad \frac{\partial}{\partial x}\left[\frac{P + \beta^2(nmc^2 + \eta)}{1 - \beta^2}\right] + \frac{\partial}{\partial(ct)}\left[\frac{\beta(P + nmc^2 + \eta)}{1 - \beta^2}\right] = 0\,,$$

$$(15)\qquad \frac{\partial}{\partial x}\left[\frac{\beta(P + nmc^2 + \eta)}{1 - \beta^2}\right] + \frac{\partial}{\partial(ct)}\left[\frac{nmc^2 + \eta + \beta^2 P}{1 - \beta^2}\right] = 0\,.$$

Taking the derivative of eq. (14) with respect to x, of eq. (15) with respect to ct and subtracting, one obtains

$$(16)\qquad \frac{\partial^2}{\partial x^2}\left[\frac{P + \beta^2(nmc^2 + \eta)}{1 - \beta^2}\right] - \frac{\partial^2}{\partial(ct)^2}\left[\frac{nmc^2 + \eta + \beta^2 P}{1 - \beta^2}\right] = 0\,.$$

We now want to apply this equation to the co-moving system in which $\beta = 0$ and, restricting our attention to first-order terms, specifically neglecting the squares of the derivatives of β, we obtain

$$(17)\qquad \frac{\partial^2 P}{\partial x^2} = \frac{\partial^2(nmc^2 + \eta)}{\partial(ct)^2}\,.$$

If we now neglect the rest energy compared to η and using the connection between η and P given by eq. (2), we get

$$\frac{1}{3}\frac{\partial^2 \eta}{\partial x^2} = \frac{\partial^2 \eta}{\partial (ct)^2}. \tag{18}$$

The general solution of this equation is any function η of the variable

$$\left(x + \frac{c}{\sqrt{3}}t\right) \quad \text{or} \quad \left(x - \frac{c}{\sqrt{3}}t\right).$$

This means that in the extreme relativistic case the sound velocity will become $c/\sqrt{3}$, a very well known result.

To obtain the Riemann relations we set $\beta = 0$ in eqs. (14) and (15) but do not neglect the derivatives of β. The result is

$$\frac{\partial P}{\partial x} + (P + E)\frac{\partial \beta}{\partial (ct)} = 0 \tag{19}$$

and

$$(P + E)\frac{\partial \beta}{\partial x} + \frac{\partial}{\partial (ct)} E = 0. \tag{20}$$

If we again use in the extreme relativistic approximation $E = \eta$ and $P = \frac{1}{3}\eta$ the equation is reduced to the pair

$$\frac{1}{3}\frac{\partial \eta}{\partial x} + \frac{4}{3}\eta\frac{\partial \beta}{\partial (ct)} = 0 \tag{21}$$

and

$$\frac{4}{3}\eta\frac{\partial \beta}{\partial x} + \frac{\partial \eta}{\partial (ct)} = 0. \tag{22}$$

This pair of equations can be solved by introducing the variable

$$\sigma = \int \frac{\sqrt{3}}{4}\frac{\partial \eta}{\eta} = \frac{\sqrt{3}}{4}\ln \eta. \tag{23}$$

If σ is introduced instead of η one obtains

$$\frac{\partial \sigma}{\partial x} + \frac{\sqrt{3}}{c}\frac{\partial \beta}{\partial t} = 0 \tag{24}$$

and

$$\frac{\partial \beta}{\partial x} + \frac{\sqrt{3}}{c}\frac{\partial \sigma}{\partial t} = 0. \tag{25}$$

Adding these two equations one finally gets

$$\frac{\partial(\sigma+\beta)}{\partial x}+\frac{\sqrt{3}}{c}\frac{\partial(\sigma+\beta)}{\partial t}=0\,. \tag{26}$$

This shows that the quantity $\sigma+\beta$ can be any function of the variable

$$\left(x-\frac{c}{\sqrt{3}}\,t\right).$$

Thus a disturbance will propagate in the forward direction with the sound velocity of $c/\sqrt{3}$ in such a manner that the quantity $(\sigma+\beta)$ is conserved. This holds, of course, only in the neighborhood of $\beta=0$.

What has just been said suffices for the following discussion. Two remarks however, are of some general interest. One is that by subtracting (25) from (24) one can obtain the result that the quantity $(\sigma-\beta)$ is unchanged when one proceeds with respect to the material with sound velocity in the negative x direction. The other is that a more general formulation of the Riemann invariant could have been obtained by introducing the more general definition of

$$\sigma=\int v_s\,\frac{\mathrm{d}E}{E+P}\,. \tag{27}$$

In this expression v_s stands for the sound velocity and the E of course includes the rest energy. By introducing this expression into eq. (19) and (20) one could have derived eq. (24) and (25) in a direct manner. Unfortunately, under these conditions the sound velocity is a more complicated function and no general integral is available for v_s. Therefore, we shall not be interested in this case.

One cannot exploit the invariance of $(\sigma+\beta)$ in the forward direction in a straightforward manner because in relativity the velocities are not additive. In fact, if you have a relative velocity β_{12} between two systems 1 and 2 and another relative velocity β_{23} between systems 2 and 3 then the relative velocity between systems 1 and 3 is given by the well-known formula

$$\beta_{13}=\frac{\beta_{12}+\beta_{23}}{1+\beta_{12}\beta_{23}}\,. \tag{28}$$

Since σ has been defined in eq. (23) as an integral it is indeed simply additive, but the β's are not. One can obtain a proper generalization of the Riemann invariant by introducing a quantity which in relativity is sometimes called the « celerity », which for small β values is the same as the velocity, but in addition, possesses the property of additivity. The celerity is the archyperbolictangent of β, or more explicitly

$$\operatorname{arctgh}\beta = \frac{1}{2}\ln\frac{1+\beta}{1-\beta}. \tag{29}$$

It is indeed easy to verify that

$$\frac{1}{2}\ln\frac{1+\beta_{13}}{1-\beta_{13}} = \frac{1}{2}\ln\frac{1+(\beta_{12}+\beta_{32})/(1+\beta_{12}\beta_{23})}{1-(\beta_{12}+\beta_{23})/(1+\beta_{12}\beta_{23})} = \tag{30}$$

$$= \frac{1}{2}\ln\frac{1+\beta_{12}+\beta_{23}+\beta_{12}\beta_{23}}{1-\beta_{12}-\beta_{23}+\beta_{12}\beta_{23}} = \frac{1}{2}\ln\frac{1+\beta_{12}}{1-\beta_{12}} + \frac{1}{2}\ln\frac{1+\beta_{23}}{1-\beta_{23}}.$$

The result is that one can make the following general statement for the approximation of the Riemann invariant in the positive x direction in the extreme relativistic case. The quantity

$$\frac{1}{2}\ln\frac{1+\beta}{1-\beta} + \frac{\sqrt{3}}{4}\ln\eta = \text{const}_{(x-(c/\sqrt{3})t)} \tag{31}$$

is propagated relative to the flow of matter with the sound velocity $c/\sqrt{3}$ in an unchanged manner in the positive x direction.

This last circumstance we have expressed by writing on the right-hand side of eq. (31) a constant with a subscript which indicates the path along which the quantity of the left-hand side is invariant. It is convenient to multiply eq. (31) by 2 and to exponentiate the quantity so obtained. This gives the result

$$\frac{1+\beta}{1-\beta}\eta^{\sqrt{3}/2} = \text{const}_{(x-(c/\sqrt{3})t)}, \tag{32}$$

that the quantity written on the left-hand side of eq. (32) is again a constant along the paths indicated by the subscript on the right-hand side.

We shall apply these results to the extreme exterior of a supernova. We shall assume that this exterior region is thin enough so that the curvature does not matter and the problem can be treated as though it were planar. Under these conditions we can see that in the outermost layers of the star in the shocked region the Riemann adiabat as given in eq. (32) is constant. In this adiabat we will have to identify η with η_1, the value prevailing in the shocked region. On the other hand, β is the relative velocity of the shocked and unshocked regions and, therefore, β in eq. (32) has to be identified with

β_2. Using eq. (8) we obtain

$$\frac{2}{\varepsilon}\eta_1^{\sqrt{3}/2} = \text{const} . \tag{33}$$

We have omitted the subscript from the constant because the equation relates to a very thin region which derives from the same neighborhood and very nearly the same value of the adiabat in the interior of the star. By using eq. (11) we can eliminate η_1 and obtain eqs. (34), (35) and (36):

$$\frac{2}{\varepsilon}\left[\frac{2n_2 mc^2}{\varepsilon}\right]^{\sqrt{3}/2} = \text{const} , \tag{34}$$

$$\frac{n_2^{\sqrt{3}/2}}{\varepsilon^{\sqrt{3}/2+1}} = \text{const} , \tag{35}$$

$$\frac{1}{\varepsilon} = \text{const}\; n_2^{-((\sqrt{3}/2)/(\sqrt{3}/2+1))} = \text{const}\; n_2^{3-2\sqrt{3}} . \tag{36}$$

We now have a relation between the small velocity difference ε and the density n_2 in the unshocked region.

After the shock has impinged, the material will adiabatically expand and at the same time it will accelerate and in this process the Riemann relation (32) will be conserved.

At the same time, one should note that in an adiabatic expansion process the energy density varies with the $\frac{4}{3}$ power of the particle density in the relativistic region

$$\eta \approx n^{\frac{4}{3}} \tag{37}$$

and therefore, the energy/particle varies with the $\frac{1}{3}$ power of the particle density

$$\frac{\eta}{n} \approx n^{\frac{1}{3}} . \tag{38}$$

Thus the energy/particle is changing with the fourth root of the energy density

$$\left(\frac{\eta}{n}\right)^4 \approx \eta . \tag{39}$$

All this holds up to the point where the energy/particle η/n in the co-moving system becomes approximately equal to mc^2.

Just after the shock has struck η/n is given according to eq. (6) by

$$\frac{\eta_1}{n_1} = \frac{mc^2}{(2\varepsilon)^{\frac{1}{2}}} = \text{const}\, n_2^{\frac{3}{2}-\sqrt{3}}\,, \tag{40}$$

where eq. (36) has been used.

During the process of adiabatic expansion η/n starts at the value given by eq. (40) and gets down to a fixed value in the neighborhood of mc^2, and will then have no further strong reason to change. The total change of η/n during the expansion process is, therefore, proportional to

$$n_2^{\sqrt{3}-\frac{3}{2}} \tag{41}$$

The total change in η is the fourth power of this quantity and according to the constancy of the quantity given in eq. (33) ε is proportional to the $\sqrt{\frac{3}{2}}$ power of η. This gives a total variation in ε by a factor

$$n_2^{2\sqrt{3}(\sqrt{3}-\frac{3}{2})} = n_2^{6-3\sqrt{3}} \tag{42}$$

Considering the value of ε before the expansion given by (36) one obtains the final value

$$\varepsilon_{\text{final}} \approx n_2^{2\sqrt{3}-3} n_2^{6-3\sqrt{3}} = n_2^{3-\sqrt{3}}\,. \tag{43}$$

The final value of ε in turn is connected with the energy of the particles which are supposed to be the cosmic rays. This energy is proportional to

$$E = \frac{mc^2}{(1-\beta^2)^{\frac{1}{2}}} \approx \varepsilon_{\text{final}}^{-\frac{1}{2}} \approx n_2^{\frac{1}{2}(\sqrt{3}-3)}\,. \tag{44}$$

Let us apply this result to the case where the outermost region of the star can be described as an atmosphere at constant temperature. In that case n_2 will vary exponentially with the height. The number of cosmic rays above a certain energy will be proportional to the total number of particles ejected from outside a given radius and this number in turn will be proportional to the density n_2 of particles at that radius. Thus the number of cosmic rays N which we can set proportional to n_2 and which will have an energy greater than the energy given by eq. (44) can be written as

$$N_{\text{energy}>E} \approx n_2 \approx E^{2/\sqrt{3}-3} = E^{-(1+\sqrt{3}/3)} = E^{-1.577}\,. \tag{45}$$

This is actually a reasonable representation of the empirical cosmic-ray spectrum.

When one compares this explanation with the one given 20 years ago by FERMI one should note that in that older explanation the number of cosmic rays was indeed proportional to a power of the energy. But the exponent of this power was left undetermined. Actually, it depended on quantities produced by statistical processes in the galaxies and a detailed check could not be made by comparison with other observations. The theory of COLGATE and JOHNSON has the advantage that it leads to a rather definite power which is in agreement with experiments. Actually, this power could be somewhat varied by changing the matter density distribution in the space near the supernova but the change so obtained does not appear to be great.

One can raise three serious objections to this cosmic-ray theory. One is that the cosmic rays of a given velocity are ejected essentially simultaneously from the surface of the supernova and these ions, together with the accelerating electrons, form a moving plasma. In the beginning this plasma has a much greater density than the plasma present in interstellar space. After the cosmic rays have travelled for approximately one light year the two plasma densities become comparable and at this point the well-known phenomenon of two-stream instability should appear. This instability may transform the ordered motion of the cosmic rays into a disorderly jumble and could give rise to effects which are hard to predict. Recent calculations of a graduate student, MCKEE at Berkeley, have given some proof that the two-stream instability actually does occur and does affect the motion of the electrons which accompany the cosmic rays. But his calculations, carried out in a simplified manner with the help of computing machines, also give reasonably strong support to the view that the heavy particles which are the essential ones in the cosmic-ray spectrum will not be affected.

A second and stronger objection is due to the fact that, when very little material remains outside the radius r at which the shock is to be formed, radiation can leak out without further collision with electrons. In the actual shock process radiation plays an extremely important part because most of the energy is contained in radiation. When the leaking occurs the shock can no longer be formed and thus there is no obvious mechanism whereby the cosmic rays of highest energy up to 10^{20} eV could be produced.

A third objection is connected with the question whether or not the hydrodynamic phenomena which we have described occur under properly adiabatic condition, that is, whether or not equilibrium and, in particular, radiation equilibrium is attained. This question is closely connected with the topic which we have mentioned. There seems to be little doubt that cosmic rays of lower energy can be explained in the manner described above. But at what limit of energy difficulties will set in is uncertain and depends to a great extent on the detailed model of the pre-supernova state of the star.

It is a further interesting question to follow the emission of energy from a supernova beyond the emission of the highest-energy particles which are presumably cosmic rays. At somewhat deeper layers within the star the original shock will be relatively weaker. Even under those conditions a lot of energy will be produced in the form of black-body radiation. As the star expands, this black-body radiation will retain its general shape but will degrade in energy-density and temperature, transforming this energy into the kinetic energy of the outer layers of material. Eventually, however, the total material per square centimeter that lies above a certain material volume will be small enough so that radiation can leak out. Calculations of these detailed processes have been carried out for several supernova models by WOOD at Livermore.

The result is that, at the very beginning, the supernova will emit in addition to the cosmic rays, some electromagnetic radiation which could be observed in the MeV region. At that time, the tail of the photon distribution that lies in the visible is so weak as to be unobservable. As time progresses, the total intensity distribution of the radiation shifts from the gamma-ray region into the X-ray region and later into the ultraviolet. Eventually, the maximum should shift into the visible region. In a model that starts from a highly dense star of approximately 10^9 cm radius this occurs after a few hours and the intensity so obtained appears to be too weak to explain the observed visible intensities of supernovae. It is generally assumed that the actual intensity is due to an afterglow produced by nuclear phenomena, particularly β-decay that was originating in the mantle surrounding the center of the supernova.

There is, however, another interesting possibility. It is conceivable that the pre-supernova state is a very extended star with a big gaseous envelope. In that case the calculations of the type which WOOD has carried out would give a delay of several days between the supernova explosion and the observed maximum of emitted visible radiation. Furthermore, in that case a greater fraction of the energy deposited by the original shock in the body of the supernova could eventually appear as visible radiation and this may suffice to explain some or perhaps even all the observed supernova intensity. This possibility has also been mentioned by the Russian authors GRASBERG and NADEZHIN in their paper «Emergence of a Shock Wave into an Extended Envelope of a Star and Supernova Explosion», Institute of Science, National Academy of Science, Moscow, 1969.

There is little doubt that the most spectacular phase of a supernova explosion is its earliest stage in which gamma rays are emitted. Unfortunately, we cannot see these gamma rays since our atmosphere provides a very effective shield. Appropriate apparatus mounted on spacecraft could observe this early phase and check on the way in which the intensity and the average wave length varies with time. This could give valuable information about the theory and would allow us to draw conclusions concerning the nature of the star in the pre-supernova state.

Another consequence of such an observation would be that we could do what the Chief Calendrical Computer in the times of the Sung Dynasty could not do. We could predict by several days the appearance of a supernova as an optically observable object. MARK and his co-workers at the Ames Research Laboratory in California are actually planning such an observation. It is not impossible that TIME magazine will be notified of a 20th century « guest star », a feat which was indeed impossible in the 11th century.

PROCEEDINGS OF THE INTERNATIONAL SCHOOL OF PHYSICS
« ENRICO FERMI »

Course XIV
Ergodic Theories
edited by P. CALDIROLA

Course XV
Nuclear Spectroscopy
edited by G. RACAH

Course XVI
Physicomathematical Aspects of Biology
edited by N. RASHEVSKY

Course XVII
Topics of Radiofrequency Spectroscopy
edited by A. GOZZINI

Course XVIII
Physics of Solids (Radiation Damage in Solids)
edited by D. S. BILLINGTON

Course XIX
Cosmic Rays, Solar Particles and Space Research
edited by B. PETERS

Course XX
Evidence for Gravitational Theories
edited by C. MØLLER

Course XXI
Liquid Helium
edited by G. CARERI

Course XXII
Semiconductors
edited by R. A. SMITH

Course XXIII
Nuclear Physics
edited by V. F. WEISSKOPF

Course XXIV
Space Exploration and the Solar System
edited by B. ROSSI

Course XXV
Advanced Plasma Theory
edited by M. N. ROSENBLUTH

Course XXVI
Selected Topics on Elementary Particle Physics
edited by M. CONVERSI

Course XXVII
Dispersion and Absorption of Sound by Molecular Processes
edited by D. SETTE

Course XXVIII
Star Evolution
edited by L. GRATTON

Course XXIX
Dispersion Relations and Their Connection with Causality
edited by E. P. WIGNER

Course XXX
Radiation Dosimetry
edited by F. W. SPIERS and G. W. REED

Course XXXI
Quantum Electronics and Coherent Light
edited by C. H. TOWNES and P. A. MILES

Course XXXII
Weak Interactions and High-Energy Neutrino Physics
edited by T. D. LEE

Course XXXIII
Strong Interactions
edited by L. W. ALVAREZ

Course XXXIV
The Optical Properties of Solids
edited by J. TAUC

Information about Courses I-XIII may be obtained from the Italian Physical Society.

Course XXXV
High-Energy Astrophysics
edited by L. GRATTON

Course XXXVI
Many-Body Description of Nuclear Structure and Reactions
edited by C. BLOCH

Course XXXVII
Theory of Magnetism in Transition Metals
edited by W. MARSHALL

Course XXXVIII
Interaction of High-Energy Particles with Nuclei
edited by T. E. O. ERICSON

Course XXXIX
Plasma Astrophysics
edited by P. A. STURROCK

Course XL
Nuclear Structure and Nuclear Reactions
edited by M. JEAN

Course XLI
Selected Topics in Particle Physics
edited by J. STEINBERGER

Course XLII
Quantum Optics
edited by R. J. GLAUBER

Course XLIII
Processing of Optical Data by Organisms and by Machines
edited by W. REICHARDT

Course XLIV
Molecular Beams and Reaction Kinetics
edited by CH. SCHLIER

Course XLV
Local Quantum Theory
edited by R. JOST

Course XLVI
Physics with Storage Rings
edited by B. TOUSCHEK

Course XLVII
General Relativity and Cosmology
edited by R. K. SACHS

Tipografia Compositori - Bologna - Italy